CIRCADIAN RHYTHMS AND HEALTH RESEARCH TRENDS

CIRCADIAN RHYTHMS AND HEALTH RESEARCH TRENDS

ALFREDO V. LIGNELLI
EDITOR

Nova Biomedical Books
New York

For permission to use material from this book please contact us:
Telephone 631-231-7269; Fax 631-231-8175
Web Site: http://www.novapublishers.com

Library of Congress Cataloging-in-Publication Data

Circadian rhythms and health research trends / Alfredo V. Lignelli (editor).
p. ; cm.
Includes bibliographical references and index.
ISBN-13: 978-1-60021-875-0 (hardcover)
ISBN-10: 1-60021-875-X (hardcover)
1. Circadian rhythms. I. Lignelli, Alfredo V.
[DNLM: 1. Circadian Rhythm. 2. Adaptation, Physiological. QT 167 C5765 2007]
QP84.6.C56 2007
612'.022--dc22

2007026944

Published by Nova Science Publishers, Inc. ✦ New York

Contents

Preface

A circadian rhythm is a roughly-24-hour cycle in the physiological processes of living beings, including plants, animals, fungi and cyanobacteria. The term "circadian", comes from the Latin circa, "around", and dies, "day", meaning literally "about a day." The formal study of biological temporal rhythms such as daily, weekly, seasonal, and annual rhythms, is called chronobiology. In a strict sense, circadian rhythms are endogenously generated, although they can be modulated by external cues such as sunlight and temperature. Most of a person's body systems demonstrate circadian variations. The body systems with the most prominent circadian variations are the sleep-wake cycle, the temperature regulation system, and the endocrine system. The malfunctioning of a person's circadian system, or biological clock, causes circadian rhythm disorders. Circadian rhythms and their relation to health outcomes in the globalized hyper world of the 21st century are now understood to be extremely important.

Chapter I - The suprachiasmatic nucleus (SCN) and the pineal gland are critical components of a neural oscillator system implicated in the timing of a wide variety of biological processes. The circadian cycles established by this biological clock occur throughout nature and have a period of approximately 24 hours. With advancing age these daily fluctuations deteriorate, leading to disrupted cycles with a reduced amplitude. In humans, age-related changes have been described for hormonal rhythms, body core temperature, sleep-wakefulness, and several other behavioral cycles. It appears that the disruption of circadian rhythms and the increased incidence of disturbed sleep during aging are paralleled by age-related alterations in the neural and temporal organization of the circadian oscillator and a decreased photic input to the clock. The observed neuronal degeneration of the SCN-pineal complex in senescence strongly suggests an organic deterioration of the circadian oscillator during aging and supports the idea that damage to the biological clock is the underlying anatomical substrate for the clinically often-observed disturbances in circadian rhythmicity in mood disorders and Alzheimer's disease. Recent studies furthermore indicate that illuminance and photoperiod may be considered as potential environmental factors controlling the functional activity of the circadian timing system. If light is the effective Zeitgeber in man, as it is in many other organisms, disturbances in processing of light information during aging or in some neurodegenerative and affective diseases may have profound effects on the timing of a variety of physiological and behavioral activities, including sleep. In view of the assumed sensitivity of the biological clock to changes in day length and light intensity research is now focussing on whether the synchronizing and antidepressant effects of bright light exposure in sleep disturbances and mood disorders can be improved by applying the therapy in temporal coherence with the circadian cycles underlying the neuronal activity of the biological clock.

Chapter II - Circadian, about 24-hour, rhythms are universal features of most organisms on earth. In mammals, a master circadian clock located in the suprachiasmatic nuclei (SCN) of the anterior hypothalamus controls circadian rhythmicity. The SCN synchronizes peripheral clocks and extra-SCN clocks in the brain to allow the animal to express coherent behavioral and physiological rhythms that are entrained to the ambient circadian cycle. Important advances have been made in the last couple of decades in the understanding of the molecular mechanisms underlying circadian cycling at the cellular level. One aspect that has been largely overlooked in this research field is the adaptive significance of circadian organization to the animal under natural conditions. In this chapter, the authors will review what is currently known about the potential evolutionary benefits of circadian control of behavior and physiology, with a focus on humans and other mammals. The next challenge, the authors believe, will be to connect the advances in molecular understanding to the circadian challenges an organism faces in its natural circadian environment. The authors end with a discussion of how to test the adaptive significance of mammalian circadian organization. The authors goal is to elicit discussion about the importance of understanding the significance of circadian rhythms under natural conditions and how these rhythms adapt an individual to its local natural environment. In the context of a literature review, the authors outline why this understanding has great potential to contribute to novel approaches to understanding human health and treating disease.

Chapter III - Circadian, about 24-hour, rhythms are universal features of most organisms on earth. In mammals, a master circadian clock located in the suprachiasmatic nuclei (SCN) of the hypothalamus controls circadian rhythmicity. The SCN synchronizes peripheral clocks and extra SCN clocks in the brain to allow the animal to express coherent behavioral and physiological rhythms that are entrained to the ambient environmental circadian cycle. The strongest timing cue for entrainment is light. However non-photic cues, such as activity, food, and social cues, also can entrain the SCN. In this chapter, the authors will review what is currently known about how non-photic cues affect the circadian system in mammals and what types of neural and physiological mechanisms may be involved. We will also discuss how this current understanding can guide the development of research paradigms to study whether non-photic cues can improve human health and to what extent non-photic cues might contribute to health problems.

Chapter IV - *Purpose:* To elucidate the underlying pathophysiology of nocturnal polyuria (NP) in older adults, their diurnal diuretic pattern was investigated, including its possible association with blood pressure and natriuresis.

Materials and Methods: A total of 46 nocturic outpatients were enrolled in study 1. Using frequency-volume charts, patients were divided into two groups: an NP group with a mean age of 68.8 years with nocturnal urine production of more than 10 ml per kg body weight (n=34), and another group with a mean age of 63.5 years without NP (n=12). Eleven non-nocturic healthy adults with a mean age of 60.2 years served as controls. Furthermore, plasma arginine vasopressin (pAVP) release on 5 % hypertonic saline loading was evaluated.

In study 2, eight inpatients with a mean age of 69.3 years with NP and mobility disabilities were subjected to diuretic-rate measurement every 3 hours, where an association between diurnal diuretic rates and the urinary solute excretion was studied.

Results: The NP group showed a greater diuretic rate during the night than in the daytime compared to the non-NP nocturic and control groups. Although the NP group had a significantly higher systolic blood pressure than the non-NP group, both groups showed a similar altered circadian rhythm of pAVP release with a lack of nocturnal surge. Moreover, nocturnal and daytime diuretic rates were correlated with systolic blood pressure only in the NP group, positively and negatively, respectively. Neither pAVP nor brain natriuretic peptide were correlated with nocturnal or daytime diuretic rates. An overall upward shift was shown by a linear function relating pAVP levels to increasing plasma sodium changes using saline loading, although there was wide individual variability. Even in study 2, a reversed diuretic pattern was seen, where the urinary sodium excretion profile was similar to subjects' diuretic pattern, and a significant correlation between the diuretic rates and urinary sodium levels was found ($r=0.58$, $p=0.0003$). The ratios of night or early morning to daytime urinary sodium excretion were significantly increased in patients compared to controls ($p<0.001$), although those with NP had a greater 24-hour total urine volume than controls.

Conclusion: Patients with NP seemed to be predisposed to a reversed diuretic rhythm and nocturnal increase in urinary sodium excretion, where increased systolic blood pressure may be associated. NP may be predicted by the increased ratios of night or early morning to daytime urinary sodium excretion.

Chapter V - Feeding and light are the two main synchronizers of human circadian clocks. In adults, the oscillator located in the suprachiasmatic nuclei may be the center of the network and the unique oscillator entrainable by light / dark cycles. In contrast with the central clock, the peripheral clocks could not be entrained by sunlight but are under the dependence of the central clock to perceive the photic signal through humoral factors like glucocorticoids, transforming growth factor alpha or prokineticin-2. Peripheral clocks may be entrained by signals issued by the SCN but could be uncoupled under conflictual conditions to follow the entrainment of synchronizers more relevant to their homeostasis. Circadian rhythms are crucial to the acquisition of long-term homeostasis by tuning individual cellular oscillators with environmental cues. The fetus (and consequently the neonate) may be entrained not only by maternal signals induced by the light / dark cycles but also by the rhythmicity of nutrient supplies from the mother which may entrain rhythmic variations of blood parameters, of locomotor activities and of the body's rectal temperature. At birth, premature neonates are exposed to a timeless environment and have to organize their rhythmicities to achieve the organ's maturation. Studies on rat neonates and primate babies propose two main synchronizers: the light / dark cycles driving the suprachiasmatic nuclei oscillator and a synchronizer related to feeding and delayed of 12 hours. Molecular data obtained on murine embryos are in favor of the hypothesis that babies of Mammals are born with a functional circadian system but have to tune their clocks with environmental synchronizers. The ontogenic development of the circadian system can be modified or imprinted by changing environmental conditions, the main consequence being on nursing of preterm infants who are reared in the relatively « timeless » conditions of the neonatal intensive care nursery. However, direct comparisons between laboratory mammals and humans are difficult because of divergent milk compositions, maternal care or maturity of the babies as well as the difficulty in obtaining biopsies.Without a proper non-invasive methodology to recover mucosal cells, there is no information about development of infant peripheral clocks, and the acquisition of rhythms are studied on sleep/rest actigraphy, electroencephalograms, heart beat and body temperature. Exfoliated cells may constitute surrogates of tissue biopsies so it is useful to obtain cells from the gastrointestinal area with some relevance to the clocks' biology. A better understanding of the acquisition of circadian rhythms by gastrointestinal epithelium may be relevant to studies on metabolic syndrome which has been related to clock genes' dysfunction and on the imprinting of neonates metabolism circuitry with potential health consequences during adult life.

Chapter VI - A relationship between autonomic nervous system function and insulin resistance has been identified and it appears very complex. The central biological clock in the hypothalamus uses the autonomic network to generate and organize metabolic rhythms. The sympathetic nervous system modulates both hepatic glucose production and the glucose uptake in peripheral tissue. Some studies demonstrated that the increase of plasma insulin level was related to increased urinary and plasma norepinephrine. Other studies, performed by microneurography, have shown that acute hyperinsulinemia increases sympathetic activity in muscle nerves Analysis of heart rate variability (HRV), a non invasive technique used to assess autonomic neural regulation, has been successfully introduced to investigate the influence of autonomic neuromodulation in several pathophysiologic conditions Several studies performed by spectral analysis of HRV have shown sympathetic hyper-activity in insulin resistant subjects with normoglycemia.

The authors have evaluated, by HRV analysis in 24-hours ECG recording, the circadian rhythm of autonomic nervous system function in insulin resistant subjects with normal glucose regulation.

Chapter VII - Administration of melatonin in variable doses resulting in supraphysiological or physiological night-time blood levels of melatonin has been used to improve sleep quality. Melatonin is a potent endogenous radical scavenger and has also been available as a food additive. The ultimate goal of melatonin use is to increase the subsequent daytime well-fare and activity. Melatonin is found in all body fluids, and it is also excreted in milk at a distinct diurnal rhythm. Night-time milk provides low doses of melatonin, which do not affect daytime blood melatonin concentrations. The aim of this study was to monitor objective sleep quality and daytime physical activity in healthy middle-aged people consuming melatonin-rich night-time milk daily or during intake of a pharmacological dose of melatonin. The study was performed in 4 two-week periods in 12 healthy middle-aged subjects. During the first phase of the double blind trial the subjects received either normal commercial milk or melatonin-rich night-time milk with their meals providing 5-20 ng melatonin daily for 2 weeks and during the second phase normal commercial milk and a placebo or 2 mg melatonin before bed-time. Activity and sleep were continuously recorded by an actigraph during the second week of each experimental period. Sleep efficiency indicated by a decrease in the number of awakenings improved significantly ($p< 0.05$) during the night-milk period compared to the normal milk period. Similar improvement was not found with the 2 mg daily melatonin treatment when compared to placebo. No significant changes were observed in daytime activity between the treatment groups. However, significantly increased daytime activity was seen in 7 out of the 12 subjects during the night-milk period compared to the normal milk period while the 2 mg melatonin treatment increased daytime activity compared to placebo only in 5 subjects. Objective recording of daytime activity may be an alternative method to monitor the effect of low supply of melatonin. The authors' findings suggest that the benefits of melatonin may be obtained with far lower doses than used until now and without a risk of daytime drowsiness.

Chapter VIII - Cardiovascular variability results from a number of associated factors, intrinsic, pseudorhythmic and transient. Although cardiovascular variability arises from many different influences, the most consistent "external" modulation is breathing.

Recognition that the respiratory sinus arrhythmia (RSA) was mediated by efferent vagal activity acting on the sinus node led investigators to attempt to quantify the fluctuations in R-R intervals that were related to breathing. Power spectral analysis of R-R intervals permits calculation amount of variance of a signal at a given frequency. Given that vagal modulation of sinus node activity occurs at the respiratory frequency, measuring the variance or power in the R-R interval time series at the respiratory frequency provides a pure index of how much the vagus is modulating sinus node activity. Low frequency (LF) waves are affected by both sympathetic and parasympathetic systems. They are correlated with baroreflex gain, although other rhythms also are closely related to baroreflex activity. Baroreflex activity modulates blood pressure and projections from the baroreflexes stimulate the hypothalamus and play a central role in general autonomic homeostasis. Cortical effects of baroreflex activity reflecting emotional effects have been shown. High frequency (HF) waves usually coincide with RSA, the increase and decrease in heart rate produced by respiratory cycle. RSA is produced by a combination of respiration-induced biochemical changes, changes in intrathoracic pressure and central vagal stimulation. Although RSA amplitude is often considered to be an index of vagal tone, it can be dissociated from vagal influence on tonic heart rate.

To test the modulatory effect of respiration on circadian rhythm of the cardiovascular variability, we have analyzed the heart rate variability (HRV) in healthy zen practitioners with 5 years of experience. Practice of zen was characterized by a rhythmic and slow breathing. HRV analysis has shown a reduction in HF component, an increase in LF component and LF/HF ratio during the practice of zen. These results were reported also in night time, when zen practitioners were sleeping.

In conclusion, repeated training to slow down breathing reduces the spontaneous breathing rate with a long term effects on the cardiovascular control mechanisms. Indeed, when respiration slows to about 6 cycles/min, as in zen practitioners, in the frequency range of the spontaneous LF oscillation, the cardiovascular fluctuations become maximal. The changes in the breathing rate might modify the chemoreflex and the continuos practice in slow breathing can reduce chemoreflex. This change in the automonic control of respiration can be permanent with a resetting of endogenous circulatory rhythms.

Chapter IX - As for many human tissues and organs, the presence of circadian rhythms on eye parameters has been largely demonstrated over the last years. The interest on the role of 24-hour fluctuations in eye parameters interestingly grew up as basic and clinical research progressively put lights on the pathogenetic features of many ocular diseases..

Up-to-now, the main fields of interest is represented by the differences in the 24-hour intraocular pressure (IOP) curve between normal and glaucomatous patients with and without treatment. In fact, a different circadian effect has been shown when comparing comparing medical, laser and surgical treatment of glaucoma; anti-glaucomatous eyedrops are also characterised by peculiar mechanisms of action which reflect on the different effects on 24-hour IOP curve flattening power. Moreover, IOP assessment is influenced by corneal thickness, which is also influenced by circadian fluctuations. Finally, data on 24-hour changes in ocular perfusion and in melatonin shading have been published very recently.

The aims of the present article is to review the published literature in order to ascertain the role of circadian rhythms, both in healthy and pathological eyes, and to investigate their pathogenetic and clinical impact in eye diseases.

Chapter X - The effects of circannual changes in photoperiod and environment on reproductive performance in the pig are reviewed. The ancestral pig is a short day seasonal breeder and therefore seasonal effects on fertility of the pig may be regarded as physiological by nature. In principle, the breeding season in the pig is though to be controlled by the central nervous system, synchronized by the ambient photoperiod and further modified by other environmental cues such as availability of food and ambient temperature. Like in other short day seasonal breeders, breeding season in the pig is favoured in the mid winter in order to provide the offspring with best chances to survive in the following spring four months later. Under hard circumstances when food is difficult to find and temperatures fall too far below the thermal comfort zone, the breeding season of the pig may be delayed. In the modern pig industry, several parameters of reproductive performance indicate seasonal fluctuation in reproductive efficiency corresponding well to the seasonality of breeding seen in the wild for this species. Within the industry, this phenomenon is known as seasonal infertility and is characterised by lower pregnancy rates, lower proportion of females in oestrus when expected and delayed puberty when the pigs are kept in long days. Seasonal infertility is seen as an indicator period of the year, revealing the level of reproductive management and stress in a herd. Low pregnancy rates are likely due to a seasonal disruption of pregnancy, which is economically important and may be caused by a hormonal mechanism. Central effects of the pineal melatonin may suppress LH pulsatility and thereby reduce the pituitary support for CL, retarding the embryonic development and interfering with the physiological mechanism for establishment of pregnancy. Detrimental seasonal effects on reproduction may be avoided by adequate lighting programs involving alternation of short and long days, by providing female pigs with abundant feed, by using breed records for genetic selection, by using of cooling systems where the upper thermal comfort zone is exceeded and by providing the pig with adequate shelter, bedding material and freedom to move under cold conditions.

Chapter XI - Through centuries the man's life was dependent on the twenty-four hour cycle of day and night. The factor that seriously limited undertaking of work on a larger scale in the night time was lack of proper lighting. Invention of the bulb by Thomas Alva Edison gradually resulted in usage of the rest of the twenty-four hours for working. For that phenomenon the period of the First World War was crucial, when the considerable needs of the armaments industry contributed to the mass introduction of the night work in Europe. In this chapter, the authors talk about different problems of the "24-hour society". The part of workers are engaged on the non-standard working hours (near shift and night work - part time work, weekend work, varying working hours, split shifts, seasonal work, on-call work and compressed work week). The shift work is a reality for about 25 percent of the North American working population.

Chapter XII - The majority of the researches on circadian rhythms at the cellular and tissue level were done in mammalian species such as rodents, and there is significant knowledge accumulated about it, especially on the circadian rhythm in the cell proliferation and migration in the small intestinal epithelium that will be subject of this chapter. For this proposal the authors divided the chapter in the following topics: 1) Circadian variation in cell proliferation and migration in the small intestine of young and adult animals; 2) General aspects of the circadian variation regulation in cellular and tissue level; 3) Role of the circadian variation in the cell proliferation during growth of the small intestine and 4) perspectives.

In: Circadian Rhythms and Health Research Trends
Editor: Alfredo V. Lignelli, pp. 1-30 ISBN: 978-1-60021-875-0

Chapter I

The Human Biological Clock: From Genes to Chronotherapy

Michel A. Hofman*
Netherlands Institute for Neuroscience
Amsterdam, the Netherlands

Abstract

The suprachiasmatic nucleus (SCN) and the pineal gland are critical components of a neural oscillator system implicated in the timing of a wide variety of biological processes. The circadian cycles established by this biological clock occur throughout nature and have a period of approximately 24 hours. With advancing age these daily fluctuations deteriorate, leading to disrupted cycles with a reduced amplitude. In humans, age-related changes have been described for hormonal rhythms, body core temperature, sleep-wakefulness, and several other behavioral cycles. It appears that the disruption of circadian rhythms and the increased incidence of disturbed sleep during aging are paralleled by age-related alterations in the neural and temporal organization of the circadian oscillator and a decreased photic input to the clock. The observed neuronal degeneration of the SCN-pineal complex in senescence strongly suggests an organic deterioration of the circadian oscillator during aging and supports the idea that damage to the biological clock is the underlying anatomical substrate for the clinically often-observed disturbances in circadian rhythmicity in mood disorders and Alzheimer's disease. Recent studies furthermore indicate that illuminance and photoperiod may be considered as potential environmental factors controlling the functional activity of the circadian timing system. If light is the effective Zeitgeber in man, as it is in many other organisms, disturbances in processing of light information during aging or in some neurodegenerative and affective diseases may have profound effects on the timing of a variety of physiological and behavioral activities, including sleep. In view of the assumed sensitivity of the biological clock to changes in day length and light intensity research is

* Correspondence: M.A. Hofman. Ph.D. Netherlands Institute for Neurosciene, Meibergdreef 47, 1105 BA Amsterdam. THE NETHERLANDS. Telephone number: +31-20-5665500. Fax number: +31-20-6961006. E-mail: M.Hofman@nin.knaw.nl

now focussing on whether the synchronizing and antidepressant effects of bright light exposure in sleep disturbances and mood disorders can be improved by applying the therapy in temporal coherence with the circadian cycles underlying the neuronal activity of the biological clock.

Keywords: Suprachiasmatic nucleus - Pineal gland - Human brain - Hypothalamus - Circadian oscillator - Clock genes - Biological clock - Neuroplasticity - Sexual differentiation - Aging - Alzheimer's disease.

1. Introduction

Many organisms experience circadian oscillations in various biological processes (e.g., neuroendocrine, autonomic, cardiovascular, sleep-wake cycle). Circadian rhythms enable organisms to anticipate periodic changes in the environment and are consequently important adaptive mechanisms. In mammals, these circadian cycles are regulated by an endogenous clock, the central component of which resides in the suprachiasmatic nucleus (SCN) of the hypothalamus.

Since the discovery of the SCN as the site of the master circadian pacemaker, many attempts have been undertaken to unravel the mechanisms underlying its endogenous circadian rhythmicity. In particular, lesion and transplantation experiments of the SCN and *in vitro* slice studies have provided firm evidence for its biological clock characteristics (for reviews, see Buijs et al., 1996; Van Esseveldt et al., 2000; Panda et al., 2002). Lesioning the SCN results in a disappearance of most circadian rhythms and makes the animal arrhythmic, while transplantation of fetal SCN tissue may restore circadian rhythmicity in such lesioned animals. When the SCN is removed from the brain and maintained in a slice preparation, the neurons continue to generate circadian rhythms in electrical activity, secretion and gene expression. Consistent with its role in the temporal organization of circadian processes, investigations in rodents and non-human primates suggest that the SCN is also involved in the seasonal timing of reproduction, sexual behavior and energy metabolism (for a review, see Hofman, 2004).

Studies in humans also seem to support the notion of the SCN being the principal neural substrate that organizes and coordinates circadian rhythms. Clinically documented disruption of circadian behavior shows involvement of the SCN region (Schwartz et al., 1986; Cohen and Albers, 1991) and age-related decrements in circadian timekeeping have been attributed to the observed neuronal degeneration of the SCN in senescence (Swaab et al., 1985, 1996; Moore, 1991; Hofman, 2000; Hofman and Swaab, 2006). It appears that photic information may have a synchronizing effect on the clock mechanism of the SCN by inducing changes in the functional activity of groups of neurons.

Increasing age and a variety of diseases, however, may impair many of these functions and may have deleterious effects on the neuronal organization and biological activity of the clock. In humans, age-related changes have been described for hormonal rhythms, body core temperature, sleep-wakefulness, and several other behavioral cycles. It appears that the disruption of circadian rhythms and the increased incidence of disturbed sleep during aging are paralleled by age-related alterations in the neural and temporal organization of the circadian oscillator and a decreased photic input to the clock.

The observed neuronal degeneration of the SCN-pineal complex in senescence strongly suggests an organic deterioration of the circadian oscillator during aging and supports the idea that damage to the biological clock is the underlying anatomical substrate for the clinically often-observed disturbances in circadian rhythmicity in mood disorders and Alzheimer's disease (Hofman, 2000; Skene and Swaab, 2003; Hofman and Swaab, 2006). Recent studies furthermore indicate that illuminance and photoperiod may be considered as potential environmental factors controlling the functional activity of the circadian timing system. In this chapter, recent data on the circadian organization of the SCN-pineal complex in humans are discussed in relation to aging and Alzheimer's disease.

2. Organization of the Human SCN

2.1. Neuronal Organization

The human SCN is a small collection of parvocellular neurons in the basal part of the anterior hypothalamus, just dorsal to the optic chiasm on either side of the third ventricle (Figure 1). The bilateral SCN in humans is about 1 mm^3 in volume and contains a total number of neurons close to 100,000 (Hofman et al., 1988; Hofman and Swaab, 2002). Based on differences in morphology, afferent inputs and output projections, the SCN can be divided into a dorsomedial part or "shell" and a ventrolateral part or "core" (Mai et al., 1991; Hofman et al., 1996; Moore and Silver, 1998; Antle and Silver, 2005). Neurons in the dorsomedial region of the SCN, for example, are small and poor in dendritic arbor and contain arginine vasopressin (AVP) and neurotensin, whereas those in the ventrolateral region of the nucleus are relatively large with extensive denditric arbors and contain vasoactive intestinal polypeptide (VIP), gastrin-releasing peptide, neurotensin, neuropeptide-Y, substance-P and calbindin (Mai et al., 1991; Moore, 1992; Romijn et al., 1999; Moore et al., 2002; Figure 2). It has been established that both AVP- and VIP-expressing neurons send efferents to the rest of the brain, constituting the main output pathway of the clock (Dai et al., 1997; Buijs and Kalsbeek, 2001). Many neurons in both regions of the SCN contain γ-aminobutyric acid (GABA), an inhibitory neurotransmitter in the brain (Moore et al., 2002; Saper et al., 2005).

The ventrolateral subdivision of the SCN receives retinal input from the retino-hypothalamic tract (RHT) and secondary visual projections from the intergeniculate leaflet of the lateral geniculate complex via the geniculo-hypothalamic tract (GHT). The RHT terminals contain glutamate, as well as pituitary adenylate cyclase-activating polypeptide, which code chemically for 'light' and 'darkness' information, respectively, while the GHT has been characterized in several rodents by the neuropeptide-Y and GABA content of its neurons (Van Esseveldt et al., 2000; Shirakawa et al., 2001; Hannibal, 2002; Moore et al., 2002).

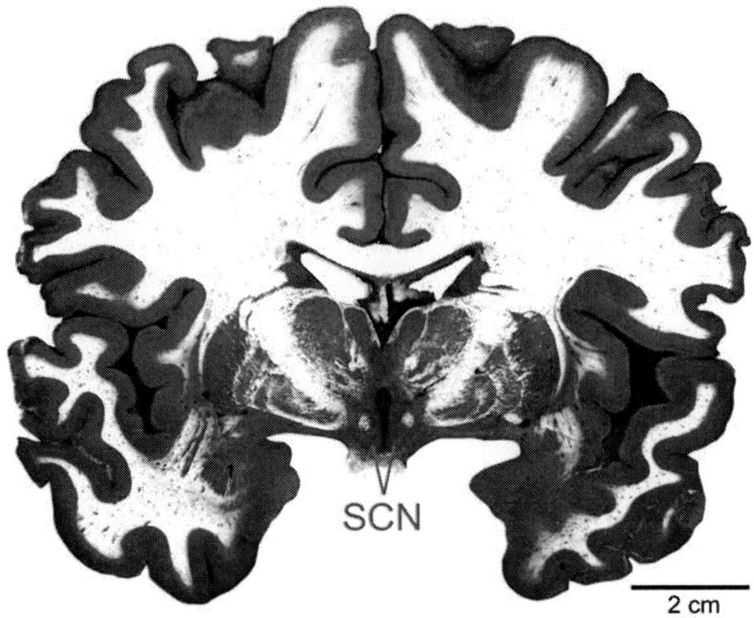

Figure 1. Coronal section through the human brain at the rostral end of the diencephalon to show the anterior part of the hypothalamus with the suprachiasmatic nucleus (SCN).

2.2. Functional Organization

The 24 h oscillations of biological processes observed in a broad spectrum of organisms are controlled by internal clocks. Although these rhythms are generated endogenously, they do not function in isolation from their surroundings. Rather, they are entrained by external cues so they can anticipate the natural temporal variations of the environment. Classically, the expression of circadian rhythms is thought to have three fundamental components: input pathways that transmit environmental cues to the circadian clock, the clock itself, which generates and coordinates the biological rhythm(s) and output pathways that transmit the clock's information regarding phase and periodicity to the rest of the brain and body (Buijs and Kalsbeek, 2001; Kalsbeek and Buijs, 2002, 2006; Hofman, 2004).

The biological clock is synchronized to the environmental light-dark cycle by photic information from the retina to the SCN. These afferent fibers predominantly terminate in the ventral part of the SCN, where they make synaptic contacts with VIP neurons (Reuss, 1996). The VIP subdivision of the SCN is therefore thought to play a prominent role in the mediation of photic information to the circadian timing system. The direct neuronal pathway from the retina to the SCN also exists in humans, as was shown by staining degenerating neurons in patients who had incurred optic nerve damage prior to death (Sadun et al., 1984) and by *in vitro* postmortem tracing studies (Dai et al., 1998).

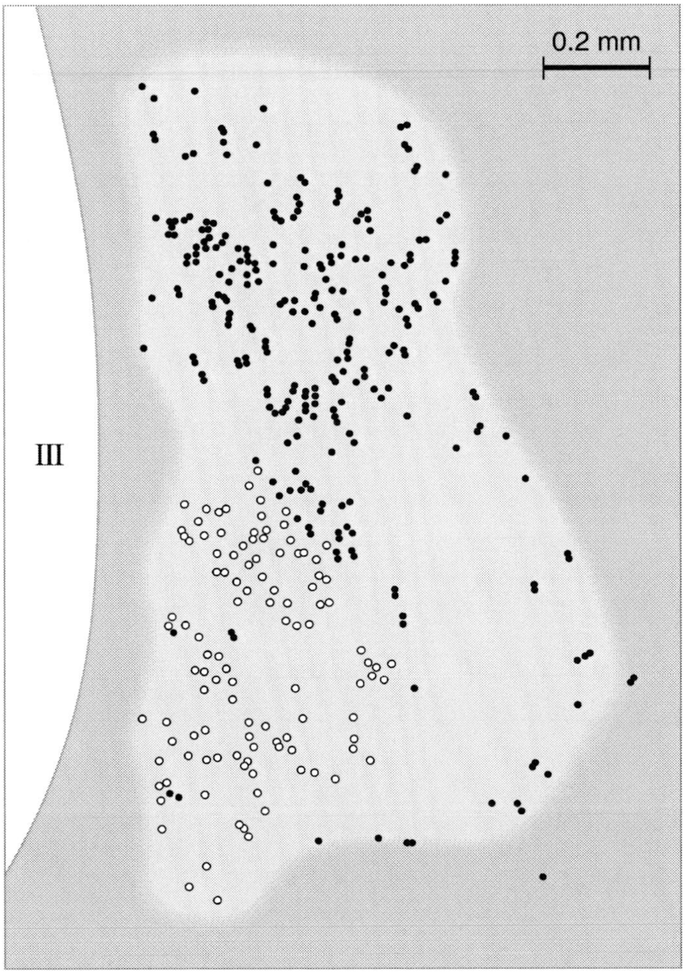

Figure 2. Distribution of arginine vasopressin (AVP)-expressing and vasoactive intestinal polypeptide (VIP)-expressing neurons in the human suprachiasmatic nucleus (SCN). Drawing showing the distribution of the AVP neurons (dots) and VIP neurons (open circles) through the mid-portion of the SCN. III = third ventricle. Modified from Hofman (2000).

2.3. Temporal Organization

In humans and other species living under semi-natural lighting conditions, little is known about the role of VIP neurons in the temporal organization of the biological clock. Totally blind people often lack the entraining effects of light, as do many subterrestrial species, and may show free-running temperature, cortisol and melatonin rhythms. They may also suffer from sleep disturbances (Sack et al., 1992). Surprisingly, some blind people maintain circadian entrainment and show light-induced suppression of melatonin secretion, despite the apparent total lack of pupillary light reflexes and with no conscious perception of light (Czeisler, 1995). It has been proposed that in these patients the retino-hypothalamic pathway

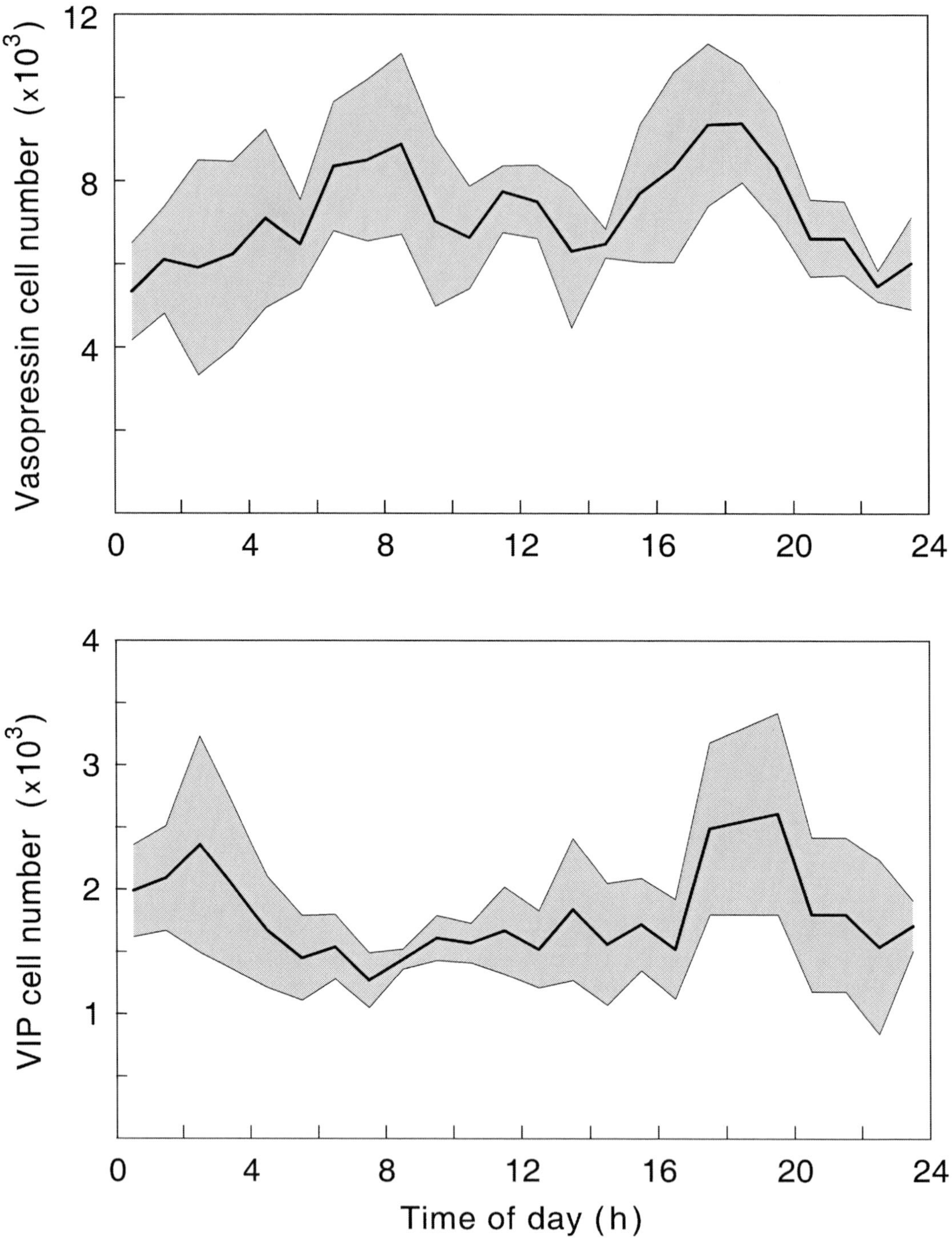

Figure 3. Circadian oscillations in the biological activity of the AVP- and VIP-expressing neurons in the human suprachiasmatic nucleus (SCN). Note the asymmetrical, bimodal waveform of the cycles. The data are represented by mean±SEM values. Reproduced from Hofman (2003).

that passes through the SCN would still be intact, but that the exact nature of the circadian photoreception in these patients is unknown. It might be of practical importance to distinguish these patients, since in this group enucleation might cause recurring insomnia and other symptoms associated with the loss of circadian rhyhthmicity.

In the past decade, several studies have been conducted to determine whether variations in light intensity and photoperiod affect the morphology and neuronal activity of the mammalian SCN, by studying brains obtained at autopsy (for reviews, see Hofman and Swaab, 1992, 2002; Swaab et al., 1996). The AVP- and VIP-expressing neurons in the human SCN, for example, were found to exhibit distinct circadian rhythms with an asymmetrical, bimodal waveform (Hofman and Swaab, 1993; Hofman, 2000, 2003; Figure 3). The AVP cycle has a peak in the early morning, a lower plateau during the day, a second peak in the late afternoon, and a decline beginning in the early evening, leading to a nadir around midnight. The VIP cycle shows a peak in the middle of the night, a lower plateau beginning in the late night and lasting for about 12 hours, and a second peak in the late afternoon, followed by a sharp decline in the early evening.

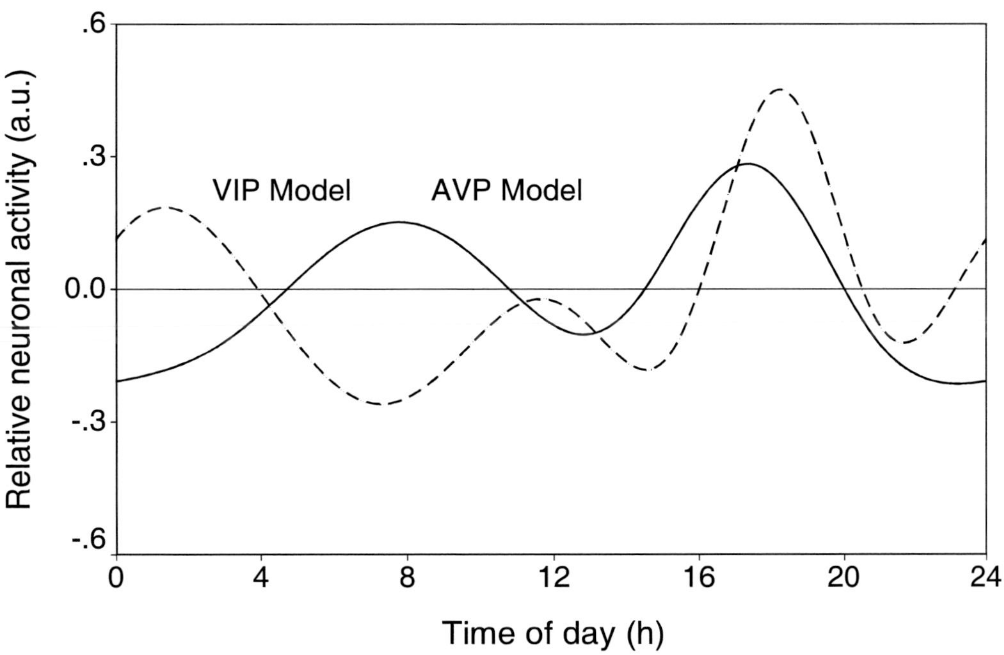

Figure 4. The nonlinear periodic functions describing the circadian AVP and VIP cycles in the human suprachiasmatic nucleus (SCN). The model of the AVP cycle reaches an estimated maximum at the end of the afternoon (tmax = 17.3 h) and a minimum around midnight (tmin = 23.2 h). The model of the VIP cycle also reaches its maximum at the end of the afternoon (tmax = 18.2 h), but has its minimum in the early morning (tmin = 7.3 h), considerably later than the minimum of the AVP cycle. Reproduced from Hofman (2003).

Time series analysis, furthermore, indicated that the circadian cycles in the SCN can be adequately described by a model consisting of nonlinear periodic functions, which could be decomposed into mono- and diphasic cycles, with periods of 24-hours and 12-hours, respectively (Hofman, 2003; Figure 4). The demonstration of two significantly different, but temporally linked, output profiles strongly suggest that the circadian clock in the SCN consists of a multioscillator system.

2.4. Multioscillator Circadian System

While it is still difficult to come up with a definitive "circuit diagram" for the biological clock, a recently proposed theory, based on anatomical and biochemical studies, considers the clock as a multioscillator system, consisting of coupled, independently operating subdivisions located in the core and shell of the SCN (Moore and Silver 1998; Leak and Moore 2001). The inputs and outputs of these two regions also appear to be distinct, indicating that the nature of the signal conveyed to areas receiving core and shell projections varies as a function of the subdivision from which the innervation is derived.

It is likely that the period of the rhythmic output from the core and shell is the same, as the distinct effector sites of these subdivisions need to be entrained to the same sidereal time. By contrast, differential inputs to the core and shell should control phase, amplitude, or waveform of their rhythms separately. The specific neurological organization and information flow in the biological clock might explain why the circadian profiles of the AVP are somewhat different from those of the VIP cycle. Although the exact nature of the biological clock and the means by which it is synchronized to the 24-hour photic cycle is still elusive, the mathematical models describing the circadian rhythms in the human clock suggest that both the core and the shell of the SCN contain circadian oscillators. From rodent studies we know that the dorsomedial part of the SCN indeed contains a circadian pacemaker. The ventromedial part of the SCN, on the other hand, receiving primary and secondary visual afferents, may reflect the external light-dark cycle, rather than endogenous oscillator activity.

The notion of separate circadian pacemakers in the SCN was recently supported by a differential entrainment of two components in the circadian rhythm of SCN neuronal activity to the light-dark cycle (Jagota et al., 2000). Also the circadian rhythm of *c-fos* expression and that of *Per1* and *Per2* expression, two clock genes implicated in the molecular mechanisms of the circadian pacemaker, were found to be slightly different between the dorsomedial and ventromedial regions of the SCN (Schwartz et al. 2000; Dardente et al. 2002). It does not necessarily mean that the AVP- and VIP-expressing cells in the SCN are pacemaker cells. In fact, the role of neurotransmitters and neuromodulators in the generation of circadian rhythms is unclear. So far, none of the peptides found in the SCN has been reported to play a crucial role in the circadian oscillator system. Instead, some of the peptide genes may act as clock-controlled genes and their transcripts may drive the circadian output of the clock (for reviews, see Dunlap 1999; Cermakian and Sassone-Corsi 2000; Hastings and Maywood, 2000; King and Takahashi, 2000; Hastings et al. 2003). In other words, circadian oscillations in neuropeptide levels in the SCN are probably part of the output system of the circadian clock, and are not an intrinsic element of the circadian pacemaker.

It is evident that there must be some kind of coupling for the two components of the oscillating system in the SCN to maintain their mutual phase relationship under free-running conditions. There are various ways how the coupling of these oscillators may be envisaged, depending on whether the oscillators are restricted to different populations of neurons in the SCN or that they are colocalized in the same neuron. In the coupling interaction, it is likely that there is some degree of asymmetry in the strengths of the two oscillators (see e.g., Daan et al. 2001; Wehr 2001; Hofman 2003). Cell-to-cell communication is clearly necessary for conveying inputs to and outputs from the SCN and may be involved in ensuring the high precision of the observed rhythms. Although at present the details of how time of day is encoded by the molecular clockwork are not known it is clear that there are properties in the behaviors of the clock genes that could be interpreted as circadian oscillators.

From a functional perspective the main attraction of the dual oscillator model is that it uniquely counts for the adjustment of circadian organization to season and latitude and that the phase relationship of these oscillators or the temporal organization of the expression of their genes reflect the photoperiod, and thus time of year, to which an animal has been exposed (Daan et al. 2001; Hofman 2003; 2004). This dual oscillator model also has obvious implications for the evolution of circadian timing. The rotation of the earth provides two precise timing signals, sunrise and sunset, rather than a single one. It might well be that life that evolved on its surface has developed ways to exploit both signals, for the fine tuning of temporal organization over the 24-hours, as well as for adapting its physiology and behavior to the seasonal fluctuations in photoperiod.

3. Molecular Genetics of Circadian Timing

3.1. Clock Genes and the Suprachiasmatic Nucleus

As internal systems, circadian clocks arise from the expression and temporally regulated activity of specific genes and gene products. The past decade has witnessed rapid breakthroughs in our understanding of how the circadian clock functions at the cellular and molecular levels (for reviews, see King and Takahashi, 2000; Reppert and Weaver, 2001; 2002; Hastings and Herzog, 2004). Individual SCN neurons express self-sustained circadian oscillations driven by autoregulatory transcription-translation feedback loops. The period of this process is approximately 24 hours, giving rise to the circadian rhythm. Beginning with the molecular cloning of the mouse *Clock* gene at least six genes now have been identified as putative components of the mammalian clock: *Per1*, *Per2*, *Cry1*, *Cry2*, *Clock* and *Bmal1*. Additional genes potentially involved in output pathways from and input pathways to the circadian clock have been identified. Oscillations of mRNA transcripts of these clock genes have been an important element of the study of biological clocks.

Based on the oscillation patterns of the *Period* genes within the SCN (high during the day, low during the night), the mammalian clock was predicted to be a "day-active" clock, in which the oscillating components of the clock are at higher levels during daytime than during the night. However, analysis of the phase response curve (PRC) to light shows that the circadian clock is more greatly affected at nighttime than during the day (daytime is within a "dead zone" during which light exposure has little or no effect on the phase of the clock) (see e.g. King and Takahashi, 2000). Thus, during the dead zone of the PRC, the oscillating clock molecules are abundant, and during the active phase of the PRC, these molecules are scarce.

What follows from this is the prediction that this clock would be phase shifted (and thus entrained) not by degradation of clock molecules following zeitgeber exposure but rather by the induced expression of clock molecules.

Although phase information is essential for a functioning clock, multiple clock components may work in concert to invest this phase information in the cycling activity of one or few elements of the system. In this scenario, all of the components are important for the presence of a circadian rhythm and thus should be considered as elements of the clock, even though only a subset directly reflects the current phase of the clock. Such a scenario appears to be the case for the mammalian circadian clock, in which mRNA abundance of some components exhibit high-amplitude circadian oscillations while others have little or no apparent oscillations over the 24-hour cycle. The nearly 24-hour oscillation is thought to depend on delays between transcription, translation, and subsequent transcriptional repression, but how these delays are established is only beginning to be understood. Insights into the steps modulating synthesis, turnover, and localization of key mRNAs and proteins will inevitably modify the model for molecular timekeeping. The molecular loop must broadcast its timing signal to the SCN neuron and beyond that to brain and body. This is achieved, in part, by the temporally regulated expression of a series of clock-controlled genes.

3.2. Organizing Principles of Cellular Clocks

The human SCN is not simply a collection of 100,000 clock cells: rather, it is a heterogeneous structure composed of multiple functional compartments. Indeed recent studies challenge the view that the SCN is functionally homogeneous (see e.g., Hastings and Herzog, 2004; Antle and Silver, 2005). Some SCN cells are not endogenously rhythmic with respect to clock gene expression; only some cells receive direct retinal input and immediately express clock genes following photic stimulation. Furthermore, individual clock cells within the SCN exhibit various phases and free-running periods (Nakamura et al., 2001; Quintero et al., 2003; Yamaguchi et al., 2003). These phenomena are difficult to explain by coupling alone. An alternative 'network' model suggests that it is precisely this heterogeneity of the SCN that is integral to keeping the population of oscillators ticking in a coherent fashion (Antle et al., 2003). This model hypothesizes that the phases of individual oscillators are synchronized by a daily signal from a subset of SCN cells. By pulling the phases closer together, the overall output of the system remains rhythmic.

Central to this model are recent discoveries demonstrating that the morphological and peptidergic heterogeneity of the SCN reflects its functional heterogeneity. Rather than being uniformly distributed in the SCN, these intrinsically rhythmic cells are largely confined to the SCN shell, occupying roughly the same area as AVP-containing cells (Hamada et al., 2001, 2004; Yan and Okamura, 2002;). This region of the SCN receives little retinal innervation and displays delayed clock gene expression following phase-shifting light exposure (Hamada et al., 2001).

Cells in the SCN core receive direct retinal innervation and express c-fos, *Per1* and *Per2* in response to phase-shifting light pulses (Romijn et al., 1996; Yan et al., 1999; Hamada et al., 2001; Yan and Silver, 2002). It has been observed that cells in the core of the mouse SCN either lack rhythmic expression of *Per1* and *Per2* (Karatsoreos et al., 2004) or express these genes in antiphase to the expression in the SCN shell (King et al., 2003). Cells in the core oscillate in their responsiveness to photic input: although light exposure always increases firing rates in SCN neurons (Meijer et al., 1992), light induces clock gene expression in the SCN only during the night (Reppert and Weaver, 2001).

Although expression in each individual cell peaks at a particular time of the day that is stable for that cell over several cycles, peak expression for the population as a whole occurs during the middle of the day. When the dorsal third of the nucleus is surgically separated from the ventral portion that contains the core, coordinated rhythmic output persists in only the ventral portion (Yamaguchi et al., 2003). Individual cells in the dorsal portion remain rhythmic but coherent phase relationships among cells are lost, resulting in a tissue without net rhythmic output.

3.3. Clock Gene Expression Patterns

For the circadian clock to be phase-shifted or entrained by light, photic information must be relayed from the lightinduced cells to the rhythmic cells, which must respond appropriately to the resetting signal (with delays and advances to early-night and late-night light exposure, respectively). Detailed examination of the temporal and spatial patterns of gene expression reveal that *Per1* and *Per2* are regulated separately in a phase-specific and region-specific manner (Yan and Silver, 2002; 2004). Following phase-shifting light pulses, *Per* expression is first induced in the SCN core (Yan et al., 1999; Hamada et al., 2001; 2004; Kawamoto et al., 2003). In the mouse, early-night delaying light pulses induce *mPer1* and *mPer2* expression initially in the SCN core, followed later by *mPer2* expression in the SCN shell. Late-night advancing light pulses induce expression of *mPer1* but not *mPer2*. This *mPer1* expression occurs first in the core and later in the shell.

Mid-night light pulses that do not alter circadian phase induce *mPer1* expression only in the SCN core. Based on these findings, it appears that the spread of *mPer2* expression to the shell underlies phase delays, whereas a similar spread of *mPer1* underlies phase advances (Yan and Silver, 2002; 2004). Phase delays to light are impaired in *mPer2* mutant mice (Albrecht et al., 2001; Spoelstra et al., 2004) and following *Per2* antisense treatment (Wakamatsu et al., 2001). The role of *Per1* in light-induced phase shifts is less clear, although *Per1* antisense treatment also attenuates phase delays.

Cells in the SCN shell have a specific spatial arrangement. Daily rhythmic expression of *Per1*, *Per2* and AVP does not occur in all cells simultaneously, but rather spreads and recedes through the SCN (Hamada et al., 2004). Rhythmic expression of *Per1*, *Per2* and AVP starts in a small group of cells located adjacent to the third ventricle in the dorsomedial SCN and spreads ventrolaterally over 4–8 hours. It then recedes until expression is again limited to the dorsomedial region. The same pattern of expression has been observed in vitro using transgenic mice in which luciferase reports *Per1* expression (Yamaguchi et al., 2003). Expression initially occurs in the dorsomedial periventricular SCN and spreads slowly to the ventral SCN over several hours.

Blocking protein synthesis or electrical activity reveals functional aspects of SCN organization. Because the intracellular clock is driven by an autoregulatory transcription–translation feedback loop, individual clocks can all be reset to a common phase by a protein synthesis inhibitor. When the inhibitor is removed, cells begin to oscillate in phase with one another, but re-establish their original phase relationships after several cycles. Furthermore, maintaining this phase relationship requires electrical signaling, either between individual cells or between the core and the shell. When action potentials are blocked with tetrodotoxin, circadian patterns of luminescence in individual cells persist but the system as a whole becomes desynchronized (Yamaguchi et al., 2003). These data reveal that, although individual cells within the SCN can sustain oscillations, the specific phases of such oscillations are regulated by the network properties of the SCN.

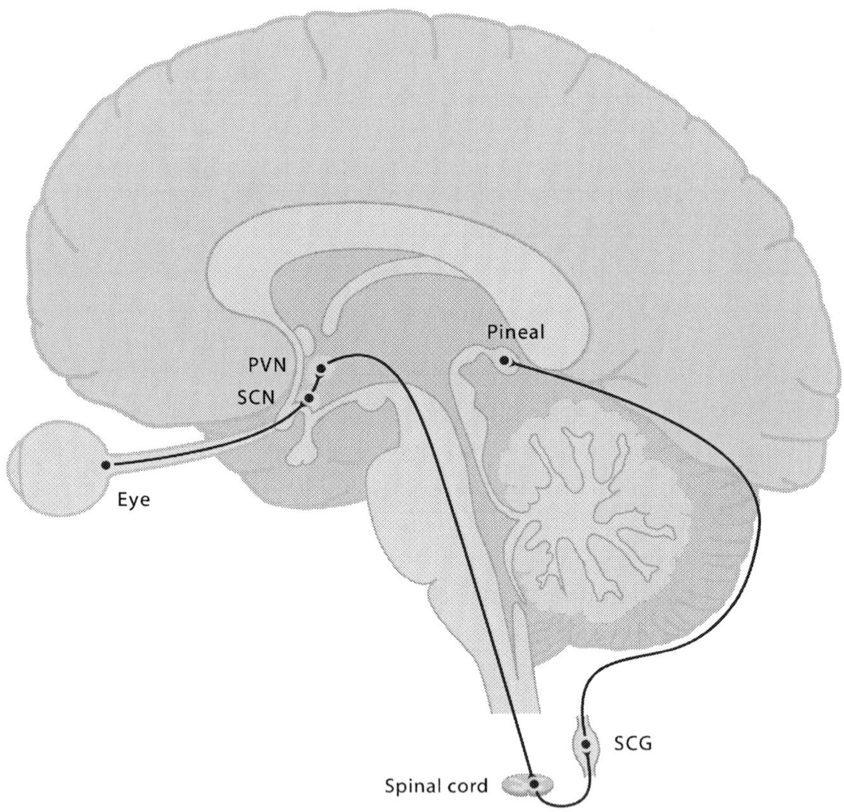

Figure 5. Diagram of the human brain (mid-sagittal section) showing the multi-synaptic pathway of the circadian timing system by which environmental light information reaches the SCN and the pineal gland. SCN = suprachiasmatic nucleus; PVN = paraventricular nucleus; SCG = superior cervical ganglion. Modified from Hofman and Swaab (1992).

4. The SCN-Pineal Complex

4.1. Circadian Timing

The pineal gland is a central structure in the circadian timing system and the major source of the hormone melatonin (Arendt, 1995). The pineal is innervated by a neural multi-synaptic pathway originating in the suprachiasmatic nucleus (SCN) of the hypothalamus (Buijs and Kalsbeek, 2001; Figure 5).

Disruption of any portion of this pathway from the SCN to the pineal gland abolishes melatonin rhythmicity. In all animals studied to date, whether they are diurnal or nocturnal, there is a day/night variation in pineal melatonin production with peak concentrations occurring during the dark phase. The circadian clock and its output rhythms are synchronized to the 24 h light/dark cycle by ocular light which is transmitted from the retina primarily via the retinohypothalamic tract (RHT) to the SCN. Although functions of this hormone in humans are mainly based on correlative observations, there is some evidence that melatonin stabilises and strengthens coupling of circadian rhythms, especially of core temperature and sleep-wake rhythms (Skene and Swaab, 2003; Claustrat et al., 2005).

4.2. Seasonal Timing

In recent years it has also become clear that in mammals the SCN and the pineal gland are the principal neural structures involved in the regulation of annual cycles (for a review see, Hofman, 2004). In fact, many of the functions that exhibit seasonal cycles in mammals, such as sexual behavior, energy metabolism, food intake, and hibernation, are regulated by this timing system in the brain. Photoperiodic information has been shown to be the strongest synchronizer of seasonal functions in most species. These findings strongly suggest that the endocrine activity of the mammalian pineal is under neural control, and receives a major input from the SCN. This means that in addition to its role as a circadian pacemaker, the SCN may also be involved in the seasonal timing of a number of physiological and behavioral processes by regulation of the photoperiod-dependent changes in melatonin secretion.

5. Biological Clock and Aging

5.1. Circadian Rhythmicity and Aging

With advancing age the circadian timing system is progressively disturbed, both in humans and other mammals, as is clearly demonstrated by a reduced amplitude and period length of circadian rhythms and an increased tendency towards internal desynchronization (for reviews, see Van Someren, 2000a). In humans, age-related changes have been described for hormonal rhythms, body core temperature, sleep-wakefulness, and several other behavioral cycles (see, e.g., Cornélissen et al., 1992; Monk and Kupfer, 2000; Touitou and Haus, 2000; Van Someren et al., 2002; Avidan, 2005). Weitzman and colleagues (1982), for example, who studied young and elderly subjects under conditions of temporal isolation, reported a reduction in amplitude and period of body temperature rhythm.

A similar reduction in the amplitude of the circadian sleep-wake cycle was found in long-term registrations of activity patterns in young and elderly men (Renfrew et al., 1987), and even more so in patients with Alzheimer's disease (Witting et al., 1990; Van Someren et al., 1996; Bhatt et al., 2005). The increase in time spent in both wakefulness during the night and daytime naps, a characteristic pattern in many elderly people, is a symptom of disruption of circadian sleep rhythms similar to the one found following experimental SCN lesions (Eastman et al., 1984). In fact, disorders of the circadian timing system during aging may first be manifested as sleep-wake pathologies (Turek et al., 1995; Hofman, 1997; Dijk et al., 2000; Van Someren, 2000b).

The question now is why the timing and consolidation of sleep and other physiological rhythms change so greatly with age? Observations in animals suggest that there are age-related changes in the properties of the circadian oscillator, which regulates the timing of sleep, body temperature and other variables within the 24-hour day. In old animals the circadian pacemaker oscillates more rapidly and with a lower amplitude than in young ones (Morin, 1988; Satinoff et al., 1993). Moreover, a change in the molecular responsiveness of the pacemaker to photic stimulation has been found in aged rats (Sutin et al., 1993).

5.2. Clock Genes and Aging

Recent experiments in rodents show that the age-related deterioration in circadian pacemaker function occurs at the level of individual cells in the SCN, possibly accounting for age-related deficits observed in the expression of behavioral rhythmicity (Aujard et al., 2001). Indeed aging is found to be correlated with decreased expression of several of the genes that make up the molecular machinery of the circadian clock, including both *Clock* and its binding partner *Bmal1* (Asai et al., 2001; Weinert et al., 2001; Kolker et al., 2003; 2004).

Kolker and colleagues, for example, reported that age alters the 24-hour expression profile of *Clock* and *Bmal1* in the SCN of hamsters (Kolker et al., 2003). However, there was no effect of age on the 24-hour profile of either *Per1* or *Per2* in constant darkness, although light induced less *Per1,* but not *Per2* in the SCN of old hamsters. Given the importance these genes have in the generation of circadian rhythms one might expect that expression of either or both of these *Per* homologs would be decreased in the SCN of old animals. Asai et al. (2001) found similar results in rats, while Yamazaki et al. (2002) found that advanced age does not dampen rhythmic expression of a *Per1*-driven luciferase in the SCN of rats. It seems that while induction of the *Per* homologs by light are altered in old age, the daily rhythm of expression of these genes is similar in the SCN of young and old rodents.

Weinert and colleagues, who investigated the expression of clock genes in the SCN of young and old laboratory mice found that the expression of *mPer1* and *mPer2* was rhythmic in both groups, with peak values at CT17 (Weinert et al., 2001). The levels of *mClock* and *mCry1* were not different depending on the time of day and did not vary with age. In contrast, an age-dependent difference was observed in the case of *mPer2* (but not *mPer1*) expression, with the maximum at CT17 significantly lower in old mice. Clock gene expression in peripheral tissues (liver and heart) also seems to be affected by aging (Claustrat et al, 2005). The liver of old rats exhibits a significant lower level of *Per* gene expression in the evening compared to middle-aged rats. In the heart of old rats similar 24-hour gene profiles have been observed, with a tendency towards a decrease of *Per* expression and an increase of *Bmal1* expression in the evening.

Furthermore, mutation experiments show that age-related changes in circadian rhythmicity occur equally in wild-type and heterogezygous *Clock* mutant mice (Kolker et al., 2004), suggesting that the *Clock* mutation does not render mice more susceptible to the effects of age on the circadian pacemaker. While they clearly have altered circadian rhythms as young adult animals, and these disruptions are exacerbated by age, the age-related changes in the rhythms of mutant mice are similar to those of wild-type mice. Together these data suggest that a mutation that has pronounced effects on the circadian timing system of young adult animals does not necessarely have a major effect on the way the circadian rhythm of locomotor activity changes with age.

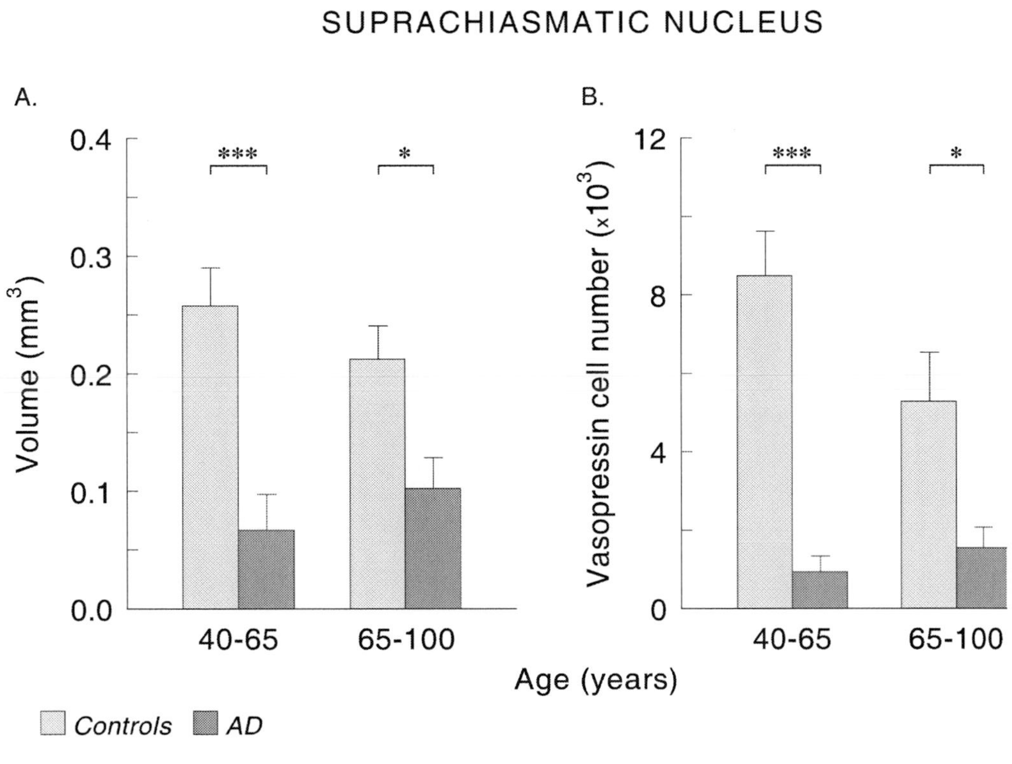

Figure 6. Effect of aging and Alzheimer's disease on the human suprachiasmatic nucleus (SCN). In Alzheimer patients, both presenile (<65 years of age) and senile (>65 years of age), the volume of the SCN (A) and the number of arginine vasopressin (AVP)-expressing neurons in the SCN (B) are significantly decreased compared to age-matched controls. Note that in presenile Alzheimer patients the number of AVP-expressing neurons is only 10% of that of controls of comparable age. Reproduced from Swaab (2004), with permission.

5.3. Suprachiasmatic Nucleus Dysfunction in Older People

The many lines of evidence of age-related decrements in circadian time-keeping in human beings (Vitiello and Prinz., 1990; Czeisler et al., 1992; Van Someren et al., 2002) and the observed neuronal degeneration of the SCN in senescence and Alzheimer patients (Swaab et al., 1985, 1996; Zhou et al., 1995; Hofman et al., 1996; 2006; Hofman, 1997; Swaab, 1999) strongly suggest an organic deterioration of the circadian oscillator.

The disruption of circadian rhythms and the increased incidence of disturbed sleep in humans during aging (Van Someren et al., 2002) are paralleled by age-related alterations in the neural organization of the SCN, a decreased photic input to the clock, and, in the SCN of Alzheimer patients, also with a dramatic decrease in peptide synthesis (Liu et al., 2000; Swaab, 2004; Figure 6), the presence of pretangles (Swaab et al., 1992; Van de Nes et al., 1998) and tangles (Stopa et al., 1999). However, diffuse amyloid plaques are only seldom noted in this nucleus (Van de Nes et al., 1998; Stopa et al., 1999). Stopa and colleagues (1999), for example, reported that the SCN of Alzheimer patients, who are frequently suffering from sleep disruption and other circadian rhythm disturbances, is severely damaged and exhibits a dramatic loss of vasopressin- and neurotensin-expressing neurons, as well as a corresponding increase in the GFAP (glial fibrillary acidic protein) stained astrocytes.

The immunocytochemical data showing decreased activity of the SCN in Alzheimer's disease (AD) have been confirmed by *in situ* hybridization (Figure 7). The total amount of AVP mRNA in the SCN of Alzheimer patients was three times lower than in age- and sex-matched controls. In addition, the AVP mRNA-expressing neurons in the SCN showed a marked day–night difference in controls under 80 years of age. The amount of AVP mRNA was more than three times higher during the day than at night, whereas no clear diurnal rhythm of AVP mRNA was observed in AD patients (Liu et al., 2000). These data support the idea that damage to the SCN is the underlying anatomical substrate for the clinically often-observed disturbances in circadian rhythmicity in Alzheimer's disease.

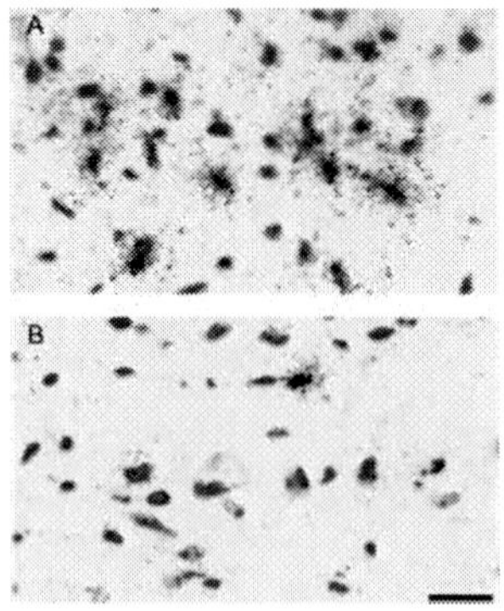

Figure 7. Thionin-counterstained emulsion autoradiograms in frontal sections (6 μm) of the human suprachiasmatic nucleus (SCN) at high magnification. Note the black silver deposits that show the presence of vasopressin (AVP) mRNA in the SCN neurons. A. Control subject. B. Alzheimer patient. Note the low number of AVP mRNA expressing neurons in the AD patient. Scale bar = 100 μm. Modified from Liu et al. (2000).

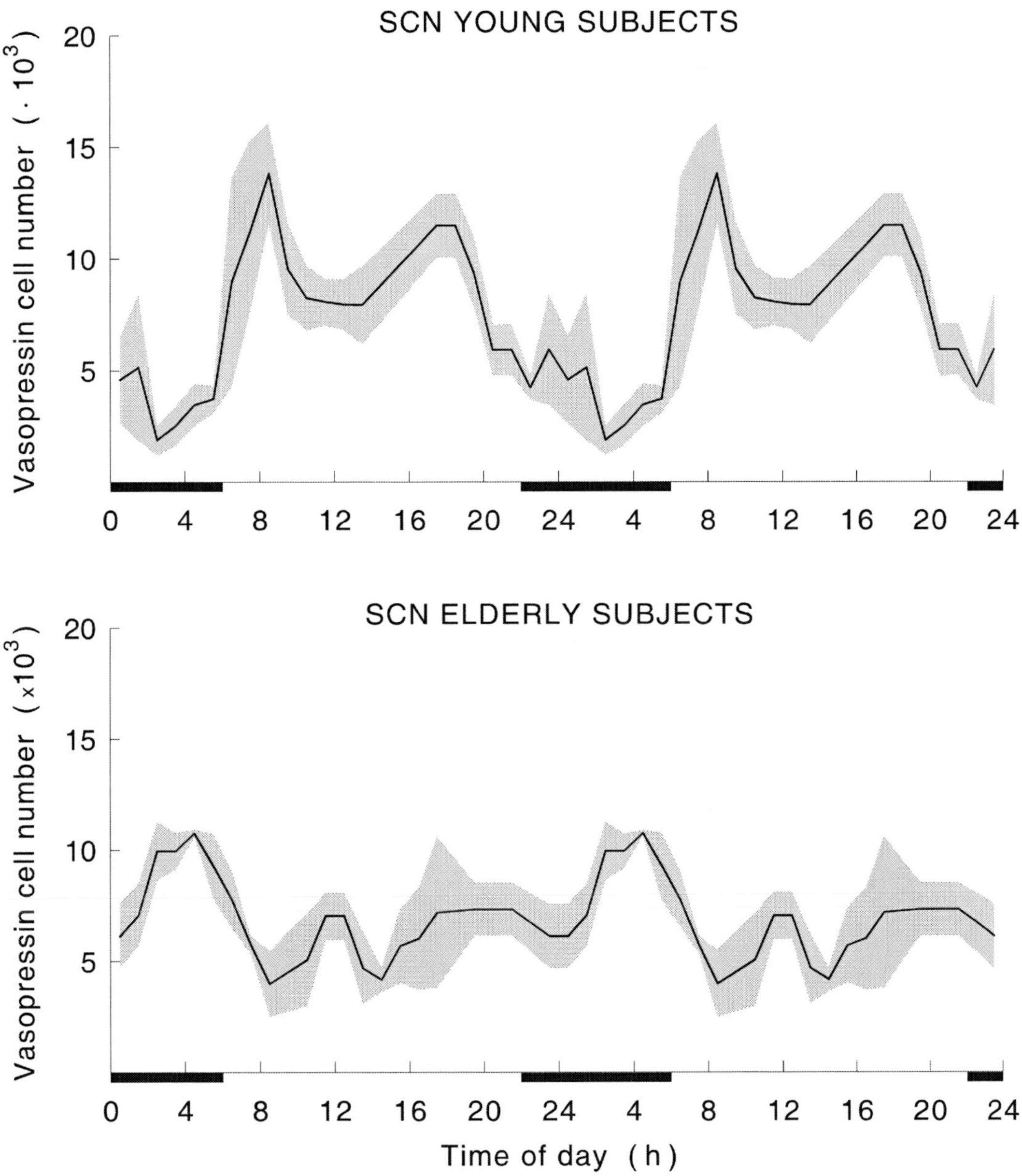

Figure 8. Circadian rhythm in the number of arginine vasopressin (AVP)-expressing neurons in the human suprachiasmatic nucleus (SCN) of (upper panel) young subjects (<50 years of age) and (lower pannel) elderly subjects (>50 years of age). The black bars indicate the night period (22:00-06:00 h). The general trend in the data is enhanced by using a smoothed double plotted curve and is represented by mean ± SEM values. Note the distinct circadian neuropeptide rhythm in the biological clock of young people compared with the disrupted rhythm in older people. Reproduced from Hofman and Swaab (1994).

Moreover, we found that the number of AVP-expressing neurons in the human SCN exhibits a marked diurnal oscillation in young (up to 50 years of age), but not in elderly people (over 50 years of age) (Hofman and Swaab, 1994; Figure 8). Whereas in young subjects low AVP-expressing neuron numbers were observed during the night and peak values during the early morning, a disrupted cycle with a reduced amplitude was found in the SCN of elderly people. Similar age-related decrements have been reported for the seasonal timing system (Hofman and Swaab, 1995).

As the number of AVP-expressing neurons probably reflects the peptidergic activity state of the cells, these findings suggest that the synthesis of AVP in the human SCN exhibits diurnal as well as seasonal fluctuations, and that the temporal organization of these rhythms becomes progressively disturbed in senescence.

Immunocytochemical studies, furthermore, provide evidence that degenerative alterations in the human SCN occur at a later phase in life than the reported functional changes in circadian organization. More frequent and prolonged awakenings and shorter sleep periods have already been found in 50- to 60-year-old subjects (Myers and Badia, 1995; Prinz et al., 2000), whereas a reduction in SCN volume and number of AVP-expressing neurons are only present from the age of 80 years onwards (Swaab et al., 1985, 1993; Hofman, 1997). Thus, the observed loss of AVP-expressing neurons in the SCN of very old people may only be a relatively late correlate of functional changes in the biological clock appearing much earlier.

5.4. Pineal Melatonin and Aging

Reduced melatonin concentrations during aging, especially nocturnal levels, have been reported in plasma, cerebro-spinal fluid (CSF) and in urine (reviewed in Skene and Swaab, 2003; Karasek, 2003; Wu and Swaab, 2005). Studies of the major urinary metabolite of melatonin, 6-sulfatoxymelatonin show that age related decrease in melatonin production occurs even as early as 20 – 30 years of age (Kennaway et al., 1999). Zhao and colleagues found that a decline of the nocturnal serum melatonin peak was only significant at the age of 60 and further declined from 70 years of age onwards (Zhao et al., 2002). Even within a fairly narrow age range (40 - 69 years), there is a significant effect of age on the daily excretion of urinary 6-sulfatoxymelatonin (Skene et al., 1990a). Although many reports indicate that melatonin levels decline with age, especially the nocturnal melatonin peak, some recent studies do not support a reduction (Zeitzer et al., 1999; Fourtillan et al., 2001).

Melatonin content in the human pineal has also been found to be reduced with age (Skene et al., 1990b; Wu and Swaab, 2005; Hofman, unpublished results; Figure 9). In another study of human postmortem pineal tissue, elderly subjects had lower pineal melatonin contents than younger subjects, but this difference was not statistically significant (Luboshitzky et al., 1998). Besides the age-related decline of melatonin production, age-related changes in the timing of the melatonin rhythm have also been reported (Duffy et al., 2002). Moreover, older subjects enter sleep and awake earlier relative to their nightly melatonin secretory episode, which indicates that aging is also associated with a change in the internal phase relationship between the sleep–wake cycle and the output of the circadian pacemaker.

In addition to the effects of normal aging various studies have shown that melatonin levels are lower in Alzheimer's disease (AD) patients compared with aged matched controls

(Skene et al., 1990b; Uchida et al., 1996; Liu et al., 1999; Mishima et al., 1999; Ohashi et al., 1999; Ferrari et al., 2000; Karasek, 2003). In patients who lack serum melatonin rhythms, clinical symptoms of delirium and sleep-wake disturbance were frequently but not always found (Uchida et al., 1996). In AD patients with disturbed sleep-wake rhythms a higher degree of irregularities in melatonin secretion have been observed (Mishima et al., 1999). Our finding that the daily variations in pineal melatonin and 5-methoxytryptophol content disappeared in AD patients (Skene et al., 1990b; Wu et al., 2003) may be linked with the clinical observations of sleep disorders and sundowning in these patients. Others have

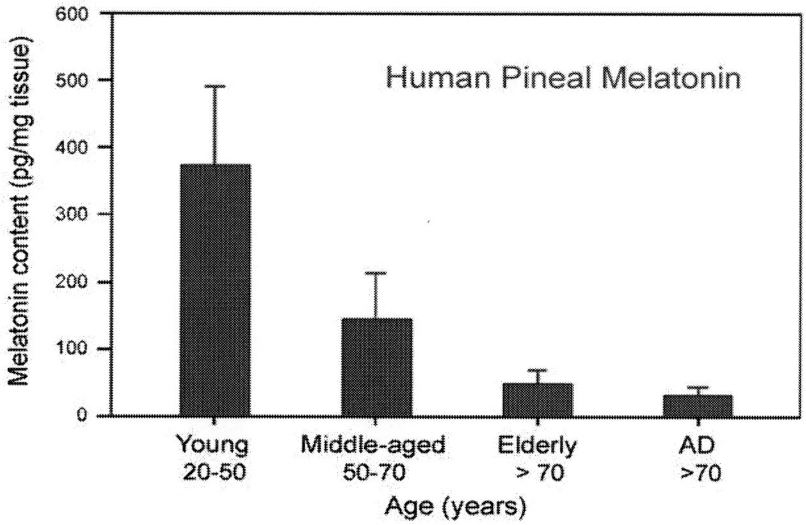

Figure 9. Effect of aging and Alzheimer's disease (AD) on the night-time melatonin level in the human pineal gland. Note the dramatic decline of the nocturnal melatonin production in middle-aged subjects compared to that of young subjects and the further decline after the age of 70 years. Reproduced from Hofman (2008).

reported that there is a selective impairment of the nocturnal melatonin peak in dementia (Ferrari et al., 2000) or that the melatonin levels are increased in AD patients during daytime and that these patients do not react to bright light (Ohashi et al., 1999).

5.5. Sex, Circadian Clock and Aging

In humans, during normal aging, the number of peptidergic neurons in the SCN deteriorate in a sexually dimorphic way (for a review, see Swaab, 1999). The number of VIP-expressing neurons in the SCN of women, for example, did not show any age-related changes as opposed to the neurons expressing AVP, whereas in men a complex pattern of changes was observed of VIP-expressing neurons with advancing age. Between 10 and 40 years, the male SCN contained twice as many VIP neurons as the female, but a subsequent decrease in the number of male VIP neurons between 40 and 65 years of age resulted in fewer VIP neurons in men than in women. After 65 years of age the sex difference remained just short of significance. A significant decrease in the number of VIP-expressing neurons in the SCN was found in female presenile Alzheimer patients only (Zhou et al., 1995).

With respect to the sex differences in age-related changes in VIP neurons it is of interest to mention that there are differences between healthy aged women and men as far as entrained circadian temperature rhythms are concerned that suggest that aging may affect the circadian timing system in a sexually dimorphic way. The acrophase of body temperature was phase-advanced by an average of 1.25 h in the aged women compared to the age-matched men. Women woke up earlier and slept for shorter periods of time (Moe et al., 1991). In addition, the sex differences (Swaab and Hofman, 1995), the presence of sex hormone receptors (Kruijver and Swaab, 2002) and the sexual orientation related differences in the human SCN (Swaab and Hofman, 1990, 1995), reinforce the idea on the possible involvement of this nucleus in sexual behavior or reproduction.

Many findings indicate that changes in the SCN and other parts of the circadian timing system underlie some of the sleep disturbances among elderly people (Dijk et al., 2000; Hofman, 2000; Van Someren, 2000b; Hofman and Swaab, 2006). Based on the hypothesis that increased stimulation of the brain can improve or even restore the decreased neuronal activity (Swaab et al., 2002), it has been demonstrated in aged rats that both overt sleep-wake rhythms (Witting et al., 1993) and cell function in the SCN (Lucassen et al., 1995) can be restored by enhancing the light stimulus the circadian timing system normally uses for synchronization of the sleep-wake rhythm. There is evidence that a similar degree of plasticity exists in the human circadian timing system. Improvement of the sleep-wake rhythm of elderly people has been demonstrated by application of a variety of potent modulators of the circadian timing system, like bright light, melatonin and physical activity (Van Someren et al., 2002; Meijer and Schwartz, 2003). Contrary to treatment with hypnotics, the improvement of sleep following these treatments is without adverse effects and even results in improvement of mood, performance, daytime energy, and quality of life.

Conclusion

It is now generally accepted that the suprachiasmatic nucleus (SCN) is the principal neural structure that mediates circadian rhythms in mammals, including man. Consistent with its role in the circadian organization of physiological processes and behavior, recent investigations suggest that measurable biological activities within the SCN, such as glucose utilization, neuronal electrical activity, and protein synthesis exhibit circadian rhythms (Klein et al., 1991; Van Esseveldt et al., 2000).

Animal studies have shown that circadian rhythmicity is disrupted during senescence at various levels of biological organization, including an age-related reduction in the amplitude of the rest-activity cycle (Turek et al., 1995; Weinert, 2000). Analogous to these animal data, age-related alterations have also been described for hormonal rhythms, temperature rhythms, sleep-wakefulness rhythms, as well as for various other physiological and behavioral rhythms in humans (Touitou and Haus, 1992, 2000; Turek et al., 1995; Van Someren, 2000b; Dijk et al., 2000; Van Someren et al., 2002). It appears that aging not only affects the amplitude but also the frequency of circadian rhythms, particularly in dementia (Witting et al., 1990; Turek et al., 1995). Generally, dissociation of internal synchronization of circadian rhythms seems to occur more frequently in the elderly (Touitou and Haus, 1992, 2000). Such age-related changes in the circadian period may have major consequences for the synchronzation of the biological clock with external Zeitgebers as well as for the temporal organization of the multioscillatory system.

The fact that the internal synchronization of various rhythms appears to be disturbed in elderly humans, in addition to alterations in the rhythmicity itself, points towards the circadian oscillator as the site of origin of these age-related changes. The observed reduction in the amplitude of the AVP rhythm of the SCN in elderly people is in accordance with this view, and with findings that the human pineal gland, as part of the circadian timing system, shows similar alterations with an age-related decrease in melatonin synthesis and increased pineal calcium deposits (Skene et al., 1990a; Skene and Swaab, 2003; Wu et al., 2003; Wu and Swaab, 2005). Moreover, the circadian melatonin rhythm becomes increasingly unstable in older people. Similar disorders of the pineal biosynthetic activity have been shown in the aging hamster and gerbil (Reiter, 1995). These findings suggest that degeneration of the SCN-pineal complex could well be the neural substrate for the disrupted circadian rhythms, which have been reported in elderly subjects, demented patients and in depression (Zhou et al., 2001). Experimental studies, for example, show that the free-running period of the rest-activity cycle is directly proportional to the residual volume of the SCN in partly lesioned animals, and such animals show a pronounced reduction in the amplitude and phase advance of daily activity, drinking and eating rhythms (Turek et al., 1995).

In conclusion, the age-related functional changes in circadian organization in humans may be associated with subtle degenerative alterations in the SCN and other parts of the clock, leading to disruptions of the circadian system. In the present review it was shown that this may happen relatively early in life, well before any dramatic neuronal atrophy of the biological clock becomes manifest. Furthermore, the present findings indicate that illuminance and photoperiod may be considered as potential environmental factors controlling the functional activity of the human SCN. If light is the effective Zeitgeber in man, as it is in many organisms, disturbances in processing of light information during aging or in some neurodegenerative and affective diseases may have profound effects on the timing of a variety of physiological and behavioral activities, including sleep (see e.g., Meijer and Schwartz, 2003). In view of the assumed sensitivity of the biological clock to changes in day length and light intensity it would be interesting to investigate whether the synchronizing and antidepressant effects of bright light exposure in sleep disturbances and mood disorders can be improved by applying the therapy in temporal coherence with the circadian cycles underlying the neuronal activity of the biological clock.

References

Albrecht, U., Zheng, B., Larkin, D., Sun, Z.S. and Lee, C.C. (2001). *mPer1* and *mPer2* are essential for normal resetting of the cicadian clock. *Journal of Biological Rhythms*, 16, 100-104.

Antle, M.C. and Silver, R. (2005). Orchestrating time: arrangements of the brain circadian clock. *Trends in Neurosciences*, 28, 145-151.

Antle, M.C., Foley D.K., Foley N.C. and Silver R. (2003). Gates and oscillators: a network model of the brain clock. *Journal of Biological Rhythms*, 18, 339-350.

Asai, M., Yoshinobu, Y., Kaneko, S., Mori, A., Nikaido, T., Moriya, T., Akiyama, M. and Shibata, S. (2001). Circadian expression of *Per* gene mRNA expression in the suprachiasmatic nucleus, paraventricular nucleus, and pineal body of aged rats. *Journal of Neuroscience Research*, 66, 1133-1139.

Aujard, F., Herzog, E.D. and Block, G.D. (2001). Circadian rhythms in firing rate of individual suprachiasmatic nucleus neurons from adult and middle-aged mice. *Neuroscience*, 106, 255–261.

Avidan, A.Y. (2005). Sleep in the geriatric patient population. *Seminars in Neurology*, 25, 52-63.

Bhatt, M.H., Podder, N. and Chokroverty, S. (2005). Sleep and neurodegenerative diseases. *Seminars in Neurology*, 25, 39-51.

Buijs, R.M. and Kalsbeek, A. (2001). Hypothalamic integration of central and peripheral clocks. *Nature Reviews/ Neuroscience*, 2, 521-526.

Buijs, R.M., Kalsbeek, A., Romijn, H.J., Pennartz, C.M.A. and Mirmiran, M. (Eds.) (1996). Hypothalamic Integration of Circadian Rhythms. *Progress in Brain Research*, Vol. 111. Elsevier, Amsterdam.

Cermakian N. and Sassone-Corsi P. (2000). Multilevel regulation of the circadian clock. *Nature Reviews/ Molecular Cell Biology*, 1, 59-67.

Claustrat, F., Fournier, I., Geelen, G., Brun, J., Corman, B. and Claustrat, B. (2005). Aging circdian clock gene expression in peripheral tissue in rats. *Pathologie-Biologie (Paris),* 53, 257-260.

Cohen, R.A. and Albers, H.E. (1991). Disruption of the human circadian and cognitive regulation following a discrete hypothalamic lesion: a case study. *Neurology*, 41, 726-729.

Cornélissen, G., Haus, E. and Halberg, F. (1992). Chronobiologic blood pressure assessment from womb to tomb. In: Touitou, Y., Haus, E. (Eds.). *Biological Rhythms in Clinical and Laboratory Medicine*. Springer Verlag, Berlin, pp. 428-452.

Czeisler, C.A., Dumont, M., Duffy, J.F., Steinberg, J.D., Richardson, G.S., Brown, E.N., Sanchez, R., Rios, C.D. and Ronda, J.M. (1992). Association of sleep-wake habits in older people with changes in output of circadian pacemaker. *Lancet*, 340, 933-936.

Czeisler, C.A. (1995). The effect of light on the human circadian pacemaker. In: Waterhouse, J.M. (Ed.). *Circadian Clocks and their Adjustment*, Ciba Foundation Symposium 183, John Wiley, Chichester, pp. 254-302.

Daan S., Albrecht U., Van der Horst G.T.J., Illnerová H., Roenneberg T., Wehr T.A. and Schwartz W.J. (2001). Assembling a clock for all seasons: are there M and E oscillators in the genes? *Journal of Biological Rhythms*, 16, 105-116.

Dai, J., Swaab, D.F. and Buijs, R.M. (1997). Distribution of vasopressin and vasoactive intestinal polypeptide (VIP) fibers in the human hypothalamus with special emphasis on suprachiasmatic nucleus efferent projections. *Journal of Comparative Neurology*, 383, 397-414.

Dai, J., Swaab, D.F., Van der Vliet, J. and Buijs, R.M. (1998). Postmortem tracing reveals the organization of hypothalamic projections of the suprachiasmatic nucleus in the human brain. *Journal of Comparative Neurology*, 400, 87-102.

Dardente H., Poirel V-J., Klosen P., Pévet P. and Masson-Pévet M. (2002). *Per* and neuropeptide expression in the rat suprachiasmatic nuclei: compartmentalization and differential cellular induction by light. *Brain Research*, 958, 261-271.

Dijk, D.J., Duffy, J.F. and Czeisler, C.A. (2000). Contribution of circadian physiology and sleep homeostasis to age-related changes in human sleep. *Chronobiology International*, 17, 285-311.

Duffy, J.F., Zeitzer, J.M., Rimmer, D.W., Klerman, D.B., Dijk, D.J. and Czeisler, C.A. (2002). Peak of circadian melatonin rhythm occurs later within the sleep of older subjects. *American Journal of Physiology. Endocrinology and Metabolism*, 282, E297-E303.

Dunlap J.C. (1999). Molecular bases for circadian clocks. *Cell*, 96, 271-290.

Eastman, C.I., Mistlberger, R.E. and Rechtschaffen, A. (1984). Suprachiasmatic nuclei lesions eliminate circadian temperature and sleep rhythms in the rat. *Physiology and Behavior*, 32, 357-368.

Ferrari E., Arcaini, A., Gornati, R., Pelanconi, L., Cravello, L., Fioravanti, M., Solerte, M.B. and Maqri, F. (2000). Pineal and pituitary-adrenocortical function in physiological aging and in senile dementia. *Experimental Gerontology*, 35, 1239-1250.

Fourtillan, J.B., Brisson, A.M., Fourtillan, M., Ingrand, L., Decourt, J.P. and Girault, J. (2001). Melatonin secretion occurs at a constant rate in both young and older men and women. *American Journal of Physiology. Endocrinology and Metabolism*, 280, E11-E22.

Hamada T., Antle M.C. and Silver R. (2004). Temporal and spatial expression patterns of canonical clock genes and clock-controlled genes in the suprachiasmatic nucleus. *European Journal of Neuroscience*,19, 1741-1748.

Hamada, T., LeSauter J., Venuti J.M. and Silver R. (2001). Expression of *Period* genes: rhythmic and nonrhythmic compartments of the suprachiasmatic nucleus pacemaker. *Journal of Neuroscience*, 21, 7742-7750.

Hannibal, J. (2002). Neurotransmitters of the retino-hypothalamic tract. *Cell and Tissue Research*, 309, 73-88.

Hastings, M.H. and Herzog, E.D. (2004). Clock genes, oscillators, and cellular networks in the suprachiasmatic nuclei. *Journal of Biological Rhythms*, 19, 400-413.

Hastings M.H. and Maywood E.S. (2000). Circadian clocks in the mammalian brain. *BioEssay*, 22, 23-34.

Hastings M.H., Reddy A.B. and Maywood, E.S. (2003). A clockwork web: circadian timing in brain and periphery, in health and disease. *Nature Reviews/ Neuroscience*, 4, 449-461.

Hofman, M.A. (1997). Lifespan changes in the human hypothalamus. *Experimental Gerontology*, 32, 559-575.

Hofman, M.A. (2000). The human circadian clock and aging. *Chronobiology International*, 17, 245-259.

Hofman, M.A. (2003). Circadian oscillations of neuropeptide expression in the human biological clock. *Journal of Comparative Physiology* A, 189, 823-831.

Hofman, M.A. (2004). The brain's calendar: neural mechanisms of seasonal timing. *Biological Reviews*, 79, 61-77.

Hofman, M.A. (2008). The human circadian timing system. In: Binder, M.D., Hirokawa, N., Windhorst, U. (Eds.). *Encyclopedia of Neuroscience*. Springer Verlag, Heidelberg (in press).

Hofman, M.A., Fliers, E., Goudsmit, E. and Swaab, D.F. (1988). Morphometric analysis of the suprachiasmatic and paraventricular nuclei in the human brain: sex differences and age-dependent changes. *Journal of Anatomy*, 160, 127-143.

Hofman, M.A. and Swaab, D.F. (1992). The human hypothalamus: comparative morphometry and photoperiodic influences. In: Swaab, D.F., Hofman, M.A., Mirmiran, M., Ravid, R., Van Leeuwen, F.W. (Eds.). The Human Hypothalamus in Health and Disease. *Progress in Brain Research*, Vol. 93. Elsevier, Amsterdam, pp. 133-149.

Hofman, M.A. and Swaab, D.F. (1993). Diurnal and seasonal rhythms of neuronal activity in the suprachiasmatic nucleus of humans. *Journal of Biological Rhythms*, 8, 283-295.

Hofman, M.A. and Swaab, D.F. (1994). Alterations in circadian rhythmicity of the vasopressin-producing neurons of the human suprachiasmatic nucleus (SCN) with aging. *Brain Research*, 651, 134-142.

Hofman, M.A. and Swaab, D.F. (1995). Influence of aging on the seasonal rhythm of the vasopressin-expressing neurons in the human suprachiasmatic nucleus. *Neurobiology of Aging*, 116, 965-971.

Hofman, M.A. and Swaab, D.F. (2002). A brain for all seasons: cellular and molecular mechanisms of photoperiodic plasticity. In: Hofman, M.A., Boer, G.J., Holtmaat, A.J.G.D., Van Someren, E.J.W., Verhaagen, J., Swaab, D.F. (Eds.). Plasticity in the Adult Brain: From Genes to Neurotherapy. *Progress in Brain Research*, Vol. 138. Elsevier, Amsterdam, pp. 255-280.

Hofman, M.A. and Swaab, D.F. (2006). Living by the clock: The circadian pacemaker in older people. *Ageing Research Reviews*, 5, 33-51.

Hofman, M.A., Zhou, J.-N. and Swaab, D.F. (1996). Suprachiasmatic nucleus of the human brain: an immunocytochemical and morphometric analysis. *Anatomical Record*, 244, 552-562.

Jagota, A., De la Iglesia, H.O. and Schwartz, W.J. (2000). Morning and evening circadian oscillations in the suprachiasmatic nucleus *in vitro*. *Nature Neuroscience*, 3,372-376.

Kalsbeek, A. and Buijs, R.M. (2002). Output pathways of the mammalian suprachiasmatic nucleus: coding circadian time by transmitter selection and specific targeting. *Cell and Tissue Research*, 309, 109-118.

Kalsbeek, A., Palm, I.F., La Fleur, S.E., Scheer, F.A., Perreau-Lenz, S., Ruiter, M., Kreier, F., Cailotto, C. and Buijs, R.M. (2006). SCN outputs and the hypothalamic balance of life. *Journal of Biological Rhythms*, 21, 458-469.

Karasek, M. (2003). Melatonin in human aging and age-related diseases. *Experimental Gerontology*, 39, 1723-1729.

Karatsoreos, I.N., Yan, L., LeSauter, J. and Silver, R. (2004). Phenotypic matters: identification of light-responsive cells in the mouse suprachiasmatic nucleus. *Journal of Neuroscience*, 24, 68-75.

Kawamoto, K., Nagano, M., Kanda, F., Chihara, K., Shigeyoshi, Y. and Okamura, H. (2003). Two types of VIP neuronal components in rat suprachiasmatic nucleus. *Journal of Neuroscience Research*, 74, 852-857.

Kenneway, D.J., Lushington, K., Dawson, D., Lack, L., Van den Heuvel, C. and Rogers, N. (1999). Urinary 6-sulfatoxymelatonin secretion ang aging: new results and a critical review of the literature. *Journal of Pineal Research*, 27, 210-220.

King, D.P. and Takahashi, J.S. (2000). Molecular genetics of circadian rhythms in mammals. *Annual Review of Neuroscience*, 23, 713-742.

King, V.M., Chahad-Ehlers, S., Shen, S., Harmar, A.J., Maywood, E.S. and Hastings, M.H. (2003). A hVIPR transgene as a novel tool for the analysis of circadian function in the mouse suprachiasmatic nucleus. *European Journal of Neuroscience*, 17, 822-832.

Klein, D.C., Moore, R.Y. and Reppert, S.M. (Eds.) (1991). *Suprachiasmatic Nucleus; The Mind's Clock*. Oxford University Press, New York.

Kolker, D.E., Fukuyama, H., Huang, D.S., Takahashi, J.S., Horton, T.H. and Turek, F.W. (2003). Aging alters circadian and light-induced expression of clock genes in golden hamsters. *Journal of Biological Rhythms*, 18, 159-169.

Kolker, D.E., Vitaterna, M.H., Fruechte, E.M., Takahashi, J.S. and Turek, F.W. (2004). Effects of age on circadian rhythms are similar in wild-type and heterozygous *Clock* mutant mice. *Neurobiology of Aging*, 25, 517-523.

Kruijver, F.P.M. and Swaab, D.F. (2002). Sex hormone receptors are present in the human suprachiasmatic nucleus. *Neuroendocrinology*, 75, 296-305.

Leak, R.K. and Moore, R.Y. (2001). Topographic organization of suprachiasmatic nucleus projection neurons. *Journal of Comparative Neurology*, 433, 312-314.

Liu, R.-Y., Zhou, J.-N., Van Heerikhuize, J., Hofman, M.A. and Swaab, D.F. (1999). Decrease melatonin levels in postmortem cerebrospinal fluid in relation to aging, Alzheimer's disease and apolipoprotein E *epsilon*-4/4 genotype. *Journal of Clinical Endocrinology and Metabolism,* 84, 323-327.

Liu, R.-Y., Zhou, J.-N., Hoogendijk, W.J.G., Van Heerikhuize, J., Kamphorst, W., Unmehopa, U.A., Hofman, M.A. and Swaab, D.F. (2000). Decreased vasopressin gene expression in the biological clock of Alzheimer's disease patients with and without depression. *Journal of Neuropathology and Experimental Neurology*, 59, 314-322.

Luboshitzky, R., Yanai, D., Shen-Orr, Z., Israeli, E., Herer, P. and Lavie, P. (1998). Daily and seasonal variations in the concentration of melatonin in the human pineal gland. *Brain Research Bulletin*, 47, 271-276.

Lucassen, P.J., Hofman, M.A. and Swaab, D.F. (1995). Increased light intensity prevents the age-related losss of vasopressin-expressing neurons in the rat suprachiasmatic nucleus. *Brain Research*, 693, 261–266.

Mai, J.K., Kedziora, O., Teckhaus, L. and Sofroniew, M.V. (1991). Evidence for subdivisions in the human suprachiasmatic nucleus. *Journal of Comparative Neurology*, 305, 508-525.

Meijer, J.H., Rusak, B. and Ganshirt, G. (1992). The relation between light-induced discharge in the suprachiasmatic nucleus and phase shifts of hamster circadian rhythms. *Brain Research*, 598, 257-63.

Meijer, J.H. and Schwartz, W.J. (2003). In search of the pathways for light-induced pacemaker resetting in the suprachiasmatic nucleus. *Journal of Biological Rhythms*, 18, 235-249.

Mishima, K., Okawa, M., Shimizu, T. and Hishikawa, Y. (2001). Diminished melatonin secretation in the elderly caused by insufficient environmental illumination. *Journal of Clinical Endocrinology and Metabolism*, 86, 129-134.

Moe, K.E., Vitiello, M.V., Larsen, L.H. and Prinz, P.N. (1995). Cognitive processes and sleep disturbances: Sleep–wake patterns in Alzheimer's disease: relationships with cognition and function. *Journal of Sleep Research*, 4, 15–20.

Monk, T.Y. and Kupfer, D.J. (2000). Circadian rhythms in healthy aging – effects downstream from the pacemaker. *Chronobiology International*, 17, 355-368.

Moore, R.Y. (1991). Disorders of circadian function and the human circadian timing system. In: Klein, D.C, Moore, R.Y., Reppert, S.M., (Eds.). *Suprachiasmatic Nucleus: The Mind's Clock*. Oxford University Press, New York, pp. 429-441.

Moore, R.Y. (1992). The organization of the human circadian timing system. In: Swaab, D.F., Hofman, M.A., Mirmiran, M., Ravid, R., Van Leeuwen, F.W. (Eds.). The Human Hypothalamus in Health and Disease. *Progress in Brain Research*, Vol. 93. Elsevier, Amsterdam, pp. 101-117.

Moore, R.Y. and Silver, R. (1998). Suprachiasmatic nucleus organization. *Chronobiology International*, 15, 475-487.

Moore, R.Y., Speh, J.C. and Leak, R.K. (2002). Suprachiasmatic nucleus organization. *Cell and Tissue Research*, 309, 89-98.

Morin, L.P. (1988). Age-related changes in hamster circadian period, entrainment, and rhythm splitting. *Journal of Biological Rhythms*, 3, 237-248.

Myers, B.L. and Badia, P. (1995). Changes in circadian rhythms and sleep quality with aging: Mechanisms and interventions. *Neuroscience and Biobehavioral Reviews*, 19, 553-571.

Nakamura, W., Honma, S., Shirakawa, T. and Honma, K. (2001). Regional pacemakers composed of multiple oscillator neurons in the rat suprachiasmatic nucleus. *European Journal of Neuroscience*, 14, 666-674.

Ohashi, Y., Okamoto, N., Uchida, K., Iyo, M., Mori, N. and Morita, Y. (1999). Daily rhythm of serum melatonin levels and effect of light exposure in patients with dementia of the Alzheimer's type. *Biological Psychiatry*, 45, 1646-1652.

Panda, S., Hogenesch, J.B. and Kay, S.A. (2002). Circadian rhythms from flies to human. *Nature*, 417, 329-335.

Prinz, P.N., Bailey, S.L. and Woods, D.L. (2000). Sleep impairments in healty seniors: roles of stress, cortisol, and interleukin-1 beta. *Chronobiology International*, 17, 391-404.

Quintero, J.E., Kuhlman, S.J. and McMahon, D.G. (2003). The biological clock nucleus: a multiphasic oscillator network regulated by light. *Journal of Neuroscience*, 23, 8070-8076.

Reiter, R.J. (1995). The pineal gland and melatonin in relation to aging: summary of the theories and of the data. *Experimental Gerontology*, 30, 199-212.

Renfrew, J.W., Pettigrew, K.D. and Rapoport, S.I. (1987). Motor activity and sleep duration as a function of age in healthy men. *Physiology and Behavior*, 41, 627-634.

Reppert, S.M. and Weaver, D.R. (2001). Molecular analysis of mammalian circadian rhythms. *Annual Reviews of Physiology*, 63, 647-676.

Reppert, S.M. and Weaver, D.R. (2002). Coordination of circadian timing in mammals. *Nature*, 418, 935-941.

Reuss, S. (1996). Components and connections of the circadian timing system in mammals. *Cell and Tissue Research*, 285, 353-378.

Romijn, H.J., Sluiter, A.A., Pool, C.W., Wortel, J. and Buijs, R.M. (1996). Differences in colocalization between Fos and PHI, GRP, VIP, and VP in neurons of the rat suprachiasmatic nucleus after a light stimulus during the phase-delay versus the phase-advance period of the night. *Journal of Comparative Neurology*, 372, 1-8.

Romijn, H.J., Van Uum, J.F., Emmering, J., Goncharuk, V. and Buijs, R.M. (1999). Colocalization of VIP and AVP in neurons of the paraventricular, supraoptic and suprachiasmatic nucleus. *Brain Research*, 832, 47-53.

Sack, R.L. and Lewy, A.J. (2001). Circadian rhythm sleep disorders: lessons from the blind. *Sleep Medicine Reviews*, 5, 189-206.

Sadun, A.A., Schaechter, J.D. and Smith, L.E.H. (1993). A retinohypothalamic pathway in man: light mediation of circadian rhythms. *Brain Research*, 302, 371-377.

Saper, C.B., Lu, J., Chou, T.C. and Gooley, J. (2005). The hypothalamic integrator for circadian rhythms. *Trends in Neurosciences*, 28, 152-157.

Satinoff, E., Li, H., Tcheng, T.K., Liu, C., McArthur, A.J., Medanic, M. and Gillette, M.U. (1993). Do the suprachiasmatic nuclei oscillate in old rats as they do in young ones? *American Journal of Physiology*, 265, R1216-R1222.

Schwartz, W.J., Bosis, N.A. and Hedley-Whyte, E.T. (1986). A discrete lesion of ventral hypothalamus and optic chiasm that disturbed the daily temperature rhythm. *Journal of Neurology*, 233, 1-4.

Schwartz, W.J., Carpino, Jr A., De la Iglesia, H.O., Baler, R., Klein, D.C., Nakabeppu, Y. and Aronin, N. (2000). Differential regulation of *fos* family genes in the ventrolateral and dorsomedial subdivisions of the rat suprachiasmatic nucleus. *Neuroscience*, 98: 535-547.

Shirakawa, T., Honma, S. and Honma, K. (2001). Multiple oscillators in the suprachiasmatic nucleus. *Chronobiology International*, 18, 371-387.

Skene, D. and Swaab, D.F. (2003). Melatonin rhythmicity: effect of age and Alzheimer's disease. *Experimental Gerontology*, 38, 199-206.

Skene, D.J., Bojkowski, C.J., Currie, J.E., Wright, J., Boulter, P.S. and Arendt, J. (1990a). 6-sulphatoxymelatonin production in breast cancer patients. *Journal of Pineal Research*, 8, 269-276.

Skene, D.J, Vivien-Roels, B., Sparks, D.L., Hunsaker, J.C., Pevet, P., Ravid, R. and Swaab, D.F. (1990b). Daily variation in the concentrations of melatonin and 5-methoxytryptophol in the human pineal gland: effect of age and Alzheimer's disease. *Brain Research*, 528, 170-174.

Spoelstra, K., Albrecht, U., Van der Horst, G.T., Brauer, V. and Daan, S. (2001). Phase responses to light pulses in mice lacking functional *Per* or *Cry* genes. *Journal of Biological Rhythms*, 19, 518-529.

Stopa, E.G., Volicer, L., Kuo-Leblanc, V., Harper, D., Lathi, D., Tate, B. and Satlin, A. (1999). Pathologic evaluation of the human suprachiasmatic nucleus in severe dementia. *Journal of Neuropathology and Experimental Neurology*, 58, 29-39.

Sutin, E.L., Dement, W.C., Heller, H.C. and Kilduff, T.S. (1993). Light-induced gene expression in the suprachiasmatic nucleus of young and aging rats. *Neurobiology of Aging*, 14, 441-446.

Swaab, D.F. (1999). Biological rhythms in health and disease: the suprachiasmatic nucleus and the autonomic nervous system. In: Vinken, P.G., Bruyn, G.W., Appenzeller, O. (Eds.). *Handbook of Clinical Neurology*, Vol. 74. Elsevier, Amsterdam, pp. 467–521.

Swaab, D.F. (2004). The Human Hypothalamus: Basic and Clinical Aspects. Part II: Neuropathology of the Human Hypothalamus and Adjacent Brain Structures. *Handbook of Clinical Neurology*, vol. 80, Elsevier, Amsterdam.

Swaab, D.F. and Hofman, M.A. (1990). An enlarged suprachiasmatic nucleus in homosexual men. *Brain Research*, 537, 141-148.

Swaab, D.F. and Hofman, M.A. (1995). Sexual differentiation of the human hypothalamus in relation to gender and sexual orientation. *Trends in Neurosciences*, 18, 264-270.

Swaab, D.F., Fliers, E. and Partiman, T.S. (1985). The suprachiasmatic nucleus of the human brain in relation to sex, age and senile dementia. *Brain Research*, 342, 37-44.

Swaab, D.F., Dubelaar, E.J., Hofman, M.A., Scherder, E.J.A., Van Someren, E.J.W. and Verwer, R.W.H. (2002). Brain aging and Alzheimer's disease; use it or lose it. In: Hofman, M.A., Boer, G.J., Holtmaat, A.J.G.D. Van Someren, E.J.W., Verhaagen, J., Swaab, D.F. (Eds.). Plasticity in the Adult Brain: From Genes to Neurotherapy. *Progress in Brain Research,* Vol. 138. Elsevier, Amsterdam, pp. 343-373.

Swaab, D.F., Grundke-Iqbal, I., Iqbal, K., Kremer, H.P., Ravid, R. and Van de Nes, J.A. (1992). Tau and ubiquitin in the human hypothalamus in aging and Alzheimer's disease. *Brain Research*, 590, 239–249.

Swaab, D.F., Hofman, M.A., Lucassen, P.J., Purba, J.S., Raadsheer, F.C. and Van de Nes, J.A. (1993). Functional neuroanatomy and neuropathology of the human hypothalamus. *Anatomy and Embryology* (Berlin), 187, 317-30.

Swaab, D.F., Van Someren, E.J.W., Zhou, J.N. and Hofman, M.A. (1996). Biological rhythms in the human life cycle and their relationship to functional changes in the suprachiasmatic nucleus. In: Buijs, R.M., Kalsbeek, A., Romijn, H.J, Pennartz, C.M.A., Mirmiran, M. (Eds.). Hypothalamic Integration of Circadian Rhythms. *Progress in Brain Research*, Vol. 111. Elsevier, Amsterdam, pp. 349-368.

Touitou, Y. and Haus, E. (1992). Biological rhythms and aging. In: Touitou, Y., Haus, E., (Eds.). *Biological Rhythms in Clinical and Laboratory Medicine*. Springer, Berlin, pp. 188-207.

Touitou, Y. and Haus, E. (2000). Alterations with aging of the endocrine and neuroendocrine circadian system in humans. *Chronobiology International*, 17, 369-390.

Turek, F.W., Penev, P., Zhang, Y., Van Reeth, O. and Zee, P. (1995). Effects of age on the circadian system. *Neuroscience and Biobehavioral Reviews*, 19, 53-58.

Uchida, K., Okamoto, N., Ohara, K. and Morita, Y. (1996). Daily rhythms of serum melatonin in patients with dementia of degenerate type. *Brain Research*, 717, 154-159.

Van de Nes, J.A., Kamphorst, W., Ravid, R. and Swaab, D.F. (1998). Comparison of beta-protein/A4 deposits and Alz-50-stained cytoskeletal changes in the hypothalamus and adjoining areas of Alzheimer's disease patients: amorphic plaques and cytoskeletal changes occur independently. *Acta Neuropathologica* (Berlin) 96, 129–138.

Van Esseveldt, K.E., Lehman, M.N. and Boer, GJ. (2000). The suprachiasmatic nucleus and the circadian time-keeping system revisited. *Brain Research Reviews*, 33, 44-77.

Van Someren, E.J.W. (Ed.) (2000a). Circadian Rhythms and Sleep in Human Aging. *Chronobiology International* (Special Issue), Vol. 17.

Van Someren, E.J.W. (2000b). More than a marker: interaction between the circadian regulation of temperature and sleep, age-related changes and treatment possibilities. *Chronobiology International*, 17, 313-354.

Van Someren, E.J.W., Hagebeuk, E.E.O., Lijzenga, C., Scheltens, P., De Rooij, S.E., Jonker, C., Pot, A.M. and Mirmiram, M. (1996). Circadian rest-activity rhythm disturbances in Alzheimer's disease. *Biological Psychiatry*, 40, 259-270.

Van Someren, E.J.W., Riemersma, R.F. and Swaab, D.F. (2002). Functional plasticity of the circadian timing system in old age: light exposure. In: Hofman, M.A., Boer, G.J., Holtmaat, A.J.G.D., Van Someren, E.J.W., Verhaagen, J., Swaab, D.F. (Eds.). Plasticity in the Adult Brain: From Genes to Neurotherapy. *Progress in Brain Research*, Vol. 138. Elsevier, Amsterdam, pp. 205-231.

Vitiello, M.V. and Prinz, P.N. (1990). Sleep and sleep disorders in normal aging. In: Thorpy, M.J., (Ed.). *Handbook of Sleep Disorders, Neurological Diseases and Therapy*. Marcel Dekker, New York, pp. 139-151.

Wakamatsu, H., Takahashi, S., Moriya, T., Inouye, S.T., Okamura, H., Akiyama, M. and Shibata, S. (2001). Additive effect of mPerl and mPer2 antisense oligonucleotides on light-induced phase shift. *NeuroReport*, 12, 127-131.

Wehr, T.A. (2001). Photoperiodism in humans and other primates: evidence and implications. *Journal of Biological Rhythms*, 16, 348-364.

Weinert, D. (2000). Age-dependent changes of the circadian system. *Chronobiology International*, 17, 261-284.

Weinert, H., Weinert, D., Schurov, I., Maywood, E.S. and Hastings, M.H. (2001). Impaired expression of the *mPer2* circadian clock gene in the suprachiasmatic nuclei of aging mice. *Chronobiology International*,18, 559–565.

Weitzman, E.D., Moline, M.L., Czeisler, C.A. and Zimmerman, J.C. (1982). Chronobiology of aging: temperature, sleep-wake rhythms and entrainment. *Neurobiology of Aging*, 3, 299-309.

Witting, W., Kwa, J.H., Eikelenboom, P., Mirmiran, M. and Swaab, D.F. (1990). Alterations in the circadian rest-activity rhythm in aging and Alzheimer's disease. *Biological Psychiatry*, 27, 563-572.

Witting, W., Mirmiran, M., Bos, N.P. and Swaab, D.F. (1993). Effect of light intensity on diurnal sleep-wake distribution in young and old rats. *Brain Research Bulletin*, 30, 157-162.

Wu, Y.-H., Feenstra, M.G.P., Zhou, J.-N., Liu, R.-Y., Toranó, J.S., Van Kan, H.J.M., Fischer, F., Ravid, R. and Swaab, D.F. (2003). Molecular changes underlying reduced pineal melatonin levels in Alzheimer's disease: alterations in preclinical and clinical stages. *Journal of Clinical Endocrinology and Metabolism*, 88, 5898-5906.

Wu, Y.-H. and Swaab, D.F. (2005). The human pineal gland and melatonin in aging and Alzheimer's disease. *Journal of Pineal Research*, 38, 145-152.

Yamaguchi, S., Isejima, H., Matsuo, T., Okura, R., Yagita, K., Kobayashi, M. and Okamura, H. (2003). Synchronization of cellular clocks in the suprachiasmatic nucleus. *Science,* 302, 1408-1412.

Yamazaki, S., Straume, M., Tei, H., Sakaki, Y., Menaker, M. and Block, G.D. (2002). Effects of aging on central and periheral mammalian clocks. *Proceedings of the National Academy of Science USA*, 99, 10801-10806.

Yan, L. and Okamura, H. (2002). Gradients in the circadian expression of *Per1* and *Per2* genes in the rat surpachiasmatic nucleus. *European Journal of Neuroscience*, 15, 1153-1162.

Yan, L. and Silver, R. (2002). Differential induction and localization of *mPer1* and *mPer2* during advancing and delaying phase shifts. *European Journal of Neuroscience*, 16, 1531-1549.

Yan, L. and Silver, R. (2004). Resetting the brain clock: time-course and localization of mPER1 and mPER2 protein expression in suprachiasmatic nuclei during phase shifts. *European Journal of Neuroscience*, 19, 1105-1109.

Yan, L., Takekida, S., Shigeyoshi, Y. and Okamura, H. (1999). *Per1* and *Per2* gene expression in the rat suprachiasmatic nucleus; circadian profile and the compartment-specific response to light. *Neuroscience,* 94, 141-150.

Zeitzer, J.M., Daniels. J.E., Duffy, J.F., Klerman, E.B., Shanahan, T.L., Dijk, D.J. and Czeisler, C.A. (1999). Do plasma melatonin concentrations decline with age? *American Journal of Medicine*, 107, 432-436.

Zhao, Z.Y., Xie, Y., Fu, Y.R., Bogdan, A. and Touitou, Y. (2002). Aging and the circadian rhythm of melatonin: a cross-sectional study of Chinese subjects 30-110 yr of age. *Chronobiology International*, 19, 1171-1182.

Zhou, J.-N., Hofman, M.A. and Swaab, D.F. (1995). VIP neurons in the human SCN in relation to sex, age, and Alzheimer's disease. *Neurobiology of Aging*, 16, 571-576.

Zhou, J.-N., Riemersma, R., Unmehopa, U.A., Hoogendijk, W.J., Van Heerikhuize, J.J., Hofman, M.A. and Swaab, D.F. (2001). Alterations in arginine vasopressin neurons in the suprachiasmatic nucleus in depression. *Archives of General Psychiatry*, 58, 655-662.

In: Circadian Rhythms and Health Research Trends
Editor: Alfredo V. Lignelli, pp. 31-55

ISBN: 978-1-60021-875-0

Chapter II

The Adaptive Significance of Circadian Rhythms and Its Relevance to Human Health

Abel Bult-Ito* and Marina R. Castillo

Behavioral and Evolutionary Neuroscience Laboratory,
Alaskan Basic Neuroscience Program,
Institute of Arctic Biology, University of Alaska Fairbanks,
Fairbanks, Alaska 99775-7000

Abstract

Circadian, about 24-hour, rhythms are universal features of most organisms on earth. In mammals, a master circadian clock located in the suprachiasmatic nuclei (SCN) of the anterior hypothalamus controls circadian rhythmicity. The SCN synchronizes peripheral clocks and extra-SCN clocks in the brain to allow the animal to express coherent behavioral and physiological rhythms that are entrained to the ambient circadian cycle. Important advances have been made in the last couple of decades in the understanding of the molecular mechanisms underlying circadian cycling at the cellular level. One aspect that has been largely overlooked in this research field is the adaptive significance of circadian organization to the animal under natural conditions. In this chapter, we will review what is currently known about the potential evolutionary benefits of circadian control of behavior and physiology, with a focus on humans and other mammals. The next challenge, we believe, will be to connect the advances in molecular understanding to the circadian challenges an organism faces in its natural circadian environment. We end with a discussion of how to test the adaptive significance of mammalian circadian organization. Our goal is to elicit discussion about the importance of understanding the significance of circadian rhythms under natural conditions and how these rhythms adapt an individual to its local natural environment. In the context of a literature review, we

* Correspondence should be addressed to Abel Bult-Ito at ffab@uaf.edu or Institute of Arctic Biology, P.O. Box 757000, University of Alaska Fairbanks, Fairbanks, Alaska 99775-7000; phone: 907-474-7158; fax: 907-474-6967.

outline why this understanding has great potential to contribute to novel approaches to understanding human health and treating disease.

Keywords: *circadian rhythms, adaptation, human health, sleep, clock genes.*

The Goal of This Review

Our goal is to elicit discussion about the importance of understanding the significance of circadian (about 24-hour) rhythms under natural conditions and how these rhythms adapt an individual to its local natural environment. In the context of a literature review, we outline why this understanding has great potential to contribute to novel approaches to understanding human health and treating disease.

Introduction to Circadian Rhythms

The Prevalence of Circadian Organization

The organization of physiology and behavior in circadian rhythms is a near-universal characteristic of life on earth. Unicellular organisms, i.e., prokaryotes and eukaryotes, and multicellular organisms, i.e., plants, animals, and fungi, all display endogenous circadian rhythms when placed in constant environmental conditions (Pittendrigh 1960, 1993). Most of these rhythms are thought to have evolved as an adaptation to the 24-hour light-dark cycle that result from the daily rotation of the earth around its own axis, although very few studies have attempted to test this hypothesis.

The Mammalian Master Circadian Clock

In vitro studies of suprachiasmatic nuclei (SCN) firing rates (Meijer and Rietveld, 1989) and SCN lesions followed by transplantation of fetal SCN donor tissue into the third ventricle (Ralph *et al.*, 1990) have been instrumental in unequivocally demonstrating that the SCN is the endogenous light-entrainable master circadian clock in mammals (Weaver, 1998). The two SCN are located in the anterior hypothalamus on top of the optic chiasm, are separated by the bottom of the third ventricle, and contain about 10,000 cells each (Takahashi *et al.*, 2001). In transplant studies, circadian characteristics of the donor, such as free-running period, are transferred to the host with the transplanted SCN graft (Ralph *et al.*, 1990). In addition, circadian rhythms of aged animals can be restored with fetal SCN grafts (Cai *et al.*, 1997a, 1997b; Hurd *et al.*, 1995).

Peripheral and Extra-SCN Clocks

Besides the SCN, functional circadian clocks are also found in many peripheral tissues (Balsalobre, 2002; Balsalobre *et al.*, 1998; Damiola *et al.*, 2000; Hara *et al.*, 2001; Le Minh *et al.*, 2001; Rutter *et al.*, 2002; Stokkan *et al.*, 2001; Yamazaki *et al.*, 2000) and extra-SCN brain regions (Wakamatsu *et al.*, 2001). Under normal conditions, the SCN entrains these other circadian clocks so that they have phase relationships that are optimally organized during the circadian cycle according to their particular functions. Although non-SCN clocks share common molecular feedback loops for the generation of circadian rhythms with the SCN (Reick *et al.*, 2001), they respond differently to the same environmental or physiological cues (Rutter *et al.*, 2002). The master circadian clock in the SCN responds rapidly to changes in light-dark schedule (Takamure *et al.*, 1991; Yamazaki *et al.*, 2000) but not to changes in non-photic schedules, such as feeding schedules, in a light-dark cycle (Brinkhof *et al.*, 1998; Damiola *et al.*, 2000; Hara *et al.*, 2001; Le Minh *et al.*, 2001; Shibata *et al.*, 1983; Stokkan *et al.*, 2001; Wakamutsu *et al.*, 2001). In contrast, the circadian clock in the liver responds rapidly to changes in feeding schedule but very slowly to changes in the light-dark schedule. Clocks in the kidney, heart, pancreas, skeletal muscle and lung adjust gradually to changes in either food or lighting conditions (Damiola *et al.*, 2000; Stokkan *et al.*, 2001; Yamazaki *et al.*, 2000). Glucocorticoid rhythms shift quickly with scheduled feeding (Krieger, 1974) and may be part of the mechanism for peripheral clock entrainment (Le Minh *et al.*, 2001). Peripheral tissues such as liver, kidney and heart express glucocorticoid receptors (Rosenfeld *et al.*, 1988, 1993) and exposure to glucocorticoid analogs resets the phase of circadian gene expression in these tissues (Balsalobre *et al.*, 2000). However, the SCN does not express glucocorticoid receptors (Rosenfeld *et al.*, 1988, 1993) and, therefore, remains unaffected by glucocorticoid treatment (Balsalobre *et al.*, 2000).

The Molecular Circadian Clock

Transcriptional feedback loops within mammalian clock cells are the molecular basis of the 24-hour cycling mechanism underlying circadian rhythmicity (Darlington *et al.*, 1998; Dunlap, 1999; Reppert and Weaver, 2001; Van der Horst *et al.*, 1999; Wilsbacher and Takahashi, 1998). Heterodimers of CLOCK and BMAL1 proteins activate transcription of *Period* (*Per*) and *Cryptochrome* (*Cry*) genes. Subsequently, Cry proteins negatively regulate CLOCK: BMAL1-mediated gene transcription, closing the molecular feedback loops, while PER2 protein positively regulates *BMAL1* gene expression (Allada *et al.*, 2001; Bae *et al.*, 2001; Okamura *et al.*, 2002; Reppert and Weaver, 2001; Rutter *et al.*, 2002). This molecular circadian clock regulates the expression of clock-controlled genes in the SCN, e.g., arginine-vasopressin (Jin *et al.*, 1999; Silver *et al.*, 1999). The molecular basis of circadian regulation is now being elucidated, but how circadian rhythms contribute to adaptation to local environments and health is still a fledgling field.

The Food-Entrainable Circadian Clock

A number of mammalian species, e.g., mice, rats, hamsters, and marsupials, show anticipatory activity before food is presented in near 24-hour feeding or caloric restriction schedules in a light-dark cycle as well as constant dark. This anticipatory activity persists during days of total food deprivation. When food is presented during the light period to nocturnal animals in a light-dark cycle, the nocturnal bout of activity is considerably shortened, although these animals continue to become active at lights off (Boulos and Terman, 1980; Cassone and Stephan, 2002; Holmes and Mistlberger, 2000; Kennedy *et al.*, 1995; Mistlberger, 1993, 1994; Rosenwasser *et al.*, 1984; White and Timberlake, 1994). Studies in which food-anticipatory behavior persisted after SCN lesion provide evidence that the food-entrainable circadian clock is distinct from the light-entrainable clock in the SCN (Abe and Rusak, 1992; Boulos *et al.*, 1980; Boulos and Terman, 1980; Comperatore and Stephan, 1990; Marchant and Mistlberger, 1997; Mistlberger, 1992, 1993; Ruis *et al.*, 1989; Stephan, 1981, 1983, 1984, 1989a, 1989b, 1992a, 1992b; Stephan *et al.*, 1979; Walcott and Tate, 1996), although its location is currently unknown (Stephan, 2002). Lesions of the parabrachial region (Davidson *et al.*, 2000), or the ventromedial hypothalamus in one study (Inouye, 1982) but not another (Mistlberger and Rechtschaffen, 1984), reduced or abolished food-anticipatory activity. However, whether these regions are integral parts of the food-entrainable clock is still unclear. Nevertheless, the parabrachial region has been proposed as one pathway by which scheduled feeding might influence the SCN (Cassone and Stephan, 2002). Others have suggested that the paraventricular thalamus facilitates entrainment to a scheduled daily palatable meal in rats (Mendoza *et al.*, 2005), while Landry *et al.* (2007) have shown through lesion studies that the paraventricular thalamus is not necessary for the expression of food-anticipatory activity.

Behavioral and SCN Entrainment to Non-Photic Timing Cues Such as Scheduled Feeding

Nocturnal animals are normally inactive during the light period of a light-dark cycle. If food is made available only during the light period, food-anticipatory activity occurs. However, rhythmic gene expression in the SCN and the light-entrainable activity component controlled by the SCN do not shift but remain entrained by the light-dark cycle. As a result, the SCN is effectively uncoupled from peripheral food-entrainable clocks (Brinkhof *et al.*, 1998; Damiola *et al.*, 2000; Hara *et al.*, 2001; Le Minh *et al.*, 2001; Shibata *et al.*, 1983; Stokkan *et al.*, 2001; Wakamutsu *et al.*, 2001). Scheduled feeding and caloric restriction does not entrain free-running light-entrainable locomotor activity rhythms in several studies (Abe *et al.*, 1989; Aschoff *et al.*, 1983; Aschoff and Van Goetz, 1986; Boulos *et al.*, 1989; Boulos and Terman, 1980; Gibbs, 1979; Kennedy *et al.*, 1995; Moore, 1980; O'Reilly *et al.*, 1986; Stephan, 1986a), but does phase shift this activity component approximately 1-3 hours (Challet *et al.*, 1998a; 1998b; Mendoza et al., 2005; Mistlberger, 1991). In other studies, the light-entrainable activity component entrains to SF in a subset of animals (Abe *et al.* 1989;

Abe and Rusak, 1992; Castillo *et al.*, 2004; Jilge, 1991; Kennedy *et al.*, 1991, 1996; Stephan, 1986b, 1986c).

The likelihood of the free-running light-entrainable activity component being entrained by scheduled daily feeding in DD also appears to be species dependent. Hamsters typically show behavioral entrainment to daily scheduled feeding in constant darkness (Mistlberger, 1993) and mice are moderately likely to entrain (Castillo *et al.*, 2004; Holmes and Mistlberger, 2000; Marchant and Mistlberger, 1997), while rats do not typically entrain (Mistlberger, 1993). The phase of the molecular clock in the SCN was not determined in any of these studies. However, using Per2 protein expression as a marker of SCN phase, we were the first to shown that scheduled feeding entrains the wheel-running activity rhythms, body temperature rhythms (Castillo and Bult-Ito, 2007; Chapter 3 in this volume), and the mouse SCN in constant dark conditions (Castillo *et al.*, 2004). This finding has recently been confirmed in mice (Abe *et al.*, 2007) and rats (Caldelas *et al.*, 2005; Lamont *et al.*, 2005). Other non-photic timing cues, e.g., scheduled forced daily treadmill running (Marchant and Mistlberger, 1996), voluntary wheel-running (Edgar and Dement, 1991), and timed daily injections of saline (Sumova *et al.*, 1994), melatonin (Slotten *et al.*, 2000), or triazolam (Van Reeth and Turek, 1989), can also entrain the master circadian clock. For a discussion of the relevance of non-photic entrainment of the circadian system to human health, please see Castillo and Bult-Ito (2007; Chapter 3 in this volume).

Adaptive Significance of Circadian Organization

The empirical evidence that establishes the adaptive significance of circadian organization is very sparse (Johnson, 2005; Sharma, 2003), although the evolutionary significance of circadian rhythms is almost universally accepted. In the context of the evolution of circadian traits and following evolutionary biology principles, Johnson (2005) defines adaptation as "an aspect of the phenotype that is the product of evolution by natural selection in a particular environmental context and represents a solution to some challenge presented by the environment," and fitness as "a measure of reproductive success and the passing on of genes." We adopt these definitions in our discussions below on the adaptive significance of circadian organization.

Reproductive Fitness

Whether circadian organization confers reproductive fitness benefits has to date not been establish for any mammals. Carl H. Johnson's group has done some elegant competition studies in the asexual cyanobacterium with strains that have different free-running periods of 22, 25, and 30 hours. In culture, each strain overtakes the other strains if its free-running period is similar to the length of the environmental light-dark cycle (Ouyang *et al.*, 1998; Woelfle *et al.*, 2004). In constant light conditions, the strains performed similarly and cultures remained mixed, which indicates that the difference in fitness under light-dark cycles of various lengths is most likely due to differences in the circadian clock (Johnson, 2005).

Survival

Lesioning of the SCN in chipmunks resulted in arrhythmicity or abnormal activity rhythms and significantly reduced survival in their natural environment compared to sham operated animals and non-manipulated controls (DeCoursey *et al.*, 2000; DeCoursey and Krulas, 1998). DeCoursey *et al.* (2000) deduced that the decreased survival was due to weasel predation. Weasels may have been able to cue into increased activity levels in the burrow at night in the SCN-lesioned chipmunks. SCN lesions in antelope ground squirrels resulted in increased above ground activity at night, compared to intact animals. This increased nighttime activity resulted in increased mortality due to predation by a feral cat compared to the intact squirrels (DeCoursey, 2004; DeCoursey *et al.*, 1997). In the golden-mantled ground squirrel, SCN lesions resulted in aberrant hibernation patterns that were energetically expensive and resulted in lower survival rates due to increased risk of starvation immediately after emergence from hibernation in the spring compared to intact animals (DeCoursey, 2004). To truly establish that circadian activity patterns controlled by the SCN have adaptive value, follow-up studies will have to measure reproductive fitness, i.e., the number of offspring surviving to reproduce themselves, of SCN lesioned and intact animals (Johnson, 2005).

Hurd and Ralph (1998) showed that heterozygous *tau* mutant golden hamsters with a free-running period of 22 hours had a shorter lifespan in a 14:10 hour light-dark cycle than the homozygous wild type hamsters, while SCN transplants that restored robust circadian rhythms in old hamsters (Hurd *et al.*, 1995) extended their life by about 4 months. However, homozygous *tau* mutant hamsters with a free-running period of 20 hours had a longer lifespan in constant darkness than the homozygous wild type animals (Oklejewicz and Daan, 2002). In other studies (Menaker and Vogelbaum, 1993; Ruby *et al.*, 1998; Sato and Kawamura, 1984; Vitaterna *et al.*, 1994), lesions of the SCN did not affect survival in the laboratory. This indicates that SCN-controlled circadian organization is not necessary for survival per se, but that the changes in circadian activity patterns as a result of SCN lesions under natural conditions may expose the animals to adverse circumstances, such as increased predation (DeCoursey, 2004) that leads to subsequent reduction in survival. None of these studies addressed reproductive fitness.

Latitudinal Clines in Circadian Organization

The environmental light-dark cycles vary considerably along a latitudinal cline, with large changes in day length during spring and fall in polar regions, e.g., 10 minutes each day on the North Slope in Alaska at 70°N, and near constant 12:12 hour light-dark cycles throughout the year at the equator. In addition, polar environments experience near constant darkness in winter and near constant light conditions in summer. As a result, animals face different circadian challenges at different latitudes, which may require circadian adaptations through natural selection. Consequently, elucidation of the effects of latitude on circadian function has received considerable attention.

Some Arctic and Antarctic species appear to have decreased robustness of circadian clock function. The Svalbard ptarmigan at 70°N become arrhythmic in activity, feeding behavior, and plasma melatonin levels under constant lighting conditions (Reierth and Stokkan, 1998a, 1998b; Reierth *et al.*, 1999). However, other birds, such as the Lapland Longspurs at 68°N in Alaska, maintained circadian plasma melatonin and testosterone (males) and 17-β estradiol (females) rhythms throughout the polar year (Hau *et al.*, 2002). In Antarctic Adélie Penguins at 64°S, daily plasma prolactin and corticosterone patterns are arrhythmic under constant natural light conditions, when assessing the population as a whole (Vleck and Van Hook, 2002). Whether individual penguins retained circadian cycles was not assessed in this study.

Reindeer from Svalbard and northern Norway loose their circadian organization of activity during the polar winter and polar summer and become ultradian (Van Oort *et al.*, 2005). In addition, the precision of timing of circadian activity, i.e., the variation in onset, midpoint, and offset of activity, is generally larger at higher latitudes in mammals as well as birds (Daan and Aschoff, 1975). The increased propensity for loss of circadian organization at higher latitudes under constant light conditions has also been shown in the northern red-backed vole, *Clethrionomys rutilus*, and several other rodent species (Swade and Pittendrigh, 1967). We have confirmed these findings in the red-backed vole in the laboratory. *C. rutilus* from Fairbanks, Alaska, at 64°N were bred for several generations in the laboratory and 50% to 70% loose circadian organization of wheel-running activity in constant dark (Fig. 1) and constant light conditions, respectively. Most of these voles become ultradian with periods of about 4-6 hours, but some become arrhythmic (Tavernier *et al.*, 2004). The loss of circadian organization is quite rare in rodents. For example, the majority of subterranean mole rats, *Georychus capensis*, *Heterocephalus glaber* (65%), and *Spalax ehrenbergi* (50-80%), display circadian rhythms in light-dark cycles and in constant darkness or constant light, although under natural conditions they are exposed to light infrequently (Ben-Sclomo *et al.*, 1995; Goldman *et al.*, 1997; Lovegrove and Muir, 1996; Riccio and Goldman, 2000a, 2000b; Tobler *et al.*, 1998). Whether the circadian attributes at high latitudes reflect a fundamental difference in circadian clock function, are due to particular clock gene alleles, or are environmentally induced remains to be elucidated.

We have proposed two hypotheses to explain why circadian rhythms in *C. rutilus* may be less robust at higher latitudes (Tavernier *et al.*, 2004) and these can be extended to other mammals as well. 1. Food resources are limited during the dark Arctic winter and loss of circadian rhythmicity may be an adaptation to exploit limited resources around the clock. Similarly, in summer the short Arctic breeding season puts large energetic demands on the parents and being able to forage around the clock when food is abundant may confer fitness benefits. 2. During the Arctic spring and fall, day length changes rapidly and keeping up with these changes to entrain to the ambient light-dark cycle, e.g., for predator avoidance or maintaining coherent circadian physiology, may require a circadian system that is less robust and more responsive to light. We have shown in laboratory mice that less robustly organized circadian rhythms are correlated with larger light-induced phase shifts (Amy *et al.*, 2000; Bult *et al.*, 1993, 2001; Castillo *et al.*, 2005; Yan *et al.*, 2003).

Latitudinal clines in circadian organization have also been established for invertebrates, such as nemobiine crickets (Shimizu and Masaki, 1997) and the fruit fly *Drosophila*. Circadian eclosion rhythms in *D. subobscura* are latitude dependent (Lankinen, 1993) and in *D. auraria* at high latitude these rhythms become arrhythmic in constant darkness (Pittendrigh *et al.*, 1991).

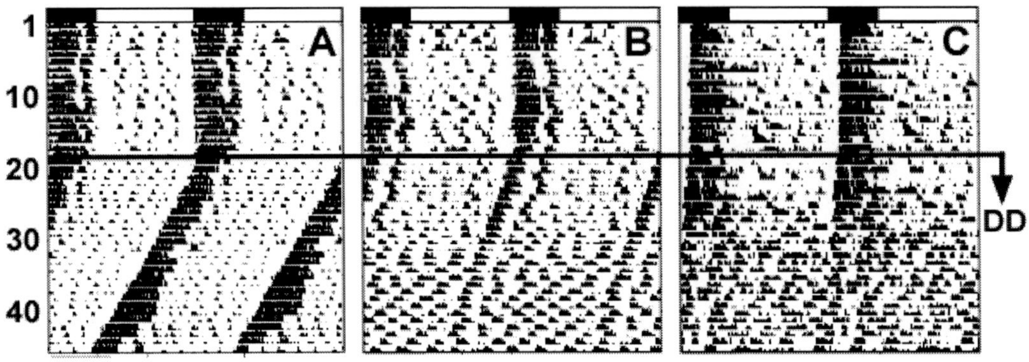

Figure 1. Representative double plotted wheel-running actograms of a red-backed vole that maintained circadian rhythmicity (A), a vole that developed an ultradian rhythm (B), and a vole that became arrhythmic (C) after four weeks in constant darkness (DD). Nineteen days of a 16-8 hour light-dark cycle, top white and black bars, respectively, followed by 28 days of DD. Days are shown on the y-axis and 48 hours are shown on the x-axis. The small black bars represent the number of wheel revolutions in five-minute bins. Wheel-running activity data were collected by computer with the VitalView data collection system (Mini-Mitter Co., Inc., Bend, OR). Wheel-running data were analyzed with Vitalview and Actiview (Mini-Mitter Co, Inc.), Matlab Version 7.1 (The Mathworks, Inc. Natick, MA), and Clock Lab (Actimetrics, Evanston, IL) software programs. Circadian and ultradian rhythms and arrhythmicity were confirmed with a chi-squared periodogram (not shown). As is evident in A-C, in a light-dark cycle, red-backed voles are predominantly nocturnal, as we have shown previously (Tavernier *et al.*, 2004).

The amplitude of the circadian clock in *D. auraria* is temperature dependent and appears to be larger at higher latitudes (Pittendrigh *et al.*, 1991). In addition, midday egg laying in *D. melanogaster* (Allemand and David, 1976), and midday activity in *D. ananassae* (Joshi, 1999) and 11 *Drosophila* species (Simunovic and Jaenike, 2006) are positively correlated with latitude. Furthermore, the amplitude of the phase response curve to light is decreased and free-running periods are increased at higher latitudes (Pittendrigh and Takamura, 1989). Many of these latitudinal traits are affected by genetic factors as they are maintained under standardized laboratory conditions. Latitudinal variation in the clock gene *period* in *Drosophila* (Costa *et al.*, 1992; Costa and Kyriacou, 1998; Sawyer *et al.*, 1997) may be more important for temperature compensation of the circadian clock (Peixoto *et al.*, 1998; Sawyer *et al.*, 1997) than that it explains the latitudinal cline in circadian organization. Interestingly, the latitudinal clines in the amplitude of the circadian clock in *Drosophila* are contrary to the findings in mammals and birds (see above).

In humans, urine melatonin concentrations are weakly but positively correlated with latitude (Wetterberg *et al.*, 1999). Considering that urine samples were collected during the night and that melatonin concentrations are very low during the day (Pandi-Perumal *et al.*, 2006), this result suggests higher amplitude rhythms at higher latitudes. Circadian processes have been implicated in seasonal affective disorder (SAD) (Klerman, 2005; Lewy *et al.*, 2006; Wirz-Justice, 2006). SAD was first reported to be more common at higher latitudes (Rosen *et al.*, 1990), but a latitudinal cline has since only been supported weakly (Magnússon, 2000; Michalak and Lam, 2002) or not at all (Blazer *et al.*, 1998; Levitt and Boyle, 2002; Magnússon and Axelsson, 1993; Magnússon and Stefánsson, 1993). Melatonin rhythms are phase delayed in SAD patients (Thalén *et al.*, 1995) and individuals with sleep disturbances (Bratlid and Wahlund, 2003). Morning bright light, which greatly reduces melatonin levels (Pandi-Perumal *et al.*, 2006), alleviates SAD symptoms (Lewy *et al.*, 1998; Postolache and Oren, 2005; Sack *et al.*, 1990), improves sleep (Hansen et al., 1987; Lingjaerde *et al.*, 1985; Winkler *et al.*, 2005), and normalizes melatonin rhythms (Hansen *et al.*, 1987; Thalén *et al.*, 1995). Environmental factors other than light (Axelsson *et al.*, 2002a, 2002b) and genetic factors (Madden *et al.*, 1996; Ozaki *et al.*, 1995; Stefánsson *et al.*, 1994) also play an important role in SAD.

The weak support for or lack of a latitudinal cline in SAD in previous studies might be related to the particular populations that were studied. For example, Winnipeg, Manitoba, (50°N) residents wholly from Icelandic descent had significantly lower rates of SAD than a random sample of non-Icelandic Winnipeg residents, which revealed an important genetic effect (Axelsson *et al.*, 2002b). This finding was used as an argument against a latitudinal cline in SAD prevalence. However, could resistance to SAD have evolved in populations that had lived at high latitude for many generations, which would still be present in descendents of Icelandic people who lived at 63.4°N to 66.5°N? The study by Rosen *et al.* (1990) was based on populations along the east coast of the United States of America, mostly comprised of people who had been in the US for only a few generations, were of diverse backgrounds, and were generally not descendents of high-latitude peoples. Therefore, these people may have been more susceptible to developing SAD at higher latitudes.

Desynchronization of internal circadian rhythms from the external environmental day-night cycle appears to play an important role in the development of mood disorders, such as SAD (Wirz-Justice, 2006), and sleep disorders (Roldan and Ang, 2006), such as associated with shift work (Schwartz and Roth, 2006) and jetlag (Richardson, 2005). Desynchronization is more likely if the internal circadian system cannot keep up with seasonal changes in the external environmental light-dark cycle. We hypothesize that SAD may be caused by an inability of the circadian system to keep up with changes in day length through phase shifts because of a robust circadian organization that is not very sensitive to light. As a result, internal circadian organization may be uncoupled from the environmental cycle and result in aberrant physiological cycles. This may lead to changes in brain neurotransmitter systems and the development of psychiatric disorders, such as SAD (Wirz-Justice, 2006).

This hypothesis leads to two main predictions: 1. SAD patients have more robust circadian rhythms when they do not have SAD symptoms compared to healthy matched controls, and 2. Populations that have lived at high latitudes for many generations, e.g., the original circumpolar peoples, are expected to have low rates of SAD and exhibit less robust circadian organization than populations that have lived at high latitudes for only a few generations or circumpolar peoples with high SAD rates. Other hypotheses have been proposed for SAD, such as the photoperiod and phase-shift hypotheses (Sohn and Lam, 2005).

Human Diurnal Preference

The human *Per3* clock gene is polymorphic (Archer *et al.*, 2003; Ebisawa *et al.*, 2001) and diurnal preference, i.e., whether people are morning or evening types, is correlated with a human *Per3* polymorphism in a UK population (Archer *et al.*, 2003). Pereira *et al.* (2005) suggested a latitudinal cline in *Per3* polymorphisms in Brazil, but this was not confirmed in populations from Europe, Africa, Asia, and Australia, and the *Per3* polymorphisms appear not to have been under natural selection (Nadkarni *et al.*, 2005). These findings are in agreement with another study in Brazil in which the distribution of morning and evening types does not correlate with latitude, but may have been influenced by local social habits instead (Benedito-Silva *et al.*, 1998).

A C/T human *clock* polymorphism at position 3111 is associated with evening preference and delayed sleep time in a Japanese (Mishima *et al.*, 2005) and Caucasian (Katzenberg *et al.*, 1998) population, while not confirmed in another Caucasian population (Robilliard *et al.*, 2002). One polymorphism in the human *Per1* gene, i.e., an A to G synonymous substitution at position 2548, is not associated with diurnal preference in normal adults (Katzenberg *et al.*, 1999), while another *Per1* polymorphism, i.e., a T to C synonymous substitution at position 2434, is more frequent in subjects with extreme morningness compared to subjects with extreme eveningness (Carpen *et al.*, 2006). Interestingly, this latter polymorphism is not associated with delayed sleep phase syndrome (Carpen *et al.*, 2006). Familial advanced and delayed sleep phase syndromes can be viewed as extreme manifestations of diurnal preference, but the molecular mechanisms controlling this diverse range of sleep disorders appears to be different (Tafti *et al.*, 2005).

Familial Advanced Sleep Phase Syndrome in Humans

Familial advanced sleep phase syndrome, an autosomal dominant trait, is associated with a mutation in the human *Per2* gene (Toh *et al.*, 2001), a mutation in the human *casein kinase Iδ* gene (Xu *et al.*, 2005), and a mutation unrelated to the human *Per2* gene mutation identified by Toh *et al.* (2001) (Satoh *et al.*, 2003). This circadian syndrome is characterized by individuals getting up very early in the morning, while going to bed very early in the evening (Jones *et al.*, 1999). Transgenic mice expressing this human *Per2* mutation show 4-6 hour phase advanced locomotor activity rhythms (Xu *et al.*, 2007), which is similar to the advanced phase angle of entrainment to the light-dark cycle observed in humans (Jones *et al.*, 1999). This advanced phase angle of entrainment was due to a shortening of the free-running period, as measured in constant darkness, by 2-3 hours compared to wild type mice (Xu *et al.*, 2007), which is also similar to the human phenotype (Jones *et al.*, 1999). Similar circadian characteristics are found in cycling fibroblasts carrying the human *Per2* mutation (Vanselow *et al.*, 2006). These mutations affect the phosphorylation state of the Per2 protein *in vitro*, which affect the speed by which the molecular circadian clock runs through the transcription and translational feedback loops. As a result, the endogenous period of the clock is shortened

(Vanselow *et al.*, 2006; Xu *et al.*, 2007). These studies confirm the causal relationship between the human clock genes *Per2* and human *casein kinase Iδ* and the phenotype observed in human familial advanced sleep phase syndrome. The circadian system appears to be directly involvement of in this human circadian disorder. To what extent familial advanced sleep phase syndrome can be considered a maladaptive trait remains to be determined.

Familial Delayed Sleep-Phase Syndrome in Humans

Per3 polymorphisms, a missense polymorphism (Ebisawa *et al.*, 2001) and a gene-length polymorphism (Pereira *et al.*, 2005), are associated with familial delayed sleep-phase syndrome. Familial delayed sleep-phase syndrome in humans is another example of the connection between clock genes that control circadian cycling at the molecular level to a human sleep disorder that greatly changes the temporal expression of behavioral and physiological circadian rhythms. To what extent familial delayed sleep phase syndrome has fitness consequences remains to be determined.

Fatal Familial Insomnia in Humans

Fatal familial insomnia is an extreme circadian and sleep disorder that results in death and is caused by a mutation in the prion protein gene (Gambetti *et al.*, 2003; Montagna *et al.*, 2003). Because this disorder is not due to a change in the molecular circadian clock mechanism, we will not discuss it here in detail.

Testing the Adaptive Significance of Mammalian Circadian Organization

Although clock gene mutations with large phenotypic effects have been instrumental in the elucidation of the molecular clock mechanism (see the Molecular Circadian Clock section), they are not very useful in determining the adaptive significance of circadian organization under natural conditions. Clock gene alleles with small phenotypic effects are more likely to affect adaptations to local circadian environments, because mutations with large phenotypic effects are usually deleterious and have no chance to spread through or be maintained in the population (Crow, 1997). Therefore, identification of natural clock gene allelic variation most likely will reveal novel polymorphisms. These polymorphisms will allow for the further refinement of the molecular clock mechanism and provide insights into pathways that may eventually lead to therapeutic strategies. We believe that natural variation in clock gene alleles and circadian phenotypes are an untapped resource that may eventually explain the majority of circadian disorders and related human health problems and normal variation. Most of the circadian disorders and normal variation in circadian characteristics are probably due to variation in multiple clock genes as most complex traits have a polygenic basis. Environmental factors also play an important role and will have to be an integral part of the investigations to obtain a comprehensive understanding.

Reproductive fitness measurements as a consequence of circadian organization and variation in clock gene alleles has to be the ultimate goal to confirm the adaptive significance of circadian rhythms (Johnson, 2005). However, this approach may often not be feasible in mammals, with generally long generation times, and also in humans, with a number of ethical and moral constraints. Therefore, we would like to propose a set of experimental approaches that in aggregate will address the adaptive significance of circadian organization in mammals.

The elucidation of the adaptive significance of mammalian circadian organization requires expertise from divergent fields of research. Therefore, an interdisciplinary approach that includes, but is not limited to, circadian biologists, molecular biologists, physiologists, neuroscientists, evolutionary biologist, ecologists, and for human studies epidemiologists, medical care providers, and psychologists, is a must.

Latitudinal Clines

If phenotypic circadian rhythm characteristics in humans and other mammals are under natural selection, we would expect that populations that have lived in the same region for hundreds of generations to be optimally adapted to the circadian characteristics of their local environment. These adaptations would have evolved within the genetic constraints of the specific populations, such as phylogenetic history and available clock gene allelic variation. Latitudinal clines are very useful tools to elucidate circadian adaptations because of the large variation in circadian challenges from equatorial to polar environments. Identification of latitude dependent clock gene allelic variation that correlates with circadian phenotypic variation will be the most direct approach.

Mammalian Studies

To our knowledge, no latitudinal clines have been determined for mammalian clock gene allelic variation and the elucidation of latitude dependent circadian phenotypes have been sparse, anecdotal, and limited in latitudinal range (see the Latitudinal Clines in Circadian Organization section). The latitudinal cline should optimally be done in the same species (Garland and Adolph, 1994), include a large latitudinal range, and be replicated at different locations on earth. For example, the northern red-backed vole, *Clethrionomys rutilus*, is a non-hibernating rodent species (Tavernier *et al.*, 2004) with a worldwide distribution in the northern hemisphere (Stenseth, 1985) and a statewide distribution in Alaska with a latitudinal range of 12° (70°N, North Slope, to 58°N, Juneau). Day length changes during spring and fall are two-fold different between the North Slope (10 minutes) and Juneau (5 minutes). A closely related species, the southern red-backed vole *Clethrionomys gapperi*, extends from Juneau, Alaska, to New Mexico (34°N) (Stenseth, 1985), potentially extending this latitudinal cline to 36°.

House mice are widely distributed throughout the world as they have followed human habitation. Their suitability for studies of latitudinal clines has been demonstrated. For example, on a latitudinal cline along the east coast of the United States of America (USA), wild-caught mice from Maine (north) build larger nests than mice from Florida (south) after 1-3 generations of breeding in the laboratory (Lynch, 1992). A similar latitudinal cline was found in mice from Colorado and Texas in the USA (Plomin and Manosevitz, 1974). Consequently, nest-building behavior appears to have been under natural selection.

Laboratory studies on reproductive measures in mice bidirectionally selected for nest-building behavior also confirm that nest building is a behavioral adaptation to cold ambient temperatures (Bult and Lynch, 1996, 1997, 2000; Lynch, 1980).

Human Studies

Indigenous peoples who have lived in the same location for hundreds of generations may have evolved circadian characteristics that adapt them to specific local circadian environments. Elucidating clock gene allelic variation in indigenous peoples along a latitudinal cline has the potential of identifying polymorphisms that have been or continue to be subject to natural selection.

Another important approach is to evaluate circadian characteristics and circadian disorders along a latitudinal cline in peoples who have lived in a particular region for only a few generations. This approach has the potential to identify clock gene allelic variation that makes people more vulnerable to circadian disturbances in particular circadian environments. Comparing people, who have grown up, or have ancestors from, equatorial regions and now live in polar regions (or visa versa), to indigenous populations may be a powerful approach if studying an entire latitudinal cline is prohibitive. The advantage of studying humans is that they live in virtually every habitable region of the world and therefore the within species comparison can be very powerful.

Circadian Rhythms and Disease

Vulnerability to disease can be maladaptive if it leads to reduced fitness. Identification of clock gene polymorphisms in human circadian disorders have already been successful, such as those identified in familial advanced and delayed sleep phase syndromes (see Human Diurnal Preference section). These studies should be expanded and include other diseases, such as depression. For example, depression induced by major life events (Grandin *et al.*, 2006; Haynes *et al.*, 2006), SAD (Lewy *et al.*, 2006), and menopausal depression (Parry et al., 2006) are associated with several circadian abnormalities. Bipolar disorder, characterized by circadian abnormalities (Harvey *et al*, 2006) and alternations between manic and major depressive episodes, is associated with *Per3* and *timeless* polymorphisms (Mansour *et al.*, 2005). These findings indicate that bipolar disorder might be caused by changes in the master circadian clock itself. Alternatively, these clock genes may have functions outside of the SCN that are not related to time keeping.

Targeted Clock Gene Allele Transgenic Mice

With the advance of molecular tools, including rapid and accurate DNA sequencing, naturally occurring clock gene allelic variation with correlated differences in circadian organization can be put to rigorous molecular, behavioral, and fitness analyses using transgenic mice in a laboratory setting that express particular naturally occurring clock gene alleles. Subsequent crosses of transgenic mouse lines could investigate the effects of combinations of these clock gene alleles. One drawback of this approach is that the allelic variation might have been found in a species other than the mouse. As a result, these different alleles may function differently in the genetic background of the mouse or even specific mouse line, compared to the genetic background of the species the allelic variation was identified in. Another drawback of this approach is that these clock gene alleles may function differently in different environments, as a result of genotype by environment interactions, and the laboratory environment may not reflect the allele's normal function in the natural environment. Despite these potential experimental constraints, this approach has already been very successful. As discussed, transgenic mice expressing the *Per2* mutation associated with human familial advanced sleep phase syndrome (Toh *et al.*, 2001) showed similar advanced activity rhythms as observed in humans (Xu *et al.*, 2007).

Conclusions

Human health is usually expressed as a collection of continuous traits with diseases being identified at the high and/or low ends of the distributions. Most diseases, e.g., psychiatric disorders and cardiovascular diseases, are not due to mutations with large phenotypic effects, but are most likely influenced by a collection of genes with small effects that interact with particular environmental factors. We hypothesize that the genetic predisposition for a particular disease can be attributed to the combination of a particular set of genetic factors that are part of the normal genetic variation in a population. Natural clock gene allelic variation and associated circadian phenotypes and how these traits contribute to local adaptations are largely unexplored in mammals. We believe that this is truly an untapped resource that may eventually explain the majority of circadian disorders and related human health problems. We therefore strongly advocate for increased support and effort in this field of research.

References

Abe H, Kida M, Tsuji K, Mano T 1989 Feeding cycles entrain circadian rhythms of locomotor activity in CS mice but not in C57BL/6J mice. *Physiol. Behav*. 45:397-401.

Abe H, Honma S, Honma K-I 2007 Daily restricted feeding resets the circadian clock in the suprachiasmatic nucleus of CS mice. *Am. J. Physiol.* 292:R607-R615.

Abe H, Rusak B 1992 Anticipatory activity and entrainment of circadian rhythms in Syrian hamsters exposed to restricted palatable diets. *Am. J. Physiol.* 263:R116-24.

Allada R, Emery P, Takahashi JS, Rosbash M 2001 Stopping time: the genetics of fly and mouse circadian clocks. *Annu. Rev. Neurosci.* 24:1091-1119.

Allemand R, David, JR 1976 The circadian rhythm of oviposition in *Drosophila melanogaster*: a genetic latitudinal cline in wild populations. *Experientia* 32:1403-1405.

Amy SP, Chari R, Bult A 2000 Fos in the suprachiasmatic nucleus of house mouse lines that reveal a different phase-delay response to the same light pulse. *J. Biol. Rhythms*. 15:95-102.

Archer SN, Robilliard DL, Skene DJ, Smits M, Williams A, Arendt J, and von Schantz M 2003 A length polymorphism in the circadian clock gene *Per3* is linked to delayed sleep phase syndrome and extreme diurnal preference. *Sleep* 26:413-415.

Aschoff J, Von Goetz C 1986 Effects of feeding cycles on circadian rhythms in squirrel monkeys. *J. Biol. Rhythms*. 1:267-276.

Aschoff J, Von Goetz C, Honma K 1983 Restricted feeding in rats: effects of varying feeding cycles. *Z Tierpsychol* 63:91-111.

Axelsson J, Káradóttir R, Karlsson MM 2002a Differences in prevalence of seasonal affective disorder that are not explained by either genetic or latitude differences. *Int. J. Circumpolar Health* 61:17-20.

Axelsson, J, Stefánsson JG, Magnússon A, Sigvaldason H, Karlsson MM 2002b Seasonal Affective Disorders: The relevance of Icelandic and Icelandic-Canadian evidence to ætiologic hypotheses. *Can. J. Psychiatry* 47:153-158.

Bae K, Jin X, Maywood ES, Hastings MH, Reppert SM, Weaver DR 2001 Differential functions of mPer1, mPer2, and mPer3 in the SCN circadian clock. *Neuron.* 30:525-536.

Balsalobre A 2002 Clock genes in mammalian peripheral tissues. *Cell Tissue Res.* 309:193-199.

Balsalobre A, Damiola F, Schibler U 1998 A serum shock induces circadian gene expression in mammalian tissue culture cells. *Cell* 93:917-919.

Balsalobre A, Brown SA, Marcacci L, Tronche F, Kellendonk C, Reichardt HM, Schütz G, Schibler U 2000 Resetting of circadian time in peropheral tissues by glucocorticoid signaling. *Science* 289:2344-2347.

Benedito-Silva AA, Menna-Barreto L, Farias Alam M, Rotenberg L, Moreira LFS, Menezes AAL, Pereira da Silva H, Marques N 1998 Latitude and social habits as determinants of the distribution of morning and evening types in Brazil. *Biol. Rhythm. Res.* 29:591–597.

Ben-Sclomo R, Ritte U, Nevo E 1995 Activity pattern and rhythm in the subterranean mole rat superspecies *Spalax ehrenbergi*. *Behav. Genet* 25:239-245.

Blazer DG, Kessler RC, Swartz MS 1998 Epidemiology of recurrent major and minor depression with a seasonal pattern. The National Comorbidity Survey. *Br. J. Psychiatry* 172:164-167.

Boulos Z, Frim DM, Dewey LK, Moore-Ede MC 1989 Effects of restricted feeding schedules on circadian organization in squirrel monkeys. *Physiol. Behav*. 45:507-515.

Boulos Z, Rosenwasser AM, Terman M 1980. Feeding schedules and the circadian organization of behavior in the rat. *Behav. Brain Res.* 1:39-65.

Boulos Z, Terman M 1980 Food availabiltiy and daily biological rhythms. *Neurosci. Biobehav. Rev*. 4:119-131.

Bratlid T, Wahlund B 2003 Alterations in serum melatonin and sleep in individuals in a sub-arctic region from winter to spring. *Int. J. Circumpolar Health*. 62:242-254.

Brinkhof MWG, Daan S, Strubbe JH 1998 Forced dissociation of the food- and light-entrainable circadian rhythms of rats in a skeleton photoperiod. *Physiol. Behav.* 65:225-231.

Bult A, Hiestand L, VanderZee EA, Lynch CB 1993 Circadian rhythms differ between selected mouse lines: a model to study the role of vasopressin neurons in the suprachiasmatic nuclei. *Brain Res. Bull* 32:623-627.

Bult A, Kobylk ME, Van der Zee, EA 2001 Differential expression of protein kinase C βI (PKCβI) but not PKCα and PKCβII in the suprachiasmatic nucleus of selected house mouse lines, and the relationship to arginine-vasopressin. *Brain Res.* 914:123-133.

Bult A, Lynch CB 1996 Multiple selection responses in house mice bidirectionally selected for thermoregulatory nest-building behavior: crosses of replicate lines. *Behav. Genet* 26:439-446.

Bult A, Lynch CB 1997 Nesting and Fitness: Lifetime reproductive success in house mice bidirectionally selected for thermoregulatory nest-building behavior. *Behav. Genet* 27:231-240.

Bult A, Lynch CB 2000 Breaking through artificial selection limits of an adaptive behavior in mice and the consequences for correlated responses. *Behav. Genet*, in press.

Cai A, Lehman MN, Lloyd JM, Wise PM 1997a Transplantation of fetal suprachiasmatic nuclei into middle-aged rats restores diurnal Fos expression in host. *Am. J. Physiol.* 272:R422-428.

Cai A, Scarbrough K, Hinkle DA, Wise PM 1997b Fetal grafts containing suprachiasmatic nuclei restore the diurnal rhythm of CRH and POMC mRNA in aging rats. *Am. J. Physiol.* 273:R1764-1770.

Caldelas I, Feillet CA, Dardente H, Eclancher F, Malan A, Gourmelen S, Pevet P, Challet E 2005 Timed hypocaloric feeding and melatonin synchronize the suprachiasmatic clockwork in rats, but with opposite timing of behavioral output. *Eur. J. Neuroscience* 22:921-929.

Carpen JD, von Schantz M, Smits M, Skene DJ, Archer SN 2006 A silent polymorphism in the *PER1* gene associates with extreme diurnal preference in humans. *J. Hum. Genet* 51:1122-1125.

Cassone VM, Stephan FK 2002 Central and peripheral regulation of feeding and nutrition by the mammalian circadian clock: implications for nutrition during manned space flight. *Nutrition* 18:814-819.

Castillo, MR, Bult-Ito, A 2007 Non-photic effects on the circadian system and how they relate to human health. Ed. A. V. Lignelli, *Circadian rhythms and health research trends*. Nova Science Publishers, Inc., Hauppauge, NY, Chapter 3.

Castillo MR, Hochstetler KJ, Greene DM, Firmin SI, Tavernier RJ, Raap DK, Bult-Ito A 2005 Circadian rhythm of core body temperature in two laboratory mouse lines. *Physiol. Behav.* 86:538-545.

Castillo MR, Hochstetler KJ, Tavernier RJ Jr, Greene DM, Bult-Ito A 2004 Entrainment of the master circadian clock by scheduled feeding. *Am. J. Physiol.* 287:R551-R555.

Challet E, Bernard DJ, Turek FW 1998a Lesions of glucose-responsive neurons impair synchronization effects of caloric restriction in mice. *Brain Res.* 801:244-250.

Challet E, Solberg LC, Turek FW 1998b Entrainment in calorie-restricted mice: conflicting zeitgebers and free-running conditions. *Am. J. Physiol.* 274:R1751-R1761.

Comperatore CA, Stephan FK 1990 Effects of vagotomy on entrainment of activity rhythms to food access. *Physiol. Behav.* 47:671-678.

Costa R, Kyriacou CP 1998 Functional and evolutionary implications of natural variation in clock genes. *Cur. Opinion Neurobiol.* 8:659-664.

Costa R, Peixoto AA, Barbujani G, Kyriacou CP 1992 A latitudinal cline in *Drosophila* clock gene. *Proc. R . Soc. Lond* B 250:43-49.

Crow JF 1997 The high spontaneous mutation rate: is it a health risk? *Proc. Natl. Acad. Sci. USA* 94:8380-8386.

Daan S, Aschoff J 1975 Circadian rhythms of locomotor activity in captive birds and mammals: their variations with season and latitude. *Oecologia* 18:269-316.

Damiola F, Le Minh N, Preitner N, Kornmann B, Fleury-Olela F, Schibler U 2000 Restricted feeding uncouples circadian oscillators in peripheral tissues from the central pacemaker in the suprachiasmatic nucleus. *Genes Devel.* 14:2950-2961.

Darlington TK, Wager-Smith K, Ceriani MF, Staknis D, Gekakis N, Steeves TDL, Weitz CJ, Takahashi JS, Kay SA 1998 Closing the circadian loop: CLOCK-induced transcription of its own inhibitors per and tim. *Science* 280:1599-1603.

Davidson AJ, Cappendijk SL, Stephan FK 2000 Feeding-entrained circadian rhythms are attenuated by lesions of the parabrachial region in rats. *Am. J. Physiol.* 278:R1296-1304.

DeCoursey PJ 2004 Diversity of Function of SCN Pacemakers in Behavior and Ecology of Three Species of Sciurid Rodents. *Biol. Rhythm. Res.* 35:13-33.

DeCoursey PJ, Krulas JR 1998 Behavior of SCN-lesioned chipmunks in natural habitat: a pilot study. *J. Biol. Rhythms* 13:229-244.

DeCoursey PJ, Krulas JR, Mele G, Holley DC 1997 Circadian performance of suprachiasmatic nuclei (SCN)-lesioned antelope ground squirrels in a desert enclosure. *Physiol. Behav.* 62:1099-1108.

DeCoursey PJ, Walker JK, Smtih SA 2000 A circadian pacemaker in free-living chipmunks: essential for survival? *J. Comp. Physiol.* 186:169-180.

Dunlap JC 1999 Molecular basis for circadian clocks. *Cell* 96:271-290.

Ebisawa T, Uchiyama M, Kajimura N, Mishima K, Kamei Y, Katoh M, Watanabe T, Sekimoto M, Shibui K, Kim K, Kudo Y, Ozeki Y, Sugishita M, Toyoshima R, Inoue Y, Yamada N, Nagase T, Ozaki N, Ohara O, Ishida N, Okawa M, Takahashi K, Yamauchi T 2001 Association of structural polymorphisms in the human *period3* gene with delayed sleep phase syndrome. *EMBO Rep.* 2:342-346.

Edgar DM, Dement WC 1991 Regularly scheduled voluntary exercise synchronizes the mouse circadian clock. *Am. J. Physiol.* 261:R928-33.

Gambetti P, Parchi P, Chen SG 2003 Hereditary Creutzfeldt-Jakob disease and fatal familial insomnia. *Clin. Lab. Med.* 23:43-64.

Garland T, Jr, Adolph SC 1994 Why not to do two-species comparative studies: limitations on inferring adaptation. *Physiol. Zool.* 67:797-828.

Gibbs FP 1979 Fixed interval feeding does not entrain the circadian pacemaker in blind rats. Am J Physiol 236:R249-R253.

Goldman BD, Goldman SL, Riccio AP, Terkel J 1997 Circadian patterns of locomotor activity and body temperature in blind mole rats, *Spalax ehrenbergi*. *J. Biol. Rhythms* 12:348-361.

Grandin LD, Alloy LB, Abramson LY 2006 The social Zeitgeber theory, circadian rhythms, and mood disorders: review and evaluation. *Clin. Psychol. Rev.* 26:679-694.

Hansen T, Bratlid T, Lingjaerde O, Brenn T 1987 Midwinter insomnia in the sub arctic region: evening levels of serum melatonin and cortisol before and after treatment with bright artificial light. *Acta Psychiatr. Scand.* 75:428-434.

Hara R, Wan K, Wakamatsu H, Aida R, Moriya T, Akiyama M, Shibata S 2001 Restricted feeding entrains liver clock without participation of the suprachiasmatic nucleus. *Genes Cells* 6:269-278.

Harvey AG, Mullin BC, Hinshaw SP 2006 Sleep and circadian rhythms in children and adolescents with bipolar disorder. *Development Psychopathol.* 18:1147-1168.

Hau M, Romero LM, Brawn JD, Van't Hof TJ 2002 Effects of polar day on plasma profiles of melatonin, testosterone, and estradiol in high-Arctic Lapland longspurs. *Gen. Comp Endocrinol.* 126: 101-112.

Haynes PL, McQuaid JR, Ancoli-Israel S, Martin JL 2006 Disrupting life events and the sleep-wake cycle in depression. *Psychol. Med.* 36:1363-1373.

Holmes MM, Mistlberger RE 2000 Food anticipatory activity and photic entrainment in food-restricted BALB/c mice. *Physiol. Behav.* 68:655-666.

Hurd MW and Ralph MR 1998 The significance of circadian organization for longevity in the golden hamster. *J. Biol. Rhythms* 13:430-436.

Hurd MW, Zimmer KA, Lehman MN, Ralph MR 1995 Circadian locomotor rhythms in aged hamsters following suprachiasmatic transplant. *Am. J. Physiol.* 269:R958-R968.

Inouye S-HT 1982 Ventromedial hypothalamic lesions eliminate anticipatory activities of restricted daily feeding schedules in the rat. *Brain Res.* 250:183-187.

Jilge B 1991 Restricted feeding: a nonphotic zeitgeber in the rabbit. *Physiol. Behav.* 51:157-166.

Jin X, Shearman LP, Weaver DR, Zylka MJ, DeVries GJ, Reppert SM 1999 A molecular mechanism regulating rhythmic output from the suprachiasmatic circadian clock. *Cell* 96:57-68.

Johnson CH 2005 Testing the adaptive value of circadian systems. *Methods Enzymol.* 393:818-837.

Jones CR, Campbell SS, Zone SE, Cooper F, DeSano A, Murphy PJ, Jones B, Czajkowski L, Ptácek LJ 1999 Familial advanced sleep-phase syndrome: a short-period circadian rhythm variant in humans. *Nature Med.* 5:1062-1065.

Joshi DS 1999 Latitudinal variation in locomotor activity rhythm in adult Drosophila ananassae. *Can. J. Zool* 77:865–870.

Katzenberg D, Young T, Finn L, Lin L, King DP, Takahashi JS, Mignot E 1998 A CLOCK polymorphism associated with human diurnal preference. *Sleep* 21:569-576.

Katzenberg D, Young T, Lin L, Finn L, Mignot E 1999 A human period gene (HPER1) polymorphism is not associated with diurnal preference in normal adults. *Psychiatric Genet* 9:107-109.

Kennedy GA, Coleman GJ, Armstrong SM 1991 Restricted feeding entrains circadian wheel-running activity rhythms of the kowari. *Am. J. Physiol.* 261:R819-R827.

Kennedy GA, Coleman GJ, Armstrong SM 1995 Entrainment of circadian wheel-running rhythms of the northern brown bandicoot, *Isoodon macrourus*, by daily restricted feeding schedules. *Chronobiol. Intl.* 12:176-187.

Kennedy GA, Coleman GJ, Armstrong SM 1996 Daily restricted feeding effects on the circadian activity rhythms of the stripe-faced dunnart, *Sminthopsis macroura*. *J. Biol. Rhythms* 11:188-195.

Klerman EB 2005 Clinical aspects of human circadian rhythms. *J. Biol. Rhythms* 20:375-386.

Krieger DT 1974 Food and water restriction shifts corticosterone, temperature, activity and brain amine periodicity. *Endocrinol.* 95:1195-1201.

Lamont EW, Diaz LR, Barry-Shaw J, Stewart J, Amir S 2005 Daily restricted feeding rescues a rhythm of Period2 expression in the arrhythmic suprachiasmatic nucleus. *Neurosci.* 132:245-248.

Landry GJ, Yamakawa GRS, Mistlberger RE 2007 Robust food anticipatory circadian rhythms in rats with complete ablation of the thalamic paraventricular nucleus. *Brain Res.* 1141:108-18.

Lankinen P 1993 North-south differences in circadian eclosion rhythm in European populations of *Drosophila subobscura*. *Heredity* 71:210-218.

Le Minh N, Damiola F, Tronche, G Schütz G, Schibler U 2001 Glucocorticoid hormones inhibit food-induced phase-shifting of peripheral circadian oscillators. *EMBO J.* 20:7128-7136.

Levitt AJ, Boyle MH 2002 The impact of latitude on the prevalence of seasonal depression. *Can. J. Psychiatry* 47:361-367.

Lewy AJ, Bauer VK, Cutler NL, Sack RL, Ahmed S, Thomas KH, Blood ML, Jackson JML 1998 Morning vs. evening light treatment of patients with winter depression. *Arch. Gen. Psychiatry* 55:890-896.

Lewy AJ, Lefler BJ, Emens JS, Bauer VK 2006 The circadian basis of winter depression. *PNAS* 103:7414-7419.

Lingjaerde O, Bratlid T, Hansen T 1985 Insomnia during the "dark period" in northern Norway. An explorative, controlled trial with light treatment. *Acta Psychiatr. Scand.* 71:506-512.

Lovegrove BG, Muir A 1996 Circadian body temperature rhythms of the solitary cape mole rat *Georychus capensis* (Bathyergidae). *Physiol. Behav.* 60:991-998.

Lynch CB 1980 Response to divergent selection for nesting behavior in Mus domesticus. *Genetics* 96:757-765.

Lynch CB 1992 Clinal variation in cold adaptation in *Mus domesticus*: verification of predictions from laboratory populations. *Am. Nat.* 139:1219-1236.

Madden PA, Heath AC, Rosenthal NE, Martin NG 1996 Seasonal changes in mood and behavior. The role of genetic factors. *Arch. Gen. Psychiatry* 53:47-55.

Magnússon A 2000 An overview of epidemiological studies on seasonal affective disorder. *Acta Psychiatr. Scand.* 101:176-184.

Magnússon A, Axelsson J 1993 The prevalence of seasonal affective disorder is low among descendants of Icelandic emigrants in Canada. *Arch. Gen. Psychiatry* 50:947-951.

Magnússon A, Stefánsson JG 1993 Prevalence of seasonal affective disorder in Iceland. *Arch Gen. Psychiatry* 50:941-946.

Mansour HA, Monk TH, Nimgaonkar VI 2005 Circadian genes and bipolar disorder. *Annals Med.* 37:196-205.

Marchant EG, Mistlberger RE 1996 Entrainment and phase shifting of circadian rhythms in mice by forced treadmill running. *Physiol. Behav.* 60:657-663.

Marchant EG, Mistlberger RE 1997 Anticipation and entrainment to feeding time in intact and SCN-ablated C57Bl/6J mice. *Brain Res.* 765:273-282.

Meijer JH, Rietveld WJ 1989 Neurophysiology of the suprachiasmatic circadian pacemaker in rodents. *Physiol. Rev.* 69:671-707.

Menaker M, Vogelbaum MA 1993 Mutant circadian period as a marker of suprachiasmatic nucleus function. *J. Biol. Rhythms* 8:S93–S98.

Mendoza J, Angeles-Castellanos M, Escobar C 2005 A daily palatable meal without food deprivation entrains the suprachiasmatic nucleus of rats. *Eur. J. Neurosci.* 22:2855-2862.

Michalak EE, Lam RW 2002 Seasonal affective disorder: the latitude hypothesis revisited. *Can. J. Psychiatry* 47:787-788.

Mishima K, Tozawa T, Satoh K, Saitoh H, Mishima Y 2005 The 3111T/C polymorphism of *hClock* is associated with evening preference and delayed sleep timing in a Japanese population sample. *Am. J. Med. Genet. Part B* (Neuropsychiatric Genet) 133B:101-104.

Mistlberger RE 1991 Scheduled daily exercise or feeding alters the phase of photic entrainment in Syrian hamsters. *Physiol. Behav.* 50:1257-1260.

Mistlberger RE 1992 Nonphotic entrainment of circadian activity rhythms in suprachiasmatic nuclei ablated hamsters. *Behav. Neurosci.* 106:192-202.

Mistlberger RE 1993 Effects of scheduled food and water access on circadian rhythms of hamsters in constant light, dark, and light:dark. *Physiol. Behav.* 53:509-516.

Mistlberger RE 1994 Circadian food-anticipatory activity: formal models and physiological mechanisms. *Neurosci. Biobehav. Rev.* 18:171-195.

Mistlberger RE, Rechtschaffen A 1984 Recovery of anticipatory activity to restricted feeding in rats with ventromedial hypothalamic lesions. *Physiol. Behav.* 33:227-235.

Montagna P, Gambetti P, Cortelli P, Lugaresi E 2003 Familial and sporadic fatal insomnia. *Lancet Neurol.* 2:167-176.

Moore RY 1980 Suprachiasmatic nucleus, secondary synchronizing stimuli and the central neural control of circadian rhythms. *Brain Res.* 183:13-28.

Nadkarni NA, Weale ME, von Schantz M, Thomas MG 2005 evolution of a length polymorphism in the human PER3 gene, a component of the circadian system. *J. Biol. Rhythm* 20:490-499.

Okamura H, Yamaguchi S, Yagita K 2002 Molecular machinery of the circadian clock in mammals. *Cell Tissue Res.* 309:47-56.

Oklejewicz M, Daan S 2002 Enhanced longevity in tau mutant Syrian hamsters, *Mesocricetus auratus*. *J. Biol. Rhythms* 17:210–216.

O'Reilly H, Armstrong SM, Coleman GJ 1986 Restricted feeding and circadian activity rhythms of a predatory marsupial, *Dasyuroides byrnei*. *Physiol. Behav.* 38:471-476.

Ouyang Y, Andersson CR, Kondo T, Golden SS, Johnson CH 1998 Resonating circadian clocks enhance fitness in cyanobacteria. *Proc. Natl. Acad. Sci. USA* 95:8660-8664.

Ozaki N, Ono Y, Ito A, Rosenthal NE 1995 Prevalence of seasonal difficulties in mood and behavior among Japanese civil servants. *Am. J. Psychiatry* 152:1225-1227.

Pandi-Perumal SR, Srinivasan V, Maestroni GJM, Cardinali DP, Poeggeler B, Hardeland R 2006 Melatonin. Nature's most versatile biological signal? *FEBS Journal* 273:2813–2838.

Parry, BL, Martinez LF, Maurer EL, Lopez AM, Sorenson D, Meliska CJ 2006 Sleep, rhythms and women's mood. Part II. Menopause. *Sleep Med. Rev.* 10:197-208.

Peixoto AA, Hennessy JM, Townson I, Hasan G, Rosbash M, Costa R, and Kyriacou CP 1998 Molecular coevolution within a Drosophila clock gene. *Proc. Natl. Acad. Sci. USA* 95:4475-4480.

Pereira DS, Tufik S, Louzada FM, Benedito-Silva AA, Lopez AR , Lemos NA , Korczak AL, D'Ameida V, Pedrazzoli M 2005 Association of the length polymorphism in the human *Per3* gene with the delayed sleep-phase syndrome: does latitude have an influence upon it? *Sleep* 28:29-32.

Pittendrigh CS. 1960 Circadian rhythms and the circadian organization of living systems. *Cold Spring Harb. Symp. Quant Biol.* 25:159-84.

Pittendrigh CS. 1993 Temporal organization: reflections of a Darwinian clock-watcher. *Annu. Rev. Physiol.* 55:16-54.

Pittendrigh CS, Kyner WT, Takamura T 1991 The amplitude of circadian oscillations: temperature dependence, latitudinal clines, and the photoperiodic time measurement. *J. Biol. Rhythms* 6:299-313.

Pittendrigh CS, Takamura T 1989 Latitudinal clines in the properties of a circadian pacemaker. *J. Biol. Rhythms* 4:217-235.

Plomin RJ, Manosevitz M 1974 Behavioral polytypism in wild Mus musculus. *Behav. Genet* 4:145-157.

Postolache TT, Oren DA 2005 Circadian phase shifting, alerting, and antidepressant effects of bright light treatment. *Clin. Sports Med.* 24:381-413.

Ralph MR, Foster RG, Davis FC, Menaker M 1990 Transplanted suprachiasmatic nucleus determines circadian period. *Science* 247:975-978.

Reick M, Garcia JA, Dudley C, McKnight SL 2001 NPAS2: an analog of clock operative in the mammalian forebrain. *Science* 293:506-514.

Reierth E, Stokkan KA 1998a Activity rhythm in high-arctic Svalbard ptarmigan (*Lagopus mutus hyperboreus*). *Can. J. Zool* 76:2031-2039.

Reierth E, Stokkan KA 1998b Dual entrainment by light and food in the Svalbard ptarmigan (*Lagopus mutus hyperboreus*). *J. Biol. Rhythms*. 13:393-402.

Reierth E, Van't Hof TJ, Stokkan KA 1999 Seasonal and daily variations in plasma melatonin in the high-arctic Svalbard ptarmigan (*Lagopus mutus hyperboreus*). *J. Biol. Rhythms.* 14:314-319.

Reppert SM, Weaver DR 2001 Molecular analysis of mammalian circadian rhythms. *Annu. Rev. Physiol.* 63:647-676.

Riccio AP, Goldman BD 2000a Circadian rhythms of locomotor activity in naked mole rats (*Heterocephalus glaber*). *Physiol. Behav.* 71:1-13.

Riccio AP, Goldman BD 2000b Circadian rhythms of body temperature and metabolic rate in naked mole rats. *Physiol. Behav.* 71:14-22.

Richardson GS 2005 The human circadian system in normal and disordered sleep. *J. Clin. Psychiatry* 66 (Suppl 9):3-9.

Robilliard DL, Archer SN, Arendt J, Lockley SW, Hack LM, English J, Leger D, Smits MG, Williams A, Skene DJ, von Schantz M 2002 The 3111 *Clock* gene polymorphism is not associated with sleep and circadian rhythmicity in phenotypically characterized human subjects. *J. Sleep Res.* 11:305-312.

Roldan G, Ang RC 2006 Overview of sleep disorders. *Respir. Care Clin.* 12:31-54.

Rosen LN, Targum SD, Terman M, Bryant MJ, Hoffman H, Kasper SF, Hamovit JR, Docherty PJ, Welch B, Rosenthal NE 1990 Prevalence of seasonal affective disorder at four latitudes. *Psychiatry Res.* 31:131-144.

Rosenfeld P, Van Eekelen JAM, Levine S, De Kloet ER 1988 Ontogeny of the type 2 glucocorticoid receptor in discrete rat brain regions: an immunocytochemical study. *Brain Res.* 470:119-127.

Rosenfeld P, Van Eekelen JAM, Levine S, De Kloet ER 1993 Ontogeny of corticosteroid receptors in the brain. *Cell Mol. Neurobiol.* 13:295-319.

Rosenwasser AM, Pelchat RJ, Adler NT 1984 Memory for feeding time: possible dependence on coupled circadian oscillators. *Physiol. Behav.* 32:25-30.

Ruby NF, Dark J, Heller HC, Zucker I 1998 Suprachiasmatic nucleus: role in circannual body mass and hibernation rhythms of ground squirrels. *Brain Res.* 782:63–72.

Ruis JF, Talamini LM, Buys JP, Rietveld WJ 1989 Effects of time of feeding on recovery of food-entrained rhythms during subsequent fasting in SCN-lesioned rats. *Physiol. Behav.* 46:857-866.

Rutter J, Reick M, McKnight SL 2002 Metabolism and the control of circadian rhythms. *Annu. Rev. Biochem.* 71:307-331.

Sack RL, Lewy AJ, White D, Singer C, Fireman MJ, Vandiver R 1990 Morning vs. evening light treatment for winter depression. *Arch. Gen. Psychiatry* 47:343-351.

Sato T, Kawamura H 1984 Effects of bilateral suprachiasmatic nucleus lesions on the circadian rhythms in a diurnal rodent, the Siberian chipmunk (Eutamia sibiricus). *J. Comp. Physiol.* A 155:745–752.

Satoh K, Mishima K, Inoue Y, Ebisawa T, Shimizu T 2003 Two pedigrees of familial advanced sleep phase syndrome in Japan. *Sleep* 26:416-417.

Sawyer LA, Hennessy JM, Peixoto AA, Rosato E, Parkinson H, Costa R, and Kyriacou CP 1997 Natural variation in a *Drosophila* clock gene and temperature compensation. *Science* 278:2117-2120.

Schwartz JRL, Roth T 2006 Shift work sleep disorder. Burdens of illness and approaches to management. *Drugs* 66:2357-2370.

Sharma VK 2003 Adaptive significance of circadian clocks. *Chronobiol. Intl.* 20:901-919.

Shibata S, Liou SY,Ueki S, Oomura Y 1983 Effects of restricted feeding on single neuron activity of suprachiasmatic neurons in rat hypothalamic slice preparation. *Physiol. Behav.* 31:523-528.

Shimizu T, Masaki S 1997 Geographical and species variation in circadian rhythm parameters in nemobiine crickets. *Physiol. Entomol.* 22:83-93.

Silver R, Sookhoo AI, LeSauter J, Stevens P, Jansen HT, Lehman MN 1999 Multiple regulatory elements result in regional specificity in circadian rhythms of neuropeptide expression in mouse SCN. *NeuroReport* 10:3165-3174.

Simunovic A, Jaenike J 2006 Adaptive variation among *Drosophila* species in their circadian rhythms. *Evol. Ecol. Res.* 8: 803–811.

Slotten HA, Pitrosky B, Pevet P 2000 Entrainment of rat circadian rhythms by melatonin does not depend on the serotonergic afferents to the suprachiasmatic nuclei. *Brain Res.* 876:10-16.

Sohn C-H, Lam RW 2005 Update on the biology of seasonal affective disorder. *CNS Spectr.* 10:635-646.

Stenseth NC 1985 Geographic distribution of *Clethrionomys* species. *Ann. Zool Fennici.* 22:215-219.

Stefánsson JG, Magnússon A, Karlsson MM, Axelsson J 1994 Low prevalence of seasonal affective disorders in Icelanders and Canadians of Icelandic descent. In Circumpolar Health 1993: Proceedings of the 9th International Congress on Circumpolar Health. Arctic Medical Research, vol. 53, suppl. 2. Edited by Pétursdóttir G, Sigur SB, Karlsson MM, Axelsson *J. Reykjavík*: 491-492.

Stephan FK 1981 Limits of entrainment to periodic feeding in rats with suprachiasmatic lesions. *J. Comp. Physiol.* 143:410-410.

Stephan FK 1983 Circadian rhythm dissociation induced by periodic feeding in rats with suprachiasmatic lesions. *Behav. Brain Res.* 7:81-98.

Stephan FK 1984 Phase shifts of circadian rhythms in activity entrained to food access. *Physiol. Behav.* 32:663-671.

Stephan FK 1986a Coupling between feeding- and light-entrained circadian pacemakers in the rat. *Physiol. Behav.* 38:537-544.

Stephan FK 1986b The role of period and phase in interactions between feeding- and light-entrainable circadian rhythms. *Physiol. Behav.* 36:151-158.

Stephan FK 1986c Interaction between the light- and feeding-entrainable circadian rhythms in the rat. *Physiol. Behav.* 38:127-133.

Stephan FK 1989a Entrainment of activity to multiple feeding times in rats with suprachiasmatic lesions. *Physiol. Behav.* 46:489-497.

Stephan FK 1989b Forced dissociation of activity entrained to T cycles of food access in rats with suprachiasmatic lesions. *J. Biol. Rhythms* 4:467-479.

Stephan FK 1992a Resetting of a circadian clock by food pulses. *Physiol. Behav.* 52:997-1008.

Stephan FK 1992b Resetting of a feeding-entrainable circadian clock in rats. Physiol Behav 52:985-995.

Stephan FK, 2002 The “other” circadian system: food as a zeitgeber. *J. Biol. Rhythms.* 17:284-292.

Stephan FK, Swann JM, Sisk CL 1979 Entrainment of circadian rhythms by feeding schedules in rats with suprachiasmatic lesions. *Behav. Neural. Biol.* 25:545-554.

Stokkan K-A, Yamazaki S, Tei H, Sakaki Y, Menaker M 2001 Entrainment of the circadian clock in the liver by feeding. *Science* 291:490-493.

Sumova A, Ebling FJ, Maywood ES, Herbert J, Hastings MH 1994 Non-photic circadian entrainment in the Syrian hamster is not associated with phosphorylation of the transcriptional regulator CREB within the suprachiasmatic nucleus, but is associated with adrenocortical activation. *Neuroendocrinol.* 59:579-589.

Swade RH, Pittendrigh CS 1967 Circadian locomotor rhythms of rodents in the arctic. *Am. Nat.* 101:431-466.

Tafti M, Maret S, Dauvilliers Y 2005 Genes for normal sleep and sleep disorders. *Annals Med.* 37:580-589.

Takahashi JS, Turek FW, Moore RY 2001 Handbook of Behavioral Neurobiology, volume 12: *Circadian Clocks*. New York, Kluwer Academic/Plenum Publishers.

Takamure M, Murakami N, Takahashi K, Kuroda H, Etoh T 1991 Rapid reentrainment of the circadian clock itself, but not the measurable activity rhythms, to a new light-dark cycle in the rat. *Physiol. Behav.* 50:443-449.

Tavernier RJ, Largen AL, Bult-Ito A 2004 Circadian organization of a subarctic rodent, the northern red-backed vole (*Clethrionomys rutilus). J. Biol. Rhythms* 19:238-247.

Thalén BE, Kjellman BF, Morkrid L, Wetterberg L 1995 Melatonin in light treatment of patients with seasonal and nonseasonal depression. *Acta Psychiatr. Scand.* 92:274-284.

Tobler I, Herrmann M, Cooper HM, Negroni J, Nevo E, Achermann P 1998 Rest activity rhythm of the blind mole rat *Spalax ehrenbergi* under different lighting conditions. *Behav. Brain Res.* 96:173-183.

Toh KL, Jones CR, He Y, Eide EJ, Hinz WA, Virshup DM, Ptácek LJ, Fu Y-H 2001 An hPer2 phosphorylation site mutation in familial advanced sleep phase syndrome. *Science* 291:1040-1043.

Van der Horst GTJ, Muijtjens M, Kobayashi K, Takano R, Kanno S-I, Takao M, DeWilt J, Verkerk A, Eker APM, VanLeenen D, Buijs R, Bootsma D, Hoeijmakers JHJ, Yasui A 1999 Mammalian Cry1 and Cry2 are essential for maintenance of circadian rhythms. *Nature* 398:627-630.

Van Oort BEH, Tyler NJC, Gerkema MP, Folkow L, Schytte Blix A, Stokkan K-A 2005 Circadian organization in reindeer: these Arctic animals abandon their daily rhythms when it is dark all day or light all night. *Nature* 438:1095-1096.

Van Reeth O, Turek FW 1989 Administering triazolam on a circadian basis entrains the activity rhythm of hamsters. *Am. J. Physiol.* 256:R639-R645.

Vanselow K, Venselow JT, Westermark PO, Reischl S, Maier B, Korte T, Herrmann A, Herzel H, Schlosser A, Kramer A 2006 Differential effects of PER2 phosphorylation: molecular basis for the human familial advanced sleep phase syndrome (FASPS). *Genes* and Dev 20:2660-2672.

Vitaterna MH, King DP, Chang A-M, Kornhauser JM, Lowrey PL, McDonald JD, Dove WF, Pinto LH, Turek FW, Takahashi JS 1994 Mutagenesis and mapping of a mouse gene, Clock, essential for circadian behaviour. *Science* 264:719–725.

Vleck CM, Van Hook JA 2002 Absence of daily rhythms of prolactin and corticosterone in Adélie Penguins under continuous daylight. *Condor* 104:667-671.

Wakamatsu H, Yoshinobu Y, Aida R, Moriya T, Akiyama M, Shibata S 2001 Restricted-feeding-induced anticipatory activity rhythm is associated with a phase-shift of the expression of mPer1 and mPer2 mRNA in the cerebral cortex and hippocampus but not in the suprachiasmatic nucleus of mice. *Eur. J. Neurosci.* 13:1190- 1196.

Walcott EC and Tate BA 1996 Entrainment of aged, dysrhythmic rats to a restricted feeding schedule. *Physiol. Behav.* 60:1205-1208.

Weaver DR, 1998 The suprachiasmatic nucleus: a 25-year retrospective, *J. Biol. Rhythms* 13:100-112.

Wetterberg L, Bratlid T, von Knorring L, Eberhard G, Yuwiler A 1999 A multinational study of the relationships between nighttime urinary melatonin production, age, gender, body size, and latitude. *Eur. Arch. Psychiatry Clin. Neurosci.* 249:256–262.

White W, Timberlake W 1994 Two meals in the active period of the rat both entrain food-anticipatory activity. *Physiol. Behav.* 56:17-25.

Wilsbacher LD, Takahashi JS 1998 Circadian rhythms: molecular basis of the clock. *Curr. Opinion Genet Dev.* 8:595-602.

Winkler D, Pjerk E, Praschak-Rieder N, Willeit M, Pezawas L, Konstantinidis A, Stastny J, Kasper S 2005 Actigraphy in patients woth seasonal affective disorder and healthy control subjects treated with light therapy. *Biol. Psychiatry* 58:331-336.

Wirz-Justice A 2006 Biological rhythm disturbances in mood disorders. *Int. Clin. Psychopharmacol* 21:S11-S15.

Woelfle MA, Ouyang Y, Phanvijhitsiri K, Johnson CH 2004 The adaptive value of circadian clocks: an experimental assessment in cyanobacteria. *Curr. Biol.* 14:1481-1486.

Xu Y, Padiath QS, Shapiro RE, Jones CR, Wu SC, Saigoh N, Saigoh K, Ptácek LJ, Fu Y-H 2005 Functional consequences of a *CKIδ* mutation causing familial advanced sleep phase syndrome. *Nature* 434:640-644.

Xu Y, Toh KL, Jones CR, Shin J-Y, Fu Y-H, Ptácek LJ 2007 Modeling of a human circadian mutation yields insights into clock regulation by PER2. *Cell* 128:59-70.

Yamazaki S, Numano R, Abe M, Hida A, Takahashi R, Ueda M, Block GD, Sakaki Y, Menaker M, Tei H. 2000 Resetting central and peripheral circadian oscillators in transgenic rats. *Science* 288:682-685.

Yan L, Hochstetler KJ, Silver R, Bult-Ito A. 2003 Relationship between Phase Shifts and *Per* Gene Expression in Mouse Suprachiasmatic Nucleus. *NeuroReport* 14:1247-1251.

In: Circadian Rhythms and Health Research Trends
Editor: Alfredo V. Lignelli, pp. 57-74

ISBN: 978-1-60021-875-0

Chapter III

Non-Photic Effects on the Circadian System and How They Relate to Human Health

Marina R. Castillo and Abel Bult-Ito*
Behavioral and Evolutionary Neuroscience Laboratory,
Alaskan Basic Neuroscience Program,
Institute of Arctic Biology, P.O. Box 757000,
University of Alaska Fairbanks, Fairbanks,
Alaska 99775-7000, U.S.A.

Abstract

Circadian, about 24-hour, rhythms are universal features of most organisms on earth. In mammals, a master circadian clock located in the suprachiasmatic nuclei (SCN) of the hypothalamus controls circadian rhythmicity. The SCN synchronizes peripheral clocks and extra SCN clocks in the brain to allow the animal to express coherent behavioral and physiological rhythms that are entrained to the ambient environmental circadian cycle. The strongest timing cue for entrainment is light. However non-photic cues, such as activity, food, and social cues, also can entrain the SCN. In this chapter, we will review what is currently known about how non-photic cues affect the circadian system in mammals and what types of neural and physiological mechanisms may be involved. We will also discuss how this current understanding can guide the development of research paradigms to study whether non-photic cues can improve human health and to what extent non-photic cues might contribute to health problems.

---

* Correspondence should be addressed to Abel Bult-Ito, Ph.D., Behavioral and Evolutionary Neuroscience Laboratory, Alaskan Basic Neuroscience Program, Institute of Arctic Biology, P.O. Box 757000, University of Alaska Fairbanks, Fairbanks, Alaska 99775-7000. E-mail: ffab@uaf.edu.

Introduction

Several decades of research validate the organization of biological systems in time as well as in space. Generation and maintenance of the circadian time structure is provided by the not yet fully understood molecular mechanisms of circadian oscillators and the integration of multiple cellular clocks into functional multicellular organisms. Photic and non-photic external signals from the surroundings sustain coordination of circadian oscillators with the environment. To what extent circadian rhythms are adaptive is discussed in Bult-Ito and Castillo (2007; Chapter 2 of this volume).

The relationship between circadian rhythmicity and human health has become increasingly clear. Therapeutic approaches intended to restore circadian rhythmicity are expected to gain significance for the improvement of human health. Individual non-photic stimuli although not as strong as photic stimuli for phase shifting the circadian system, show great therapeutic potential. We believe that further elucidation of additive effects among combinations of different non-photic inputs and the interactions with photic stimuli will contribute to this potential. Our current understanding of the molecular mechanisms of circadian rhythms as well as experimental evidence suggests the therapeutic potential of non-photic signals.

Molecular Mechanisms of Circadian Oscillators

Circadian rhythms are understood in general as endogenous periodic fluctuations in physiology, metabolism, and behavior expressed by living organisms. Even when kept away from light most eukaryotes still organize their behavior and physiology around a 24 h day. These endogenously generated rhythms share similar characteristics across phyla, i.e., high precision, synchronization to 24 h light-dark schedules, and temperature compensation (Van Gelder *et al.*, 2003). In fact, a common molecular mechanism for the generation of circadian rhythms is well established. The core clock mechanism is consistently composed of a negative transcriptional feed back loop in which positive transcription factors activate the expression of their own inhibitors. While specific transcription factors vary among different organisms the common transcriptional feed back loop remains the same (Balsalobre, 2002; Dunlap *et al.,* 1999; Rutter *et al.,* 2002; Van Gelder *et al.,* 2003; Hastings and Herzog, 2004).

In mammals, the positive arm of the circadian clock is specified by a heterodimeric transcription factor composed of the products of the *Clock* and *BMAL 1* (also called *MOP3*) genes. This heterodimer binds to promoter sequences activating the expression of negative components of the clock, the *Per* (*Per1, Per2, Per3*) and *Cry (Cry1, Cry2)* genes (Balsalobre, 2002; Hastings and Herzog, 2004; Hastings *et al.*, 2003; Shearman *et al.*, 2000). Transcription continues until the levels of these proteins accumulate in the nucleus in sufficient levels as to repress the activation by binding to the CLOCK/BMAL1 heterodimer. Then when levels of PER and CRY proteins fall transcription begins again. Functional analogs of Clock are expressed in distinct patterns in mammalian tissues, such as NPAS2 (also called MOP4) an analog of Clock expressed in the forebrain but not in the SCN (Green and Menaker, 2003; Reick *et al.,* 2001). The NPAS2:BMAL1 heterodimer has been shown to inhibit *BMAL1* gene expression (Abe *et al.*, 1998). Since BMAL1 is an obligate component of the activating arm of the oscillator and *BMAL1* mRNA levels vary rhythmically as a function of the light dark (LD) cycle, the inhibitory action of Clock:BMAL1 on *BMAL1* gene expression reinforces the continued oscillation of the clock (Abe et al, 1998; Reick *et al.*, 2001; Rutter *et al.*, 2001). The periodic oscillation of the clock is believed to result from delays between transcription, translation, and subsequent transcriptional inhibition. Details on how these delays are established still remain to be elucidated. A clue to these mechanisms has been put forward by studies of organisms with mutated clock genes. Following a classical biochemical pathway of targeting proteins for degradation, phosphorylation of PER protein targets it for ubiquitination and proteosomal degradation. A mutation in the casein kinase1ε gene leads to hypophosphorylation of PER and a shortened circadian period in the Syrian hamster (Lowry *et al.*, 2000). In humans, mutations leading to hypophosphorylation of PER proteins are associated with advanced-phase sleep syndrome (Toh *et al.*, 2001) and polymorphism of phosphorylation sites has been linked to delayed sleep syndrome and extreme diurnal preference (Archer *et al.*, 2003). In addition, posttranslational modifications (Camacho *et al.*, 2001; Keesler *et al.*, 2000; Vielhaber *et al.*, 2000) as well as complexing of PER proteins with each other and with CRY likely influence the rhythmicity of transcription and translation of the clock (Kume *et al.*, 1999; Yagita *et al.*, 2000).

While the main components of the mammalian molecular clock have been sketched out, how the rhythmic fluctuation in gene expression of the clock is translated into rhythmicity of biochemical and physiological processes is less understood. Experimental evidence points toward the existence of different classes of clock controlled genes. Analysis of gene expression in *Per* mutant mice has revealed that some biochemical pathways are under the control of *Per1* or *Per2* while others are dependent on both (Zheng *et al.* 2001). Clock controlled pathways likely feed back directly or indirectly to the clock. Some clock controlled outputs such as vasopressin can amplify and/or transduce periodicity but are not required for time keeping (Ingram *et al.*, 1996; Tousson and Meissl, 2004).

Photic Entrainment

While a basic molecular mechanism provides for the generation of intrinsic circadian rhythmicity, the period of the rhythm must be coupled or entrained to the environmental light schedule. In mammals as well as other clock containing organisms, the cycle length of the circadian oscillator generating intrinsic rhythmicity is not exactly 24 hours. Therefore, it must be synchronized to the external environment. A clock is only useful if it can run concurrently to environmental time so that it can be used to anticipate events. The circadian clock is therefore adjusted everyday by exposure to daylight. As a result, circadian clock function is synchronized to the environment. This process is called entrainment and it was originally defined as the process by which a circadian clock resets its phase after application of extrinsic stimuli.

In unicellular and some small organisms light can directly modify cellular gene expression and entrain circadian rhythms, i.e., in *Drosophila* Cryptochrome acts as a cell-autonomous photopigment to transduce light energy into specific degradation of Timeless protein, inducing a phase shift in circadian rhythms (Van Gelder *et al.*, 2003). In mammals, light cannot directly reach and entrain the majority of circadian rhythms, i.e., the brain and internal organs. Although isolated cells from different tissues, e.g., liver, kidney, lung, heart, pancreas, and forebrain are competent circadian oscillators (Damiola *et al.*, 2000; Rutter *et al.*, 2002) their rhythms in the intact animal are coordinated by the suprachiasmatic nucleus (SCN) of the brain. The SCN is the master circadian clock in mammals, a self sustained cellular clock that coordinates circadian rhythmicity in the rest of the body likely by means of direct neural and/or diffusible substances, as demonstrated by lesion/transplant studies (Hurd *et al.*, 1995; Silver *et al.*, 1996). Light stimuli as sensed by the mammalian eye are transmitted to the SCN mainly via the retinohypothalamic tract (RHT) (Morin, 1994) by glutamatergic neurons (Ebling, 1996) and the peptide PACAP as a neuromodulator (Hannibal, 2002). Activation of ionotropic glutamate receptors mediates Ca^{+2} influx leading to targeted clock gene activation (Ding *et al.*, 1994). This mechanism represents a well-established pathway by which light reaches and influences the circadian clock (Albrecht and Oster, 2001).

Sunlight developed as the strongest natural zeitgeber for biological clocks and it is without a doubt still so. Living organisms adjust their circadian clocks to the environmental light-dark schedule imposed by the sun. The clock genes *Per1* and *Per2* are rhythmically expressed in the SCN with maximum expression around the midday (Hastings *et al.*, 2003). Components of the main oscillatory process are also implicated in synchronization to light. In mammals the expression of clock genes *Per1* and *Per2* can be modified by light which in turn modifies the state of the clock. Light synchronizes circadian rhythms largely by action during the subjective night. Activation of *Per* genes in response to light is however not the same at all circadian times. Light pulses delivered during the circadian day when *Per* gene expression is maximal has no effect on the phase of expression of circadian genes (Hastings *et al.*, 2003; Yan and Silver, 2004). Light pulses delivered during the circadian night when expression of *Per* genes is low is associated with induction of *Per1* and *Per2* (Albrecht et., 1997; Moriya *et al.*, 2000; Shearman *et al.*, 1997; Zylka *et al.*, 1998). However detection of light early in the subjective night produces phase delays in circadian rhythms while detection of light later in the night produces phase advances. Differences in signal transduction mechanisms are believed to mediate the different response to light during the subjective day and night (Gillete and Mitchell, 2002).

Although photic stimuli are the strongest signal for entrainment of circadian rhythms entrainment to other stimuli is also possible, although not yet as thoroughly studied as photic entrainment. In humans an illustrative example of entrainment is traveling across time zones. The physiology and biochemistry of the human body are entrained to a particular lighting schedule which is reflected in the expression of intrinsic needs around a particular schedule i.e. wake up and rest time and meal times. After moving relatively quickly to a new time zone circadian rhythms must be re-entrained to the new lighting schedule. Not a simple molecular and biochemical task usually accompanied of some discomfort from being initially out of phase in the new schedule. The amount of discomfort varies among individuals and is often experienced as a desire to eat, sleep, or to be active when it is not time to do so under the new schedule. More recently, after the discovery of the molecular mechanisms underlying circadian rhythmicity the definition of entrainment was extended as to include the core principle underlying circadian rhythmicity. Entrainment is defined now as a phase change in the rhythmic expression of circadian genes (Rutter *et al.*, 2002).

Non-Photic Entrainment

Non-photic stimuli include a wide variety of stimuli such as: activity, arousal, feeding, social interactions, auditory stimuli, physical contact, dark pulses, sleep deprivation, and pharmacological treatments. The potential of these stimuli to signal the circadian clock as to modify or induce entrainment of circadian rhythms has been investigated (reviewed in Klerman *et al.*, 1998; Mistleberger and Skene, 2004; Mistleberger and Skene, 2005; Mrosovsky, 1996; Yanielli and Harrington 2004). A key conclusion in these studies is that non-photic stimuli are capable of entraining at least some circadian rhythms in the organism in the absence of photic stimuli and that in the presence of light non-photic stimuli are generally overpowered by photic signals.

In the absence of light, non-photic stimuli that increase arousal can phase advance circadian rhythms during the subjective day (Mrosovsky, 1996) and to a lesser extent induce phase delays during the subjective night (Marchant and Mistlberger, 1996). Two general pathways by which non-photic stimuli can reach the SCN from other brain areas are known. One is the geniculohypothalamic tract (GHT) using neuropeptide Y (NPY), gamma-amino butyric acid (GABA), and endorphins (Harrington, 1997) as neurotranmitters and the other one is via the raphe nuclei using the neurotransmitter serotonin (Meyer-Bernstein and Morin, 1996; Marchant *et al.*, 1997). These two systems mediate the resetting effects of non-photic inputs during the circadian day. Neurotransmitters associated with these pathways are also capable of phase advancing the SCN when administered during the circadian day either in vivo or in vitro (Harrington and Schak, 2000; Hastings *et al.*, 1998).

Phase advancing effects of non-photic stimuli during the circadian day must ultimately be mediated by a change in the phase of expression of clock related genes (Yanielli *et al.*, 2002). Evidence in support of this assertion is indeed available. NPY, serotonin, some pharmacologic agents, and physical activity decrease the normal midday rise in *Per1* and *Per2* when delivered during the subjective day (Fukuhara *et al.*, 2001; Horikawa *et al.*, 2000; Maywood *et al.*, 1999; Yokota *et al.*, 2000). Although both *Per1* and *Per2* are decreased, the decrease in *Per2* seems to be correlated with the non-photic phase shift (Yanielli *et al.*, 2002).

In a real-life context, both photic and non-photic stimuli influence the circadian system simultaneously. The final phase of the clock is likely to be a result of an interaction between photic and non-photic stimuli. Convergence of both stimuli at the molecular mechanisms of the clock influences the final outcome expressed by the circadian clock. Clear examples of these interactions are the inhibition of subjective day phase shifting actions of non-photic stimuli by light and the inhibition of the effects of light in the subjective night by non-photic inputs (reviewed in Yanielli, and Harrington, 2004). More over, the fact that these interactions are expressed at specific circadian times suggests that they might be influenced by endogenous rhythms of neurotransmitters acting at the level of the SCN. In vitro experiments using brain slice preparations support this idea. For example, the magnitude of phase delay-shifts to light or light-mimetic agents is larger in vitro than in vivo. This difference is explained by the presence of an endogenous tone of NPY in the early night in vivo that acts to reduce the phase-shifting action of light, while in vitro experiments are free of this influence (Yannielli and Harrington, 2000, 2001a, 2001b).

The interaction between photic and non-photic effects has received more attention so far than the interaction between different non-photic effects in the absence of light. Feeding schedules were recently demonstrated to entrain the SCN in mice (Abe *et al.*, 2007; Castillo *et al.*, 2004). When feeding is dissociated from the normal activity period, by allowing animals to eat only during their inactive period, two behavioral activity components result. One component entrains to and is in anticipation of onset of food availability, i.e., food anticipatory activity (FAA), which is controlled by a SCN independent food-entrainable clock. The other is the animal's normal nocturnal or light-entrainable activity (LEA) component, which is controlled by the SCN (Castillo *et al.*, 2004; Challet *et al.,* 2003). Entrainment of LEA to the feeding schedule, i.e., 6 hours of food availability during the initial 12:12 hour light-dark cycle, in the absence of light occurred slowly but at a steady pace and was substantiated both at the behavioral level by the patterns of locomotor activity and at the molecular level by the expression patterns of clock Per2 protein in the mouse SCN (Castillo *et al.*, 2004). One particular mouse line used for this experiment was demonstrated to be 100% food-entrainable (49 of 49) compared to the other mouse lines tested concurrently, in which only 4%, 57% or 39% of the mice entrained to the feeding schedule (Castillo *et al.*, 2004). Entrainment of the feeding schedule took place in constant dark and with free access to a running wheel to which the animals were habituated for several weeks before the experiment began.

After these findings, a preliminary experiment using 100% entrainable mice was conducted in our laboratory under the same conditions but using core body temperature as the circadian parameter and completely excluding running wheels from the experiment. Of the four experimental subjects available, after about 12 weeks in constant dark, one showed a core body temperature rhythm entrained to the feeding schedule (Figure 1a). Another mouse free ran through the feeding schedule and was unable to entrain (Figure 1b). The two remaining mice had not entrained to the feeding schedule after 20 weeks in constant darkness when the experiment was terminated. One of these mice was slowly but clearly approaching entrainment (Figure 1c), while the phase of the other mouse seemed to be locked 10.5 h ahead of the start of food availability (Figure 1d), a pattern not seen when animals have access to a running wheel (Castillo *et al.*, 2004).

The results from this experiment (Figure 1) compared to our previous findings (Castillo *et al.*, 2004) suggest that entrainment to the feeding schedule is enhanced by physical activity since all mice were able to entrain to the feeding schedule in about 12 weeks in the presence of a running wheel. These results suggest that scheduled feeding and voluntary physical activity, two non-photic stimuli, acted synergistically to entrain the SCN. Most studies on photic and non-photic stimuli and interactions among them on circadian rhythms have been performed in animal species other than humans (Mrosovsky, 1996; Yannielli and Harrington, 2004). Therefore, whether a combination of non-photic cues, i.e., timed meals and exercise, also promotes SCN entrainment in humans remains an interesting question to be investigated.

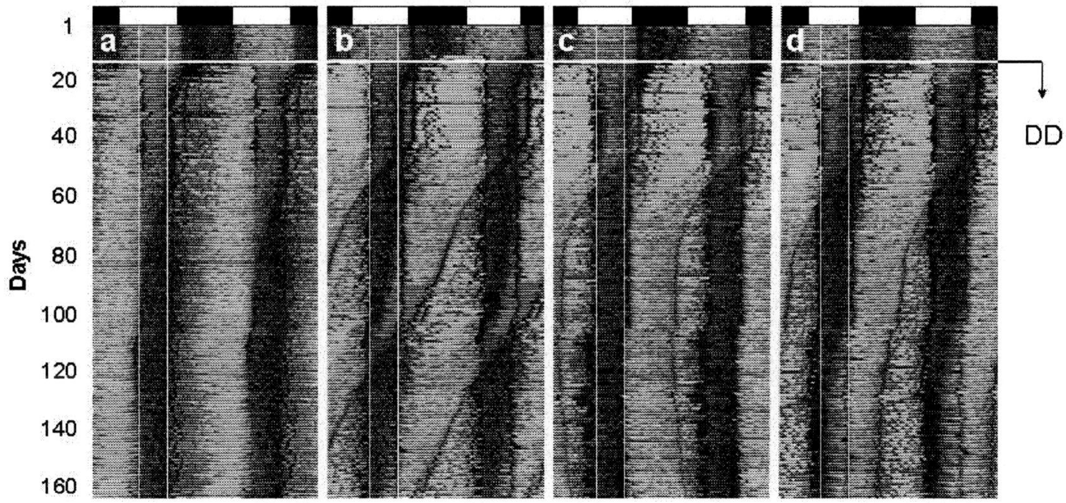

Figure 1. Core body temperature records of mice under a restricted feeding schedule in constant dark. Two weeks of a 12:12-h light-dark cycle followed by constant dark (DD). White rectangles indicate periods of food availability. Note: a 10 min shut down of the computer system at day 115 resulted in adjustment of the computer time to the daylight saving schedule, resulting in a one-hour shift to the left.

The potential of scheduled meals and sleep as means of stabilizing circadian rhythms in humans under irregular work schedules has not been completely addressed. In these studies, single randomly timed 8 h sleep or two randomly arranged 4 h sleep periods were associated with free running rhythms despite the fact that meal times were reported to be taken at similar times throughout. However, if one of the 4 h sleep periods was taken at the same time each day after one to three weeks circadian rhythms of rectal temperature and urinary potassium stabilized with periods indistinguishable from 24 h (Minors and Waterhouse, 1981). These findings suggest that non-photic stimuli, i.e., scheduled sleep, are capable of entraining human circadian rhythms. Although meal times by themselves were unable to entrain circadian rhythms (Minors and Waterhouse, 1981) they might have acted synergistically with the 4 h scheduled sleep times resulting in entrainment to the 24 h cycle. During these experiments light was under complete control of the experimental subjects. Therefore, the possibility that photic cues influenced entrainment can not be ruled out.

Perhaps the most suitable experimental model to investigate non-photic entrainment in humans is represented by totally blind subjects with free-running circadian rhythms. However, caution must be exercised because blind individuals represent a heterogeneous population in which some individuals retain non-conscious light perception capable of inducing melatonin suppression and phase shifts of circadian rhythms. Most bilaterally enucleated individuals and also some totally blind individuals with intact eyes exhibit free-running circadian rhythms despite of being exposed to daily non-photic stimuli such as scheduled activity, meals, and social interactions (Klerman et al., 1998; Lockley et al., 1997; Mistlberger and Skene, 2004; Skene et al., 1999). However, some blind individuals seem to be able to maintain synchronized circadian rhythms with their environment via non-photic cues as demonstrated in a study using blind subjects with a negative melatonin suppression response. In this study, nine out of 15 blind individuals were able to maintain synchronization of core body temperature (CBT) rhythms to the 24h day, albeit often at an atypical phase angle of entrainment. In addition, in the same study one blind individual lacking ocular photoreception was successfully entrained to a sleep-wake cycle by phase advances while living in constant near darkness (Klerman *et al.*, 1998). A common feature of phase shifting effects of non-photic cues on the circadian system is that they are not very large, in the range of several minutes per day, compared to the ones induced by photic cues, in the range of 30 minutes to several hours per day. This means in general that entrainment by non-photic stimuli is slow (Castillo *et al.*, 2004; Edgar and Dement, 1991; Honma *et al.*, 2003; Marchant and Mistlberger, 1996; Mistlberger, 2006; Miyazaki *et al.*, 2001).

Results from animal studies also support the notion that circadian rhythms can be shifted and entrained by exercise (Edgar and Dement, 1991; Marchant and Mistlberger, 1996; Mrosovsky, 1996) and indeed humans are also responsive to the phase shifting effects of exercise (Atkinson *et al.,* 2007; Baehr *et al.*, 2003; Buxton *et al.*, 1997; Buxton *et al.*, 2003; Mistlberger and Skene, 2004; Miyazaki *et al.,* 2001;) but to a lesser degree than other animal species, e.g., hamsters (Mrosovsky and Salmon 1987; Redlin and Mrosovsky 1997). Exercise of different durations and intensities produces shifts in rhythms of secretory products in humans regardless of athletic status and age (Atkinson *et al.*, 2007). In spite of difficulties in controlling for the physical condition of human research participants, characteristics of the exercise bout, and other confounding synchronizers, a phase response curve (PRC) for exercise in humans has been constructed (Baehr *et al.*, 2003; Buxton *et al.*, 1997; Buxton *et al.*, 2003). Evening exercise has a phase advancing effect if performed before the dim light melatonin onset which occurs before sleep time, and a phase delaying effect if performed during the sleep period, morning, or early afternoon (Buxton *et al.*, 2003). Most importantly, the apparent disagreement between the PCR curve for exercise and light supports the idea that exercise itself has phase shifting effects. The mechanism by which exercise resets the circadian clock remains controversial. Exercise is known to influence diverse body functions, such as increased core body temperature (Van Someren, 2003), higher levels of general excitement (Mistlberger *et al.*, 2003), increased pupil size (Harbin *et al.*, 1989), and increased hypothalamic-pituitary-adrenal stimulation (Atkinson *et al.*, 2007), all of which could influence entrainment. More research is needed to further characterize the potential use of exercise schedules to entrain circadian rhythms.

Social interactions in humans were once believed to be the main entrainment cue for circadian rhythms, representing a stronger entraining signal than light (Aschoff, 1978; Wever, 1979). However this idea was not supported by experimental evidence from a large number of reports in which humans in social contact free ran independently in the absence of regular LD cycles (for an excellent review of the literature see Mistlberger and Skene, 2004). Social interactions remain weak stimuli for the circadian system but the possibility remains that combinations of non-photic stimuli including social interactions might result in a stronger signal for entrainment.

Non-photic entrainment also includes different groups of pharmacological agents, often called chronobiotics, which possess the ability to phase shift and entrain circadian rhythms and include melatonin, stimulants and hypnotics. Melatonin is the most studied chronobiotic agent for its ability to entrain free-running circadian rhythms (Cardinali *et al.,* 2006; Lewy *et al.,* 2006). Melatonin is produced by the pineal gland and its secretion is controlled by, and at the same time indicative of, the phase and amplitude of the oscillation in the SCN. Melatonin plays a major role in controlling the sleep wake cycle (Cardinali and Pevet, 1998; Kennaway and Wright, 2002). The highest density of melatonin receptors in humans is located in the SCN and entraining free running circadian rhythms by administering melatonin is only possible if the SCN is intact (Dawson and Armstrong, 1996). Melatonin is currently the prescription of choice for the treatment of non-24-h sleep-wake disorders in the blind (Mistlberger and Skene, 2005) and has been used successfully for the treatment of jet lag (Cardinali *et al.*, 2002). Most interestingly, a combination of melatonin, light exposure and exercise hastened resynchronization of circadian rhythms after transmeridian flights comprising 12-13 hour time zones from an average of 8-10 days to 2 days (Cardinali *et al.,* 2006). These findings support the idea that photic and non-photic cues in combination with melatonin chronoterapy, are stronger entrainment stimuli in combination than alone.

Overall, the evidence available so far in support of non-photic entrainment in humans is promising and possesses potential for therapeutic applications in humans. More research is needed to elucidate the entrainment effects of different combinations of non-photic inputs and the most effective times for administration. Achieving non-photic phase shifts as well as increasing the size of the shift are both desirable for practical uses. Recent evidence indicated that exposure to constant light for one or two nights can potentiate the effects of serotonin agonists in Hamsters (Knoch *et al.*, 2006; Mistlberger, 2006), opening up the possibility that similar potentiating mechanisms could be functional for different combinations of entrainment signals in humans. Systematic research studies in this area are warranted.

Non-Photic Stimuli and Health

Circadian rhythms are usually taken for granted and go unnoticed until derangements occur. Since cellular function, the basis of life, is organized in time as well as in space it is not surprising then that alterations in circadian rhythmicity have conspicuous consequences. Aberrant circadian rhythmicity is believed to contribute either as a cause or correlate to the symptoms experienced in a number of conditions, i.e. sleep disturbances (Richardson, 2005), aging (Martin *et al.*, 2006; Martin *et al.*, 2007; Van Someren, 2000), mood disorders (Harper *et al.,* 2004; Sullivan and Richards, 2004; Wirz-Justice, 2006), blindness (Klerman *et al.*, 1998; Sack *et al.*, 1992), jet lag (Comperatore and Krueger, 1990; Parry, 2002; Waterhouse *et al.*, 2005), and reverse shift work (Horowitz and Tanigawa, 2002; Van Dongen, 2006) among others. In this context, non-photic stimuli alone or in combination with photic stimuli become relevant to the maintenance and reinforcement of circadian rhythmicity. In some cases, when photic entrainment is compromised or not possible, i.e., when the eyes or RHT are non-functional, non-photic stimuli constitute the only possibility for SCN entrainment.

While the therapeutic application of non-photic stimuli alone or in combination with photic cues remains to be fully addressed, recent reports support benefits of this approach. Combinations of photic and non-photic stimuli including scheduled bright light, staying out of bed during the day, scheduled exercise, bed time routines, and reducing noise and light at night have proven useful in restoring rest/activity rhythms among nursing home residents making a difference in their quality of life (Martin *et al.*, 2006; Martin *et al.*, 2007). More research is needed to validate these results using other circadian parameters indicative of clock function such as body temperature and melatonin rhythms.

A most recent and unexpected link between circadian rhythms and health revealed that humans and animals with disrupted circadian rhythmicity are more likely to develop tumors and cancer (Lee, 2006; Levi, 2006b). Tumors in mice grow more rapidly and cause death sooner if circadian rhythms are disrupted by SCN ablation or under a jet lag model (Filipski *et al.*, 2002; Filipski *et al.*, 2005; Filipski *et al.*, 2006). Nurses under night shifts have higher probability of developing breast cancer (Schernhammer *et al.*, 2001) or colon cancer (Schernhammer *et al.*, 2003) than their peers who never worked at night. In cancer patients, compromised circadian rhythmicity correlated with poor performance status and tumor burden. In patients with metastatic colorectal cancer, the circadian rest/activity rhythm of the patient is a strong indicator of tumor response to treatment and restoring circadian rhythmicity might improve their prognosis (Levi, 2006a). The antitumor potential of circadian rhythmicity as a therapeutic tool remains to be validated. The possibility that not only one but different combinations of photic and non-photic stimuli could be used to restore or reinforce rhythmicity according to patient needs deserves further attention.

Circadian Delivery of Medication

Circadian rhythmicity and entrainment allow prediction of metabolic function. Synchronized individuals display circadian rhythms with predictable times of peaks and troughs in physiological parameters. These rhythms have an impact on the pharmacology and tolerance of drugs, i.e., circadian dosing times influence tolerance and efficacy of anticancer drugs irrespective of injection route or single/repeated dosing (Levi, 2006b). Animal studies and clinical trials in humans confirm that optimal timing of chemotherapy leads to decreased toxicity and allows delivery of a more intense therapy with improved efficacy outcomes (Levi, 2006a; 2006b). Circadian-timed maintenance therapy is currently a consensus treatment for childhood acute lymphoblastic leukemia (Schmiegelow *et al.,* 1997) and has also shown excellent results in the treatment of gynecological and genitourinary cancers (Kobayashi *et al.*, 2002), lung cancer (Focan, 2002), and colorectal cancer (Giacchetti, 2002). Likely, future general approaches in medication dosage and administration will not only depend on the weight and sometimes the age of the patient but also on the status of his/her circadian rhythms. From this perspective, combinations of non-photic and photic stimuli that promote circadian rhythmicity represent potential therapeutic tools to assist ambulatory patients as well as bed-ridden individuals with disrupted circadian rhythmicity due to illness.

Conclusions

Biological systems are structured not only in space but also in time. The integrity of the circadian timing structure is critical for appropriate biological and cognitive function. Maintenance of circadian rhythmicity in agreement with the environment is mediated by photic as well as non-photic signals inevitably converging at the master circadian clock. While photic stimuli are the strongest timing signals for the circadian system, enough evidence at the molecular and behavioral level supports entrainment by non-photic stimuli alone or in combination with photic cues. Possible potentiating and/or additive effects among different non-photic and photic stimuli as well as execution of their corresponding phase response curves deserves further attention for its likely applications to human health. Increasing evidence correlates circadian rhythmicity with human wellbeing, susceptibility to diseases, and responses to therapy. In this direction therapeutic approaches aimed toward increasing circadian rhythmicity are expected to gain significance for the improvement of human health. Precisely in this context non-photic stimuli purport their greatest therapeutic potential with the additional asset of not adding to the pharmacological burden of the patient.

References

Abe, H. Honma, S. and Honma, K-I. (2007). Daily restricted feeding resets the circadian clock in the suprachiasmatic nucleus of CS mice. *American journal of physiology*, 292, R607-R615.

Abe, H. Honma, S. Namihira, M, Tanahashi, Y. Ikeda, M, and Honma, K. (1998). Circadian rhythm and light responsiveness of BMAL1 expression, a partner of mammalian clock gene Clock, in the suprachiasmatic nucleus of rats. *Neuroscience letters*, 258(2), 93-96.

Albrecht, U. and Oster, H. (2001). The circadian clock and behavior. *Behavioural brain research*, 125(1-2), 89-91.

Albrecht, U. Sun, ZS. Eichele, and G. Lee, CC. (1997). A differential response of two putative mammalian circadian regulators, mper1 and mper2 to light. *Cell*, 91(7), 1055-1064.

Archer, S. N. Robilliard, D. L. Skene, D. J. Smits, M. Williams, A. Arendt, J. and von Schantz, M. (2003). A length polymorphism in the circadian clock gene Per3 is linked to delayed sleep phase syndrome and extreme diurnal preference. *Sleep,* 26(4), 413-415.

Aschoff, J. (1978). Features of circadian rhythms relevant for the design of shift schedules. *Ergonomics,* 21, 739-754.

Atkinson, G. Edwards, B. Reilly, T. and Waterhouse, J. (2007). Exercise as a synchronizer of human circadian rhythms: an update and discussion of the methodological problems. *European journal of applied physiology*, 99(4), 331-341.

Balsalobre, A. (2002). Clock genes in mammalian peripheral tissues. *Cell and tissue research*, 309, 193-199.

Baehr, E. K. Eastman, C. I. Revelle, W. Olson, S. H. Wolfe, L. F. and Zee, P. C. (2003). Circadian phase-shifting effects of nocturnal exercise in older compared with young adults. *American journal of physiology*, 284(6), R1542-R1550.

Bult-Ito, A. and Castillo, M. R. (2007). The adaptive significance of circadian rhythms and its relevance to human health. Ed. A. V. Lignelli, *Circadian rhythms and health research trends*. Nova Science Publishers, Inc., Hauppauge, NY, Chapter 2.

Buxton, O. M. Lee, C. W. L'Hermite-Baleriaux, M. Turek, F. W. and Van Cauter, E. (2003). Exercise elicits phase shifts and acute alterations of melatonin that vary with circadian phase. *American journal of physiology,* 284(3), R714-724.

Buxton, O. M. L'Hermite-Balériaux, M. Hirschfeld, U. and Cauter, E. (1997). Acute and delayed effects of exercise on human melatonin secretion. *Journal of biological rhythms,* 12(6), 568-574.

Cardinaly, D. P. Bortman, G. P. Liotta, G. Pérez-Lloret, S. Albornoz, L. E. Cutrera R. A. Batista, J. and Ortega-Gallo, P. (2002). A multifactorial approach employing melatonin to accelerate resynchronization of sleep-wake cycle after a 12 time-zone westerly transmeridian flight in elite soccer athletes. *Journal of pineal research*, 32(1), 41-46.

Cardinali, D. P. Furio, A. M. Reyes, M. P. and Brusco, L. I. (2006). The use of chronobiotics in the resynchronization of the sleep-wake cycle. Cancer causes and control, 17, 601-609.

Cardinali, D. P. and Pevet, P. (1998). Basic aspects of melatonin action. *Sleep medicine reviews,* 2(3), 175-190.

Castillo, M. R. Hochstetler, K. J. Tavernier, R. J. Greene, D. M. and Bult-Ito A. (2004). Entrainment of the master circadian clock by scheduled feeding. *American journal of physiology*, 287, R551-R555.

Camacho, F. Cilio, M. Guo, Y. Virshup, D. M. Patel, K. Khorkova, O. Styren, S. Morse, B. Yao, Z. and Keesler, G. A. (2001). *Human kasein kinase 1δ phosphorylation of human circadian clock proteins Period 1 and 2.* Federation of European biochemical societies letters, 489, 159-165.

Challet, E. Caldelas, I. Graff, C. and Pevet, P. (2003). Synchronization of the molecular clockwork by light- and food-related cues in mammals. *Biological chemistry*, 384, 711-719.

Comperatore, C. A. Krueger, G. P. (1990). Circadian rhythm desynchronosis, jet lag, shift lag, and coping strategies. *Occupational medicine,* 5(2), 323-341.

Damiola, F. Le Minh, N. Preitner, N. Kornmann, B. Fleury-Olela, F. and Schibler, U. (2000). Restricted feeding uncouples circadian oscillators in peripheral tissues from the central pacemaker in the suprachiasmatic nucleus. *Genes and development,* 14, 2950-2961.

Dawson, D. Armstrong, S. M. (1996). Chronobiotics-drugs that shift rhythms. *Pharmacology and therapeutics*, 69, 15-36.

Ding, J.M. Chen, D. Weber, E. T. Faiman, L. E. Rea, M. A. and Gillette, M. U. (1994). Resetting the biological clock: mediation of nocturnal circadian shifts by glutamate and NO. *Science*, 266(5191), 1713-1717.

Dunlap, J. C. Loros, J. J. Liu, Y. and Crosthwaite, S. K. (1999). Eukaryotic circadian systems: cycles in common. *Genes to cells*, 4, 1-10.

Ebling, F. J. (1996). The role of glutamate in the photic regulation of the suprachiasmatic nucleus. *Progress in neurobiology*, 50(2-3), 109-132.

Edgar, D. M. and Dement, W. C. (1991). Regularly scheduled voluntary exercise synchronizes the mouse circadian clock. *American journal of physiology*, 261, R928-R933.

Filipski, E. Innominato, P. F. Wu, M. Li, X. M. Iacobelli, S. Xian, L. J. and Lévi, F. (2005). Effects of light and food schedules on liver and tumor molecular clocks in mice. *Journal of the national cancer institute*, 97(7), 507-517.

Filipski, E. King, V. M. Li, X. Granda, T. G. Mormont, M. C. Liu, X. Claustrat, B. Hastings, M. H. and Lévi, F. (2002). Host circadian clock as a control point in tumor progression. *Journal of the national cancer institute*, 94(9), 690-697.

Filipski, E. Li, X. and Lévi, F. (2006). Disruption of circadian coordination and malignant growth. *Cancer causes and control*, 17(4), 509-514.

Focan, C. (2002) Chronobiological concepts underlying the chronotherapy of human lung cancer. *Chronobiology international*, 19(1), 253-273.

Fukuhara, C. Brewer, J. M. Dirden, J. C. Bittman, E. L. Tosini, G. and Harrington, M. E. (2001) Neuropeptide Y rapidly reduces Period 1 and Period 2 mRNA levels in the hamster suprachiasmatic nucleus. *Neuroscience letters*, 314(3), 119-122.

Giacchetti, S. (2002). Chronotherapy of colorectal cancer. *Chronobiology international*, 19(1), 207-219.

Gillette, M. U. and Mitchell, J. W. (2002) Signaling in the suprachiasmatic nucleus, selectively responsive and integrative. *Cell and tissue research*, 309(1), 99-107.

Green, C. B. and Menaker, M. (2003). Clocks on the brain. *Science*, 301, 319-320.

Hannibal, J. (2002). Neurotransmitters of the retino-hypothalamic tract. *Cell and tissue research,* 309(1), 73-88.

Harbin, G. Durst, L. and Harbin, D. (1989). Evaluation of oculomotor response in relationship to sports performance. *Medicine and science in sports and exercise*, 21(3), 258-262.

Harper, D. G. Stopa, E. G. McKee, A. C. Satlin, A. Fish, D. and Volicer, L. (2004). Dementia severity and Lewy bodies affect circadian rhythms in Alzheimer disease. *Neurobiology,* 25(6), 771-781.

Harrington, M. E. (1997). The ventral lateral geniculate nucleus and the intergeniculate leaflet: interrelated structures in the visual and circadian systems. *Neuroscience and biobehavioral reviews*, 21(5), 705-27.

Harrington, M. E. and Schak, K. M. (2000) Neuropeptide Y phase advances the in vitro hamster circadian clock during the subjective day with no effect on phase during the subjective night. *Canadian journal of physiology and pharmacology*, 78(2), 87-92.

Hastings, M. H. Duffield, G. E. Smith, E. J. Maywood, E. S. and Ebling, F. J. (1998). Entrainment of the circadian system of mammals by nonphotic cues. *Chronobiology international,* 15(5), 425-445.

Hastings, M. H.and Herzog, E. D. (2004) Clock genes, oscillators, and cellular networks in the suprachiasmatic nuclei. *Journal of biological rhythms*, 19(5), 400-413.

Hastings, M. H. Reddy, A. B. Garabette, M. King, V. M. Chahad-Ehlers, S. O'Brien, J. and Maywood, E. S. (2003). Expression of clock gene products in the suprachiasmatic nucleus in relation to circadian behaviour. *Molecular clocks and light signaling*. Wiley, Chichester (Novartis Foundation Symposium 253) 203-222.

Honma, K-I. Hashimoto, S. Nakao, M. and Honma, S. (2003). Period and phase adjustments of human circadian rhythms in the real world. *Journal of biological rhythms*, 18(3), 261-270.

Horikawa, K. Yokota, S. Fuji, K. Akiyama, M. Moriya, T. Okamura, H. and Shibata, S. (2000). Nonphotic entrainment by 5-HT1A/7 receptor agonists accompanied by reduced Per1 and Per2 mRNA levels in the suprachiasmatic nuclei. *Journal of neuroscience*, 20(15), 5867-5873.

Horowitz, T. S. and Tanigawa, T. (2002). Circadian-based new technologies for night workers. *Industrial health*, 40(3), 223-236.

Hurd, M. W. Zimmer, K. A. Lehman, M. N. and Ralph, M. R. (1995). Circadian locomotor rhythms in aged hamsters following suprachiasmatic transplant. *American journal of physiology,* 269(5) Pt 2, R958-968.

Ingram, C. D. Snowball, R. K. and Mihai, R. (1996). Circadian rhythm of neuronal activity in suprachiasmatic nucleus slices from the vasopressin-deficient Brattleboro rat. *Neuroscience*, 75(2), 635-41.

Keesler, G. A. Camacho, F. Guo, Y. Virshup, D. M, Mondadori, C. and Yao, Z. (2000). Phosphorylation and destabilization of human period 1 clock protein by human casein kinase 1ε. *Neuroreport*, 11, 951,955.

Kennaway, D. J. and Wright, H. (2002). Melatonin and circadian rhythms. *Current topics in medicinal chemistry*, 2(2), 199-209.

Klerman, E. B. Rimmer, D. W. Dijk, D. J. Kronauer, R. E. Rizzo, J. F 3rd. and Czeisler, C. A. (1998). Nonphotic entrainment of the human circadian pacemaker. *American journal of physiology,* 274(4) Pt 2, R991-996.

Knoch, M. E. Siegel, D. Duncan, M. J. and Glass, J. D. (2006). Serotonergic mediation of constant light-potentiated nonphotic phase shifting of the circadian locomotor activity in Syrian Hamsters. *American journal of physiology*, 291, R180-R181.

Kobayashi, M. Wood, P. A. and Hrushesky, W. J. M. (2002). Circadian chemotherapy for gynecological and genitourinary cancers. *Chronobiology international*, 19(1), 237-251.

Kume, K. Zylka, MJ. Sriram, S. Shearman, L. P. Weaver, D. R. Jin, X. Maywood, E. S. Hastings, M. H. and Reppert, S. M. (1999). mCRY1 and mCRY2 are essential components of the negative limb of the circadian clock feedback loop. *Cell*, 98, 193-205.

Lee, C. C. (2006). Tumor suppression by the mammalian *Period* genes. *Cancer causes and control,* 17, 525-530.

Levi, F. (2006a). Chronotherapeutics, the relevance of timing in cancer therapy. *Cancer causes and control,* 17, 611-621.

Levi, F. (2006b). Circadian rhythms play role in cancer research. *Journal of the national cancer institute,* 98(12), 806-807.

Lewy, A. Emens, J. Jackman, A. Yuhas, K. (2006). Circadian uses of melatonin. *Chronobiology international*, 23(1-2), 403-412.

Lockley S. W. Skene, D. J. Arendt, J. Tabandeh, H. Bird, A. C. and Defrance, R. (1997). Relationship between melatonin rhythms and visual loss in the blind. *Journal of clinical endocrinology and metabolism*, 82, 3763-3770.

Lowrey, P. L. Shinomura, K. Antoch, M. P. Yamazaki, S. Zemenides, P. D. Ralph, M. R. Menaker, M. and Takahashi, J. S. (2000). Positional syntenic cloning and functional characterization of the mammalian circadian mutation tau. *Science,* 288, 483-491.

Marchant, E. G. and Mistlberger, R. E. (1996). Entrainment and phase shifting of circadian rhythms in mice by forced treadmill running. *Physiology and Behavior*, 60, 657-663.

Marchant, E. G. Watson, N. V. and Mistlberger, R. E. (1997). Both neuropeptide Y and serotonin are necessary for entrainment of circadian rhythms in mice by daily treadmill running schedules. *Journal of neuroscience*, 17(20), 7974-7987.

Martin, J. L. Marler, M. R. Harker, J. O. Josephson, K. R. and Alessi, C. A. (2007). A multicomponent nonpharmacological intervention improves activity rhythms among nursing home residents with disrupted sleep/wake patterns. *The journal of gerontology series A: biological sciences and medical sciences*, 62(1), 67-72.

Martin, J. L. Webber, A. P. Alam, T. Harker, J. O. Josephson, K. R. and Alessi, C. A. (2006). Daytime sleeping, sleep disturbance, and circadian rhythms in the nursing home. *American journal of geriatric psychiatry*, 14(2), 121-129.

Maywood, E. S. Mrosovsky, N. Field, M. D. and Hastings M. H. (1999). Rapid down-regulation of mammalian period genes during behavioral resetting of the circadian clock. *Proceedings of the National Academy of Sciences of the United States of America*, 96(26), 15211-15216.

Meyer-Bernstein, E. L. and Morin, L. P. (1996). Differential serotonergic innervation of the suprachiasmatic nucleus and the intergeniculate leaflet and its role in circadian rhythm modulation. *Journal of neuroscience*, 16(6), 2097-111.

Minors, D. S. and Waterhouse, J. M. (1981). Anchor sleep as a synchronizer of rhythms on abnormal routines. *International journal of chronobiology*, 7(3), 165-188.

Mistlberger, R. E. (1993). Effects of scheduled food and water access on circadian rhythms of hamsters in constant light, dark, and light-dark. *Physiology and behavior*, 53, 509-516.

Mistlberger, R. E. (2006). Illuminating serotonergic gateways for strong resetting of the mammalian circadian clock. *American journal of physiology*, 291(1), R177-R179.

Mistlberger, R. E. Antle, M. C. Webb, I. C. Jones, M. Weinberg, J. and Pollock, M. S. (2003) Circadian clock resetting by arousal in Syrian hamsters, the role of stress and activity. *American journal of physiology*, 285(4), R917-R925.

Mistlberger, R. E. and Skene, D. J. (2004). Social influences on mammalian circadian rhythms, animal and human studies. *Biological reviews of the Cambridge Philosophical Society*, 79(3), 533-556.

Mistlberger, R. E. and Skene, D. J. (2005). Nonphotic entrainment in humans? *Journal of biological rhythms,* 20(4), 339-352.

Miyazaki, T. Hashimoto, S. Masubuchi, S. Honma, S. and Honma K.-I. (2001). Phase-advance shifts of human circadian pacemaker are accelerated by daytime physical exercise. *American journal of physiology*, 281, R197-R205.

Morin, LP. (1994). The circadian visual system. *Brain research reviews*, 19(1), 102-27.

Moriya, T. Horikawa, K. Akiyama, M. and Shibata, S. (2000). Correlative association between N-methyl-D-aspartate receptor-mediated expression of period genes in the suprachiasmatic nucleus and phase shifts in behavior with photic entrainment of clock in hamsters. *Molecular pharmacology*, 58(6), 1554-1562.

Mrosovsky, N. (1996). Locomotor activity and non-photic influences on circadian clocks. *Biological reviews of the Cambridge Philosophical Society*, 71(3), 343-372.

Mrosovsky, N. and Salmon, P. A. (1987). A behavioural method for accelerating re-entrainment of rhythms to new light-dark cycles. *Nature*, 330(6146), 372-373.

Parry, B. L. (2002). Jet lag: minimizing it's effects with critically timed bright light and melatonin administration. *Journal of molecular microbiology and biotechnology*, 4(5), 463-466.

Redlin, U. and Mrosovsky, N. (1997). Exercise and human circadian rhythms, what we know and what we need to know. *Chronobiology international*, 14(2), 221-229.

Reick, M. Garcia, J. A. Dudley, C. and McKnight, S. L. (2001). NPAS2, An analog of clock operative in the mammalian forebrain. *Science*, 293, 506-509.

Richardson, G. S. (2005). The human circadian system in normal and disordered sleep. *Journal of clinical psychiatry*, 66(9), 3-9.

Rutter, J. Reick, M. and McKnight, S. L. (2002). Metabolism and the control of circadian rhythms. *Annual reviews of biochemistry*, 71, 307-331.

Rutter, J. Reick, M. Wu, L.and McKnight, S. L. (2001). Regulation of clock and NPAS2 DNA binding by the redox state of NAD cofactors. *Science*, 293, 510-514.

Sack, R. L. Lewy, A. J. Blood, M. L. Keith, L. D. and Nakagawa, H. (1992). Circadian rhythms abnormalities in totally blind people: incidence and clinical significance. *Journal of clinical endocrinology and metabolism*, 75(1), 127-134.

Schernhammer, E. S. Laden, F. Speizer, F. E. Willett, W. C. Hunter, D. J. Kawachi, I.and Colditz, G. A. (2001). Rotating night shifts and risk of breaast cancer in women participating in the nurses' health study. *Journal of the national cancer institute*, 93(20), 1563-1568.

Schernhammer, E. S. Laden, F. Speizer, F. E. Willett, W. C. Hunter, D. J. Kawachi, I. Fuchs, C. S. and Colditz, G. A. (2003). Night-shift work and risk of colorectal cancer in the nurses' health study. *Journal of the national cancer institute*, 95(11), 825-828.

Schmiegelow, K. Glomstein, A. Kristinsson, J. Salmi, T. Schroder, H. and Bjork, O. (1997). Impact of morning versus evening schedule for oral methotrexate and 6-mercaptopurine on relapse risk for children with acute lymphoblastic leukemia. Nordic Society for Pediatric Hematology and Oncology (NOPHO). *Journal of pediatric hematology/oncology,* 19(2), 102-109.

Shearman, L. P. Sriram S. Weaver D. R. Maywood, E. S. Chaves, I. Zheng, B. Kume, K. Lee ,C. C. van der Horst, G. T. J. Hastings, M. and Reppert, S. M. (2000). Interacting molecular loops in the mammalian circadian clock. *Sciene,* 288, 1013-1019.

Shearman, L. P. Zylka, M. J. Weaver, D. R. Kolakowski, L. F Jr. and Reppert, S. M. (1997). Two period homologs, circadian expression and photic regulation in the suprachiasmatic nuclei. *Neuron,* 19(6), 1261-1269.

Silver, R. LeSauter, J. Tresco, P. A. and Lehman, M. N. (1996). A diffusible coupling signal from the transplanted suprachiasmatic nucleus controlling circadian locomotor rhythms. *Nature*, 382(6594), 810-813.

Skene, D. J. Lockley, S. W. James, K. and Arendt, J. (1999). Correlation between urinary cortisol and 6-sulphatoxymelatonin rhythms in field studies of blind subjects. *Clinical endocrinology,* 50, 715-719.

Sullivan, S. C. and Richards, K. C. (2004). Predictors of circadian sleep-wake rhythm maintenance in elders with dementia. *Aging and mental health*, 8(2), 143-152.

Toh, K. L. Jones, C. R. He, Y. Eide, E. J. Hinz, W. A. Virshup, D. M. Ptacek, L. J. and Fy Y. H. (2001). An hPer2 phosphorylation site mutation in familial advanced sleep phase syndrome. *Science*, 291, 1040-1043.

Tousson, E. and Meissl, H. (2004). Suprachiasmatic nuclei grafts restore the circadian rhythm in the paraventricular nucleus of the hypothalamus. *Journal of neuroscience*, 24(12), 2983-2988.

Van Dongen, H. P. (2006). Shift work and inter-individual differences in sleep and sleepiness. *Chronobiology international*, 23(6), 1139-1147.

Van Gelder, R. N. Herzog, E. D. Schwartz, W. J. and Taghert, P. H. (2003). Circadian rhythms: In the loop at last. *Science*, 300, 1534-1535.

Van Someren, E. J. W. (2000). Circadian rhythms and sleep in human aging. *Chronobiology international,* 17(3), 233-243.

Van Someren, E. J. W. (2003). Thermosensitivity of the circadian timing system. *Sleep and biological rhythms*, 1, 55-64.

Vielhaber, E. Eide, E. J. Rivers, A. Gao, Z. H. and Virshup, D. M. (2000). Nuclear entry of the circadian regulator mPer1 is controlled by mammalian casein kinase 1ε. *Molecular and cellular biology*, 20, 4888-4899.

Waterhouse, J. Nevill, A. Finnegan, J. Williams, P. Edwards, B. Kao, S. Y. and Reilly, Y. (2005). Further assessments of the relationship between jet lag and some of its symptoms. *Chronobiology international*, 22(1), 121-136.

Wever, R. A. (1979). *The circadian system of man*. Springer-Verlag, New York-Heidelberg-Berlin.

Wirz-Justice, A. (2006). Biological rhythm disturbances in mood disorders. *International clinical psychopharmacology,* 21(1), S11-S15.

Yagita, K. Yamaguchi, S. Tamanini, F. van der Horst, G. T. J. Hoeijmakers, J. H. J. Yasui, A. Loros, J. J. Dunlap, J. C. and Okamura, H. (2000). Dimerization and nuclear entry of mPER proteins in mammalian cells. *Genes and development*, 14, 1353-1363.

Yan, L. and Silver, R. (2004). Resetting the brain clock, time course and localization of mPER1 and mPER2 protein expression in suprachiasmatic nuclei during phase shifts. *European journal of neuroscience*, 19(4), 1105-1109.

Yannielli, P. and Harrington, M. E. (2000) Neuropeptide Y applied in vitro can block the phase shifts induced by light in vivo. *Neuroreport*, 11, 1587-1591.

Yannielli, P. and Harrington, M. E. (2001a) The neuropeptide Y Y5 receptor mediates the blockade of "photic like" NMDA-induced phase shifts in the golden hamster. *Journal of Neuroscience*, 14, 5367-5373.

Yannielli, P. and Harrington, M. E. (2001b). Neuropeptide Y in the mammalian circadian system, effects on light induced circadian responses. *Peptides*, 22(3), 547-556.

Yannielli, P. and Harrington, M. E. (2004). Let there be "more" light, enhancement of light actions on the circadian system through non-photic pathways. *Progress in neurobiology*, 74(1), 59-76.

Yannielli, P. C. McKinley Brewer, J. and Harrington, M. E. (2002). Is novel wheel inhibition of per1 and per2 expression linked to phase shift occurrence? *Neuroscience*, 112(3), 677-685.

Yokota, S. I. Horikawa, K. Akiyama, M. Moriya, T. Ebihara, S. Komuro, G. Ohta, T. and Shibata, S. (2000). Inhibitory action of brotizolam on circadian and light-induced per1 and per2 expression in the hamster suprachiasmatic nucleus. *British journal of pharmacology*, 131(8), 1739-47.

Zheng, B. Albrecht, U. Kaasik, K. Sage, M. Lu, W. Vaishnav, S. Li, Q. Sun, Z. S. Eichele, G. Bradley, A. and Lee, C. C. (2001). Nonredundant roles of the mPer1 and mPer2 genes in the mammalian circadian clock. *Cell,* 105, 683-694.

Zylka, M. J. Shearman, L. P. Weaver, D. R, and Reppert, S. M. (1998). Three period homologs in mammals, differential light responses in the suprachiasmatic circadian clock and oscillating transcripts outside of brain. *Neuron*, 20(6), 1103-1110.

In: Circadian Rhythms and Health Research Trends
Editor: Alfredo V. Lignelli, pp. 75-97
ISBN: 978-1-60021-875-0

Chapter IV

Diuretic Pattern in Older Adults with Nocturnal Polyuria and Its Association with Blood Pressure and Nocturnal Natriuresis

Osamu Natsume*
Department of Urology, Nara Prefectural Rehabilitation Center, Nara, Japan

Abstract

Purpose: To elucidate the underlying pathophysiology of nocturnal polyuria (NP) in older adults, their diurnal diuretic pattern was investigated, including its possible association with blood pressure and natriuresis.

Materials and Methods: A total of 46 nocturic outpatients were enrolled in study 1. Using frequency-volume charts, patients were divided into two groups: an NP group with a mean age of 68.8 years with nocturnal urine production of more than 10 ml per kg body weight (n=34), and another group with a mean age of 63.5 years without NP (n=12). Eleven non-nocturic healthy adults with a mean age of 60.2 years served as controls. Furthermore, plasma arginine vasopressin (pAVP) release on 5 % hypertonic saline loading was evaluated. In study 2, eight inpatients with a mean age of 69.3 years with NP and mobility disabilities were subjected to diuretic-rate measurement every 3 hours, where an association between diurnal diuretic rates and the urinary solute excretion was studied.

Results: The NP group showed a greater diuretic rate during the night than in the daytime compared to the non-NP nocturic and control groups. Although the NP group had a significantly higher systolic blood pressure than the non-NP group, both groups

* Address: Oaza-ta 722, Tawaramoto-cho, Shiki-gun, Nara 636-0393, Japan. Tel: +81-7443-2-0200. Fax: +81-7443-2-0207. e-mail: osnatsume@hotmail.com
Address correspondence to: Osamu Natsume, Department of Urology, Nara Medical University, Shijo-cho 840, Kashihara City, Nara 634-8522, Japan. Tel: +81-7442-2-3051. Fax: +81-7442-2-9282.

showed a similar altered circadian rhythm of pAVP release with a lack of nocturnal surge. Moreover, nocturnal and daytime diuretic rates were correlated with systolic blood pressure only in the NP group, positively and negatively, respectively. Neither pAVP nor brain natriuretic peptide were correlated with nocturnal or daytime diuretic rates. An overall upward shift was shown by a linear function relating pAVP levels to increasing plasma sodium changes using saline loading, although there was wide individual variability. Even in study 2, a reversed diuretic pattern was seen, where the urinary sodium excretion profile was similar to subjects' diuretic pattern, and a significant correlation between the diuretic rates and urinary sodium levels was found ($r=0.58$, $p=0.0003$). The ratios of night or early morning to daytime urinary sodium excretion were significantly increased in patients compared to controls ($p<0.001$), although those with NP had a greater 24-hour total urine volume than controls.

Conclusion: Patients with NP seemed to be predisposed to a reversed diuretic rhythm and nocturnal increase in urinary sodium excretion, where increased systolic blood pressure may be associated. NP may be predicted by the increased ratios of night or early morning to daytime urinary sodium excretion.

Keywords: *nocturia, nocturnal polyuria, arginine vasopressin, natriuresis, blood pressure.*

Abbreviations and Acronyms

NP	Nocturnal polyuria
AVP	Arginine vasopressin
pAVP	Plasma arginine vasopressin
uAVP	Urinary arginine vasopressin
Posm	Plasma osmolality
Uosm	Urinary osmolality
BNP	Brain natriuretic peptide
Bw	Body weight
TUV	Total urine volume
NUV	Nocturnal urine volume
DUV	Daytime urine volume

Introduction

Nocturnal urinary frequency (Nocturia) is a common urinary tract symptom in both males and females [1-4], and has been indicated to increase with aging. The overall prevalence of nocturia was estimated to be approximately 50% or more in people over 50 years old when defined as at least one or more voids at night, and 12 to 30% even in the 50s when defined as at least two or more voids at night [1-3]. In a Japanese nationwide survey of nocturia prevalence, Homma et al. reported that only a minority of those affected sought treatment, regardless of the high prevalence and impact of nocturia [5]. Nocturia has been considered to be potentially associated with a multifocal etiology [6, 7], greatly affecting the quality of life; in particular, due to it being linked to not only sleep disturbance but increased mortality[8]. To date, nocturnal polyuria (NP), reduced bladder capacity, or a combination of the two, have been recognized as major components responsible for nocturia.

Although an association between an altered circadian rhythm of antidiuretic hormone, plasma arginine vasopressin (pAVP), and a diuretic pattern in older adults with nocturia or NP has recently been a field of intense research, it still remains unclear. Therefore, further study on the relationship of NP with AVP and cardiovascular–related peptides such as natriuretic peptides, including an analysis regarding whether they are determinants for increasing NP, may allow for a better understanding of NP with aging.

To date, hemodynamic change with aging has been well documented to be likely to accompany disturbed profiles in solute excretion [9], and interestingly, many investigators have demonstrated increased nocturnal diuresis with a concomitant increase of urinary sodium excretion in NP [10-14]. There is also a report that increased nocturnal natriuresis and diuresis are associated with elevations of natriuretic peptide in the elderly [12]. Therefore, the implications of this research may be important in understanding the pathophysiology of NP. Namely, these findings may suggest the interactive contribution of hemodynamic change or sympathetic nerve activity as well as osmotic control on the activity of vasopressin neurons in NP. Because a hypertonic saline infusion test [15-18] can provide us with some important information concerning not only pAVP responsiveness to the increase of plasma sodium but also volume–pressure mediated pAVP release, its application to adult NP patients might be useful in the assessment of the osmotic response or to facilitate of understanding of the underlying pathophysiology. In this study, using a comprehensive evaluation method, clinical investigations were conducted to disclose potential etiological factors underlying the disturbed diuretic rhythm in older adult patients with NP, particularly emphasizing the possible relevance of blood pressure in increasing nocturnal diuresis in nocturnal polyuria. Moreover, whether the urinary sodium excretion profiles of patients were related to increased nocturnal urine production was also investigated compared to non-nocturic healthy older adult controls.

Materials and Methods

Study 1: Comparison of Diurnal pAVP Release, Brain Natriuretic Peptide Concentrations, Blood Pressure and Chart Variables in Adult Nocturic Patients with and without NP

Subjects

From April 2004 to March 2007, among 60 patients with nocturia with an average of one or more voids per night, a total of 46 patients (31 males and 15 females) aged from 50 to 87 years (mean: 67.4 years), were enrolled in this study based on the records of their frequency-volume charts, and we evaluated the circadian variation in pAVP. NP was defined as a nocturnal urine volume (NUV) exceeding 10 ml per kg of body weight (Bw) according to the classification of nocturia proposed by Homma et al. [19]. Patients were segregated into an NP group consisting of 34 patients, and a non-NP group of 12 patients. A group of 11 non-nocturic healthy volunteers ranging from 47 to 79 years old (mean: 60.2 years, 10 males and 1 female) served as the control group, excluding evaluating diurnal changes in pAVP and plasma osmolality (Posm). All patients completed the chart record for a mean of 3.4 days (range: 1 to 5 days) in the NP group and 3.0 days (range: 1 to 5 days) in the non-NP group. Patients were excluded if they were receiving diuretic therapy or had evidence of overt congestive cardiac failure, liver dysfunction, renal dysfunction (serum creatinine □1.2 mg/dl, or a 24-hour creatinine clearance <70ml/min), diabetes mellitus, interstitial cystitis, urinary tract infection, urinary incontinence frequent enough to interfere with urinary output measurement, a 24-hour total urine volume (TUV) in excess of 40 ml/kgBw, and invariably more than 50 ml of postvoid residual urine measured using abdominal ultrasonography. Patients with a habitual fluid intake of more than 100ml at night for reasons of health and shift workers were excluded. Blood pressure values were obtained as part of the hospital assessment. Patients with a systolic blood pressure of □ 140 mmHg, diastolic blood pressure of □ 90 mmHg, or both were considered to be hypertensive. Patients were treated as hypertensive if they were taking antihypertensive medication. The mean blood pressure of three representative measurements taken during the study period in a recumbent position during the day was obtained and expressed as the mean value. Eleven healthy adult control subjects, who usually don't show nocturia and take no medication, completed the frequency-volume chart for one day at least. Both patients and control subjects were given full information, and informed consent was obtained.

Chart Analysis

In this study, the standard terminology of the International Continence Society [7] was used. The terms of daytime (or diurnal) and nighttime (nocturnal) were herein defined as the periods when they recorded that they were awake and asleep, respectively. During the recording period of the frequency-volume chart, patients didn't undergo strict fluid intake restriction except for the following advice: namely, prior to the recording, the patients were advised not to drink more than what was sufficient to satisfy their thirst to avoid excessive fluid intake. NUV was defined as total nocturnal urine production on the frequency-volume chart, including the first voided volume after rising from bed. Diuresis rates of diurnal and nocturnal urine production per kgBw were calculated as the sum of the voided volume per kgBw during the periods corresponding to daytime urine volume (DUV) and NUV, respectively, expressed as per minute urine production, i.e., ml/kg/min.

Blood and Urine Sampling

Prior to admission, pAVP, Posm, and peripheral blood count measurements, a standard chemistry panel, and electrocardiograms were performed routinely. At admission, peripheral blood for measuring pAVP and Posm at 2 PM, 2 AM, and 6 AM were taken in the recumbent position in all subjects excluding two patients. In addition, plasma specimens and timed single voided urine obtained at approximately 6 AM were subjected to measurements of the plasma brain natriuretic peptide (BNP) concentration, and urinary AVP (uAVP) [12, 20], creatinine (uCr), and osmolality (Uosm) in the early morning, respectively.

Evaluation of Osmotic Response in pAVP Release

To examine pAVP responsiveness of the hypothalamo-posterior pituitary system to osmolality changes with the increase of plasma sodium in patients with NP, a hypertonic saline infusion test was conducted [15, 16, 18]. Five % hypertonic saline infusion was perfomed at a rate of 0.05 ml/kg/min for 120 minutes in the recumbent position for 30 patients in the NP group and 7 patients in the non-NP group, respectively. A plasma sample was collected to examine pAVP concentrations related to changes in Posm at the initiation of the test, and after 1 and 2 hours. Obtained results were compared using a linear function and scattergram relating pAVP to Posm described by Robertson et al.[15] and Oiso et al.[16].

Analytical Method for Plasma and Urine Specimens

Plasma samples were collected to conduct an assay for pAVP related to Posm and BNP in a recumbent position at 6 AM following overnight hospitalization. The pAVP and uAVP concentrations were measured using a highly-sensitive radioimmunoassay (normal range: 0.3 to 4.2 pg/ml, AVP radioimmunoassay kit, Mitsubishi Kagaku Iatron, Inc.) at Ikagaku Corp. (Kyoto, Japan). The coefficients of variation of the assay were 5.9 % interassay and 9.7 % intraassay. The lower detectable assay limit was 0.15 pg/ml. BNP concentrations were measured using an enzyme immunoassay (normal range: 18.4 pg/ml or less, E test-TOSOH□). The coefficients of variation of the assay were 2.6 % interassay and 3.2 % intraassay. The lower detectable assay limit was 2.3 pg/ml. Posm and Uosm were determined using the freezing point depression method at the same laboratory. The levels of uAVP were adjusted to $pg/mgCr^{-1}$ by the urinary creatinine level.

Statistical Analysis

All results are expressed as the mean ± standard error (SE) unless otherwise indicated. The chi-square test was used for making comparisons between proportions of the two groups. A paired or unpaired Student's t-test was used to compare continuous variables between the groups. For statistical analysis of circadian variation for pAVP and Posm, a two-way repeated measures analysis of variance (ANOVA) was performed to compare intra- and intergroup differences. $P< 0.05$ was considered significant.

Study 2: Relationship between Diuresis and Natriuresis in Adults with NP

This study was conducted in order to ascertain patients' reversed diuretic pattern for bladder care based on a regular schedule following catheter removal, and identify the association between diurnal diuresis and natriuresis in rehabilitated inpatients with NP. A total of 8 patients with NP, consisting of 5 males and 3 females aged 51 to 82 years (mean: 69.3), were enrolled to assess their diurnal diuresis and natriuresis pattern because they were found to exhibit a noticeable increase of NUV, and their levels of excreted urinary electrolytes and Uosm were compared to those of healthy control subjects. These patients were hospitalized for rehabilitation due to mobility disabilities following stroke for three patients, incomplete spinal cord injury for four and multiple sclerosis for one. Herein, NUV was calculated as the sum of urine production from 9 PM through 6 AM, because the ward timetable consisted of lights-out at 9 PM, awakening at 6 AM, and regular hospital meals were taken at 7:30 AM, 12 AM (noon), and 6 PM in our hospital. Patients required bladder care including intermittent catheterization or pad exchange based on a regular schedule and had difficulty in recording their frequency-volume charts due to mobility disabilities. All remaining patients, excluding one case who was indwelled transiently with a catheter only during the assessment period, had had a transurethral catheter indwelled for urinary drainage. Therefore, they were assessed regarding their diurnal diuretic pattern in study 2 prior to bladder care following catheter removal. Circadian variations of diuretic rates and urinary solute excretion were measured by dividing the consecutive 48-hour recording period into separate 3-hour periods (from period 1 from 6 AM through 9 AM, to period 8 from 3 AM through 6 AM) and urine was collected 3-hourly throughout the 48-hour periods from an indwelling catheter bag. Urinary samples were collected to estimate urinary sodium, chloride, potassium, and Uosm on the first day at the end of each period of 12 AM to 3 PM (period 3), 6 PM to 9 PM (period 5), 0 AM to 3 AM (period 7), and 3 AM to 6 AM (period 8), respectively. Similarly, plasma samples were collected at the same periods excluding period 5. The levels of urinary sodium, chloride, and potassium were measured by an ion-selective electrode method, and the level of each electrolyte was adjusted to $mmol/mgCr^{-1}$ by the urinary creatinine level. A linear regression analysis was used to test the associations between diuresis rates and levels of excreted electrolytes, and between diuresis rates and Uosm in urine collected in periods 3, 5, 7, and 8. For the statistical analysis of circadian variation over time in urinary sodium, chloride, potassium, and Uosm, a one-way repeated measures ANOVA was performed. For comparison of the night-to-day ratios for the levels of urinary electrolytes and Uosm, the same data as those in controls used for study 1 were employed here again, where the urinary storage time predominantly including period 3 was treated as period 3, and the time of the first morning void as period 8 based on the frequency-volume chart records in controls. Other study conditions and the analytical method including statistical analysis were similar to those in study 1.

Results

Patient Characteristics in Study 1

Table 1 lists the clinical characteristics of the subjects in study 1. Nocturic patients with NP were older than controls ($p<0.01$), although there was no significant difference between those with and without NP. Approximately half of those with and without NP had hypertension. There was no significant difference in the prevalence of stroke and the taking of hypnotics between those with and without NP.

Comparison of the Summarized Baseline Data

Table 2 summarizes the demographic features and each frequency-volume chart variable in study 1. The nocturic groups with and without NP and the control group had a similar body mass index. The nocturic patients with NP showed a higher systolic blood pressure than those without NP ($p=0.029$) and controls ($p<0.0005$), although there was no significant difference in respect to the mean and diastolic blood pressure between the groups. The nocturic patients with NP tended to have a greater TUV and TUV per kgBw compared to those without NP and controls. The non-NP and control groups showed higher diuretic rates in the daytime than during the night (P=0.017 and P=0.023, respectively); in contrast, the NP group showed higher diuretic rates during the night than in the daytime (p<0.0001) (Figure 1).

Table 1. Characteristics of the nocturic group and the non-nocturic healthy control group in study 1

	Nocturia group		Control group (n=11)
	NP (n=34)	non-NP (n=12)	
Gender			
Male	21	10	10
Female	13	2	1
Age	68.8 (50 - 87)*	63.5 (50 - 76)	60.2 (47 - 79)
Hypertensive patients	18	6	0
Treated patients	17	5	0
Normotensive patients	16	6	0
Taking hypnotics	8	4	0
Patients with stroke episodes	9	4	0

The range is presented in a parenthesis.
* p<0.01 compared to controls.

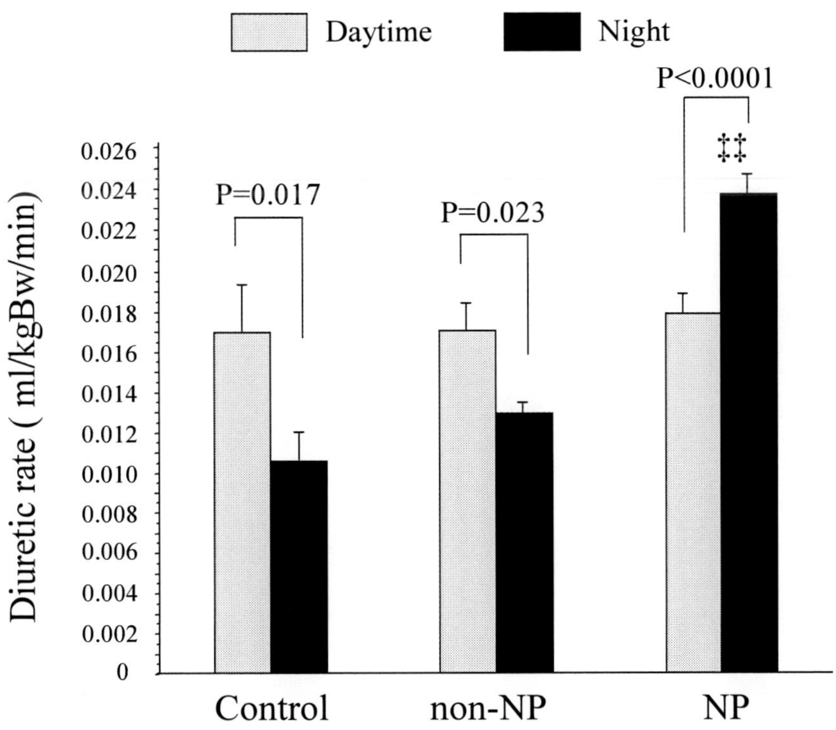

Figure 1. Comparison of diuretic rates in the daytime and during the night in the patient group with and without nocturnal polyuria and controls in study 1. ‡‡p<0.0001 compared to non-NP and control groups.

Table 2. Summary of demographic features and frequency-volume chart variables in study 1

	Nocturia group					Control group (n=11)	
	NP (n=34)		non-NP (n=12)				
	Mean	(Range)	Mean	(Range)	p value	Mean	(Range)
Body weight (kg)	57.6*	(37.0 - 80.1)	60.6	(55.0 - 74.0)	n.s.	65.2	(47.5 - 80.0)
Body mass index	22.3	(15.6 - 28.4)	23.1	(19.1 - 27.9)	n.s.	23.6	(19.5 - 26.4)
Mean blood pressure (mmHg)	95	(81 - 108)	94.3	(78 - 114)	n.s.	91	(72 - 102)
Systolic blood pressure (mmHg)	132***	(110 - 148)	124	(104 - 144)	0.029	119	(100 - 130)
Diastolic blood pressure (mmHg)	79	(64 - 96)	80	(60 - 99)	n.s.	76	(57 - 89)
Nocturnal voids	2.8	(1.0 - 4.5)	1.8	(1.0 - 3.0)	0.0017	0	
Nocturnal polyuria index§	0.47‡ ‡	(0.31 - 0.70)	0.34*	(0.22 - 0.52)	<0.0001	0.25	(0.12 - 0.35)
Maximum voided volume (ml)	344	(160 - 600)	277**	(200 - 360)	0.045	384	(250 - 600)
Maximum voided volume/body weight (ml/kg)	6	(3.0 - 10.9)	4.6†	(3.3 - 6.3)	0.0098	5.9	(4.3 - 7.9)
Total urine volume (ml)	1631*	(916 - 2387)	1331	(903 - 1830)	0.011	1365	(1070 - 1731)
Total urine volume/body weight (ml/kg)	28.4‡ ‡	(19.8 - 35.4)	22.2	(15.8 - 32.7)	0.0002	21.4	(14.6 - 33.2)
Daytime urine volume (ml)	876	(360 - 1420)	895	(483 - 1333)	n.s.	1025	(730 - 1325)
Daytime urine volume/body weight (ml/kg)	15.3	(6.1 - 23.3)	14.9	(8.5 - 24.2)	n.s.	16.2	(11.5 - 27.9)
Daytime diuretic rate (ml/kg body weight/min)	0.018	(0.006 - 0.028)	0.017	(0.011 - 0.029)	n.s.	0.017	(0.010 - 0.033)
Daytime single voided volume (ml)	156	(74 - 308)	154	(89 - 200)	n.s.	185	(142 - 292)
Nocturnal urine volume (ml)	755‡ ‡	(466 - 1247)	436*	(340 - 560)	<0.0001	339	(130 - 488)
Nocturnal urine volume/body weight (ml/kg)	13.1‡ ‡	(10.2 - 17.7)	7.3**	(4.6 - 9.7)	<0.0001	5.2	(2.6 - 7.7)
Nocturnal diuretic rate (ml/kg body weight/min)	0.024‡ ‡	(0.017 - 0.038)	0.013	(0.008 - 0.016)	<0.0001	0.011	(0.006 - 0.018)
Nocturnal single voided volume (ml)	223*	(103 - 415)	179‡	(118 - 254)	n.s.	292	(130 - 400)
Brain natriuretic peptide (pg/ml)	35.9	(5.1 - 156)	23.6	(4.3 - 83.7)	n.s.	13.5	(4.2 - 33.3)
Plasma arginine vasopressin (pg/ml)¶	1.6	(0.2 - 7.4)	1.8	(0.4 - 4.4)	n.s.		
Urinary arginine vasopressin ($pg/mg \cdot Cr^{-1}$)¶	22.1#	(2.1 - 85.5)	27.8	(7.8 - 53.3)	n.s.	34.7#	(15.9 - 86.2)
Plasma osmolality (mosm/l)¶	285	(271 - 297)	285	(279 - 289)	n.s.		
Urine osmolality (mosm/l)¶	493**	(90 - 768)	535	(160 - 927)	n.s.	678	(353 - 949)

§The ratio of nocturnal urine volume over total urine volume.

¶Data at sampling time of 6 AM in the nocturia group and at the first void after awakening in the control group are presented.

*p<0.05, **p<0.005, ***p<0.0005 compared to controls.

†p<0.01, ‡p<0.001, ‡ ‡ p<0.0001 compared to controls.

#Data missing for 1 patient.

The range is presented in a parenthesis.

Both the NP and non-NP groups had a lower mean nocturnal single voided volume compared to controls (p<0.05 and p<0.001, respectively), although there was no significant difference regarding the mean daytime single voided volume between the three groups. Particularly, the nocturic patients with NP showed a significantly smaller nocturnal single voided volume compared to those of controls, although both groups showed similar maximum voided volumes and maximum voided volumes per kgBw (p<0.05) (Figure 2). Nocturnal frequency was significantly higher in patients with than in those without NP (p=0.0017), although there was no difference in nocturnal single voided volume between the groups. In both the nocturic groups with and without NP, nocturnal voids were negatively correlated with the mean nocturnal (r= - 0.68, p<0.0001 and r= - 0.82, p=0.0005, respectively) and daytime single voided volume (r= - 0.47, p=0.0042 and r= - 0.60, p=0.037, respectively), respectively. The BNP and uAVP levels in the early morning didn't differ between the three groups (Table 2). Those with NP showed a significantly lower Uosm in the early morning compared to controls (p<0.005).

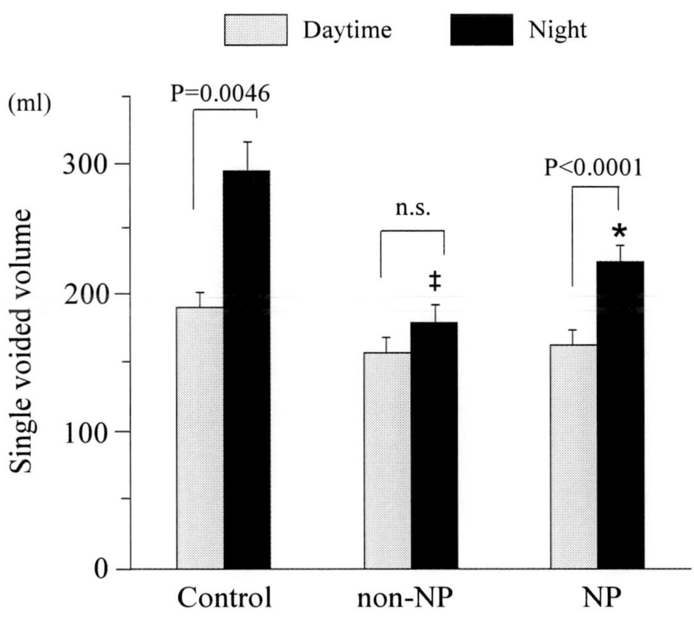

*p<0.05 and ‡<0.001 compared to the control group.

Figure 2. Comparison of single voided volume in the daytime and during the night in the patient group with and without nocturnal polyuria and controls.

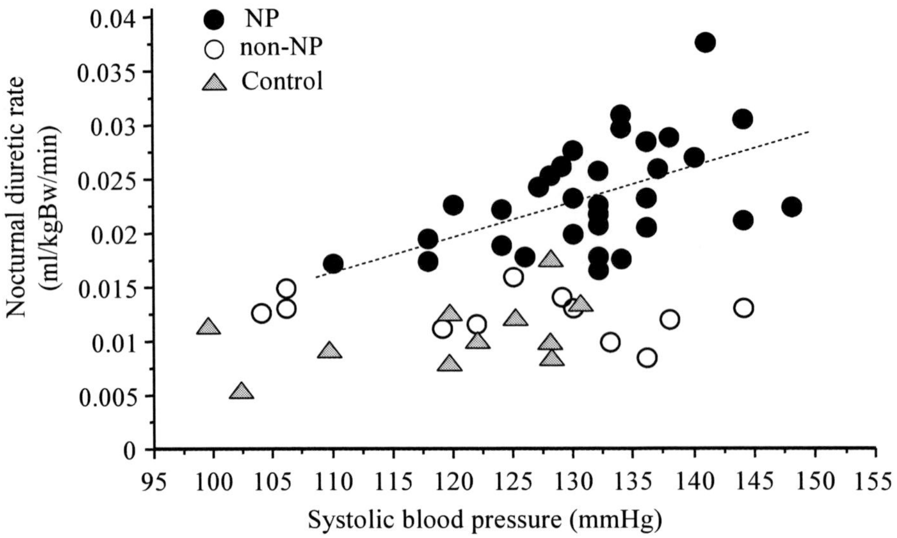

Figure 3A. Relationship of the systolic blood pressure and nocturnal diuretic rate in nocturic patients with and without nocturnal polyuria and non-nocturic healthy controls. A significant correlation described by a linear regression expressed using a dotted line overall was seen in nocturic patients with nocturnal polyuria (closed circles) (r=0.48, p=0.0037), although there was individual variation. However, no significant correlation was found in nocturic patients without nocturnal polyuria (open circles) and non-nocturic healthy controls (gray triangles).

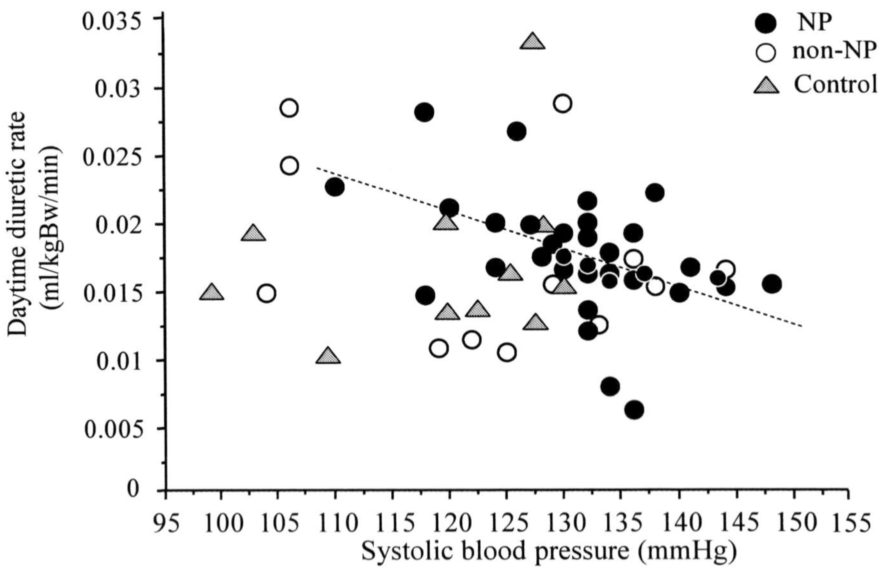

Figure 3B. Relationship of the systolic blood pressure and daytime diuretic rate in nocturic patients with and without nocturnal polyuria and non-nocturic healthy controls. A significant negative correlation described by a linear regression expressed using a dotted line overall was seen in nocturic patients with nocturnal polyuria (closed circles) (r= - 0.45, p=0.0072), although there was individual variation. No significant correlation was found in nocturic patients without nocturnal polyuria (open circles) and controls (gray triangles).

Relationship between Diuretic Rate and Blood Pressure

In the NP group, a significant positive correlation between the systolic blood pressure and nocturnal diuretic rate was seen overall, although there was individual variation (r=0.48, p=0.0037) (Figure 3A). Moreover, a significant negative correlation between the systolic blood pressure and daytime diuretic rate was found overall. (r= - 0.45, p=0.0072) (Figure 3B).

The mean blood pressure was also significantly correlated with the nocturnal diuretic rate (r=0.47, p=0.0041) and daytime diuretic rate (r= - 0.41, p=0.014), whereas diastolic blood pressure was correlated with neither parameters. On the other hand, no significant correlation between the diuretic rate and systolic, mean, and diastolic blood pressure was found in the non-NP nocturic and control groups.

Correlation between Diuresis Rate and pAVP and BNP Concentrations, and Body Mass Index

In both the NP and non-NP groups, both the levels of pAVP 6 AM in the early morning and those at 2 AM were not associated with nocturnal diuretic rates. On the other hand, the levels of pAVP at 6 AM were negatively correlated with daytime diuretic rates (r= - 0.40, p=0.023), and those at 2 AM had a negative non-significant association with those rates (r= - 0.34, p=0.054) only in the NP group and not the non-NP group. In addition, the BNP levels revealed a non-significant association with nocturnal diuresis rates in the NP group (r=0.33, p=0.054), whereas there was no correlation between BNP and daytime diuresis rates. The non-NP and control groups also showed no significant correlation between BNP levels and nocturnal and daytime diuresis rates. Moreover, the BNP levels didn't differ significantly among the three groups (Table 2), and the NP group revealed a significantly higher nocturnal diuresis rate than the remaining groups (Figure 1). There was no association between the body mass index and nocturnal and daytime diuresis rates in the three groups.

Circadian Variation in p AVP Release and Posm in NP and Non-NP Groups

Both the NP and non-NP groups showed no circadian variation with respect to diurnal changes of Posm (Figure 4). There was no significant difference in Posm levels between the groups. Altered circadian variation was identified for pAVP release in both groups, with no significant difference between them (Figure 5).

Results of 5% Hypertonic Saline Infusion Test

A 5% hypertonic saline infusion test was conducted in 30 of 34 patients in the NP group and 7 of 12 in the non-NP group. A linear function described by a formula, overall pAVP = 0.50 (Posm - 286), was obtained in the NP group, whereas overall pAVP = 0.39 (Posm - 285) in the non-NP group. When compared to the results of previous reports using healthy adults, the NP group seemed to show a more upward shift overall, although wide individual variation was noted regarding pAVP responsiveness to Posm changes in both groups (Figure 6). Representative response curves of four subjects in the NP group are shown in Figure 7, where it was suggested that individuals showing a variety of sensitivities (slope) of osmoreceptors to Posm changes from a decrease to an increase might be present among the nocturnal polyuria population.

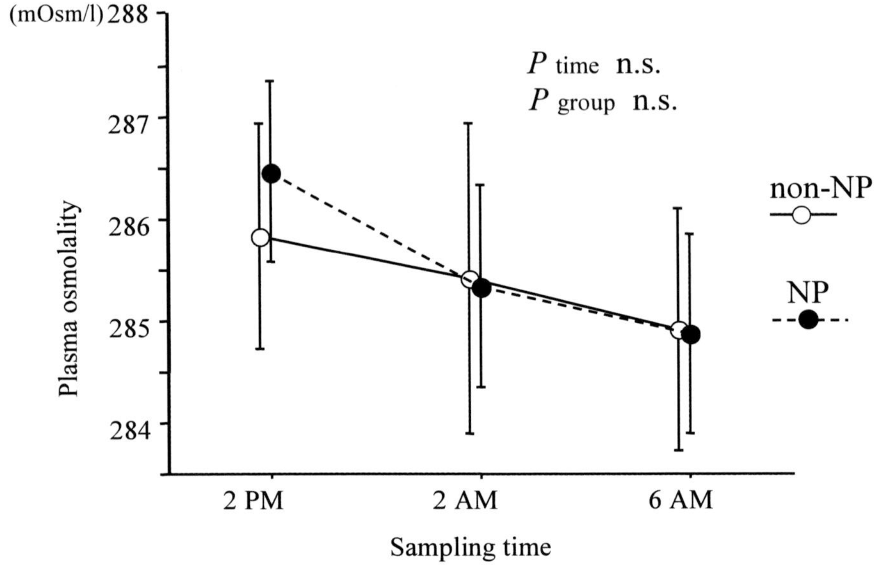

Figure 4. Diurnal changes of plasma osmolality in the patient group with and without nocturnal polyuria in study 1. Both groups showed no significant circadian variation. There was no significant difference in plasma osmolality levels between the groups.

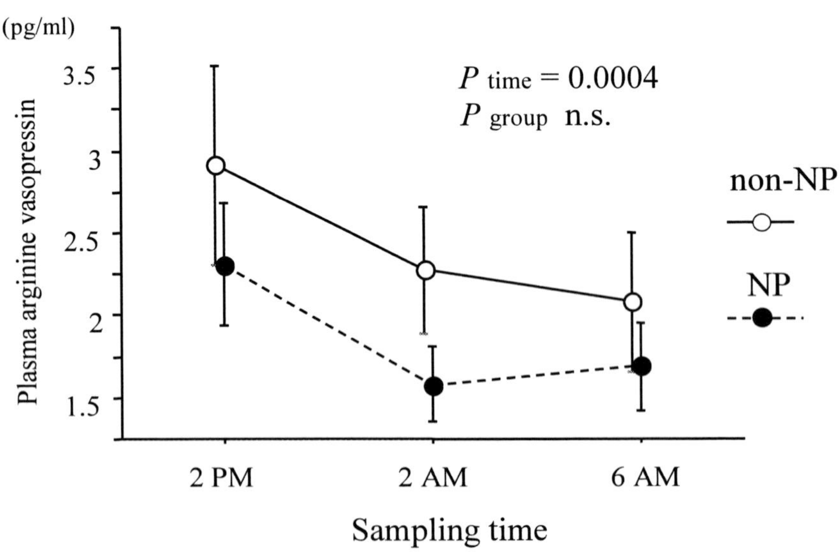

Figure 5. Circadian variation of plasma AVP in the patient group with and without nocturnal polyuria in study 1. Altered circadian variation was identified regarding plasma AVP in both groups, with no significant difference between the groups.

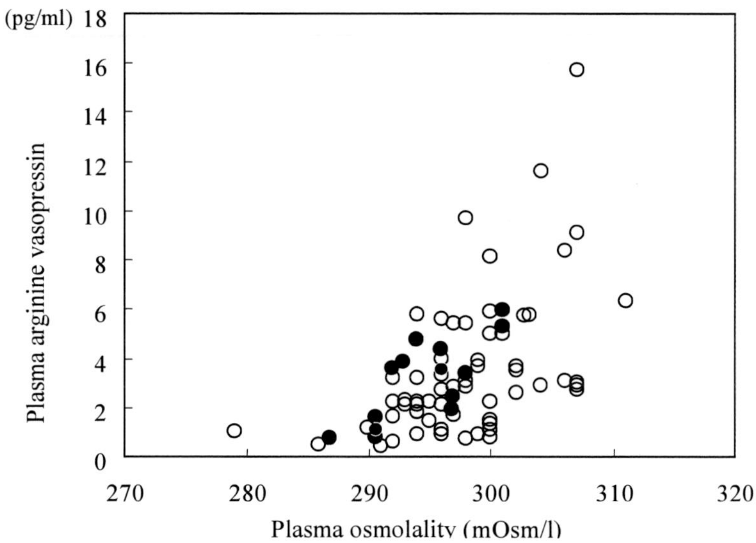

Figure 6. A scattergram relating the plasma AVP release response to plasma osmolality change after 5% hypertonic saline loading examined in 30 of 34 nocturic patients with nocturnal polyuria (open circles) and in 7 of 12 nocturic patients without nocturnal polyuria (closed circles) in study 1. The plasma AVP and osmolarity values showed wide individual variation, and particularly, a slow rise in the relationship of plasma AVP and osmolality among the nocturic patients with nocturnal polyuria appeared.

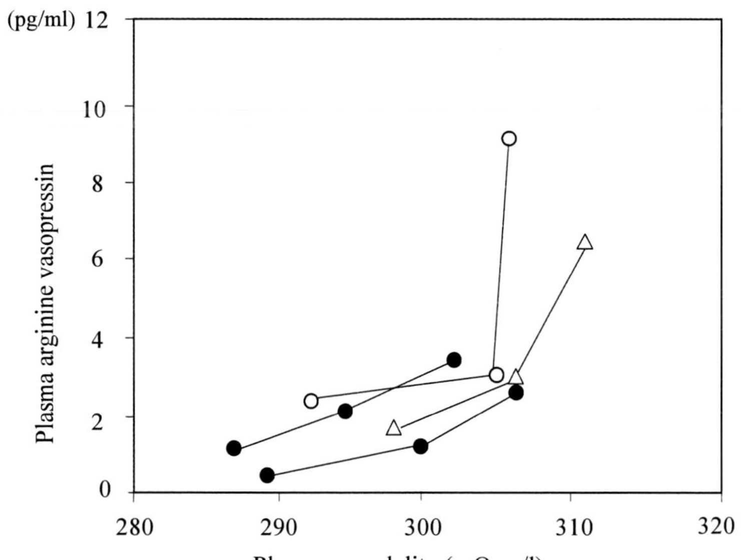

Figure 7. Representative results of a plasma AVP response curve to 5% hypertonic saline loading in patients with nocturnal polyuria in study 1, where a variety of sensitivities (slope) of osmoreceptors to plasma osmotic change was indicated. A case indicated by triangles showed an upward shift overall, with a sensitivity considered as almost normal. However, a case indicated by open circles showed an upward shift overall, with an increased sensitivity and the resetting of the threshold of plasma AVP

release. Two cases indicated by closed circles were considered to exhibit a decreased sensitivity, resulting in a more flattened response curve.

Study 2: Baseline Parameters and 24-Hour Diuresis Changes in Inpatients with NP in Study 2

The mean TUV, DUV, and NUV of eight subjects in study 2 were 1747 ml (ranged from 1480 to 2205), 825 ml (ranged from 1480 to 2205), and 921 ml (ranged from 520 to 1190), respectively. The mean TUV in the patient group was significantly greater compared to that in controls (1364 ml) ($p=0.002$). The mean BNP and uAVP levels were 21.9 pg/ml (ranged from 5.6 to 71.9) and 17.3 pg/mgCr^{-1} (ranged from 5.6 to 71.9), respectively. Nocturnal diuretic rates in the patient group were significantly greater than their daytime diuretic rates (0.033 vs. 0.018, respectively, $p=0.0006$), and significantly differed when compared to those in controls (0.033 vs. 0.011, respectively, $p<0.0001$), although there was no difference in the mean daytime diuretic rate between patients and controls. Circadian changes in the 3-hour diuresis rate obtained from inpatients with NP are shown in Figure 8 with the diurnal change of pAVP. A single peak in the diuretic rates was seen from 9 PM to 6 AM (periods 6 to 8), and a nadir during 6 AM to12 AM (periods 1 and 2). There was also a reversed diurnal pattern of pAVP noted overall, as in study 1.

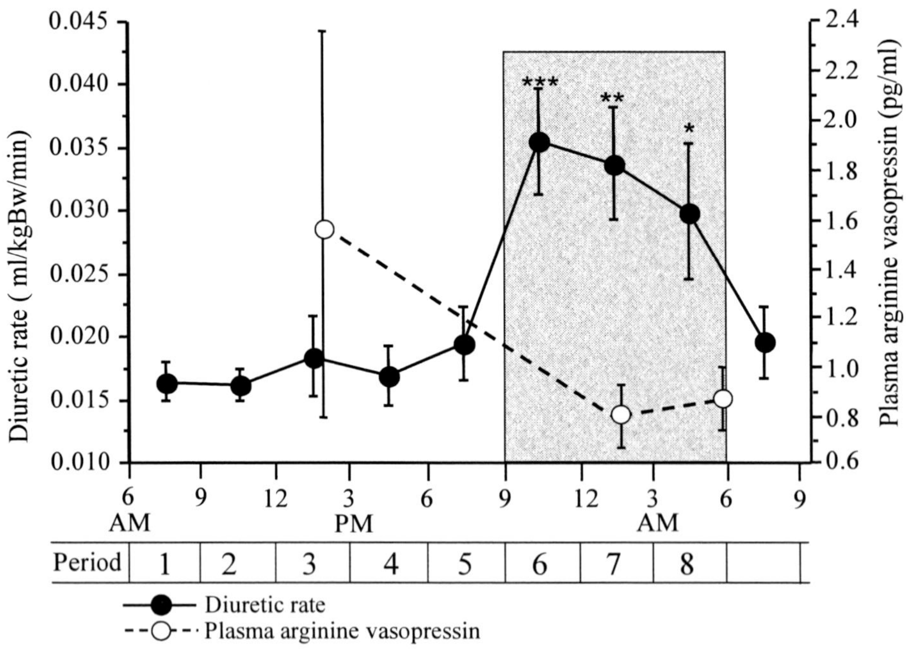

Figure 8. Circadian changes in the 3-hour diuresis rate of inpatients with NP in study 2 are shown with the circadian change of plasma AVP. A nadir was seen during 6 AM to12 AM (periods 1 and 2). *P<0.05, * *P<0.005, and * * *P<0.001 compared to nadir in period 2.

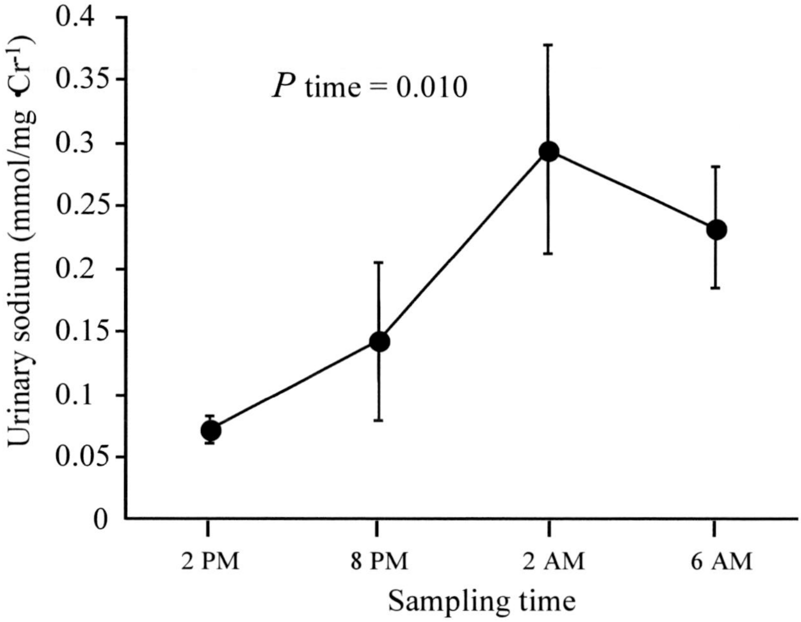

Figure 9. Circadian variation of urinary excreted sodium in inpatients with nocturnal polyuria in study 2. A significant circadian variation was seen with a change similar to the diurnal diuretic pattern shown in figure 8.

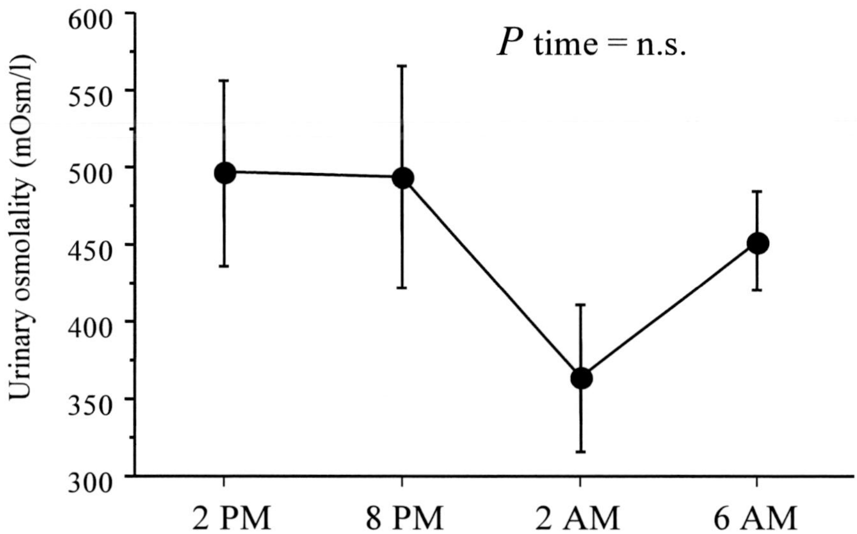

Figure 10. Circadian variation of urinary osmolality in inpatients with nocturnal polyuria in study 2. No significant circadian variation was seen, although a nadir was observed which corresponded to around the peak of diuretic rates.

Diurnal Change of Urinary Excretion of Electrolytes and Uosm

Each diurnal change based on the levels of urinary electrolytes and Uosm obtained in periods 3, 5, 7, and 8 was evaluated. Circadian variation similar to diurnal diuretic pattern was noted in urinary sodium (p=0.010) (Figure 9). Urinary chloride also showed a trend toward circadian variation with an increase during the nighttime (p=0.056). On the other hand, urinary potassium didn't exhibit circadian variation. Uosm also revealed no significant variation, although a low nadir was noted in period 7 at night (Figure 10).

Relationship between Diuresis Rates and Urinary Excreted Electrolytes, and Uosm

A linear regression analysis relating diuresis rates to levels of urinary electrolytes obtained in periods 3, 5, 7, and 8 in the patient group was conducted. A significant correlation was found between diuresis rates and excereted urinary sodium (r=0.58, p=0.0003) and Uosm (r= - 0.57, p=0.0005), whereas no significant correlation was found between diuresis rates and urinary chloride, nor between diuresis rates and urinary potassium.

Comparison of Night-to-Day Ratio of the Levels of Urinary Electrolytes and Uosm between Inpatients with NP and Controls

Night-(period 8)-to-day (period 3) ratios of urinary sodium, chloride, potassium, and Uosm in the patient group were 3.46 (p<0.0001), 3.87 (p<0.0001), 1.44 (p=0.013), and 1.01 (p=0.76), respectively, and were significantly higher, excluding Uosm, than those in controls (Figure 11). Similarly, when compared to data in controls using those in period 7 instead of period 8 as data at night, similar results were seen, where the ratios were 4.04 (p<0.0003) for sodium, 3.47 (p=0.0078) for chloride, 1.24 (p=0.017) for potassium, and 0.87 (p=0.30) for Uosm. Namely, urinary electrolyte excretion was significantly higher at night than in the daytime compared to those in controls.

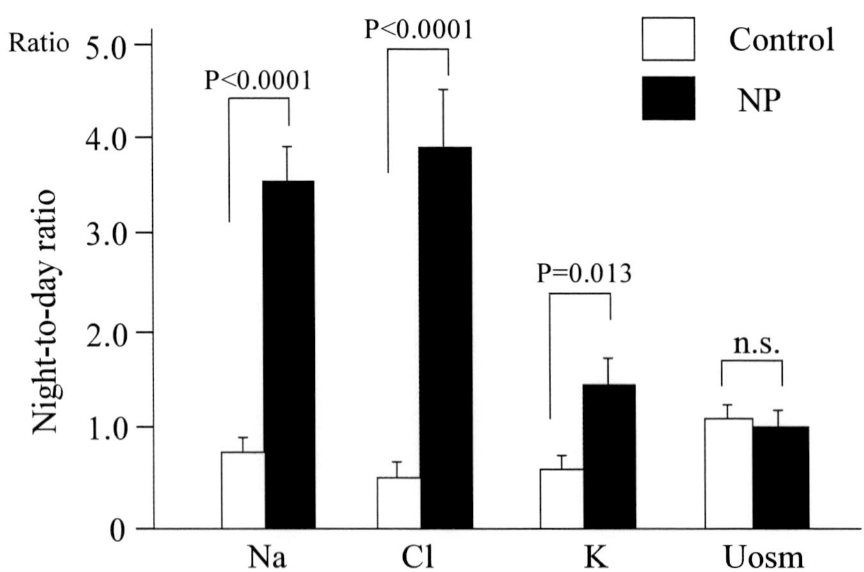

Figure 11. Comparison of the early morning-to-day ratios for urinary excreted sodium (Na), chloride (Cl), potassium (K), and urinary osmolality (Uosm) between the patient group with nocturnal polyuria and the healthy control group.

Discussion

Nocturia is a common lower urinary tract symptom in older adults, particularly around 50 years or more, and tends to increase with aging. Therefore, nocturia has recently become a field of intense research. Evaluation of the diurnal diuretic pattern using a frequency-volume chart, in addition to history taking and physical examination, is available as the initial diagnostic step of nocturia, which has been recognized to be multifactorial, because fundamental information possibly related to nocturia may underly the records. However, it seems to be not rare for urologists to encounter that it is not always easy for patients to precisely record their frequency-volume chart repeatedly. Accordingly, a convenient alternative parameter related to the underlying pathophysiology of nocturia is desirable. To date, hemodynamic change with aging has been well documented to be likely to accompany disturbed profiles in urinary solute excretion [9] and the rhythmicity of the sleep hormone, melatonin [21]. Interestingly, many investigators have demonstrated that higher nocturnal diuresis in NP is concomitantly associated with increased solute excretion [9-13]. Also, it was reported that administration of melatonin, which has a potential cardioprotective effect [22], is associated with improvements in nocturia-related problems [21]. Thus, such a hemodynamic change with aging and NP may share common underlying pathophysiologic conditions. The hypertonic saline infusion test, which is able to identify not only an individual's osmotic responsiveness of AVP release to plasma osmotic changes but also, concurrently, an individual's hemodynamic or non-osmotic condition, may be available in intractable patients with NP.

This study identified that patients with NP were likely to have a greater 24-hour total urine volume than non-NP patients and healthy control subjects overall. This finding was consistent with previous studies [13, 23]. It was also confirmed that patients with NP showed significantly higher diuretic rates during the night than in the daytime, that is, a reversed diuretic pattern overall, contrary to the control group. Then, the increase of nocturnal diuresis in such patients was concomitantly associated with an increase of urinary sodium excretion. These findings also agreed with results obtained by previous investigations [10, 11, 13]. Asplund found that systolic blood pressure was associated with an increase in the nocturnal diuresis rate, but not the daytime diuresis rate [24]. This study also found that systolic blood pressure was not only positively associated with the increase in the nocturnal diuresis rate but also negatively with the daytime diuresis rate. Thus, the possible relevance of blood pressure to nocturia or NP in older adults may have an implication, as described previously[11, 18, 24-27]; however, further investigation is required.

To date, it has been suggested that concentrations of pAVP are undetectable or low during the night in nocturic patients, and their circadian rhythm of pAVP release tends to become blunted with aging with or without NP [11, 12, 28]. More recently, Graugaard-Jensen et al. reported that circadian variation with a nocturnal surge of pAVP release was seen only in individuals without nocturia [27]. The present study also confirmed that an altered circadian variation without a nocturnal surge in pAVP release was observed in both NP and non-NP nocturic patients, with a lower nadir during the night rather than in the daytime overall. Then, it is of interest that systolic blood pressure in the NP group was significantly higher than that in the non-NP group. Furthermore, Graugaard-Jensen et al. showed a negative correlation of the nocturnal pAVP with NUV per kgBw in nocturic persons[27]. In this study, however, it was found that nocturnal pAVP had no association with nocturnal diuretic rates in the nocturic group with or without NP. The present findings seem to suggest the reduced primary importance of pAVP in the increase of nocturnal urine output, although it is unclear why this discrepancy may arise.

A linear function obtained from each value relating pAVP release responsiveness to plasma osmolarity changes during a 5% hypertonic saline infusion test in the present NP population revealed an upward shift overall compared to results of previous reports conducted in healthy adults [15, 16]. These findings may suggest an osmoregulatory inhibition of pAVP release, resulting in the resetting of the threshold. With regard to this point, some investigators have described that pAVP release could be inhibited by a possible involvement of the central nervous system in the physiological responses to a hypervolemic condition or an increased blood pressure [15, 17, 29, 30]. Thus, an overall upward shift of pAVP responsiveness obtained from hypertonic saline loading in NP patients seems to partly suggest that the hypervolemic condition or increased sympathetic nerve activity in addition to a diminished sensitivity of osmoregulatory function itself may potentially underly the population examined, and cause an inhibited pAVP release to decrease water retention. Therefore, pAVP may play a less important etiologic role in adult NP patients, excluding patients with a poor rise in pAVP due to osmotic stimulation. This point of view is consistent with that of Johnson et al., who found no association between NP and both nighttime and early morning pAVP levels [31]. Another recent cross sectional study of ours also confirmed that the mean blood pressure and systolic blood pressure were determinants for the increasing nocturnal diuresis rate and the decreasing daytime diuresis rate in patients with NP, whereas pAVP was neither a determinant for an increasing nocturnal diuretic rate nor a decreasing daytime diuretic rate for them [26]. Taken together, therefore, a reversed diuretic pattern seen in older adult patients with NP might represent a compensatory adjustment to decreased diuresis in the daytime.

Regarding a possible contribution of natriuretic peptides to NP, to date, there has also been intense research [11, 12]. Kaufman found the lack of a correlation between renal urinary output responses and plasma atrial natriuretic peptide levels[32]. A more recent investigation also reported that BNP had no direct intrarenal hemodynamic effects in hypertensive patients [33]. The present study also identified no significant correlation between BNP levels and nocturnal diuresis rates nor between the BNP levels and daytime diuresis rates. Moreover, although the BNP levels didn't differ significantly among the three groups in study 1, the NP group revealed a significantly higher nocturnal diuresis rate than the remaining groups. Also, our previous analysis using a multiple regression model confirmed that BNP was neither a determinant for increasing nocturnal diuresis rates nor for decreasing daytime diuretic rates [26]. Therefore, the prevalence of NP patients with increased BNP levels may not necessarily indicate its etiological importance in NP, but that there is an overlap of the patient population with nocturia and long-standing cardiovascular changes with age.

AVP release from the posterior lobe of the pituitary gland is known to be principally under osmotic regulation, and additionally, is also altered by nonosmotic stimuli originating from carotid and aortic baroreceptors, or volume receptors in the left atrium. The response to AVP-release stimuli from the medulla to the supraoptic and paraventricular nuclei in the hypothalamus, where they are integrated, is tonically inhibited, mediating the secretion of AVP in response to a decrease in the plasma volume and hypotension [29, 30, 34, 35]. On the other hand, an increase in the plasma volume or elevation of blood pressure inhibits p AVP release by the reverse mechanisms as mentioned above, resulting in an alteration of the osmotic threshold for pAVP [15, 17, 30]. An overall upward shift of pAVP responsiveness to plasma osmotic changes seen by hypertonic saline loading for older adults with nocturia may reflect such a volume or baroreceptor–mediated modulatory mechanism of the osmotically sensitive release of AVP.

Furthermore, it has been suggested that a principal transmitter, AVP, may not only act as a feedback regulator of endogenous rhythmicity within the suprachiasmatic nucleus (SCN), which is known to be the biological clock of mammals [36], but also function in the nucleus as a part of the neuronal output system projecting to the rest of the central nervous system, where it potentially regulates a number of physiological rhythms interactively with some other neurotransmitters [37, 38]. In addition, Yan L, and his coworkers indicated that the cellular mechanism for biological clock control may exist in the extra-SCN brain areas such as the cerebral cortex and the caudate-putamen [39]. An aspect of alternation of such an interactive cellular function within the SCN and extra-SCN regions may also contribute to an age-related decay or reverse of the circadian rhythm of diuresis and natriuresis. Recently, it was shown that the daily rhythm of AVP expression in the SCN is possibly modified with aging by examining age-related decay in primates [40]. Goncharuk et al. revealed that immunohistochemical staining for the main neuronal populations, including the vasopressin neurons of SCN, was reduced by more than 50 % in primary hypertensive patients compared with controls [41]. These age-related pathologies might account for an altered diurnal rhythm of diuresis and AVP release in older adults or the elderly, which is likely to be blunted with aging, differently from what is found in healthy children [42], young adults [43], and the elderly without nocturia [27, 28]. It is very interesting to note that neuronal loss of the SCN in patients with multiple system atrophy results in the altered circadian rhythm of plasma AVP release linked to nocturnal polyuria [44]. Further investigations concerning these biological clocks may be useful for understanding the underlying pathophysiology of diuresis and natriuresis in NP patients.

In healthy persons, nocturnal diuresis is generally less than that during the day, which is due largely to a decrease in the excretion of solutes [9-14]. According to the works of Dyer et al. and more previous investigators, it has been found that water and solute excretion in normal persons generally reaches a maximum sometime around midday, with a minimum toward the end of the sleep period, and there were several diseases with a reversed diurnal cycle regarding sodium excretion including hypertension [9-14]. The present study also confirmed a similar reversed diurnal diuresis and natriuresis pattern in patients with NP compared to non-nocturic healthy controls. Moreover, in study 2, the diurnal profile of urinary sodium excretion seemed to be in harmony with the diurnal change of the diuretic rate in patients with NP, and then, the ratios of night- or early morning-to-day of excreted sodium were significantly higher in patients than controls. When the NP group in study 1 was subdivided into the two groups based on gender because of a small sample size in study 2, no significant difference was identified in the ratios between males and females (2.2 vs. 1.6, respectively). Therefore, the ratio could be a useful and convenient parameter to potentially represent underlying conditions in patients with NP, including those with a difficulty in recording of a frequency-volume chart.

Conclusion

Older adults with NP were predisposed to be associated with a reversed circadian diuretic pattern with a concomitant increase in urinary sodium excretion at night rather than in the daytime. Then, nocturnal and daytime diuretic rates were correlated positively and negatively with systolic blood pressure, respectively, only in patients with NP. Although both the nocturic groups with NP and non-NP showed an altered circadian variation with a diminished nocturnal plasma AVP release without a nocturnal surge overall, the former revealed a significantly higher systolic blood pressure than the latter on average. The relationship of pAVP and Posm using hypertonic saline loading also suggested that NP patients were predisposed to hypervolemic or hypertensive conditions and that their pAVP release might be inhibited by a volume- or pressure-related regulatory mechanism, in addition to a diminished osmotic responsiveness by an overall upward shift of pAVP and Posm values. Neither pAVP at midnight nor in the early morning were associated with nocturnal diuretic rates. Taken together, the low pAVP levels at night may not always play an etiological role in NP, but reflect a regulatory effect to adjust the body fluid and sodium balance, excluding patients with a flattened response curve to osmotic change. Further studies are needed in the future. The increased ratios of night- or early morning-to-day of excreted urinary sodium may be useful as a potential indicator to represent the underlying pathophysiology of patients with NP.

Acknowledgments

The author gratefully acknowledges the technical staff, Masataka Hidenoi, Etsuyo Morishita, and Motomi Kimura for their help in biochemical measurement of the collected specimens.

References

[1] Yoshimura, K., Terada, N., Matsui, Y., Terai, A., Kinukawa, N. and Arai, Y. (2004) Prevalence of and risk factors for nocturia: analysis of a health screening program. *Inter. J. Urol., 11,* 282-287.

[2] Tikkinen, K.A.O., Tammela, T.L.J., Huhtala, H. and Auvinen, A. (2006) Is nocturia equally common among men and women? A population based study in Finland. *J. Urol., 175,* 596-600.

[3] Liew LC, Tiong HY, Wong ML, PngDC, Tan JK. (2006) A population study of nocturia in Singapore. *BJU Inter., 97,* 109-112.

[4] Hakkinen JT, Hakama M, Shiri R, Auvinen A, Tammela TLJ, Koskimaki J. (2006) Incidence of nocturia in 50 to 80-year-old finnish men. *J. Urol., 176,* 2541-2545.

[5] Homma Y, Yamaguchi O, Hayashi K, et al. (2006) Epidemiologic survey of lower urinary tract symptoms in Japan. *Urology, 68,* 560-564.

[6] van Kerrenbroeck, P., Abrams, P., Chaikin, D., Donovan J., Fonda, D., Jackson, S., et al. (2002) The standardization of terminology in nocturia: report from the standardization subcommittee of the International Continence Society. *Neurourol. Urodyn., 21,* 179-183.

[7] Weiss JP, Blaivas JG. (2002) Nocturnal polyuria versus overactive bladder in nocturia. *Urology, 60,* (Suppl 5A), 28-32.

[8] Asplund, R (1999) Mortality in the elderly in relation to nocturnal micturition. *BJU Int., 84,* 297-301.

[9] Dyer AR, Stamler R, Grimm R et al. (1987) Do hypertensive patients have a different diurnal pattern of electrolyte excretion? *Hypertension, 10,* 417-424.

[10] Kirkland JL, Lye M, Levy DW, Banerjee AK. (1983) Patterns of urine flow and electrolyte excretion in healthy elderly people. *Br. Med. J., 287,* 1665-1667.

[11] Matthiesen TB, Rittig S, NØrgaard JP, Pedersen EB, Djurhuus JC. (1996) Nocturnal polyuria and natriuresis in male patients with nocturia and lower urinary tract symptoms. *J. Urol., 156,* 1292-1299.

[12] Carter PG, Cannon A, McConnell AA, Abrams P. (1999) Role of atrial natriuretic peptide in nocturnal polyuria in elderly males. *Eur. Urol., 36,* 213-220.

[13] Robertson GL, Norgaard JP. (2002) Renal regulation of urine volume: potential implications for nocturia. *BJU Inter., 90,* Supple 3, 7-10.

[14] Chang S, Lin ATL, Chen K, Chang LS. (2006) Multifactorial nature of male nocturia. *Urology, 67,* 541-544.

[15] Robertson, G.L., Shelton, R.L. and Athar, S. (1976) The osmoregulation of vasopressin. *Kidney International, 10,* 25-37.

[16] Oiso, Y., Iwasaki, Y., Yamauchi, K., Takatsuki, K. Tomita, A., Kamoi, K. et al. (1986) Clinical assessment of posterior pituitary function by direct measurement of plasma vasopressin levels duing hypertonic saline infusion. *Nippon Naibunpi Gakkai Zasshi 62,* 608-618.

[17] Quillen EW, Cowwley AWJr. (1983) Influence of volume changes on osmolality-vasopressin relationships in conscious dogs. *Am. J. Physiol., 244,* H73-H79.

[18] Natsume O. (2006) A clinical investigation of nocturnal polyuria in patients with nocturia: a diurnal variation in arginine vasopressin secretion and its relevance to mean blood pressure. *J. Urol., 176,* 660- 664.

[19] Homma, Y., Yamaguchi, O., Kageyama, S., Nishizawa, O., Yoshida, M. and Kawabe, K. (2000) Nocturia in the adult: classification on the basis of largest voided volume and nocturnal urine production. *J. Urol., 163,* 777-781.

[20] Hirayama A, Fujimoto K, Akiyama T, Hirao Y. (2006) Decrease in nocturnal urinary levels of arginine vasopressin in patients with nocturnal polyuria. *Urology, 68,* 19-23.

[21] Drake MJ, Mills IW, Noble JG. (2004) Melatonin pharmacotherapy for nocturia in men with benigh prostatic enlargement. *J. Urol., 171,* 1199-1202.

[22] Sewerynek E. (2002) Melatonin and the cardiovascular system. *Neuro. Endocrinol. Lett., 23,* Suppl 1, 79-83.

[23] Rembratt A, Weiss JP, Robertson GL. (2001) Pathogenesis of nocturia in the elderly: relationship of functional bladder capacity to nocturnal urine output. *J. Urol., 165,* Supple, 250.

[24] Asplund R. (2002) Diuresis pattern, plasma vasopressin and blood pressure in healthy elderly persons with nocturia and nocturnal polyuria. Neth. *J. Med., 60,* 276-280.

[25] Bulpit CJ, Fletcher AE, Thijs L et al. (1999) Symptoms reported by elderly patients with isolated systolic hypertension: baseline data from the SYST-EUR trial. *Age Ageing, 28,*15-22.

[26] Natsume O. (2007) Diuretic pattern in adults with nocturnal polyuria: The possible contribution of blood pressure to the worsening of nocturnal polyuria. *Int. J. Urol., 14,* 822-27.

[27] Graugaard-Jensen C, Ritting S, Djurhuus C. (2006) Nocturia and circadian blood pressure profile in healthy elderly male volunteers. *J. Urol., 176,* 1034-1039.

[28] Asplund, R. and Åberg, H (1991) Diurnal variation in the levels of antidiuretic hormone in the elderly. *J. Inter. Med., 229,* 131-134.

[29] Robertson GL, Athar S. (1976) The interaction of blood osmolality and blood volume in regulating plasma vasopressin in man. *J. Clin. Endocrinol. Metab., 42,* 613-620.

[30] Grindstaff RR Cunningham JT. (2001) Cardiovascular regulation of vasopressin neurons in the supraoptic nucleus. *Exp. Neuro., 171,* 219-226.

[31] Johnson TM, □, Miller M, Pillion DJ and Ouslander JG. (2003) Arginine vasopressin and nocturnal polyuria in older adults with frequent nighttime voiding. *J. Urol., 170,* 480-484.

[32] Kaufman S. (1990) Renal and hormonal responses to prolonged atrial stretch. *Am. J. Physiol., 258,* R1286-R1290.

[33] van der Zander K, Houben AJHM, Kroon AA et al. (2003) Does brain natriuretic peptide have a direct renal effect in human hypertensives? *Hypertension, 41,* 119-123.

[34] Blessing, W.W. and Willoughby, J.O. (1985) Inhibiting the rabbit caudal ventrolateral medulla prevents baroreceptor-initiated secretion of vasopressin. *J. Physiol., 367,* 253-265.

[35] Gieroba, Z.J., Fullerton, M.J., Funder, J.W. and Blessing, W.W. (1992) Medullary pathways for adrenocorticotropic hormone and vasopressin secretion in rabbits. *Am. J. Physiol., 262,* R1047-R1056.

[36] Moore RY. (1996) Entraint pathway and the functional organization of the circadian system. *Prog. Brain Res, 111,* 103-119.

[37] Ingram CD, Ciobanu R, Coculescu IL, Tanasescu R, Coculescu M, Mihai R. (1998) Vasopressin neurotransmission and the control of circadian rhythms in the suprachiasmatic nucleus. *Prog Brain Res, 119,* 351-364.

[38] Buijs RM, Hermes MH, Kalsbeek A. (1998) The suprachiasmatic nucleus-paraventricular nucleus interactions: a bridge between the neuroendocrine and autonomic nervous system. *Prog. Brain Res., 119,* 365-382.

[39] Yan L, Miyake S, Okamura H. (2000) Distribution and circadian expression of dbp in SCN and extra-SCN areas in the mouse brain. *J. Neurosci. Res., 59,* 291-295.

[40] Cayetanot F, Bentivoglio M, Aujard F. (2005) Arginine-vasopressin and vasointestinal polypeptide rhythms in the suprachiasmatic nucleus of the mouse lemur reveal aging-related alterlations of circadian pacemaker neurons in a non-human primate. *Eur. J. Neurosci., 22,* 902-910.

[41] Goncharuk, V.D., Van Heerikhuize, J., Dai, J., Swaab, D.F. and Buijs, R.M. (2001) Neuropeptide changes in the suprachiasmatic nucleus in primary hypertension indicate functional impairment of the biological clock. *J. Comp. Neurol., 431,* 320-330.

[42] Rittig, S., Knudsen, U.B., NØrgaard, J.P., Pedersen, E.B. and Djurhuus, J.C. (1989) Abnormal diurnal rhythm of plasma vasopressin and urinary output in patients with enuresis. *Am. J. Physiol., 256,* F664-F671.

[43] George, C.P.L., Messerli, F.H., Genest, J., Nowaczynski, W., Boucher, R., Kuchel, O. and et al. (1975) Diurnal variation of plasma vasopressin in man. *J. Clin. Endocrinol. Metab., 41,* 332-338.

[44] Ozawa, T., Oyanagi, K, Tanaka, H, Horikawa, Y, et al. (1998) Suprachiasmatic nucleus in a patient with multiple system atrophy with abnormal circadian rhythm of arginine-vasopressin secretion into plasma. *J. Neurol. Sci., 154,* 116-121.

In: Circadian Rhythms and Health Research Trends
Editor: Alfredo V. Lignelli, pp. 99-116

ISBN: 978-1-60021-875-0

Chapter V

Acquisition of Circadian Rhythms by Neonates: A Review on the Potentials of Exfoliated Cells to Study Gastrointestinal Tract Maturation and Clockwork

Bertrand Kaeffer*
UMR 1280 Phan,
INRA and University of Nantes
CHU - Hôtel DIEU HNB1
Place Alexis Ricordeau
44093 Nantes Cedex 1 (France, EU)

Abstract

Feeding and light are the two main synchronizers of human circadian clocks. In adults, the oscillator located in the suprachiasmatic nuclei may be the center of the network and the unique oscillator entrainable by light / dark cycles. In contrast with the central clock, the peripheral clocks could not be entrained by sunlight but are under the dependence of the central clock to perceive the photic signal through humoral factors like glucocorticoids, transforming growth factor alpha or prokineticin-2. Peripheral clocks may be entrained by signals issued by the SCN but could be uncoupled under conflictual conditions to follow the entrainment of synchronizers more relevant to their homeostasis. Circadian rhythms are crucial to the acquisition of long-term homeostasis by tuning individual cellular oscillators with environmental cues. The fetus (and consequently the neonate) may be entrained not only by maternal signals induced by the light / dark cycles but also by the rhythmicity of nutrient supplies from the mother which may entrain rhythmic variations of blood parameters, of locomotor activities and of the body's rectal temperature. At birth, premature neonates are exposed to a timeless environment and

* Email: bertrand.kaeffer@nantes.inra.fr

have to organize their rhythmicities to achieve the organ's maturation. Studies on rat neonates and primate babies propose two main synchronizers: the light / dark cycles driving the suprachiasmatic nuclei oscillator and a synchronizer related to feeding and delayed of 12 hours. Molecular data obtained on murine embryos are in favor of the hypothesis that babies of Mammals are born with a functional circadian system but have to tune their clocks with environmental synchronizers. The ontogenic development of the circadian system can be modified or imprinted by changing environmental conditions, the main consequence being on nursing of preterm infants who are reared in the relatively « timeless » conditions of the neonatal intensive care nursery. However, direct comparisons between laboratory mammals and humans are difficult because of divergent milk compositions, maternal care or maturity of the babies as well as the difficulty in obtaining biopsies.Without a proper non-invasive methodology to recover mucosal cells, there is no information about development of infant peripheral clocks, and the acquisition of rhythms are studied on sleep/rest actigraphy, electroencephalograms, heart beat and body temperature. Exfoliated cells may constitute surrogates of tissue biopsies so it is useful to obtain cells from the gastrointestinal area with some relevance to the clocks' biology. A better understanding of the acquisition of circadian rhythms by gastrointestinal epithelium may be relevant to studies on metabolic syndrome which has been related to clock genes' dysfunction and on the imprinting of neonates metabolism circuitry with potential health consequences during adult life.

Keywords: *neonate, premature, clock, stomach, intestine, exfoliated cells.*

Introduction

Circadian clocks are playing a central role in integrating different periodic signals from the environment (day / night cycles, feeding, social and physical activities) or in the control of life processes (cell division, differentiation, apoptosis). Their main function is to adapt the organism to external conditions, and they entrain many processes including day / night cycles, body temperature regulation, neuroendocrine and digestive functions. Circadian rhythmicity of the organism relies on a network of clocks (i. e. oscillators): one for each organ or tissue functioning as an entity. The day/night cycles are the main sychronizer for the organism. Lighting cues are captured and integrated by melanopsin cells of the retina (Fu et al., 2005) which, according to the diurnal or nocturnal nature of the mammalian species, reset the central clock located in the suprachiasmatic nuclei. A fast view of the system is to imagine a web of equivalent clocks driven by a central one located in suprachiasmatic nuclei (SCN). The SCN clock is the center of the network and the unique oscillator entrained by light cycles. All others are entrained by cues from the central clock but are able to uncouple to respond to their main synchronizers. Peripheral clocks are local networks built up on the tuning of independent cell oscillators in tissues and organs. Central and peripheral clocks are characterized by a stable oscillatory period of around 24 hours and entrainability (the sensitivity of the oscillation's phase to periodic cue i. e. gating). Limit cycle oscillations are stable to minor perturbations of regulator concentrations and they persist over time even when the organism is not subjected to periodic driving forces (e. g. light exposure, feeding rhythms). In addition, clocks must be buffered against changes in temperature which typically speed up or slow down macromolecular association but the mechanism of such a regulation is still unknown (Ruby et al., 1999; Brown et al., 2002; Tsuchiya et al., 2003). Among synchronizers, feeding which includes the temporal window of food access and the nature (or composition) of the meal is considered as powerful as the alternance of day and night (Pardini and Kaeffer, 2006; Stratman and Schibler, 2006). However gating and uncoupling of peripheral clocks with the central one are largely unexplored and need new methods to address the question. Circadian rhythmicity is also dependent on a lot of individual parameters like sex (Abizaid et al., 2004) or adolescence status (Carskadon et al., 1998) as well as age of birth (full-term or preterm infants). At birth, full-term neonates have organized rhythms necessary to heart beat and maintenance of body temperature (Weinert et al., 2005) but preterm infants are discharged to parents only when they show consistent weight gain, maintain normal body temperature in an open crib, and are with mature cardiorespiratory function (Kelly et al., 2006a, b, c). In that respect problems of rhythmicity might be crucial in preterm infants who have to develop their organs in order to survive. Neonates are believed to harbor clock gene compound as their suprachiasmatic nuclei are organized from mid-gestation. Although the molecular and genetic determinants of clock ontogenesis are very complex, significant advances in the understanding of events contributing to clockwork development have been made. However there is no information about the peripheral clocks of preterm infants due to the difficulty of obtaining biopsies.

In this chapter, the current concept and studies in ontogenesis of clocks are introduced and a new strategy using exfoliated cells to study peripheral clocks of the stomach and intestine is proposed.

Ontogenesis of Clocks

Development of peripheral clocks depends on maturation of the organ housing the clock as well as on maturation of the molecular clockwork. The first appearance of molecular oscillations might be thus highly organ-specific and/or species-specific. A clear-cut example is the difference between oscillations of clock genes in the heart (Sakamoto et al., 2002) and the liver (Sladek et al., 2007). Before discussing data about clock genes, a brief molecular description of clockwork is needed. The mammalian genome encodes at least a dozen genes directly involved in the regulation of the feedback loops constituting the circadian clock. The positive limb of the loop is driven by two transcription factors, CLOCK (King et al., 1997; Gekakis et al., 1998; or a close analog called NPAS2 (Reick et al., 2001)) and BMAL1 (Gekakis et al., 1998; Hogenesch et al., 1998) which are basic Helix-Loop-Helix proteins with a PER-ARNT-SIM (PAS) domain; the equilibrium being post-translationally regulated by small ubiquitin modified proteins on BMAL1 through CLOCK-induced activity (Cardone et al., 2005). A first negative limb involves the protein regulators PERIOD 1 (Sun et al., 1997; Tei et al., 1997), PERIOD 2 (Albrecht et al., 1997; Shearman et al., 1997) or PERIOD 3 (Zylka et al., 1998). These proteins do not directly bind to DNA but they have a PAS domain to build complexes with CRYPTOCHROME 1 or 2 (van der Horst et al., 1999; Griffin et al., 1999, Kume et al., 1999). Other proteins like NONO and WDR5 are able to modulate PERIOD activity (Brown et al., 2005). The proteins CASEIN KINASE I epsilon or delta are involved in the turn-over of PERIOD molecules by phosphorylating these proteins and regulating their nuclear translocation (Shearman et al., 2000; Vielhaber et al., 2000; Camacho et al., 2001; Eide et al., 2002; Akashi et al., 2002; Lee et al., 2004). A second negative limb involves members of the retinoic acid-related orphan nuclear receptors, RORA (Sato et al., 2004) and REV-ERB-ALPHA (Preitner et al., 2002), which repress and modulate bmal1 expression. An additional loop of regulation driven by two Differentiated Embryon Chrondrocyte (DEC) proteins (Honma et al., 2002; Hamaguchi et al., 2004; Sato et al., 2004b) has been proposed. These oscillator systems control the activity of clock-controlled-genes (CCGs) that function in the rate-limiting steps of various biological pathways (Kornman et al., 2001; Storch et al., 2002; Ueda et al., 2002; Panda et al., 2002; Duffield et al., 2003).

In absence of any environmental cue, the protein PERIOD2 is believed to reboost the circadian cycle by stimulating the transcription of *bmal1* gene leading to the dimerization with CLOCK and to the reinduction of a new circadian cycle (Oishi et al., 1998a, Oishi et al., 1998b, Oishi et al., 2003, Bae et al., 2001, Zheng et al., 1999, Zheng et al., 2001, Lee et al., 2001, Yu et al., 2002). The circadian regulation of *period2* gene has also been linked *in vivo* to tumor suppression and DNA damage response (Fu et al., 2002) or the regulation of bone formation by leptin (Fu et al., 2005). *Period1* and *2* genes are molecular candidates to study the adaptive response of the cellular metabolism with environmental stressors and they have been associated with the signaling pathway allowing a cell to anticipate and integrate changes in its normal environment. Recently no relation between period2 gene mutation and food anticipatory activity has been found but if anticipation of mealtime is abolished in mutant mice, the peripheral synchronization by feeding cycles was unaffected (Feillet et al., 2006).

In that respect, the promotor of *period1* gene may be a sensor and an integrator of physiological variations in the organism with the aim to adjust the clock to the environment (Motzkus et al., 2000, Motzkus et al., 2002). Recently, Yamamoto et al., 2005 have shown that the promoter of *period1* gene contains at least a glucocorticoid-responsive element along with E-boxes for the binding of CLOCK:BMAL1 heterodimers related to the rhythmic expression of transcripts and with cAMP-responsive elements for *period1* transient expression. The regulation of *period1* gene is considered crucial for a quick adaptation to variations of environmental parameters (Brown et al., 2005a, Brown et al., 2005b) and may be modulated by food intake (Scheving et al., 2000).

Fetal Circadian Rhythmicity

Fetal circadian rhythmicity would be entrained not only by maternal cues induced by the day / night cycles but also by daily rhythms of nutrients from mother inducing rhythmic variations of blood metabolites, of motor activity and of body temperature (Weinert et al., 2005). Most data in the field have been obtained from laboratory rodents (mouse or rat pups), lambs and non-human primates.

During the murine development as early as at the embryonic day 19, all clock genes (*period1, period2, cryptochrome1, bmal1, clock*) are expressed and circadian rhythmicity has been found at postnatal day 3 for *period1*, *period2*, *cryptochrome1*, *bmal1* but not *clock*, the circadian rhythmicity of the latter being detected only at postnatal day 10 (Sladek et al., 2004). In small rats, the main model in neonatalogy, the beginning of rhythmic oscillations in liver may begin as soon as the second day after birth for rev-erb-alpha and bmal1 transcripts to reach a complete organization at day 30 (Sladek et al., 2007).

In primates, mother and infant are in synchrony or at birth out of phase. Observations realized on brain sections of human fetus or born-dead and on newborns of primates suggest that there is a neuronal zone organized as suprachiasmatic nuclei fully functional at mid-gestation (Rivkees et al., 2003). Likewise, circadian rhythmicity has been detected in fetus of ewe by 2-deoxy-glucose incorporation (Seron-Ferré et al., 2002).

Neonates and Clocks

Although it is known that the primate circadian pacemaker begins oscillating during gestation, robust circadian rhythms in activity or humoral substances are not generally apparent until 2 months after birth. The human suprachiasmatic nucleus increases its cell number and size around the time of birth according to enumeration performed on 41 subjects ranging in age from 27 weeks of gestation up to 30 years (Swaab et al., 1990). Moreover it has been shown that melatonin rhythmicity develops more slowly in preterm infants than in full-term infants, and infants born after an Intra Uterine growth restriction are the last to develop rhythms (Kennaway, 2002). However, in human and nonhuman primate newborns, day-night differences in rest-activity patterns can be detected shortly after birth, with more absolute movement occurring during the day than the night. Day-night differences in rest-activity patterns in synchrony with the light-dark cycle are generally not apparent until approximately 20 days after discharge of preterm infants from the hospital (Rivkees et al., 2003).

Preterm infants are taken in charge in Neonatal Intensive Care Unit to ensure survival and psychomotor and physiological delayed development. In such units, infants are frequently maintained in dim arhythmic light (1-5 lux) during the day. This practise is justified to preserve a cosy ambiance and to mimick uterus environment. Neonatalogists using the Neonatal Individualized Developmental Care Assessment Program (Als et al., 1994) are suggesting that infants should be dark-reared. However, some studies propose to modify this procedure to use cycled lighting. The lighting policies in the Neonatal Intensive Care Unit are still unresolved and under investigation (Haumont and Hansen, 2006). Continuous bright light has been in use to facilitate observations and care with some concerns regarding the consequences on child's health. However, negative effects previously associated with continuous bright light, including increased activity, decreased sleep, and more apnea and bradycardia, were only observed at higher illumination levels (400-900 lux) than the standard nursery light levels (around 200-225 lux). More studies were realized to evaluate effects of a Cycled Light in the Neonatal Intensive Care Unit using continuous bright light for comparison. Infants with a mean gestational age of 28 weeks gained more weight, were fed orally sooner, and had fewer ventilator days when they were in a Cycled Light environment (Miller et al., 1995). In convalescent preterm infants (mean gestational age higher than 31 weeks), Mann et al (1986) found that those in Cycled Light exhibited more weight gain and sleep 2 to 3 months after discharge and spent less time feeding in the nursery than infants in constant light. Likewise, under cycled lighting, difference in gain weight have been reported with a net gain for neonates of less than 31 weeks' postmenstrual age (Brandon et al., 2002) and no gain for neonates between 32 and 34 weeks' postmenstrual age (Rivkees et al., 2004). Anyway, Rivkees et al., 2004 are proposing that during the 2 weeks before discharge to home, cycled lighting is preferable to continuous dim lighting in the care of premature infants. Rivkees et al (2004) have used actigraphy to monitor rest-activity patterns for extended periods, and found that circadian phase can be detected in infants who are exposed to cycled lighting as early as a 37 weeks' postmenstrual age. These observations are opening speculations about nursing (adaptation of time of meals for each child or at fixed time of meal with or without a rhythmic lighting) which could have a deep influence on development of central circadian clock as well as on the maturation of gastrointestinal tract of premature infants with short-term and long-term consequences on general health.

Studies with invasive methodologies have been realized in laboratory rodents to explore synchronizing potentials of light and feeding. Light-induced c-fos expression in adult rats is restricted to a certain circadian time, namely the subjective night, when light pulses also cause phase shifts of circadian rhythmicity. Thus the circadian « gating » in rats is a prerequisite of light-dark entrainment: light exposure at any time of the day induced high c-Fos production in the SCN of 3-day-old rat pups, whereas at 10 days of age a circadian « gating » was present (Sumova et al., 2006). Sensitivity of the SCN clock to stress diminishes with postnatal age (Takahashi et al., 2001). When pups are deprived of their mothers during the light phase, i.e. at the time when they usually suckle milk, the rhythmic SCN expression of Perl and Per2 genes is completely phase-reversed within six days (Kornhauser et al., 1992). The complete absence of the mother at the time when pups are usually fed may be, however, a strong stressor for pups altering expression of stress related genes, such as corticotropine releasing hormone, glucocorticoid receptor and AVP (Ohta et al., 2003).

The capacity of non-photic synchronizers to entrain peripheral circadian clocks (and somewhat central one) has been shown on rodents by Damiola et al., 2002 and Stokkan et al., 2002. It is known from studies on human newborns that feeding can influence the daily rhythm of heart rate (Weinert et al., 1997; Weinert 2005). Feeding might not only be a masking factor but also a synchronizer, at least in the circadian range. The ontogenetic development of the circadian system can be modified or imprinted by changing environmental conditions. In premature infants, ontogenesis of these oscillators depends on functional maturity of a given organ. Inducing and synchronizing capacity of feeding on peripheral circadian clocks is unexplored in the neonate but many works have been realized in rat (Pardini and Kaeffer, 2006) among which the recent demonstration of an entraining effect of breast-feeding on the circadian rhythmicity of the clock in the liver of small rats (Sladek et al., 2007).

However, direct comparisons between laboratory mammals and humans are difficult because of divergent milk compositions, maternal care or maturity of the babies at full-term birth. A clear-cut example is the acquisition of circadian rhythms driven by light / dark cycles. The babies of rats or mice born with closed eyelids remain relatively immature up to eyes opening in contrast with the piglets or human babies born with open eyes that are considered as more mature. However rat pups are the model of choice of prematurity because small rats are born immature over the first day of life. As an animal model, they are amenable to a cheap and totally controlled environment as pup-in-cup (Yeh and Yeh, 1993). This model is used to study problems associated with prematurity (Neu et al., 2007) but according to recent results on the ontogenesis of peripheral clocks in small rats (Sumova et al., 2006), new studies to compare oscillations of clock genes (and the gating phenomenon) are needed in small rats as well as in preterm infants.

In humans, development of non-invasive methods is crucial to improve nursing care in this population by allowing the evaluation of the impact of the feeding method (with a scheduled timing or according to child's demands) as well as cycled lighting on the maturation of digestive tube of neonate. Non-invasive methods include tomography for central nervous system and metabolic markers like citrulline (Rivkees et al., 2003). However, immunonomagnetic isolation and purification have open the possibility to recover exfoliated cells from digestive epithelium allowing studies on the maturation of gastrointestinal epithelial cells.

Potentials of Exfoliated Cells to Study Peripheral Clocks

Investigation of gene expression in the gastrointestinal (GI) tract commonly relies on the analysis of tissue biopsies. However, the performance of tissue biopsies is unethical in human neonates and development of non-invasive methodology to obtain cells from GI tract is deeply needed. Exfoliated or sloughed cells are naturally lost by mammals in their environment, and have been used as surrogates for target cells to predict the response to bioactive food components (Davis et al., 2003), or to detect biomarkers of cancer in children (Holland et al., 2007) and adults (Bandelatova et al., 2002). Yet the feasibility of this approach has not been assessed for the exploration of neonatal GI tract.

Exfoliated cells have long been found to be shed in abundance in the stools of adult patients with colorectal cancer. These cells are non-apoptotic, and markers in exfoliated cells originating from the tumor are actively studied as a potential tool for cancer screening and prevention (Albaugh et al. 1992; Osborn and Ahlquist, 2005). Although the recovery of exfoliated colonocytes from the stools of normal adults has been described (Bandatelova et al., 2002), and exfoliated cells from the urinary tract of infants have been studied in the past (Robine et al., 1988), we were not aware of previous studies in preterm infants.

To gain access to cellular material isolated from preterm infants, we have adapted a technique to isolate exfoliated cells from gastric residual fluid aspirates and stools (Kaeffer et al., 2006 and 2007) allowing to realize proteomic and metabonomic analyses to evaluate gastrointestinal maturation. Gastrointestinal maturation is the acquisition by epithelial tissue of a capacity to withstand its internal conditions to progressive or sudden changes of the environment but also to anticipate happening of events reoccuring with a periodicity (Moore-Ede et al., 1986). Under that assumption, maturation implies acquisition by epithelium of nutrients transporters and water and ions exchangers (Thiesen et al., 2000) but also acquisition of a circadian rhythmicity which refer s to the concept of clock.

The presence of clock components has been demonstrated in the digestive tract (oral mucosa (Bjarnason et al., 2001), liver (Zylka et al., 1998, Sakamoto et al., 1998) and pancreas (Damiola et al., 2000). We have shown that the main molecular components (*period1, period2, clock*) of an intestinal clock are expressed by colonocytes of healthy human donors as well as by colon carcinoma cells (Pardini et al., 2005). More recently, peripheral circadian clock molecular components have been found in gastric cells and colonocytes of preterm infants (Kaeffer et al., 2006 and 2007).

Many rhythms which could be controlled by circadian clocks are observed in the intestine mainly cellular proliferation (Buchi et al., 1991, Marra et al., 1994), migration (Qiu et al., 1994) or apoptosis (Iwakiri et al., 2001). Circadian activities in the intestine are well known and concern the food intake, digestion (and associated enzymes activities), cellular proliferation, migration of epithelial cells along the physiological unit like crypts or villus, and apoptosis. Several genes have been described as regulated according to a circadian rhythmicity (Harmer et al., 2001): sodium / glucose cotransporter implicated in the absorption of dietary glucose and galactose, two transcription factors (hepatocyte nuclear factors HNF-1 alpha and beta), and uroguanylyl and guanyline which are intestinal receptors for guanylyl cyclase. Diurnal rhythmicity of intestinal transporters SGLT1 and PEPT1 is regulated by feeding (Pan et al., 2004) as well as those of innate immunity molecular components in the mouse jejunum (Froy et al., 2007). Studies on adult mice mutated for clock genes have shown that circadian clock are playing a crucial role in energy balance (Turek et al., 2005) and participate in the evolution toward a metabolic syndrome (Reddy et al., 2006). Up to now, liver has been the organ of choice to study peripheral clock functioning but a better understanding of circadian rhythms implies to integrate also complex tissue architecture and interfacing with external environment like in lung or gut.

The technical aspect of our methodology is summarized in figure 1. Our current studies demonstrate that a sufficient number of exfoliated cells can be recovered from gastric aspirates in preterm infants to measure the expression of specific genes of interest (Kaeffer et al., 2006 and 2007). This innovative approach may prove a useful, non invasive tool to assess the expression of genes of interest in the GI tract of preterm infants, as a surrogate for true biopsies. In contrast with reports on urocytograms (Robine et al., 1988), neither gestational age nor gender affected the amount of exfoliated cells lost in gastric residual fluid aspirates by premature babies. The collection of exfoliated cells would potentially be usable to delineate the time course of gene expression in infants by serial analysis of gene expression, a method already used to compare the transcriptomes of normal and colon carcinoma cells in human (Zhang et al., 1997). However, more works are needed to evaluate RNA status for RT-PCR in extracts of neonate exfoliated cells. In that respect, works on evaluation of RNA status of postmortem human brain tissue indicate that ribosomal and messenger RNA have different degrees of degradation depending on postmortem conditions and extraction procedure (Miller et al., 2004).

As collection of gastric residual fluid aspirates are performed as part of the nursing care routine every three hours in many neonatal intensive care unit (NICU), the technique may be highly relevant for time series experiments in chronobiology to reassess recent demonstration of the influence of lighting in neonatal intensive care units (Rivkees et al., 2004, Weinert 2005) on neonates gastrointestinal maturation (Pardini and Kaeffer, 2006). With stools, there is little way to improve cell sampling as feces transit is different from one child to another and as we cannot proceed like with adults during routine colonoscopy (Davidson et al., 2003). Consequently, the technique cannot be used to detect oscillations of clock genes in colonocytes but it can be used to detect expression levels of genes related to intestinal maturation.

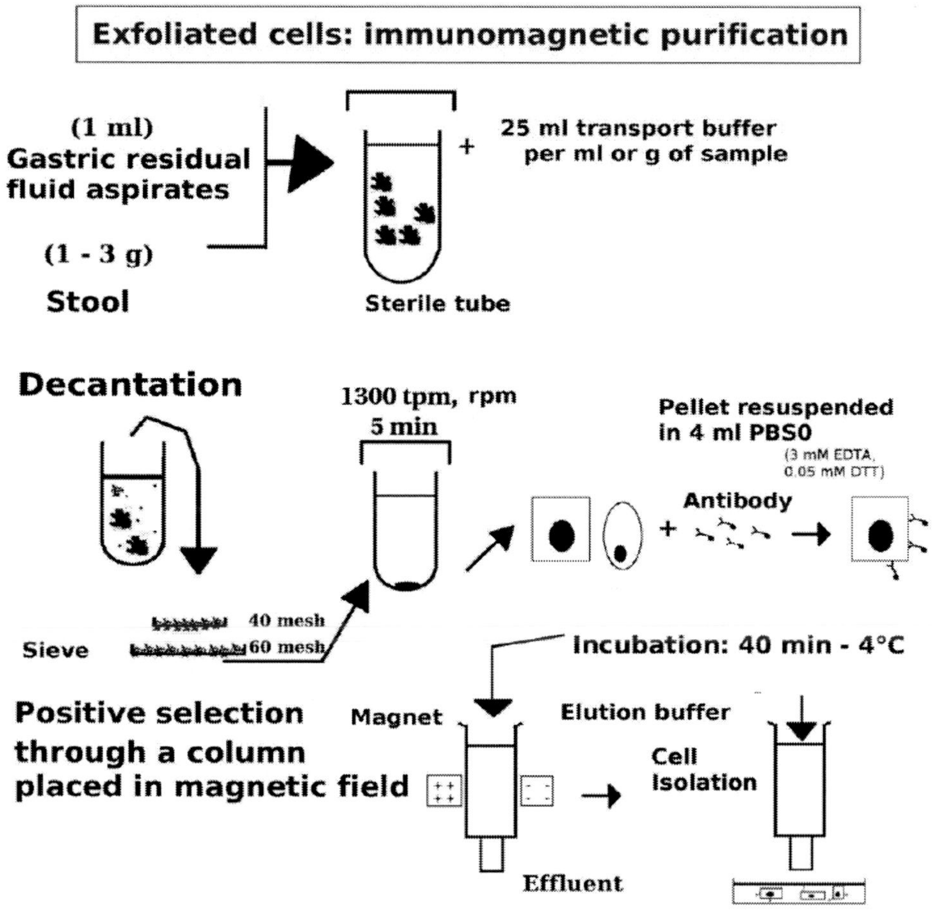

Figure 1. Isolation of exfoliated cells from gastric residual fluid aspirates and stools of preterm infants by immunomagnetic capture and purification. The method, originally described on adult fecal samples by Bandaletova et al., 2002, was adapted by Kaeffer et al., 2006 and 2007. Gastric residual fluid aspirates or stools are diluted in transport buffer (Phosphate-buffered saline solution containing 0.05 mM DTT and 1 g/l D-glucose). The next step can be realized 30 to 60 min later by decantation through a sieving device. After centrifugation, the pellet is resuspended in incubation buffer containing 3 mM EDTA and 0.05 mM DTT to allow cell dissociation and capture by antibodies directed against human epithelial cell surface antigen like monoclonal antibodies described by Moldenhauer et al., 1987.

Progressive adaptation of preterm infants to enteral feeding and lighting environment are, in intensive care units, crucial elements in the progressive adaptation to a good circadian autonomy by preterm infant (consolidated sleep / activity cycles, gastrointestinal maturity without any diarrhea). This autonomy is dependent on each organ's maturity and on adaptation potential of each infant such reasoning justify the NIDCAP system (Als et al., 1994; Sizun et al., 2006). NIDCAP is a rational technique based on close observations of infants' behavior before, during and after care.

Our new technique to isolate exfoliated cells in gastric residual fluid aspirates and stools (Kaeffer et al., 2006 and 2007) opens the way to non-invasive cellular sampling in addition to more classic studies on "gastric aspirates" and "stools" matrices by proteomics or metabonomics to obtain biological fingerprints. A strong point in favor of such non-invasive methodology is that gastric residual fluid aspirates are collected by the nurse every three to four hours. This timing is opening the way to study circadian rhythmicities in conditions similar to those of oral mucosa of healthy human volunteers (Bjarnason et al., 2000). The isolation and phenotyping of exfoliated cells from gastric residual fluid aspirates and fecal samples is opening the way to direct estimation of gastrointestinal maturation (DNA, mRNA, proteins) of preterm infants. By general methodology such as proteomic or metabonomic, we believe that one can obtain biological identities (bidimensional 2D protein map and/ or metabolomic fingerprints) with non-invasive sampling on groups of neonates. These measurements may be used to propose recommendations relating to nurse care and environment on digestive tract maturity of preterm infants and their future health (like prevention of metabolic syndrome).

Acquisition of a fully mature intestine results from a delicate balance between the cell proliferation and the cellular loss which might be revealed indirectly by different rates of exfoliation. With such general methodology we hope for new information relevant to: (1) occurrence of anionic transporters (Cl-) on the gastric cell, Na+/H+ ion exchangers on colonocyte and on the level of guanylins allowing to appreciate gastrointestinal maturity through its capacity to absorb water and, *in fine*, on its capacity to maintain hydration. (2) gastric markers of food intake (ghrelin, leptin) which are scarce in preterm infants. In adults, leptin is well-known for being synthesized by gastric cells (Bado et al., 2000) and being submitted at the plasma level, to a circadian regulation (Schoeller et al., 1997) and several studies have shown the existence of a slow-down of ghrelin levels after a meal (Kim et al., 2004; Schmidt et al., 2006). Last but not least, we may be able to follow the biological effects of the two synchronizers (lighting and meal frequency), on the acquisition of a central circadian rhythm (specialy wake / sleep cycles) by actigraphic or clinical recordings as reported by Brandon et al., 2002 and Rivkees et al., 2004.

Perspectives

A major stake of neonatalogy is to ensure a growth weight tripling infants' weight to allow a healthy development. Over the last fifty years, improvements of the care of preterm infants were the ability to provide nutrition intravenously, the capacity to maintain normal body temperature by the use of incubators and the miniaturization of blood samples needed to determine blood gases, serum electrolytes, glucose, calcium, bilirubin and other biochemical measurements (Philip, 2005). Observations realized in NIDCAP units are opening speculations about nursing (adaptation of time of meals for each child or at fixed time of meal with or without a rhythmic lighting) which could have a deep influence on development of the gastrointestinal tract of premature infants and general health. However, research in neonatalogy is hampered by several difficulties, some ethical ones. Development of non-invasive methods is crucial to improve nursing care in Neonatal Intensive Care Units. Isolation and purification of exfoliated cells from gastric fluid aspirates or stools of preterm infants open the way to appreciate maturation of the GI tract and to detect any oscillations of clock genes compounds of stomach circadian clock, hypothetically entrained by nursing care (adaptation of time of meals for each child or at fixed time of meal with or without a rhythmic lighting).

References

Abizaid A, Mezei G, Sotonyi P, Horvath TL. Sex differences in adult suprachiamastic nucleus neurons emerging late prenatally in rats. *European Journal of Neuro*science 2004, 19: 2488-2490.

Akashi M, Tsuchiya YT, Yoshino T, Nishida E. Control of intracellular dynamics of mammalian Period proteins by casein kinase I epsilon (CKIe) and CKIdelta in cultured cells. *Molecular and Cellular Biology* 2002, 22: 1693-1703.

Albaugh, GP, Iyengar V, Lohani A, Malayeri M, Bala S, Nair PP 1992 Isolation of exfoliated colonic epithelial cells, a novel non-invasive approach to the study of cellular markers. *Int. J. Cancer* 52: 347-350.

Albrecht U, Sun ZS, Eichele G, Lee CC. A differential response of two putative mammalian circadian regulators *mper1* and *mper2*, to light. *Cell* 1997, 91: 1055-1064.

Als H., Lawhon G, Duffy FH, McAnulty GB, Gibes-Grossman R, Blickman JG. Individualized developmental care for the very low-birth-weight preterm infant. Medical and neurofunctional effects. *JAMA* 272: 853-858; 1994.

Antonov J, Goldstein DR, Oberli A, Baltzer A, Pirotta M, Fleischmann A, Altermatt HJ, Jaggi R 2005 Reliable gene expression measurements from degraded RNA by quantitative real-time PCR depend on short amplicons and a proper normalization. *Laboratory Investigation* 1-11.

Bado A, Levasseur S, Attoub S, Kermorgant S, Laigneau JP, Bortoluzzi MN, Moizo L, Lehy T, Guerre-Millo M, Le Marchand-Brustel Y, Lewin MJ. The stomach is a source of leptin. *Nature.* 1998, 394:790-3.

Bae K, Jin X, Maywood ES, Hastings MH, Reppert SM, Weaver DR. Differential functions of mPer1, mPer2, and mPer3 in the SCN circadian clock. *Neuron* 2001, 30: 525-536.

Bandaletova T, Bailey N, Bingham SA, and Loktionov A 2002 Isolation of exfoliated colonocytes from human stools as a new technique for colonic cytology. *APMIS* 110: 239-246.

Brandon, D. H., Holditch-Davis, D., Belyea, M. Preterm infants born at less than 31 weeks' gestation have improved growth in cycled light compared with continuous near darkness. *Journal of Pediatrics* 140: 192-199; 2002.

Brown SA, Fleury-Olela F, Nagoshi E, Hauser C, Juge C, Meier CA, Chicheportiche R, Dayer J-M, Schibler U. The Period length of fibroblast circadian gene expression varies widely among human individuals. *PLOS Biology* 2005b, 3: 1813-1818.

Brown SA, Ripperger J, Kadener S, Fleury-Olela F, Vilbois F, Rosbash M, Schibler U. PERIOD1-associated proteins modulate the negative limb of the mammalian circadian oscillator. *Science* 2005a, 308: 693-696.

Brown SA, Zumbrunn G, Fleury-Olela F, Preitner N, Schibler U. Rhythms of mammalian body temperature can sustain peripheral circadian clocks. *Current Biology* 2002, 12: 1574-1583.

Buchi KN, Moore JG, Hrushesky WJM, Sothern RB, Rubin NH. Circadian rhythm of cellular proliferation in the human rectal mucosa. *Gastroenterology* 1991, 101:410-415.

Camacho F, Cilio M, Guo Y, Virshup DM, Patel K, Khorkova O, Styren S, Morse B, Yao Z, Keesler GA. Human casein kinase Idelta phosphorylation of human circadian clock proteins period 1 and 2. *FEBS Letters* 2001, 489: 159-165.

Cardone L, Hirayama J, Giordano F, Tamaru T, Palvimo JJ, Sassone-Corsi P. Circadian clock control by SUMOylation of BMAL1. *Science* 2005, 309: 1390-1394.

Carskadon MA, Wolfson AR, Acebo C, Tzischinsky O, Seifer R. Adolescent sleep patterns, circadian timing, and sleepiness at a transition to early school days. *Sleep* 1998, 21: 871-881.

Damiola F, Le Minh N, Preitner N, Kornmann B, Fleury-Olela F, Schiebler U. Restricted feeding uncouples circadien oscillators in peripheral tissues from the central pacemaker in the suprachiasmatic nucleus. *Genes and development* 2000, 14: 2950-2961.

Davidson LA, Lupton JR, Miskovsky E, Fields AP, Chapkin RS 2003 Quantification of human intestinal gene expression profiles using exfoliated colonocytes: a pilot study. *Biomarkers* 8: 51-61.

Davis CD 2003 Use of exfoliated cells from target tissues to predict responses to bioactive food components. *Journal of Nutrition* 133: 1769-1772.

Duffield GE. DNA Microarray analyses of circadian timing: the genomic basis of biological time. *Journal of Neuroendocrinology* 2003, 15: 991-1002.

Eide EJ, Vielhaber EL, Hinz WA, Virshup DM. The circadian regulatory proteins BMAL1 and cryptochromes are substrates of casein kinase Iepsilon. *The Journal of Biological Chemistry* 2002, 277: 17248-17254.

Feillet CA, Ripperger JA, Magnone MC, Dulloo A, Albrecht U, Challet E. Lack of food anticipation in Per2 mutant mice. *Current Biology* 2006, 16: 2016-2022.

Froy O, Chapnik N. Circadian oscillation of innate immunity components in mouse small intestine. *Mol. Immunol.* 44: 1964-1970; 2007.

Fu L, Patel MS, Bradley A, Wagner EF, Karsenty G. The molecular clock mediates leptin-regulated bone formation. *Cell* 2005, 122: 803-815.

Fu L, Pelicano H, Liu J, Huang P, Lee CC. The circadian gene Period2 plays an important role in tumor suppression and DNA damage response in vivo. *Cell* 2002, 111: 41-50.

Fu Y, Zhong H, Wang M-H H, Luo D-G, Liao H-W, Maeda H, Hattar S, Frishman LJ, Yau K-W. Intrinsically photosensitive retinal ganglion cells detect light with a vitamin A-based photopigment, melanopsin. *PNAS* 2005, 102: 10339-10344.

Gekakis N, Staknis D, Nguyen HB, Davis FC, Wilsbacher LD, King DP, Takahashi JS, Weitz CJ. Role of the CLOCK protein in the mammalian circadian mechanism. *Science* 1998, 280:1564-1569.

Griffin EA Jr, Staknis D, Weitz CJ. Light-independent role of CRY1 and CRY2 in the mammalian circadian clock. 1999, *Science* 286: 768-771.

Hamaguchi H, Fujimoto K, Kawamoto T, Noshiro M, Maemura K, Takeda N, Nagai R, Furukawa M, Honma S, Honma K -i, Kurihara H. Expression of the gene for Dec2, a basic helix-loop-helix transcription factor, is regulated by a molecular clock system. *Biochem. J.* 2004, 382: 43-50.

Harmer SL, Panda S, Kay SA. Molecular bases of circadian rhythms. *Annu. Rev. Cell Dev. Biol.* 2001, 17: 215-253.

Haumont D, Hansen V. Neonatal development: effects of light. *In* Sizun and Browne, Research on early developmental care for preterm neonates. Edition: John Libbey, Eurotext, 2006, 33-37.

Hogenesch JB, Gu YZ, Jain S, Bradfield CA. The basic-helix-loop-helix-PAS orphan MOP3 forms transcriptionally active complexes with circadian and hypoxia factors. *Proc. Natl. Acad. Sci. USA* 1998, 95: 5474-5479.

Holland N, Harmatz P, Golden D, Hubbard A, Wu YY, Bae J, Chen C, Huen K, Heyman MB 2007 Cytogenetic damages in blood lymphocytes and exfoliated epithelial cells of children with inflammatory bowel disease. *Pediatric Research* 61: 1-6.

Honma S, Kawamoto T, Takagi Y, Fujimoto K, Sato F, Noshiro M, Kato Y, Honma K -i. Dec1 and Dec2 are regulators of the mammalian molecular clock. *Nature* 2002, 419: 841-844.

Iwakiri R, Gotoh Y, Noda T, Sugihara H, Fujimoto K, Fuseler J, Aw TY. Programmed cell death in rat intestine: effect of feeding and fasting. *Scandinavian Journal of Gastroenterology* 2001, 36: 39-47.

Kaeffer B, Des Robert C, Pagnièz A, Alexandre MC, Amarger V, Legrand A, Küster A, Piloquet H, Champ M, Le Huërou-Luron I, Rozé J-C. Cellules exfoliées du tube digestif de prématurés, mise en évidence des composants moléculaires d'une horloge circadienne périphérique.12èmes *Journées Francophones de Recherche en Néonatologie*, Paris, 14-15 décembre 2006.

Kaeffer B, Des Robert C, Alexandre MC, Pagnièz A, Amarger V, Legrand A, Küster A, Piloquet H, Champ M, Le Huërou-Luron I, Rozé J-C. Recovery of exfoliated cells from the gastro-intestinal tract of premature infants: a new tool to obtain "non-invasive biopsies"?[s] *Pediatric Research*, 2007, *in press*.

Kelly MM. Primary care issues for the healthy premature infant. *Journal of Pediatric Health Care*. 2006a, 20: 293-299.

Kelly MM. The basics of Prematurity. *Journal of Pediatric Health Care*. 2006b, 20: 238-244.

Kelly MM. The medically complex premature infant in primary care. *Journal of Pediatric Health Care*. 2006c, 20: 367-373.

Kennaway DJ. Programming of the fetal suprachiasmatic nucleus and subsequent adult rhythmicity. *Trends in Endocrinology and Metabolism* 2002, 13: 398-402.

King D, Zhao Y, Sangoram A, Wilsbacher L, Tanaka M, Antoch M, Steeves T, Vitaterna M, Kornhauser J, Lowrey P, Turek F, Takahashi J. Positional cloning of the mouse circadian clock gene. *Cell* 1997, 89:641-653.

Kornhauser JM, Nelson DE, Mayo KE, Takahashi JS. Regulation of jun-B messenger RNA and AP-1 activity by light and a circadian clock. *Science* 1992, 255: 1581-1584.

Kornmann B, Preitner N, Rifat D, Fleury-Olela F, Schibler U. Analysis of circadian liver gene expression by ADDER, a highly sensitive method for the display of differentially expressed mRNAs. *Nucleic Acids Research* 2001, 29: 11 e51.

Kume K, Zylka M, Sriram S, Shearman L, Weaver D, Jin X, Maywood E, Hasting M, Reppert S. mCRY1 and mCRY2 are essential components of the negative limb of the circadian clock feedback loop. *Cell* 1999, 98:193-205.

Lee C, Etchegaray J-P, Cagampang FRA, Loudon ASI, Reppert SM. Posttranslational mechanisms regulate the mammalian circadian clock. *Cell* 2001, 107: 855-867.

Lee C, Weaver D, Reppert SM. Direct association between mouse PERIOD and CKIepsilon is critical for a functioning circadian clock. *Molecular and Cellular Biology* 2004, 24: 584-594.

Mann NP, Haddow R, Stokes L, Goodley S, Rutter N. Effect of night and day on preterm infants in a newborn nursery: randomized trial. *Br. Med. J. (Clin. Res. Ed.),* 1986 293: 1265-1267.

Marra G, Anti ,M, Percesepe A, Armelao F, Ficarelli ,R, Coco C, Rinelli A, Vecchio F. M, D'Arcangelo E. Circadian variations of epithelial cell proliferation in human rectal crypts. *Gastroenterology* 1994, 106: 982-987.

Miller CL, Diglisic S, Leister F, Webster M, Yolken RH 2004 Evaluating RNA status for RT-PCR in extracts of postmortem human brain tissue. *Biotechniques* 36: 628-633.

Miller CL, White R, Whitman TL. The effect of cycled versus non-cycled lighting on the growth and development or preterm infants. *Infant. Behav. Dev*. 1995 18: 95.

Moldenhauer G, Momburg F, Möller P, Schwartz R, Hämmerling GJ 1987 Epithelium-specific surface glycoprotein of Mr 34,000 is a widely distributed human carcinoma marker. *Br. J. Cancer* 56: 714-721.

Moore-Ede MC. Physiology of the circadian timing system: predictive versus reactive homeostasis. *Am. J. Physiol.* 1986, 250: R735-52.

Motzkus D, Albrecht U, Maronde E. The human *PER1* gene is inductible by interleukin-6. *Journal of Molecular Neuroscience* 2002, 218: 105-109.

Motzkus D, Maronde E, Grunenberg U, Lee C, Forssmann W-G, Albrecht U. The human PER1 gene is transcriptionally regulated by multiple signaling pathways. *FEBS Letters* 2000, 486: 315-319.

Neu J, Hauser N, Douglas-Escobar M. Postnatal nutrition and adult health programming. *Seminars in Fetal and Neonatal Medicine* 2007, 12: 78-86.

Ohta H, Honma S, Abe H, Honma K. Periodic absence of nursing mothers phase-shifts circadian rhythms of clock genes in the suprachiasmatic nucleus of rat pups. *Eur. J. Neurosci.* 2003, 17: 1628-1634.

Oishi K, Miyasaki K, Kadoka K, Kikuno R, Nagase T, Atsumi G-I, Ohkura N, Azama T, Mesaki M, Yukimasa S, Kobayashi H, Iitaka C, Umehara T, Horikoshi Y, Kudo T, Shimizu Y, Yano M; Monden M, Machida K, Matsuda J, Horie S, Todo T, Ishida N.

Genome-wide expression analysis of mouse liver reveals CLOCK-regulated circadian ouput genes. *Journal of Biological Chemistry* 2003, 42: 41519-41527.

Oishi K, Sakamoto K, Okada T, Nagase T, Ishida N. Antiphase circadian expression between BMAL1 and period homologue mRNA in the suprachiamastic nucleus and peripheral tissues of rats. *Biochemical and biophysical research communications* 1998a, 253: 199-203.

Oishi K, Sakamoto K, Okada T, Nagase T, Ishida N. Humoral signals mediate the circadian expression of rat period homologue (rPer2) mRNA in peripheral tissues. *Neuroscience Letters* 1998b, 256: 117-119.

Osborn NK, Ahlquist DA 2005 Stool screening for colorectal cancer: molecular approaches. *Gastroenterology* 128: 192-206.

Pan X, Terada T, Okuda M, Inui K. The diurnal rhythm of the intestinal transporters SGLT1 and PEPT1 is regulated by the feeding conditions in rats. *Journal Nutrition* 2004 134: 2211-2115.

Panda S, Antoch M, Miller B, Su A, Schook A, Schultz P, Kay S, Takahashi J, Hogenesh J. Coordinated transcription of key pathways in the mouse by the circadian clock. *Cell* 2002, 109: 307-320.

Pardini L, Kaeffer B. Feeding and circadian clocks. *Reproduction Nutrition Development* 2006, 5: 463-480.

Pardini, L., Kaeffer, B., Trubuil, A., Bourreille, A., Galmiche, J-P. Human intestinal circadian clock: expression of clock genes in colonocytes lining the crypt. *Chronobiology International*, 2005, 6: 1313-1323.

Philip AGS. The evolution of neonatalogy. *Pediatric Research*, 2005, 58: 799-815.

Preitner N, Damiola F, Lopez-Molina L, Zakany J, Duboule D, Albrecht U, Schibler U. The orphan nuclear receptor REV-ERBalpha controls circadian transcription within the positive limb of the mammalian circadian oscillator. *Cell* 2002, 110: 251-260.

Qiu JM, Roberts SA, Potten CS. Cell migration in the small and large bowel shows a strong circadian rhythm. *Epithelial Cell Biology* 1994, 3: 137-148.

Reddy AB, Karp NA, Maywood ES, Sage EA, Deery M, O'Neill JS, Wong JKY, Chesham J, Odell M, Lilley KS, Kyriacou CP, Hasting MH. Circadian Orchestration of the Hepatic Proteome. *Current Biology* 2006, 16: 1107-1115.

Reick M, Garcia JA, Dudley C, McKnight SL. NPAS2: an analog of Clock operative in the mammalian forebrain. *Science* 2001, 293: 506-509.

Rivkees SA, Mayes L, Jacobs H, Gross I 2004 Rest-Activity patterns of premature infants are regulated by cycled lighting. *Pediatrics* 113: 833-839.

Rivkees SA. Developing circadian rhythmicity in infants. *Pediatrics* 2003, 112: 373-381.

Robine N, Relier JP, Le Bars S 1988 Urocytogram, an index of maturity in premature infants. *Biol. Neonate* 54: 93-99.

Ruby NF, Burns DE, Heller HC. Circadian rhythms in the suprachiasmatic nucleus are temperature-compensated and phase-shifted by heat pulses in vitro. *The Journal of Neuroscience* 1999, 19: 8630-8636.

Sakamoto K, Oishi K, Nagase T, Miyasaki K, Ishida N. Circadian expression of clock genes during ontogeny in the rat heart. *Neuroreport* 2002 13: 1239-1242.

Sato F, Kawamoto T, Fujimoto K, Noshiro M, Honda KK, Honma S, Honma K -i, Kato Y. Functional analysis of the basic helix-loop-helix transcription factor DEC1 in circadian regulation. *Eur. J. Biochem.* 2004b, 271: 4409-4419.

Sato TK, Panda S, Miraglia LJ, Reyes TM, Rudic RD, McNamara P, Nalk KA, FitzGerald GA, Kay SA, Hogenesch JB. A functional genomics strategy reveals Rora as a component of the Mammalian circadian clock. *Neuron* 2004, 43: 527-537.

Scheving LA. Biological clocks and the digestive system. *Gastroenterology* 2000, 119: 536-549.

Schmidt PT, Degerblad M, Lindstrom E, Sundqvist M, Naslund E, Gillberg PG, Husebye E, Theodorsson E, Hellstrom PM. Circulating ghrelin levels after food intake during different phases of the migrating motor complex in man. *Eur. J. Clin. Invest.* 2006 36: 503-8.

Schoeller DA, Cella LK, Sinha MK, Caro JF. Entrainment of the diurnal rhythm of plasma leptin to meal timing. *J. Clin. Invest.* 1997 100: 1882-1887.

Seron-Ferré M, Torres C, Parraguez VH, Vergara M, Valladares L, Forcelledo ML, Constandil L, Valenzuela GJ. Perinatal neuroendocrine regulation. Development of the circadian time-keeping system. *Molecular and Cellular Endocrinology* 2002, 186: 169-173.

Shearman L, Sriram S, Weaver D, Maywood E, Chaves I, Zheng B, Kume K, Lee C, van der Horst G, Hastings M, Reppert S. Interacting molecular loops in the mammalian circadian clock. *Science* 2000, 288: 1013-1019.

Shearman L, Zylka M, Weaver D, Kolakowki L. Jr, Reppet S. Two period homologs: circadian expression and photic regulation in the suprachiasmatic nuclei. *Neuron* 1997, 19: 1261-1269.

Shimomura H, Moriya T, Sudo M, Wakamatsu H, Akiyama M, Miyake Y, Shibata S. Differential daily expression of Per1 and Per2 mRNA in the suprachiasmatic nucleus of fetal and early postnatal mice. *Eur. J. Neurosci.* 2001 13: 687-693.

Sizun J, Browne JV, *Research on early developmental care for preterm neonates*. Edition: John Libbey, Eurotext, 2006, p 123.

Sladek M, Jindrakova Z, Bendova Z, Sumova A. Postnatal ontogenesis of the circadian clock within the rat liver. *Am. J. Physiol. Regul. Integr. Comp. Physiol.* 2007, 292:R1224-9.

Sladek M, Sumova A, Kovacikova Z, Bendova Z, Laurinova K, Illnerova H. Insight into molecular core clock mechanism of embryonic and early postnatal rat supraciasmatic nucleus. *PNAS* 2004, 101: 6231-6236.

Stokkan K-A, Yamazaki S, Tei H, Sakaki Y, Menaker M. Entrainment of the circadian clock in the liver by feeding. *Science* 2001, 291: 490-493.

Storch K-F, Lipian O, Leykin I, Viswanathan N, Davis F, Wong W, Weitz C. Extensive and divergent circadian expression in liver and heart. *Nature* 2002, 417: 78-83.

Stratmann M, Schibler U. Properties, entrainment, and physiological functions of mammalian peripheral oscillators. *J. Biol. Rhythms.* 2006, 21:494-506.

Sumova A, Bendova Z, Sladek M, El-Hennamy R, Laurinova K, Jindrakova Z, Illnerova H. Setting the biological time in central and peripheral clocks during ontogenesis. *FEBS Letters* 2006 580: 2836-2842.

Sun Z, Albrecht U, Zhuchenko O, Bailey J, Eichele G, Lee C.C. RIGUI a putative mammalian ortholog of the drosophila period gene. *Cell* 1997, 90: 1003-1011.

Swaab DF, Hofman MA, Honnebier MB. Development of vasopressin neurons in the human suprachiasmatic nucleus in relation to birth. *Dev. Brain Res.* 1990 52: 289-293.

Takahashi S, Yokota S, Hara R, Kobayashi T, Akiyama M, Moriya T, Shibata S. Physical and inflammatory stressors elevate circadian clock gene mPerl mRNA levels in the paraventricular nucleus of the mouse. *Endocrinology* 2001, 142: 4910-4917.

Tei H, Okamura H, Shigeyoshi Y, Fukuhara C, Ozawa R, Hirose M, Sakaki Y. Circadian oscillation of a mammalian homologue of the Drosophila period gene. *Nature* 1997, 389: 512-516.

Thiesen, A., Wild, G., Keelan, M., Clandinin, M. T., McBurney, M., Van Aerde, J., Thomson, A. B. R. Ontogeny of intestinal nutrient transport. *Can. J. Physiol. Pharmacol.* 2000, 78: 513-527.

Tsuchiya Y, Akashi M, Nishida E. Temperature compensation and temperature resetting of circadian rhythms in mammalian cultured fibroblasts. *Genes to Cells* 2003, 8: 713-720.

Turek, F. W., Joshu, C., Kohsaka, A., Lin, E., Ivanova, G., McDearmon, E., Laposky, A., Losee-Olson, S., Easton, A., Jensen, D. R., Eckel, R. H., Takahashi, J. S., Bass, J. Obesity and metabolic syndrome in circadian Clock mutant mice. *Science* 2005, 308: 1043-1045

Ueda H, Chen W, Adachi A, Wakamatsu H, Hayashi S, Takasugi T, Nagano M, Nakahalma K-I, Suzuki Y, Shigeyoshi Y, Hashimoto S A transcriptional factor response element for gene expression during circadian night. *Nature* 2002, 418: 534-538.

van der Horst G, Muijtjens M, Kobayashi K, Takano R, Kanno S-I, Takao M, de Wit J, Verkerk A, Eker A, van Leenen D, Buijs R, Bootsma D, Hoeijmakers J, Yasui A. Mammalian cry1 and cry2 are essential for maintenace of circadian rhythms. *Nature* 1999, 398: 627-630.

Vielhaber E, Eide E, Rivers A, Gao Z-H, Virshup D. Nuclear entry of the circadian regulator mPER1 is controlled by mammalian casein kinase I epsilon. *Molecular and Cellular Biology* 2000, 20: 4888-4899.

Weinert D 2005 Ontogenetic development of the mammalian circadian system. *Chronobiology International* 22: 179-205.

Weinert D, Sitka U, Minors D, Menna-Barreto L, Waterhouse J. Twenty-four hours and ultradian rhythmicities in healthy full-term neonates: exogenons and endogenous influences. *Biol. Rhythm Res.* 1997, 28: 441-452.

Yamamoto T, Nakahata Y, Tanaka M, Yoshida M, Soma H, Shinohara K, Yasuda A, Mamine T, Takumi T. Acute physical stress elevates mouse *Period1* mRNA expression in mouse peripheral tissues via a Glucocorticoid-responsive element. *Journal Biological Chemistry* 2005, 280: 42036-42043.

Yeh K-Y, Yeh M. Use of pup in a cup model to study gastrointestinal development: interaction of nutrition and pituitary hormones. *Journal of Nutrition* 1993, 123: 378-381.

Yu W, Nomura M, Ikeda M. Interactivating feedback loops within the mammalian clock : BMAL1 is negatively autoregulated and upregulated by CRY1 CRY2 and PER2. *Biochemical and Biophysical Research Communication* 2002, 290: 933-941.

Zhang L, Zhou W, Velculescu VE, Kern SE, Hruban RH, Hamilton SR, Vogelstein B, Kinzler KW 1997 Gene expression profiles in normal and cancer cells. *Science* 276: 1268-1272.

Zheng B, Albrecht U, Kaasik K, Sage M, Lu W, Vaishnav S, Li Q, Sun Z S, Eichele G, Bradley A, Lee C C. Nonredundant roles of the mPerl and mPer2 genes in the mammalian circadian clock. *Cell* 2001, 105: 683-94.

Zheng B, Larkin DW, Albrecht U, Sun ZS, Sage M, Eichele G, Lee CC, Bradley A. The mPER2 gene encodes a functional component of the mammalian circadian clock. *Nature* 1999, 400: 169-173.

Zylka M J, Shearman L P, Weaver D R, Reppert S M. Three period homologs in mammals: differential light responses in the suprachiamatic circadian clock and oscillating transcripts outside of brain. *Neuron* 1998, 20: 1103-1110.

In: Circadian Rhythms and Health Research Trends
Editor: Alfredo V. Lignelli, pp. 117-130
ISBN: 978-1-60021-875-0

Chapter VI

Circadian Rhythm of the Autonomic Nervous System in Quality Sleep Disorders

Antonio Perciaccante, Alessandra Fiorentini and Luigi Tubani
Department of Clinical Medicine,
University "La Sapienza",
Rome, Italy

Abstract

A relationship between autonomic nervous system function and insulin resistance has been identified and it appears very complex. The central biological clock in the hypothalamus uses the autonomic network to generate and organize metabolic rhythms. The sympathetic nervous system modulates both hepatic glucose production and the glucose uptake in peripheral tissue. Some studies demonstrated that the increase of plasma insulin level was related to increased urinary and plasma norepinephrine. Other studies, performed by microneurography, have shown that acute hyperinsulinemia increases sympathetic activity in muscle nerves Analysis of heart rate variability (HRV), a non invasive technique used to assess autonomic neural regulation, has been successfully introduced to investigate the influence of autonomic neuromodulation in several pathophysiologic conditions Several studies performed by spectral analysis of HRV have shown sympathetic hyper-activity in insulin resistant subjects with normoglycemia.

We have evaluated, by HRV analysis in 24-hours ECG recording, the circadian rhythm of autonomic nervous system function in insulin resistant subjects with normal glucose regulation.

Introduction

Sleep is an important factor contributing to health. Insomnia is the most common sleep disorder, especially in Europe and in the United States. Some epidemiological studies have reported that the prevalence of this disorder ranges from 10% to 34% [1-3].

As pointed out in the recent National Institutes of Health (NIH) Consensus and State-of-the-Science and Statement [4], insomnia may be defined as complaints of disturbed sleep in the presence of adequate opportunity and circumstance for sleep. The disturbance may consist of one or more of these features: difficulty in initiating sleep, difficulty in maintaining sleep, waking up too early.

Chronic insomnia should be distinguished from acute insomnia, which may occur in anyone at one time or another. In the last Statement [4], NIH has defined chronic insomnia as 30 days or more of the symptoms described above.

Diagnosis of insomnia is difficult and it is based primary on patient and family or caregiver-derived complaints, as determined by the clinical interview. The difficulty is related to how accurately subjects reporting sleep disorders can judge their sleep latency or period of wakefulness during the night.

Various tools have been proposed as an aid to diagnosis, but many are limited in their validation. Several questionnaires have been formulated, but there is lack of standardization. Sleep diaries can help to evaluate sleep/wake cycles. The actigraph, measuring movement to infer sleep and wake cycle, may help in the evaluation of circadian rhythm disorders, but its use has not been validated in insomnia. Multichannel polysomnography is the gold standard for sleep-disordered breathing and it may play an important role in the diagnosis of insomnia, because it is the most sensitive tool to differentiate wakefulness and sleep. Actually, the use of polysomnography in insomnia is limited because it is expensive and numerous monitoring electrodes can disrupt sleep.

Insomnia often is comorbid with other disorders as depression, cardiovascular, pulmonary and gastro-intestinal disorders. Cases of insomnia that are comorbid with other conditions have been termed "secondary insomnia" or most recently "comorbid insomnia".

Insomnia is thought to be a primary disorder in itself. "Primary insomnia" is the term used when not co-existing with associated disorders.

Whether it is a primary or secondary disorder, chronic insomnia is often associated with a wide range of adverse conditions. Insomnia appears to be associated with high health care utilization. The direct and indirect costs of chronic insomnia has been estimated at tens of billions of dollars annually.

Some studies [5-8] have evaluated the effects of insomnia on quality of life. These studies suggest that chronic insomnia is associated with a reduction in quality of life and social functioning. Further studies [9-10] have shown a relationship between chronic insomnia and work days missed and impaired work performance. A relationship between sleep loss and an impaired psychomotor and cognitive functioning has been observed [11, 12].

Recently, there has been an increasing interest in a possible impact of the quality sleep disorders on glucose metabolism.

The data from the Sleep Heart Health Study [13] have suggested that type 2 diabetes mellitus is associated with periodic breathing, a respiratory abnormality associated with abnormalities in the control of ventilation, and that sleep disordered breathing is associated with glucose intolerance and insulin resistance [14].

Population-based studies have shown that a self-reported history of snoring is independently associated with impaired glucose tolerance and incident type 2 diabetes mellitus [15, 16].

Prospective data from the Nurse Health Study [17] and a population-based Swedish study [15] have shown that habitual snoring is associated with more than twofold risk of developing type 2 diabetes mellitus over a 10-year period, independent of body mass index and other confounding variables.

In the MONICA/KORA Study [18], authors have observed a significantly positive association between difficulty in maintaining sleep and incident type 2 diabetes mellitus in men and women.

Gottlieb et al. [19] have assessed the cross-selectional relationship of usual sleep time to diabetes mellitus and impaired glucose tolerance and they have observed that a sleep duration of 6 hours or less or 9 hours or more are associated with increased prevalence of these pathologies.

A prospective study by Nilsson et al. [20] has shown that difficulties in falling asleep or regular use of hypnotics and elevated resting heart rate in middle-aged men, are associated with an increased risk of diabetes mellitus.

Recently, there has been a great interest in a possible association between sleep quality disorders and hypertension. Several studies have demonstrated that hypertension is a factor associated with insomnia [21-23].

In a recent study [24], we have examined the prevalence of hypertension and type 2 diabetes mellitus in subjects with quality sleep disorders. The prevalence of hypertension was 87.1% in "poor sleepers" subjects versus 35.1% "good sleepers" subjects. The mean values of systolic and diastolic blood pressure were higher in "poor sleepers" than in "good sleepers" subjects. Moreover, we have observed a relationship between Pittsburgh Sleep Quality Index and systolic and diastolic blood pressure values.

Gottlieb et al. [25] have observed that usual sleep duration above or below the median of 7 to less than 8 hours per night is associated with an increased prevalence of hypertension, particularly at the extreme of less than 6 hours per night. In this study, the relationship between sleep duration and hypertension was examined using categorical logistic regression with adjustment for age, sex, race, apnea-hypopnea index, and body mass index. Compared to subjects sleeping 7 to less than 8 hours per night, those sleeping less than 6 and between 6 and 7 hours per night had adjusted odds ratios for hypertension of 1.66 (95% confidence interval 1.35-2.04) and 1.19 (1.02-1.39), respectively, whereas those sleeping between 8 and 9 and 9 or more hours per night had adjusted odds ratios for hypertension of 1.19 (1.04-1.37) and 1.30 (1.04-1.62), respectively ($p < .0001$ for association of sleep duration with hypertension). These associations persisted when analyses were further adjusted for caffeine and alcohol consumption, current smoking, insomnia symptoms, depression symptoms, sleep efficiency, and prevalent diabetes mellitus or cardiovascular disease.

The results of a study by Prejbisz [26] confirmed that insomnia is common in patients with essential hypertension and indicates an association between insomnia and gender, known duration of hypertension and number of hypertensive drugs taken. Untreated essential hypertension insomniacs were characterized by less pronounced nocturnal fall in both systolic and diastolic blood pressure compared with non-insomniacs.

Cross sectional and retrospective studies suggest that disturbed sleep is associated with an increase in prevalent vascular disease [27-32]. The results from Caerphilly cohort [33] have shown that, compared with men who reported not sleep disturbance, the adjusted relative odds of an ischemic stroke were significantly increased in men with any sleep disturbance. Therefore, the risk of ischemic stroke is increased in men whose sleep is frequently disturbed, and daytime sleepiness is associated with a significant increase in ischemic heart disease events.

Studies have observed an increased risk of coronary artery disease in subjects with sleep complaints. A poor sleep quality has been associated with an increased risk of death and hospitalization because of ischemic heart disease [34]. In survey of females difficulty falling asleep predicted myocardial infarction or coronary death [35].

Mallon L et al [36] have observed that there is an association between difficulties falling asleep and coronary artery disease mortality in males.

In a 6,5 year longitudinal study, Seigrist et al [37] have reported a significant risk ratio of 1.96 for nonfatal myocardial infarction, coronary heart disease death or stroke death in poor sleeprs subjects. Similar results have been reported by Appels [38]. In the Framingham study, the incidence density ratio for non fatal myocardial infarction and coronary heart disease death was 3.9 adjusted for age, systolic blood pressure, cholesterol, diabetes, smoking and body mass index.

The succession of night and day and circadian rhythms influence life and govern several fundamental physiological functions in almost all organisms, from prokaryotes to humans. These periodic changes in environment have been internalised as an endogenous circadian clock.

Circadian rhythms are generated in pacemaker neurons, located in the suprachiasmatic nucleus in the anterior hypothalamus [39, 40].

Recently, it was shown that the endogenous rhythm is generated by a suite of clock genes forming different sets of interlocking transcriptional-translational feed-back loops. The link between transcriptional and translational events of the molecular clock and between metabolic and electric activity of the suprachiasmatic nucleus are unclear.

The main function of the central biological clock is to generate and organize physiological, hormonal and metabolic rhythms [41-43] to enable the organism to anticipate the changing environmental conditions and adapt optimally it to ambient conditions.

Peripheral tissues also contain independent clocks [44-46]. The oscillatory function of peripheral clocks in mammals is orchestrated by the suprachiasmatic nucleus.

The central biological clock uses autonomic and hormonal network to generate and organize physiological rhythms and it requires information from the external and internal environment respectively by the sensory organs [47] and spinal cord [48].

The main result of biological clock output is the behavioral sleep/wake or rest/activity cycle.

In mammals, sleep is controlled by a homeostatic process, which regulates depth of sleep, and by the circadian clock of the suprachiasmatic nucleus. Sleep deprivation is known to cause molecular an physiological changes and alteration in the timing of sleep. The data by Deboer et al [49] have shown that sleep deprivation influences not only sleep homeostatic mechanisms, but also suprachiasmatic nucleus electrical activity, resulting in reduction of circadian amplitude in the major output signal from the suprachiasmatic nucleus. Following sleep deprivation, suprachiasmatic neuronal activity was reduced to about 60% of baseline levels.

Because in humans and animals the temporal distribution of sleep and wakefulness is affected by the biological clock, it has been assumed that the daily rhythms of physiological events as body temperature, blood pressure and circulating concentrations of hormones are induced by the behavioural rhythm, that is subject to a direct control of the biological clock.

Recently, Kreier et al [50] have hypothesized that the western lifestyle, characterized by increased energy intake and decreased energy expenditure, affects the central biological clock effects the selective balance of the autonomic nervous system in different part of the body. The autonomic activity in the cardiovascular compartment and in the muscle compartment is shifted in favor of the sympathetic branch, resulting in high blood pressure and impaired glucose uptake by the muscle. Contrarily, in the visceral compartment the autonomic balance is shifted in favor of the parasympathetic branch, leading to an increased insulin secretion and growth of intra-abdominal fat tissue.

The last two decades have witnessed the recognition of a significant relationship between the autonomic nervous system and cardiovascular mortality, including sudden cardiac death [51-54]. Analysis of beat-to-beat heart rate variability (HRV) has emerged as one of the non-invasive methods to quantitatively assess cardiac autonomic activity.

The variations in heart rate may be evaluated by a number of methods. Perhaps the simplest to perform are the time domain measures. In these methods, either the heart rate at any point in time or the intervals between successive normal complexes are determined. In a continuous ECG record, each QRS complex is detected, and the so-called normal-to-normal (NN) intervals or the instantaneous heart rate is determined.

From a series of instantaneous heart rates or cycle intervals, particularly those recorded over longer periods, traditionally 24 hours, more complex statistical time domain measures can be calculated. The simplest variable to calculate is the standard deviation of the NN intervals (SDNN), that is, the square root of variance. Since variance is mathematically equal to total power of spectral analysis, SDNN reflects all the cyclic components responsible for variability in the period of recording. In many studies SDNN is calculated over a 24-hour period and thus encompasses short-term HF variations as well as the lowest-frequency components seen in a 24-hour period. SDNN correlates to total autonomic activity.

Another commonly used measure derived from interval differences includes RMSSD, the square root of the mean squared differences of successive NN intervals that correlates to parasympathetic activity.

Various spectral methods for the analysis of the tachogram have been applied since the late 1960s.

Fast Fourier Transform is the most commonly used to obtain power spectral estimates of HRV.

The total power in the frequency range (0 – 0,40 Hz) is divided into: very low frequency (VLF: < 0,40 Hz), low frequency (LF: 0,04 – 0,15 Hz modulated both by sympathetic and parasympathetic system), and high frequency (HF: 0,15 – 0,40 Hz mediated by parasympathetic system). The measurement of VLF, LF, and HF power components is usually made in absolute values of power (milliseconds squared) [55]. LF and HF may also be measured in normalized units [56, 57], which represent the relative value of each power component in proportion to the total power minus the VLF component. The representation of LF and HF in normalized units emphasizes the controlled and balanced behavior of the two branches of the autonomic nervous system.

Disagreement exists in respect to the LF component. Some studies suggest that LF, when expressed in normalized units, is a quantitative marker of sympathetic modulations; other studies view LF as reflecting both sympathetic activity and vagal activity. Consequently, the LF/HF ratio is considered by Task Force of the European Society of Cardiology and the North American Society of Pacing and Electrophysiology [55] to mirror sympathovagal balance or to reflect the sympathetic modulations.

Multiple studies have demonstrated the stability of short-term and long-term measures of HRV obtained from 24-hour ambulatory monitoring, suggesting a great stability of HRV measures derived from 24-hour ambulatory monitoring in both normal subjects [58,59] and in the postinfarction [60] and ventricular arrhythmia [61] populations.

High heart rate and low HRV, both measures of cardiac autonomic status, predict coronary disease incidence [62] and prognosis [63]. In the Whitehall II study [64], HRV has shown strong linear associations whit components of the metabolic syndrome, as waist circumference, systolic blood pressure, HDL cholesterol, triglycerides and fasting and 2-hour postload glucose. The authors claim that chronically impaired autonomic function may link social position to different components of coronary risk in the general population.

Several studies on patient population have found that an impaired HRV is associated with a higher risk of all-cause mortality in survivors of an acute myocardial infarction [65-68]. Janszky et al [69] in the Stockholm Female Coronary Risk Study have observed that low HRV is a predictor of long-term mortality amongst middle-aged women with coronary heart disease when measure 3-6 months after hospitalization for an acute coronary syndrome, even after controlling for established clinical prognostic markers. Indeed, after controlling for the independent significant predictors of mortality amongst the clinical variables, SDNN index, total power, VLF power, LF power and HF power were found to be significant predictors of all-cause mortality.

Results from population-based follow-up studies, as the Zutphen Study [70], the ARIC Study [62] and the Framingham Heart Study [71] have shown that lower HRV is associated with the risk of developing coronary heart disease. These results are have been confirmed also in individuals with diabetes. Liao et al [72] have observed a significantly association between lower HRV and increased risk of coronary heart disease mortality in diabetic subjects.

Analysis of HRV has been used to assess autonomic cardiac activity particularly and to investigate the influence of autonomic neuromodulation in several pathophysiological conditions [73-81]: an impaired heart rate variation is a marker of autonomic neuropathy[82-88].

We want to evaluate the circadian rhythm of the sympatho-vagal balance through heart rate variability analysis in subjects with chronic insomnia. Twenty-five (15 men and 10 women, age: 38 $\pm$ 5.4 years) otherwise healthy Caucasian subjects affected by chronic insomnia were recruited. Chronic insomnia was diagnosed according to the Standards of Practice Committee of the American Academy of Sleep Medicine [4]. All subjects recruited have not glucose metabolism disorders, hypertension, obesity, renal and heart failure, hyperthyroidism, hypothyroidism, dyslipidemia and coronary heart disease and they are not receiving medical treatment that potentially disturbs cardiac autonomic activity (ß-blockers, antiarrhythmic drugs, ACE-inhibitors). The control group consisted in twenty-five healthy subjects matched for sex, age, body mass index, systolic and diastolic blood pressure. Written informed consent was obtained from all participants; all the investigations were performed in accordance with the principles of the Declaration of Helsinki.

Autonomic nervous system function was evaluated by HRV analysis during 24-hour ECG recording. All Holter recordings were performed using a three-channel recorder. Cardiovascular variability was analysed in frequency domain following the recommendations of the Task Force of the European Society of Cardiology and the North American Society of Pacing and Electrophysiology [55]. Spectral estimates of R-R intervals were obtained from stationary regions free of ectopic beats and technical artefacts. Fast Fourier Transform was used to obtain power spectral estimates of HRV; total power in the frequency range (0-0.40 Hz) was divided into: very low frequency (VLF: < 0.04 Hz), low frequency (LF: 0.04-0.15 Hz), high frequency (HF: 0.15-0.40 Hz). The integral under respective power density functions were measured and expressed in absolute units (ms^2/Hz). Each spectral component was presented in normalized units (n.u.) by dividing LF power or HF power by total power minus very low frequency power (LF or HF/ total power – VLF). Sympathovagal balance was expressed by LF/HF ratio.

ECG Holter was analysed for 24 hours, at 10 minutes intervals. Artificial data and arrhythmic events were excluded. All subjects, during ECG Holter registration, spent 8 hours in bed, from 11 p.m. to 7 a.m.. The 24-hour period was divided into four periods: night (0 a.m. – 6 a.m.), morning (6 a.m. – 12 a.m.), afternoon (12 a.m. – 6 p.m.) and evening (6 p.m. – 0 a.m.). We examined HRV during 24 hours, in morning, afternoon, evening and night time. Data analyses were performed with Del Mar Reynolds, Irvine California, USA. In ten healthy subjects, reproducibility was evaluated by means of the interclass correlation coefficient (ICC), comparing baseline values with the results obtained at the fourth week ICC was > 0.7 for HRV.

Statistical analysis was performed with SPSS 13.0 (SPSS Inc., Chicago, IL, U.S.A.) for Windows XP. Normality tests were performed on all data. Parametric data are expressed as means ± SD. The significance of difference was calculated by using Student's t test.

The analysis of HRV showed increased LF u.n. and LF/HF in subjects with chronic insomnia in 24 hours (LF u.n.: 70.1± 6.6 vs 55.4 ± 7.5, $p < 0.001$; LF/HF: 3.1 ± 1.3 vs 1.9 ± 0.4, $p < 0.001$) and in night time (LF u.n.: 66.7 ± 9.8 vs 32.5 ± 5.9, $p < 0.001$; LF/HF: 2.9 ± 1.5 vs 0.7 ± 0.5, $p < 0.001$) when compared with controls. HF u.n. were lower in subjects with chronic insomnia than controls in 24 hours (24.3 ± 6.5 vs 37.6 ± 7.5, $p < 0.001$) and in night time (26.7 ± 8.1 vs 58.6 ± 6.6, $p < 0.001$). Not statistically significant difference were reported in LF u.n., HF u.n. and LF/HF in morning, afternoon and evening time.

Conclusion

Our results show an impaired circadian rhythm of heart rate variability with an arrhythmic sympathovagal balance characterized by a sympathetic hyper-activity during the night in subjects with chronic insomnia (Figure 1). This unbalance in heart rate variability may have important clinical implications.

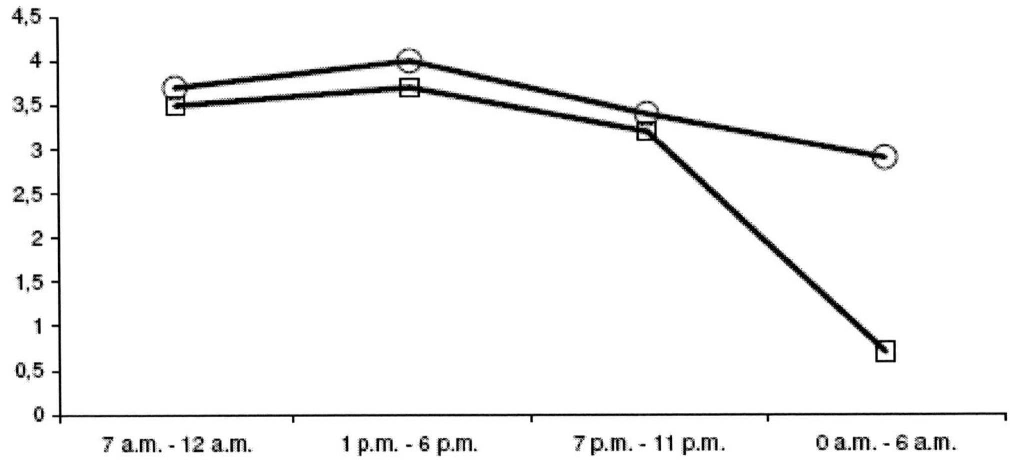

Figure 1. Circadian variation of LF/HF in subjects with chronic insomnia and in control group. White circles: group with chronic insomnia; white squares: control group.

Several studies have suggested that poor sleep is associated with an increase in risk of cardiovascular diseases [27-38]. As previously reported, impairments in heart rate variability have been associated with an higher risk of all-causes mortality and coronary heart disease [62-68, 71]. Therefore, the impaired cardiovascular autonomic activity observed in subjects with chronic insomnia may represent a mechanism involved in the association between quality sleep disorders and higher risk of cardiovascular mortality.

Moreover, the impairment in heart rate variability reflects an unbalanced and arrhythmic output toward the autonomic nervous system that follow the impaired sleep/wake rhythm characterizing the chronic insomnia.

It's possible that chronic quality sleep disorders cause an impairment in the activity of the central biological clock that organizes autonomic rhythm.

An arrhythmic circadian output of autonomic nervous system may play an important role in the relationship between chronic insomnia and its comorbidities, connecting quality sleep disorders to the pathogenesis of hypertension, diabetes mellitus and cardiovascular events.

This autonomic unbalance may explain the higher incidence and prevalence of hypertension and diabetes mellitus in subjects with chronic insomnia and it may be involved in the pathogenesis of these comorbidities [89]. Further studies are necessary to confirm this hypothesis.

References

[1] Sateia MJ, Nowell PD. Insomnia. *Lancet*. 2004; 364: 1959-1973.

[2] Partinen M, Hublin C. Epidemiology of sleep disorders. In: Kryger M, Roth T, Dement WC, editors. *Principle and practice of sleep medicine*. Philadelphia, W. B. Saunders Company; 2000, 558-580.

[3] Roth T. New developments for treating sleep disorders. *J. Clin. Psychiatry*. 2001; 62 (suppl 10): 3-4.

[4] NIH State of the Science Conference Statement on Manifestations and Management of Chronic Insomnia in Adults. National Institutes of Health State-of-the-Science Conference Statement. June 13–15, 2005.

[5] Stewart R, Besset A, Bebbington P, Brugha T, Lindesay J, Jenkins R, Singleton N, Meltzer H. Insomnia comorbidity and impact and hypnotic use by age group in a national survey population aged 16 to 74 years. *Sleep*. 2006; 29(11):1391-7.

[6] Philip P, Leger D, Taillard J, Quera-Salva MA, Niedhammer I, Mosqueda JG, Bioulac B, Gerard D. Insomniac complaints interfere with quality of life but not with absenteeism: respective role of depressive and organic comorbidity. *Sleep Med*. 2006; 7(7):585-91.

[7] Fullerton DS. The economic impact of insomnia in managed care: a clearer picture emerges. *Am. J. Manag. Care*. 2006; 12(8 Suppl):S246-52.

[8] Dixon S, Morgan K, Mathers N, Thompson J, Tomeny M. Impact of cognitive behavior therapy on health-related quality of life among adult hypnotic users with chronic insomnia. *Behav. Sleep Med*. 2006; 4(2):71-84.

[9] Leger D, Massuel MA, Metlaine A; SISYPHE Study Group. Professional correlates of insomnia. *Sleep*. 2006; 29(2):171-8.

[10] Godet-Cayre V, Pelletier-Fleury N, Le Vaillant M, Dinet J, Massuel MA, Leger D. Insomnia and absenteeism at work. Who pays the cost? *Sleep*. 2006 Feb 1;29(2):179-84.

[11] Riedel BW, Lichstein KL. Insomnia and daytime functioning. *Sleep Med. Rev*. 2000; 4(3):277-298.

[12] Edinger JD, Glenn DM, Bastian LA, Marsh GR. Slow-wave sleep and waking cognitive performance II: Findings among middle-aged adults with and without insomnia complaints. *Physiol. Behav*. 2000; 70(1-2):127-34.

[13] Resnick HE, Redline S, Shahar E, Gilpin A, Newman A, Walter R, Ewy GA, Howard BV, Punjabi NM. Diabetes and sleep disturbances. Findings from the Sleep Heart Health Study. *Diabetes Care*. 2003; 26: 702-709.

[14] Punjabi NM, Shahar E, Redline S, Gottlieb DJ, Givelber R, Resnick HE. Sleep-disorders breathing, glucose intolerance and insulin resistance. *Am. J. Epidemiol*. 2004; 160: 521-530.

[15] Elmasry A, Janson C, Lindberg E, Gislason T, Tageldin MA, Boman G. The role of habitual snoring and obesity in the development of diabetes: a 10-year follow-up study in a male population. *J. Intern. Med.* 2000; 248: 13-20.

[16] Al-Delaimy WK, Manson JE, Willet WC, Stampfer MJ, Hu FB. Snoring as a risk factor for type II diabetes mellitus: a prospective study. *Am. J. Epidemiol.* 2002; 155: 387-393.

[17] Ayas NT, White DP, Al Delaimy WK, Manson JE, Stampfer MJ, Speizer FE, Patel S, Hu FB. A prospective study of self-reported sleep duration and incident diabetes in women. *Diabetes Care*. 2003; 26 (2): 380-384.

[18] Meisinger C, Heier M, Loewel H. The MONICA/KORA Ausburg Cohort Study. Sleep disturbance as a predictor of type 2 diabetes mellitus in men and women from the general population. *Diabetologia*. 2005; 48: 235-241.

[19] Gottlieb DJ, Punjabi NM, Newman AB, Resnick HE, Redline S, Baldwin CM, Nieto FJ. Association of sleep time with diabetes mellitus and impaired glucose tolerance. *Arch. Intern. Med.* 2005; 165(8):863-7.

[20] Nilsson PM, Roost M, Engstrom G, Hedblad B, Berglund G. Incidence of diabetes in middle-aged men is related to sleep disturbances. *Diabetes Care*. 2004; 27(10):2464-9.

[21] Suka M, Yoshida K, Sugimori H. Persistent insomnia is a predictor of hypertension in Japanese male workers. *J. Occup. Health*. 2003 Nov;45(6):344-50.

[22] Bixler EO, Vgontzas AN, Lin HM, Vela-Bueno A, Kales A. Insomnia in Central Pennsylvania. J Psychosom Res. 2002; 53: 589-592.

[23] Smolensky MH, Hermida RC, Castriotta RJ, Portaluppi F. Role of sleep-wake cycle on blood pressure circadian rhythms and hypertension. *Sleep Med.* 2007 (in press).

[24] Fiorentini A, Valente R, Perciaccante A, Tubani L. Sleep's quality disorders in patients with hypertension and type 2 diabetes mellitus. *Int. J. Cardiol.* 2007;114(2):E50-2.

[25] Gottlieb DJ, Redline S, Nieto FJ, Baldwin CM, Newman AB, Resnick HE, Punjabi NM. Association of usual sleep duration with hypertension: the Sleep Heart Health Study. *Sleep*. 2006; 29(8):1009-14.

[26] Prejbisz A, Kabat M, Januszewicz A, Szelenberger W, Piotrowska AJ, Piotrowski W, Piwonski J, Makowiecka-Ciesla M, Widecka K, Patera B, Bieniaszewski L, Narkiewicz K, Tykarski A, Piejko A, Grodzicki T, Czerwienska B, Wiecek A. Characterization of insomnia in patients with essential hypertension. *Blood Press*. 2006;15(4):213-9.

[27] Liu Y, Takana H. The Fukuoka Heart Study Group. Overwork, insufficient sleep and risk of non-fatal acute myocardial infarction in Japanese men. *Occup. Environ. Med.* 2002;59:447–451.

[28] Spriggs DA, French JM , Murdy JM, Curless RH, Bates D, James OFW. Snoring increases the risk of stroke and adversely affects prognosis. *QJM* 1992;83:555–562.

[29] Shahar E, Whitney CW, Redline S, Lee ET, Newman AB, Javier Nieto F, O'Connor GT, Boland LL, Schwartz JE, Samet JM. Sleep-disordered breathing and cardiovascular disease: cross sectional results of the Sleep heart health study. *Am. J. Resp. Crit. Care Med.* 2001;163:19–25.

[30] Olson LG, King MT, Hensley MJ, Saunders NA. A community study of snoring and sleep-disordered breathing. Health outcomes. *Am. J. Resp. Crit. Care Med.* 1995;152:717–720.

[31] Koskenvuo M, Kaprio J, Telakivi T, Partinen M, Heikkila K, Sarna S. Snoring as a risk factor for ischaemic heart disease and stroke in men. *BMJ* 1987;294:16–19.

[32] Grunstein RR, Stenlof K, Hedner J, Sjostrom L. Impact of obstructive sleep aprea and sleepiness on metabolic and cardiovascular risk factors in the Swedish obese subjects (SOS) study. *Int. J. Obes.* 1995;19:410–418.

[33] Elwood P, Hack M, Pickering J, Hughes J, Gallacher J. Sleep disturbance, stroke, and heart disease events: evidence from the Caerphilly cohort. *J. Epidemiol. Community Health.* 2006 Jan;60(1):69-73.

[34] Koskenvuo M, Kaprio J, Partinen M. Poor sleep quality, emotional stress and morbidity: a six year follow-up study of 10,778 persons aged 35–59 years. In: Pakaslahti A, ed. *Stress and Psychosomatics*. Helsinki: Psychiatrica Fennica, 1986; 115–20.

[35] Eaker E, Pinsky J, Castelli W. Myocardial infarction and coronary death among women: psychological predictors from a 20-year follow-up of women in the Framingham Study. *Am. J. Epidemiol.* 1992; 135: 854–64.

[36] Mallon L, Broman JE, Hetta J. Sleep complaints predict coronary artery disease mortality in males: a 12-year follow-up study of a middle-aged Swedish population. *J. Intern. Med.* 2002;251(3):207-16.

[37] Siegrist J, Peter JH. Sleep disorders and cardiovascular risk. *Med. Klin.* (Munich). 1986; 81(12):429-32

[38] Appels A, de Vos Y, van Diest R, Hoppner P, Mulder P, de Groen J. Are sleep complaints predictive of future myocardial infarction? *Act Nerv. Super* (Praha). 1987; 29(2):147-51.

[39] Moore RY, Eichler VB. Loss of circadian adrenal corticosterone rhythm following suprachiasmatic lesions in the rat. *Brain Res.* 1972; 42: 201-206.

[40] Stephan FK, Zucker I. Circadian rhythms in drinking behaviour and locomotor activity of rats are eliminated by hypothalamic lesions. *Proc. Natl. Acad. Sci. USA.* 1972; 69: 1583-1586.

[41] Kalsbeek A, Palm IF, La Fleur SE, Scheer FAJL, Perrau-Lenz S, Ruiter M, Kreier F, Cailotto C, Buijs RM. SNC Outputs and the hypothalamic balance of life. *J. Biol. Rhythms.* 2006; 21(6): 458-467.

[42] La Fleur SE, Kalsbeek A, Wortel J, Fekkes ML, Buijs RM. A daily rhythm in glucose tolerance: a role for the suprachiasmatic nucleus. *Diabetes.* 2001; 50: 1237-1243.

[43] Buijs RM, La Fleur SE, Wortel J, Van Heyningen C, Zuiddam L, Mettenleiter TC, TC, Kalsbeek A, Nagai K, Niijima A. The suprachiasmatic nucleus balances sympathethic and parasympathetic output to peripheral organs through separate preautonomic neurons. *J. Comp. Neurol.* 2003; 464: 36-48.

[44] Buijs RM, Kalsbeek A. Hypothalamic integration of central and peripheral clocks. *Nat. Neurosci. Rev.* 2001; 2:521-526.

[45] Guo H, Brewer JM, Champhekar A, Harris RB, Bittman EL. Differential of peripheral circadian rhythms by suprachiasmatic-dependent neural signals. *Proc. Natl. Acad. Sci. USA.* 2005; 102: 3111-3116.

[46] Terazono H, Mutoh T, Yamaguchi S, Kobayashi M, Akiyama M, Udo R, Ohdo S, Okamura H, Shibata S. Adrenergic regulation of clock gene expression in mouse liver. *Proc. Natl. Acad. Sci. USA.* 2003; w100:6795-6800.

[47] Craig AD. How do you feel? Interoception: the sense of the physiological condition of the body. Nat Rev Neurosci. 2002; 3:655-666.

[48] Penicaud L, Leloup C, Lorsignol A, Alquier T, Guillod E. Brain glucose sensing mechanism and glucose homeostasis. *Curr. Opin. Clin. Nutr. Metab. Care.* 2002; 5:539-543.

[49] Deboer T, Detari L, Meijer JH. Long term effects of sleep deprivation on the mammalian circadian pacemaker. *Sleep*. 2007; 30(3):257-62.

[50] Kreier F, Yilmaz A, Kalsbeek A, Romijn JA, Sauerwein HP, Fliers E, Buijs RM. Hypothesis: shifting the equilibrium from activity to food leads to autonomic unbalance and the metabolic syndrome. *Diabetes.* 2003; 52: 2652-2656.

[51] Lown B, Verrier RL. Neural activity and ventricular fibrillation. *N. Engl. J. Med.* 1976; 294:1165-1170.

[52] Corr PB, Yamada KA, Witkowski FX. Mechanisms controlling cardiac autonomic function and their relation to arrhythmogenesis. In: Fozzard HA, Haber E, Jennings RB, Katz AN, Morgan HE, eds. *The Heart and Cardiovascular System.* New York, NY: Raven Press; 1986:1343-1403.

[53] Schwartz PJ, Priori SG. Sympathetic nervous system and cardiac arrhythmias. In: Zipes DP, Jalife J, eds. *Cardiac Electrophysiology: From Cell to Bedside.* Philadelphia, Pa: WB Saunders Co; 1990:330-343.

[54] Levy MN, Schwartz PJ, eds. *Vagal Control of the Heart: Experimental Basis and Clinical Implications.* Armonk, NY: Futura; 1994.

[55] Task Force of the European Society of Cardiology the North American Society of Pacing Electrophysiology. Heart Rate Variability. Standards of Measurement, Physiological Interpretation, and Clinical Use. *Circulation.* 1996;93:1043-1065.

[56] Pagani M, Lombardi F, Guzzetti S, Rimoldi O, Furlan R, Pizzinelli P, Sandrone G, Malfatto G, Dell'Orto S, Piccaluga E, Turiel M, Baselli G, Cerutti S, Malliani A. Power spectral analysis of heart rate and arterial pressure variabilities as a marker of sympathovagal interaction in man and conscious dog. *Circ. Res.* 1986;59:178-193.

[57] Malliani A, Pagani M, Lombardi F, Cerutti S. Cardiovascular neural regulation explored in the frequency domain. *Circulation*. 1991;84:1482-1492.

[58] Kleiger RE, Bigger JT, Bosner MS, Chung MK, Cook JR, Rolnitzky LM, Steinman R, Fleiss JL. Stability over time of variables measuring heart rate variability in normal subjects. *Am. J. Cardiol.* 1991;68:626-630.

[59] Van Hoogenhuyze DK, Weinstein N, Martin GJ, Weiss JS, Schaad JW, Sahyouni XN, Fintel D, Remme WJ, Singer DH. Reproducibility and relation to mean heart rate of heart rate variability in normal subjects and in patients with congestive heart failure secondary to coronary artery disease. *Am. J. Cardiol.* 1991;68:1668-1676.

[60] Kautzner J. Reproducibility of heart rate variability measurement. In: Malik M, Camm AJ, eds. *Heart Rate Variability.* Armonk, NY: Futura; 1995:165-171.

[61] Bigger JT, Fleiss JL, Rolnitzsky LM, Steinman RC. Stability over time of heart period variability in patients with previous myocardial infarction and ventricular arrhythmias. *Am. J. Cardiol.* 1992;69:718-723.

[62] Liao D, Cai J, Rosamond WD, Barnes RW, Hutchinson RG, Whitsel EA, Rautaharju P, Heiss G. Cardiac autonomic function and incident coronary heart disease: a population-

based case-cohort study: the ARIC study: Atherosclerosis Risk in Communities Study. *Am. J. Epidemiol.* 1997; 145: 696-706.

[63] Bigger JT, Fleiss JL, Rolnitzky LM, Steinman RC. The ability of several short-term measures of RR variability to predict mortality after myocardial infarction. *Circulation.* 1993; 88: 927-934.

[64] Hemingway H, Shipley M, Brunner E, Britton A, Malik M, Marmot M. Does autonomic function link social position to coronary risk?: The Whitehall II Study. Circulation. 2005; 111: 3071-3077.

[65] Kleiger RE, Miller JP, Bigger JT Jr. Decreased heart rate variability and its association with increased mortality after acute myocardial infarction. *Am. J. Cardiol.* 1987; 59: 256-262.

[66] Vaishnav S, Stevenson R, Marchant B, Lagi K, Ranjadayalan K, Timmis A. Relation between heart rate variability early after acute myocardial infarction and long-term mortality. *Am. J. Cardiol.* 1994; 73: 653-657.

[67] Lombardi F, Sandrone G, Pernpruder S, Sala R, Garimoldi M, Cerutti S, Baselli G, Pagani M, Malliani A. Heart rate variability a san index of sympathovagal interaction after acute myocardial infarction. *Am. J. Cardiol.* 1987; 60:1239-1245.

[68] Malik M, Farrel T, Camm AJ. Circadian rhythm of heart rate variability after acute myocardial infarction and its influence on the prognosis value of heart rate variability. *Am. J. Cardiol.* 1990; ì66: 1049-1054.

[69] Janszky I, Ericson M, Mittleman MA, Wamala S, Al-Khalili F, Schenck-Gustafsson K, Orth-Gomer K. Heart rate variability in long-term assessment in middle-aged women with coronary heart disease: The Stockholm Female Coronary Risk Study. *J. Int. Med.* 2004; 255: 13-21.

[70] Dekker JM, Schouten EG, Klootwijk P, Pool J, Swenne CA, Kromhout D. Heart rate variability from short electrocardiographic recordings predicts mortality from all causes in middle-aged and elderly men: The Zutphen Study. *Am. J. Epidemiol.* 1997; 45 (10): 899-908.

[71] Tsuji H, Larson MG, Venditti FJ Jr, Manders ES, Evans JC, Feldman CL, Levy D. Impact of reduced heart rate variability on risk for cardiac events. The Framingham Heart Study. *Circulation.* 1996 Dec 1;94(11):2850-5.

[72] Liao D, Carnethon M, Evans GW, Cascio WE, Heiss G. Lower heart rate variability is associated with the development of coronary heart disease in individuals with diabetes: The Ateherosclerosis Risk in Communities (ARIC) Study. *Diabetes.* 2002; 51: 3524-3531.

[73] Pagani M, Lombardi F, Guzzetti S, Rimoldi O, Furlan R, Pizzinelli P, Sandrone G, Malfatto G, Dell'Orto S, Piccaluga E, et al.: Power spectral analysis of heart rate and arterial pressure variabilities as a marker of sympatho-vagal interaction in man and conscious dog. *Circ. Res.* 1986, 59(2):178-93.

[74] Malliani A, Pagani M, Lombardi F, Cerutti S: Cardiovascular neural regulation explored in the frequency domain. *Circulation* 1991, 84(2):482-92.

[75] Montano N, Ruscone TG, Porta A, Lombardi F, Pagani M, Malliani A: Power spectrum analysis of heart rate variability to assess the changes in sympathovagal balance during graded orthostatic tilt. *Circulation* 1994, 90(4):1826-31.

[76] Schwartz P, La Rovere M, Vanoli E: Autonomic nervous system and sudden cardiac death: experimental basis and clinical observations for post-myocardial infarction risk stratification. *Circulation* 1992, 85:177-191.

[77] Malliani A, Schwartz P, Zanchetti A: Neural mechanism in life threatening arrhythmias. *Am. Heart J.* 1980, 200:705-715.

[78] [78] Tubani L, Baratta L, Giorgino F, Delfino M, Fiore G, Golluscio V, Giacovazzo M: Heart rate variability in cluster headache. *Ann. Ital. Med. Int.* 2003, 18(1):42-46.

[79] Baratta L, Delfino M, Fiorentini A, Martuscelli M, Tubani L: Alterations of sympathovagal bilance by heart rate variability: a rare case of adult Still's disease. *Ann. Ital. Med. Int.* 2004, 19(3):193-197.

[80] Mastrocola C, Vanacore N, Giovani A, Locuratolo N, Vella C, Alessandri A, Baratta L, Tubani L, Meco G: Twenty-four-hour heart rate variability to assess autonomic function in Parkinson's disease. *Acta Neurol. Scand.* 1999, 99(4):245-247.

[81] Laganà B, Tubani L, Maffeo N, Vella C, Makk E, Baratta L, Bonomo L: Heart rate variability and cardiac autonomic function systemic lupus erythematosus. *Lupus* 1996, 5(1):49-55.

[82] Ewing DJ, Neilson JM, Shapiro CM, Stewart JA, Reid W: Twenty four hour heart rate variability: effects of posture, sleep, and time of day in healthy controls and comparison with bedside tests of autonomic function in diabetic patients. *Br. Heart J.* 1991, 65:239-244.

[83] Ziegler D: Diabetic cardiovascular autonomic neuropathy: prognosis, diagnosis and treatment. Diabetes *Metab. Rev.* 1994, 10:339-383.

[84] Malpas SC, Maling TJB: Heart rate variability and cardiac autonomic function in diabetes. *Diabetes* 1990, 39:1177-1181.

[85] Pagani M, Malfatto G, Pierini S, Casati R, Masu AM, Poli M, Guzzetti S, Lombardi F, Cerutti S, Malliani A: Spectral analysis of heart rate variability in the assessment of autonomic diabetic neuropathy. *J. Auton. Nerv. Syst.* 1988, 23:143-15.

[86] Bianchi A, Bontempi B, Cerutti S, Gianoglio P, Comi G, Natali Sora MG: Spectral analysis of heart rate variability signal and respiration in diabetic subjects. *Med. Biol. Eng. Comput.* 1990, 28:205-211.

[87] Bellavere F, Balzani I, De Masi G, Carraro M, Carenza P, Cobelli C, Thomaseth K: Power spectral analysis of heart rate variation improves assessment of diabetic cardiac autonomic neuropathy. *Diabetes* 1992, 41:633-640.

[88] Van den Akker TJ, Koelman ASM, Hogenhuis LAH, Rompelman G: Heart rate variability and blood pressure oscillations in diabetics with autonomic neuropathy. *Automedica* 1983, 4:201-208.

[89] Kreier F, Kalsbeek A, Sauerwein HP, Fliers E, Romijn JA, Buijs RM. "Diabetes of the elderly" and type 2 diabetes in younger patients: possible role of the biological clock. *Exp. Gerontol.* 2007; 42:22-27.

In: Circadian Rhythms and Health Research Trends
Editor: Alfredo V. Lignelli, pp. 131-143
ISBN: 978-1-60021-875-0

Chapter VII

Melatonin-Rich Night-Time Milk, Sleep and Daytime Activity in Healthy Middle-Aged Subjects

Maija Valtonen[1], Leo Niskanen[2], Teuvo Koskinen[3] and Antti-Pekka Kangas[4]

[1]Institute of Applied Biotechnology, University of Kuopio, P.O. Box 1627, 70211 Kuopio, Finland
[2]Department of Medicine, Kuopio University Hospital, P.O. Box 1777, F-70211 Kuopio, Finland
[3]Department of Psychogeriatrics, Kuopio University Hospital, P.O. Box 1777, F-70211 Kuopio, Finland
[4]Institute of Applied Biotechnology, University of Kuopio, P.O. Box 1627, 70211 Kuopio, Finland

Abstract

Administration of melatonin in variable doses resulting in supraphysiological or physiological night-time blood levels of melatonin has been used to improve sleep quality. Melatonin is a potent endogenous radical scavenger and has also been available as a food additive. The ultimate goal of melatonin use is to increase the subsequent daytime well-fare and activity. Melatonin is found in all body fluids, and it is also excreted in milk at a distinct diurnal rhythm. Night-time milk provides low doses of melatonin, which do not affect daytime blood melatonin concentrations. The aim of this study was to monitor objective sleep quality and daytime physical activity in healthy middle-aged people consuming melatonin-rich night-time milk daily or during intake of a pharmacological dose of melatonin. The study was performed in 4 two-week periods in 12 healthy middle-aged subjects. During the first phase of the double blind trial the subjects received either normal commercial milk or melatonin-rich night-time milk with their meals providing 5-20 ng melatonin daily for 2 weeks and during the second phase normal commercial milk and a placebo or 2 mg melatonin before bed-time. Activity and sleep were continuously recorded by an actigraph during the second week of each experimental period. Sleep efficiency indicated by a decrease in the number of awakenings improved significantly ($p < 0.05$) during the night-milk period compared to the normal milk period. Similar improvement was not found with the 2 mg daily melatonin treatment when compared to placebo. No significant changes were observed in daytime activity between the treatment groups. However, significantly increased daytime activity was seen in 7 out of the 12 subjects during the night-milk period compared to the normal milk period while the 2 mg melatonin treatment increased daytime activity compared to placebo only in 5 subjects. Objective recording of daytime activity may be an alternative method to monitor the effect of low supply of melatonin. Our findings suggest that the benefits of melatonin may be obtained with far lower doses than used until now and without a risk of daytime drowsiness.

Introduction

The well-being of people depends on the maintenance of the endogenous daily rhythms of biological functions and the sleep-wake rhythm. Melatonin that is produced by the pineal gland at night is considered to have an important role in regulation of the sleep-wake cycle [Dawson and Encel 1993]. Although the underlying mechanisms of melatonin on sleep are unknown, it has been suggested that it exerts a sleep inducing effect through thermoregulatory mechanisms. By lowering core body temperature, melatonin reduces arousal and increases sleep-propensity. It has also been suggested that melatonin improves sleep by attenuating a hypothesized alerting signal that emanates from the suprachiasmatic nucleus and which peaks before the endogenous secretion of melatonin. [Dawson and Encel 1997].

Melatonin production from the pineal gland is very low prior to three months of age in humans. It then commences, becomes circadian, and reaches its highest nocturnal blood levels between the ages of one to three years. Nocturnal peak levels drop progressively by 80% during the remainder of childhood. In adults melatonin levels decrease additionally some 10%, mainly during senescence [Waldhauser et al.1998]. The decline of melatonin with age causes changes in circadian rhythms and in sleep quality along with behavioural changes such as a reduction in physical activity and exposure to photic stimulation [Myers and Badia 1995]. In addition, a decrease in exposure to natural light in institutions can exacerbate sleep difficulties. If there is no light passing through the eye, there is not a message to the SCN to reset the circadian rhythm. Also this may lead to diminution of melatonin excretion in old age [Haimov et al 1994].

When in sleep, we pass through two fundamentally different stages: rapid eye movement (REM) and non-REM (NREM) sleep. REM sleep, sometimes referred to as shallow sleep, accounts for about 25 % of total sleep time. REM sleep is the stage during which most dreaming occurs. During REM sleep, muscle tone decreases, rapid eye movements occur, and the electroencephalogram (EEG) shows low-amplitude, fast-frequency activity similar to that seen during wakefulness. NREM sleep accounts for the remaining 75 % of sleep time. On the basis of gradual slowing of frequency and an increase of amplitude in EEG, NREM sleep is further classified into four stages, and stages 3 and 4 are often termed slow-wave sleep or deep sleep.

Sleep is an active process regulated by the central nervous system. The most widely held theory about the function of sleep is that it serves as a period of recuperation or restoration. It has been postulated that slow wave and REM sleep have different restorative functions, slow wave sleep being important for macromolecular synthesis and REM sleep for removing the synthetic products of slow wave sleep to maintain synaptic connections. It has been suggested that during REM sleep neuronal connections in the catecholamine system are formed and that this activity is necessary to maintain cognitive function [Shapiro and Flanigan 1993].

Age-related changes in the amount and pattern of sleep and wakefulness have been well described. Older people generally spend more time in bed but less time asleep and are more easily aroused from sleep than younger persons. There is a reduction in slow-wave sleep (particularly the decline in stage 3 and 4 sleep can be nearly complete by the age of 60 years), increased night-time awakenings, and increased fragmentation of sleep by periods of wakefulness. Reductions in REM sleep and total night-time sleep also occur with age, and NREM transitions between stage 1 and 2 become more common [Gorbien 1993].

The sleep patterns that accompany advancing age suggest age-related disturbances in the circadian sleep-wake rhythm. The most striking age-related change in the sleep pattern is the repeated and frequent interruption of sleep by long periods of wakefulness. Increased night-time awakenings in the elderly are mirrored by increase in daytime fatigue, daytime napping, and the likelihood of falling asleep during the day [Carskadon et al.1982]. Age-related changes in sleep occur independent of any medical or psychiatric illnesses or primary sleep disorders [Vitello 1999].

The most melatonin is secreted into the circulation during REM sleep and the least during slow wave sleep [Birkeland 1982]. The synthesis would be stimulated by total sleep deprivation [Salin-Pascual et al.1988]. These findings suggest that melatonin is required to sustain sleep uninterrupted during the night.

It has been shown that exogenous melatonin has an overall beneficial effect on sleep quality in melatonin-deficient elderly people [Garfinkel et al.1995]. Total sleep time is usually not affected but sleep latency is decreased. With high doses of melatonin, which induce supraphysiological blood concentrations, a dose depended effect on subjective sleepiness has been demonstrated e.g. by Deacon and Arendt [1995]. An effect of melatonin on sleep has also been observed by much lower doses leading to blood levels closer to the normal melatonin concentration range at night. With the lower doses (i.e. doses not higher than 0.5 mg) the effect did not show any dose-response [Zhdanova et al.1995].

The therapeutic effects of melatonin in pharmacological doses on circadian-based sleep disorders in blind people, travellers with jet lag and patients with delayed–sleep-phase syndrome as well as the improving effect of melatonin on mood in depressed patients are well established [Dawson and Encel 1993]. The lowest dose of 0.1 mg of melatonin orally 30 min before bedtime also improved sleep in elderly people with poor sleep efficiency [Zhdanova et al. 2001]. A dose of 0.1 mg taken during the daytime raises blood melatonin concentrations to within the normal nocturnal range and has also been shown to cause hypnotic effects in normal people [Dollins et al.1994]. The lowest dose used is still ten times more than the total amount of night-time secretion of melatonin, that is, 10 μg/night in humans [Huether 1993].

Studies on doses of melatonin that do not affect the daytime blood concentration of melatonin are scarce.

Melatonin is found in all body fluids and its concentration in these parallels that in blood and reflects the secretion pattern from the pineal gland. Melatonin is also excreted in milk and a distinct melatonin rhythm has been recorded in human and bovine milk [Illnerová et al. 1993, Eriksson et al. 1998]. Our recently reported study with night-time milk which provided only 10-20 ng melatonin daily indicated that doses of this category may benefit part of elderly people by increasing their daytime activity without measurable effects on sleep [Valtonen et al. 2005]. In the present work we used actigraphy to objectively monitor sleep time and daytime physical activity in normal middle-aged subjects who consumed a moderate amount of night-time milk daily.

Material and Methods

The participants were 12 healthy independently living middle-aged subjects (8 females and 4 males, mean age 60.5 years [range 46-70]) and without specific sleep complaints. For objective recording of physical activity and sleep wrist-actigraphy was used to minimise intrusion into the life of the volunteers recruited by word-of-mouth. Participants were asked to keep their daily routines as constant as possible for the duration of the study. Each subject signed informed consent to participate. The study was approved by the local committee on ethics of human research. The subjects agreed to drink milk with all their meals. The mean milk consumption was 0.87 (±0.32 SD) l per day.

Night-time milk was obtained from the dairy Osuuskunta Maitomaa, Suonenjoki, Finland. The milk was obtained by milking cows in darkness at night [Eriksson et al.1998]. The light conditions and milking times were adjusted so that the melatonin concentration of the milk was highest possible [Valtonen et al. 2003]. It was pasteurised and the unhomogenised fat content was standardised to 1.5 % and packaged as normal commercial low fat milk containing the same 1.5 % fat. The melatonin concentrations of night-milk were determined every second week by radioimmunoassay as previously described [Eriksson et al.1998] and was 15.1±3.1 pg/ml (mean ± SD) during this study. Daytime melatonin concentration in bovine milk is close to the detection limit of the assay that is 5 pg/ml.

The study was performed in four two-week periods. During the first phase, all subjects received in a double blind fashion either normal commercial milk (period 1) or night-milk (period 2) for 2 weeks. The second phase consisted of two weeks treatment with placebo (period 3) or melatonin 2 mg per day (period 4) taken half an hour before the usual bed-time and consumption of normal milk at meals. During the second week of the different treatment periods an actigraph was worn by the subjects on their non-dominant wrist. The actigraph (Mini Motionlogger, Ambulatory Monitoring Inc. New York, USA) is a small computerised device which continuously records wrist movements over several consecutive days with negligible distortion, while subjects are living at home and sleeping in their own beds. The acticraph utilizes a sensitive motion sensor known as accelometer to monitor the occurrence and degree of motion. It produces a signal which magnitude and duration depends on the amount of motion. The activity signals are amplified and digitized and stored as activity counts. When the recording period was completed the stored movement data were transferred to a computer for reading and analysis using Act software version 6.10. This study was conducted during the late autumn months.

The statistical analysis of the subjects' data was based on the actigraphic parameters from the continuous 1-week (second week) monitoring in each experimental period. Sleep variables derived from the actigraphic data included total sleep time (TST), defined as the actual time (min) spent asleep, sleep efficiency index (SEI) as a percentage of total sleep time out of total time from sleep onset to wake time, and fragmentation of sleep as number of awakenings during the sleep and the accumulated time awake after onset of sleep (WASO). The parameters were analysed by repeated measures ANOVA and by t-test to determine treatment differences.

Daytime activity (cumulative absolute value sum m/s^2) was derived from the stored movement data reported numerically as total activity counts per minute during the whole day when awake. In addition the time waken was divided into two equal parts here called forenoon and afternoon and monitored separately (am activity and pm activity). Very large interindividual differences were evident in the level of activity of the subjects and the comparisons between treatments were done within-subjects using two pairs: Normal day-time milk (period 1) vs. night-time milk (period 2) and placebo and normal milk (period 3) vs. 2 mg melatonin and normal milk (period 4). Differences in these pairs were analyzed using repeated measures pair wise t-test. The data are presented as means and standard deviations (SD). A confidence level of 95% was used throughout all statistical tests (i.e. $p<0.05$ was considered significant).

Results

Comparison of sleep parameters derived for period 1 normal commercial milk and period 2 night-milk during the first phase revealed significant improvement in sleep efficiency index during intake of night-milk (Table 1.). It seems to be a consequence of decreased wakening times during the sleep and correspondingly decrease of WASO. The effect of night-milk on TST was not significant.

Table 1. Sleep characteristics (mean ± SD, n = 12) obtained by actigraphy and results of comparisons between treatment periods 1 vs. 2 and 3 vs. 4

Period	Wakening times/night	Wake after sleep onset min/night	Total sleep time min/night	Sleep efficiency index %
1 Normal	2.5 ± 1.5	29,1 ± 32,8	460 ± 57	93.8 ± 4.9
2 Night-milk	1.9 ± 1.6 *	21,0 ± 24,1 *	456 ± 52	95.5 ± 4.4 *
3 Placebo	2.2 ± 3.3	22,6 ± 24,7	453 ± 56	94.5 ± 4.3
4 Mel 2 mg	2.7 ± 1.0 *	29,8 ± 29,9 *	459 ± 65	93.5 ± 5.5

Significant difference between the periods *($p<0.05$).

Comparisons of sleep parameters during the second phase revealed increased wakening times and WASO during melatonin intake leading to a tendency of less efficient sleep. The increase in TST was neither significant. Daytime actigraph recordings, conducted for 6 successive days during the second week of each experimental period and reported numerically as total activity counts per minute are presented in Table 2. The differences in activity between subjects were great ($p<0.000$) and group wise comparisons during the first or second phase did not show any significant differences between the treatment periods.

Table 2. Daytime activity scores (mean ± SD, n = 12) during the different treatment periods

	Total activity counts/min	Activity am counts/min	Activity pm counts/min
1 Normal	198 ± 82	206 ± 77	189 ± 85
2 Night-milk	198 ± 84	212 ± 76	183 ± 91
3 Placebo	203 ± 83	215 ± 76	188 ± 90
4 Mel 2 mg	204 ± 81	217 ± 72	188 ± 87

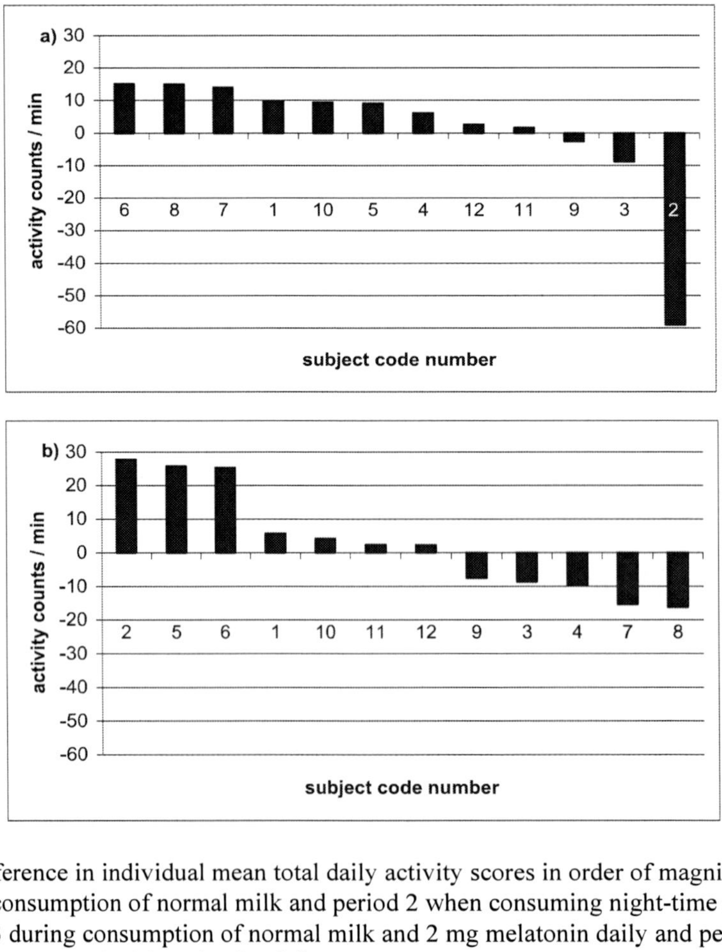

Figure 1.The difference in individual mean total daily activity scores in order of magnitude a) between period 1 during consumption of normal milk and period 2 when consuming night-time milk and b) between period 3 during consumption of normal milk and 2 mg melatonin daily and period 4 when normal milk was consumed with a placebo daily.

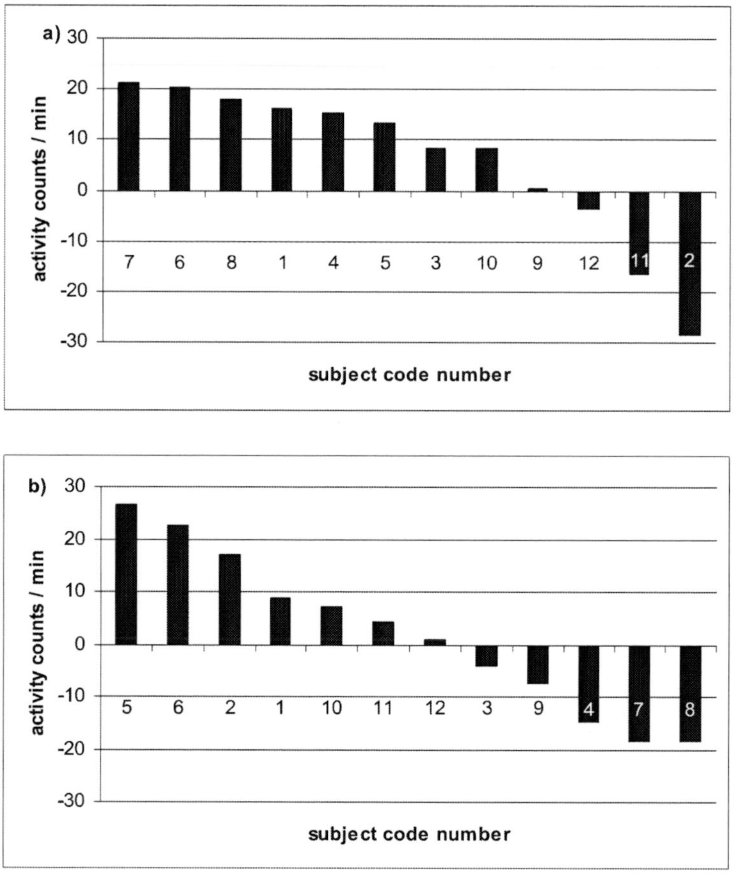

Figure 2.The difference in individual mean daily forenoon activity scores in order of magnitude a) between period 1 during consumption of normal milk and period 2 when consuming night-time milk and b) between period 3 during consumption of normal milk and 2 mg melatonin daily and period 4 when normal milk was consumed with a placebo daily.

If the activity comparisons are done within-subjects daytime activity in the first phase increased in 7 (58 %) of the subjects during period 2 when night-milk was consumed compared to period 1 (normal milk). In three subjects daytime activity decreased and in two subjects it was unchanged. In the second phase when 2 mg melatonin was taken in the evening (period 4) intake of melatonin increased daytime activity in only five subjects (42 %) but decreased it in six subjects (50 %) when compared to placebo condition.

In order to analyse the activity scores in more detail the waking hours of each day were divided in two equal parts here called forenoon (am) and afternoon (pm) and the activity counts were analysed separately . These analyses revealed that daytime activity in the first phase when the night-milk period was compared to the normal milk period increased in 9 subjects (75%) with increased activity either forenoon or afternoon or both. The increase in activity mostly occurred in forenoon (7 out of 9) subjects. While increase in activity when melatonin was taken (period 4) was seen in forenoon in 2 and both in forenoon and afternoon in 3 subjects that experienced increased daytime activity. Six subjects were more active in the morning when taking placebo than when taking melatonin. Two of the subjects (no 3 and no

9) did not increase their activity either using night-milk or melatonin and 2 of them (no 7 and 8) increased their activity when drinking night-milk but not when taking melatonin.

The difference in the mean daily total activity between period 1 and 2 (Figure 1a) and period 3 and 4 (Figure 1b) and the code number of the subject increasing or decreasing their activity when consuming night-time milk or taking melatonin are shown in order of magnitude of the response in Figure 1. The increase in activity when consuming night-time milk compared to the activity during consumption of normal milk of subjects 6, 8, 7, 1, 10, 5 and 4 is significant as well as the decrease in subjects 9, 3 and 2 (Figure 1a). The subject no 2 represents an obvious outlier response.

During treatment of melatonin there was a great increase in activity in three subjects when compared to the placebo condition subjects no 2, 5 and 6, but only a moderate increase in subjects no 1 and 10. If only forenoon is considered the response to night-time milk and melatonin is more pronounced as seen in Figure 2, which describes the difference in daily forenoon activity when period 1 and 2 are compared (Figure 2a) and when periods 3 and 4 are compared (Figure 2b).

Discussion

A broad range of doses of melatonin from 0.1 mg to several grams have been employed in treatment of insomnia [Sack et al. 1997]. Although the data are mixed it is generally thought that melatonin in doses which induce supraphysiological or physiological blood levels of melatonin improves sleep in elderly insomniacs [Haimov et al.1995, Zhdanova et al. 2001] and similar doses may also improve sleep in healthy middle-aged subjects [Attenburrow et al. 1996]. In this study it was investigated if far lower doses of melatonin in nanogram order affects sleep quality or daytime activity in normal middle-aged people. The main result in this study detected by actigraphy was the significant decrease in the number of wakening times resulting in a decrease in WASO (time awake after onset of sleep) and consequently improved sleep efficiency.

The subjects did not suffer from insomnia but because of their age some decline in their melatonin secretion was supposed [Waldhauser et al. 1998] and sleep efficiency will decrease, brief wakenings become common and WASO increases with aging.

Dawson and Encel [1997] and Sack et al. [1997] supposed that melatonin improves sleep by 1) producing corrective circadian phase shifts and thereby improving the alignment of the endogenous sleep-propensity rhythm with the desired sleep schedule. The benefit of melatonin-induced phase shifting is based on the principle that sleep quality and quantity are maximal if sleep occurs at the optimal circadian phase. Studies of Attenburrow et al. [1996], Dawson and Encel 1993], and later Pires et al. [2001] suggest a supraphysiological threshold for sleep-promoting and phase shifting effects with increased plasma melatonin concentration and decreased body core temperature somewhere between 0.1 and 1.0 mg melatonin depending on the time of administration and age and sensitivity of the subject as well as the properties of the melatonin preparation.

So the postulated effects of melatonin are improvement of sleep mostly by correcting circadian phase abnormalities and increasing sleep propensity. The low dose of melatonin that can be administered in night-time milk hardly has any synchronising effect and it is not possible to drink so much of milk that a direct sleep-promoting effect could be achieved. In the present study increased sleep efficiency resulted from decreased number of wakening times. The improvement of sleep quality is likely mediated by increased slow-wave sleep although it was not studied in detail herein.

The low physiological content of melatonin in night-milk does not cause any rise in plasma melatonin concentration but it may increase the melatonin availability and support declined endogenous melatonin secretion to affect all target systems improving in this way sleep quality. The effect may be mediated directly in the gut through receptors on the intestinal cell wall. A delayed response to melatonin and an increased efficacy with repeated treatments has been observed with pharmacological doses of melatonin [MacFarlane et al. 1991, Haimov et al. 1995]. Long-term use probably would substantiate the efficacy of night-time milk, too. The lowest effective melatonin dose that has measurable effects has not been determined and may be far below the presently considered physiological doses.

Questionnaires have been the main means to study the impact of melatonin treatment on sleep and mood. However, questionnaires are subjective and qualitative and tend to underestimate low-intensity activity associated with activities of daily living. In order to perceive also the minor chances in daily activity the effect of exogenous melatonin was monitored by actigraph recording in this study. The acticraph automatically collect and score data for sleep parameters and assess activity in any instance where quantifiable analysis of physical motion is desired. Acticraphic testing has been predominantly used in research studies to evaluate rest-activity cycles in patients with sleep disorders. Although it is not considered as a fully valid method for effective assessment of sleep disorders the practice guidelines for actigraphy established by the Standards of Practice Committee of the American Academy of Sleep Disorders, American Sleep Medicine [Littner et al. 2003] state that actigraphy testing is reliable and valid for detecting sleep in normal, healthy populations. Besides detection of rest-activity cycles actigraphy testing can be used to evaluate basic activity levels that are considered to correlate with mood and energy expenditure.

Although all day round actigraphy did not show any differences between treatment groups because of the large individual variability in physical activity of the subjects intraindividual examination revealed significant increase in daytime physical activity in more subjects drinking night-milk (7 out of the 12 subjects, 58 %) compared to the normal milk than taking melatonin 2 mg half an hour before bedtime compared to placebo (5, 42%). Although daytime activity increased in many subjects total sleep time did not increase when night-milk was consumed instead it showed a tendency to decrease. On melatonin treatment a trend in increasing sleep duration was seen. However, quantity does not compensate quality.

Improved sleep quality or increased sleep efficiency could explain the increase in daytime activity when consuming night-milk or melatonin. Studies using subjective rating of daytime alertness on melatonin conditions have indicated improved memory and mood in the elderly [Jean-Louis et al.1998] and effects on subjective feelings of well-being [MacFarlane et al.1991] and an increase in daytime activity in some elderly [Valtonen et al. 2005] confirming our results. Despite patients in the study of James et al. [1990] reported less sleep on melatonin conditions, subjective quality of sleep was reported to be improved. Studies measuring objective daytime alertness or activity on conditions of melatonin intake in the evening have, however, not been earlier reported.

Further more, we found increased activity more often before noon than afternoon when consuming night-time milk. If increased activity both before and/or afternoon together are considered on night-milk consumption, the proportion of subjects with this beneficial response is 75 %. As is true with any type of therapy, melatonin in high or low doses does not affect everyone as seen in subjects number 3 and 9 who did not increase their activity any time of the day on either night-milk or melatonin condition. The low physiological content of melatonin in night-milk although it does not cause any rise in plasma melatonin concentration may, however, increase the melatonin availability, this in turn may boost the daytime alerting signal generated by the supraschiasmatic nucleus (SCN) causing the numerous "early birds" showing increased activity in forenoon in the present study. Attenuation of this signal is proposed as a model to explain the sleep-promoting effect of melatonin in humans [Sack et al.1997]. Thus this low-dose melatonin may affect the sensitivity of the pineal gland. From the field of human endocrinology there is evidence to support that a very low physiological dose may be more sensitive than high pharmacological doses. It has been found that low-dose (only 1 μg) ACTH test is more sensitive than the conventional high-dose (250 μg) ACTH-test for the diagnosis of pituitary adrenal insufficiency [Ambrosis et al. 1998].

Increased total sleep time has not been observed in all studies with melatonin [Garfinkel et al. 1995, Jean-Louis et al. 1998]. However, improvement of sleep quality in elderly people has been reported with 2 mg melatonin taken in the evening [Garfinkel et al.1995] and sleep inducing effects have been noticed in young individuals with even lower doses (0.3 and 1.0 mg) [Sack et al. 1997]. These doses of melatonin although considered low increase plasma melatonin concentration to or well beyond the normal nocturnal level of melatonin and may cause plasma melatonin to remain elevated into the daylight hours. Sleep duration showed a trend to increase in the melatonin condition in this study. This may result in decreased alertness the next day and may be the reason for fewer subjects showing increased activity in the morning when taking 2 mg melatonin than when drinking night-milk.

The benefits of drinking milk are mostly attributed to the bioactive peptides derived from milk proteins by gastrointestinal digestion, fermentation or hydrolysis by proteolytic enzymes. The health benefits attributed to peptides are e.g. antimicrobial, antioxidative, antithrombotic, antihypertensive and immunomodulatory activities including sleep aiding effect [Korhonen and Pihlanto 2006]. They are much the same health benefits attributed to melatonin [Bubenic et al.1998]. However, the occurrence of peptides have not been shown to be more abundant in milk during the night so it can be concluded that the improvement of sleep efficiency is a consequence of increased melatonin in night-time milk. It is, however, possible that melatonin and milk borne peptides exert a synchronous effect.

Although no serious injurious effects have been reported in humans with either low or high doses of melatonin, when used for short periods of time, melatonin in doses that may increase blood melatonin concentration can have unrecognized subtle side effects when given chronically. A million litres of night-milk per year have been used as a drink with meals and in cooking in Finland for the past seven years without signs of any harmful effect.

Conclusion

Actigraphy revealed the efficacy of melatonin to improve sleep quality and increase subsequent daytime activity already in very low doses. Actigrafic monitoring of daytime performance, too, could be an alternate to define effects of melatonin as well as any risks that might be involved with long-term treatment. Doses in nanogram order that can be ingested in the night-time milk daily do not increase blood melatonin concentration and hardly can be suspected to have any detrimental side effects. Benefits of melatonin may be reached with long-term use of ultra low doses of melatonin like those in night-milk and without a risk of long-term use of pharmacological doses of melatonin.

References

Ambrosi, B., Barbetta, L., Re, T., Passini, E. and Faglia, G. (1998). The one microgram adrenocorticotropin test in assesment of hypothalamic-pituitary-adrenal function. *Eur. J. Endocrinol., 139,* 575-579.

Attenburrow, M.E.J., Cowen, P.J. and Sharpley, A.L. (1996). Low dose melatonin improves sleep in healthy middle-aged subjects. *Psychopharmacology; 126,* 179-181.

Birkeland, A.J. (1982). Plasma melatonin levels and nocturnal transitions between sleep and wakefulness. *Neuroendocrinology, 34*, 126-131.

Bubenic, B.A., Blask, D.E., Brown, G.M., Maestroni, G.J., Pang, S.F., Reiter, R.J., Viswanathan, M. and Zisapel, N. (1998). Prospects of the clinical utilization of melatonin. *Biol. Signals Recept., 7*, 195-219.

Carskadon, M.A., Brown, E.D. and Dement, W.C. (1982). Sleep fragmentation in the elderly: relationship to daytime sleep tendency. *Neurobiol. Aging., 3*, 321-327.

Dawson, D. and Encel, N.P. (1993). Melatonin and sleep in humans. Mini review. *J. Pineal. Res., 15,* 1-12.

Dawson, D. and Encel, N. P. (1997). Melatonin: Role in gating nocturnal rise in sleep propensity. *J. Biol. Rhythms., 12,* 657-665.

Deacon, S. and Arendt, J. (1995). Melatonin induced temperature suppression and its acute phase-shifting effects correlate in a dose-dependent manner in humans. *Brain Res., 688,* 77-85.

Dollins, A.B., Zhdanova, I.V., Wurtman, R.J., Lynch, H.J. and Deng, M.H.. (1994). Effect of inducing nocturnal serum melatonin concentrations in daytime on sleep, mood, body temperature, and performance. *Proc. Natl. Acad. Sci. USA, 91*, 1824-1828.

Eriksson, L., Valtonen, M., Laitinen, J.T., Paananen, M. and Kaikkonen, M. (1998). Diurnal rhythm of melatonin in bovine milk: pharmacokinetics of exogenous melatonin in lactating cows and goats. *Acta vet. scand., 39*, 301-310.

Garfinkel, D., Laudon, M., Nof, D. and Zisapel, N. (1995). Improvement of sleep quality in elderly people by controlled-release melatonin. *Lancet, 346*, 541-544.

Gorbien, M.J. (1993). Stages of sleep: Changes that occur with age. *Geriatrics, 48*, 67.

Haimov, I., Laudon, M., Zisapel, N., Souroujon, M., Nof, D., Herer, P., Tzischinsky, O. and Lavie, P. (1994). Sleep disorders and melatonin rhythms in elderly people. *BMJ, 309*, 167.

Haimov, I., Lavie, P., Laudon, M., Herer, P., Vigder, C. and Zisapel, N. (1995).. Melatonin replacement therapy of elderly insomniacs. *Sleep, 18,* 598-603.

Huether, G. (1993). The contribution of extrapineal sites of melatonin synthesis to circulating melatonin levels in higher vertebrates. *Experentia, 49,* 665-670.

Illnerová, H., Buresová, M. and Presl, J. (1993). Melatonin rhythm in human milk. *J. Clin. Endocrinol. Metab., 77*, 838-841.

James, S., Sack, D., Rosenthal, N. and Mendelson, W. (1990). Melatonin admistration in insomnia. *Neuropsychopharmacology, 3,* 19-23.

Jean-Louis, G., von Gizycki, H. and Zizi, F. (1998). Melatonin effects on sleep, mood, and cognition in elderly with mild cognitive impairment. *J. Pineal Res., 25,* 177-183.

Korhonen, H. and Pihlanto, A. (2006). Bioactive peptides: Production and functionality. *Int. Dairy J.,16*, 945-960.

Lavie, P. (1997). Melatonin: Role in gating nocturnal rise in sleep propensity. *J. Biol. Rhythms., 12,* 657-665.

Littner, M., Kushida, C.A., Anderson, W.M., Bailey, D., Berry, R.B., Davila, D.G., Hirshkowitz, M., Kapen, S., Kramer, M., Loube, D., Wise, M. and Johnson, S. (2003). Practice parameters for the role of actigraphy in the study of sleep and circadian rhythms: An update for 2002. American Academy of Sleep Medicine Practice Parameters. Standards of Practice Committee of the American Academy of Sleep Medicine. *Sleep, 26*, 337-341.

MacFarlane, J.C., Cleghorn, J.M., Brown, G.M. and Streiner, D.L. (1991). The effects of exogenous melatonin on the total sleep time and daytime alertness of chronic insomniacs: a preliminary study. *Biol. Psychiatry, 30*, 371-376.

Myers, B.L. and Badia, P. (1995). Changes in circadian rhythms and sleep quality with aging: Mechanisms and interventions. *Neurosci. Biobehav. Rev., 19*, 53-571.

Pires, M.L.N., Benedito-Silva, A.A., Pinto, L., Souza, L., Vismari, L. and Calil, H.M. (2001) Acute effects of low doses of melatonin on the sleep of young healthy subjects. *J. Pineal. Res., 31*, 326-332.

Sack, R.L., Hughes, R.J., Edgar, D.M. and Lewy, A.J. (1997). Sleep-promoting effect of melatonin: At what dose, in whom, under what conditions, and by what mechanisms. *Sleep, 20,* 908-915.

Salin-Pascual, R.J., O.rtega-Soto, H., Huerto-Delgadillo, L., Camacho-Arroyo, I., Roldán-Roldán, G. and Tamarkin, L. (1988). The effect of total sleep deprivation on plasma melatonin and cortisol in healthy human volunteers. *Sleep, 11*, 362-369.

Shapiro, C.M. and Flanigan, M.J. (1993). Function of sleep. *BMJ, 306*, 383-385.

Valtonen, M., Kangas, A.-P. and Voutilainen, M. (2003). Menetelmä melatoniinipitoisen maidon tuottamiseksi (A method to produce milk with high melatonin concentration). Finnish patent; n:o111324.

Valtonen, M., Niskanen, L., Kangas, A.-P. and Koskinen, T. (2005). Effect of melatonin-rich night-time milk on sleep and activity in elderly institutionalized subjects. *Nord. J. Psychiatry, 59*, 217-221.

Waldhauser, F., Kovacs, J. and Reiter, E. (1998). Age-related changes in melatonin levels in humans and its potential consequences for sleep disorders. *Experimental Gerontology, 33*, 759-772.

Vitello, M.V. (1999). Effective treatments for age-related sleep disturbances. *Geriatrics, 54,* 47-52.

Zhdanova, I.V., Wurtman, R.J., Lynch, H.J., Ives, J.R., Dollings, A.B., Morabito, C., Matheson, J. K. and Schomer, D.L. (1995). Sleep-inducing effect of low doses of melatonin ingested in the evening. *Clin. Pharmacol. Ther., 57*, 552-558.

Zhdanova, I.V., Wurtman, R., Matheson, J. K., Schomer, D.L.J., Regan, M.M., Taylor, J.A., Shi, J.P. and Leclair, O.U. (2001). Melatonin treatment for age-related insomnia. *J. Clin. Endocrinol. Metab., 86*, 4727-4730.

In: Circadian Rhythms and Health Research Trends
Editor: Alfredo V. Lignelli, pp. 145-157
ISBN: 978-1-60021-875-0

Chapter VIII

Modulatory Effects of Respiration on Circadian Rhythm of Heart Rate Variability

Alessandra Fiorentini, Luigi Tubani and Antonio Perciaccante
Department of Clinical Medicine,
University "La Sapienza", Rome

Abstract

Cardiovascular variability results from a number of associated factors, intrinsic, pseudorhythmic and transient. Although cardiovascular variability arises from many different influences, the most consistent "external" modulation is breathing.

Recognition that the respiratory sinus arrhythmia (RSA) was mediated by efferent vagal activity acting on the sinus node led investigators to attempt to quantify the fluctuations in R-R intervals that were related to breathing. Power spectral analysis of R-R intervals permits calculation amount of variance of a signal at a given frequency. Given that vagal modulation of sinus node activity occurs at the respiratory frequency, measuring the variance or power in the R-R interval time series at the respiratory frequency provides a pure index of how much the vagus is modulating sinus node activity. Low frequency (LF) waves are affected by both sympathetic and parasympathetic systems. They are correlated with baroreflex gain, although other rhythms also are closely related to baroreflex activity. Baroreflex activity modulates blood pressure and projections from the baroreflexes stimulate the hypothalamus and play a central role in general autonomic homeostasis. Cortical effects of baroreflex activity reflecting emotional effects have been shown. High frequency (HF) waves usually coincide with RSA, the increase and decrease in heart rate produced by respiratory cycle. RSA is produced by a combination of respiration-induced biochemical changes, changes in intrathoracic pressure and central vagal stimulation. Although RSA amplitude is often considered to be an index of vagal tone, it can be dissociated from vagal influence on tonic heart rate.

To test the modulatory effect of respiration on circadian rhythm of the cardiovascular variability, we have analyzed the heart rate variability (HRV) in healthy zen practitioners with 5 years of experience. Practice of zen was characterized by a rhythmic and slow breathing. HRV analysis has shown a reduction in HF component, an increase in LF component and LF/HF ratio during the practice of zen. These results were reported also in night time, when zen practitioners were sleeping.

In conclusion, repeated training to slow down breathing reduces the spontaneous breathing rate with a long term effects on the cardiovascular control mechanisms. Indeed, when respiration slows to about 6 cycles/min, as in zen practitioners, in the frequency range of the spontaneous LF oscillation, the cardiovascular fluctuations become maximal. The changes in the breathing rate might modify the chemoreflex and the continuos practice in slow breathing can reduce chemoreflex. This change in the automonic control of respiration can be permanent with a resetting of endogenous circulatory rhythms.

Introduction

The breathing is the most consistent modulator of cardiovascular variability. Because cardiovascular variability results from a number of associated factors intrinsic, extrinsic, pseudorhythmic and transient, the effects of respiration are complex and it affect cardiovascular variability at every level.

It is known that an intrinsic rhythmicity is present at various frequencies in the brainstem, in the efferent neurons and in the cardiac pacemaker cells. These central rhythms produce oscillations in all frequency bands, modulating the cardiac pacemaker [1-3] and it may interact with rhythms caused by baroreflex [4,5].

The respiration represent a typical pseudo-rhythmic phenomenon because it is an event that is often rhythmic, but can change in frequency and show sudden changes in response to various stimuli.

The frequency and depth of the rhythm generated by respiratory pacemaker cells, located in the brainstem, are primarily controlled by the central and peripheral chemoreflexes, but it can be varied by factors such as voluntary control, stress, temperature, exercise and by the sympathetic and parasympathetic branches of the autonomic nervous system [6].

Mechanical, gravitational, acoustic and thermal stimuli can alter the frequency and depth of respiration affecting cardiovascular system by changes in sympathetic-parasympathetic interaction [7]. Therefore, changes in respiration can affect cardiovascular variability and breathing plays a major role in the genesis of cardiovascular rhythms.

An important example is the respiratory sinus arrhythmia (RSA), that represents the respiration related fluctuations in RR intervals and is considered an index of the gain of the baroreceptor cardiac vagal reflex responses [8-10].

Some studies [1, 2, 11] have shown that RSA originate from respiration entrained oscillations in medullary neuron firing rates. These autonomic changes may be related to baroreflex stimulation caused by inspiratory increase in venous return with correlated changes in arterial blood pressure [4, 10,12] and to reflexes originating from the cardiopulmonary and lung stretch receptors [13].

An additional important role in the origin of RSA is played by the arterial baroreflex. Bernardi et al [14] have observed that during apnea, RSA is restored by stimulating the carotid baroreceptors by cycled neck suction at the frequency of the previous respiration.

Conversely, the authors have observed that RSA is considerably suppressed by neck suction phased so as to smooth out respiration-related blood pressure changes. These observations suggest that blood pressure fluctuation at neck region is the most important source of afferent information for generating RSA.

The respiration modulates heart rate variability though both autonomic nervous system by effect of the arterial baroreflex, and directly by a non-neural mechanism [15,16]. The frequency of breathing and RSA are clearly interacted [17-19]. Indeed, when frequency of breathing is to about 6 cycles/minute, then RSA and other cardiovascuklar fluctuations are maximal.

The importance of breathing pattern in cardiorespiratory control is known.

The frequency and depth of breathing affect the cardiovascular variability and circulatory system. Bernardi et al [18] have shown that, in patients with heart failure, the reduction of the breathing rate improves the ventilation/perfusion inequality and increases the resting oxygen saturation. Indeed, simple training aimed at a permanently slow breathing rate has reduced dyspnea and improved exercise performance.

The changes in the breathing rate may modify the chemoreflexes. Clinicians and experimentalists routinely estimate vagal-cardiac nerve traffic from RSA [20-22]. However, evidence suggests that sympathetic mechanism may also modulate RSA. Although sympathetic-cardiac nerve activity fluctuates at respiratory frequencies [23,24], oscillatory sympathetic stimuli do not causes appreciable heart period responses at frequencies above 0.15 Hz [25], probably because of the sluggishness of adrenergic receptor responses [26]. Taylor et al [27] have suggested that cardiac sympathetic nerve traffic reduces respiratory sinus arrhythmia by modulating driving force and thus restrains heart period oscillations at all frequencies. Thus, RSA cannot be mediated exclusively by vagal mechanisms. Indeed, sympathetic nerve activity opposes vagally mediated RR interval oscillations and this influence is exerted over high as well as low breathing frequencies. The effect of breathing frequency on RR interval fluctuations reflects primarily the kinetics of acetylcholine influences on the sinoatrial node. During rapid breathing, acetylcholine released during the expiratory phase is reduced and, because the time for hydrolysis of acetylcholine is 1.5-2.0 seconds [28,29], responses to acetylcholine released during one expiration merge with responses to acetylcholine released during the next expiration. At slow breathing frequencies, the acetylcholine release is increased during expiration and, because the time of hydrolysis is costant, sinoatrial responses summate and inhibit sinoatrial node firing.

Taylor et al [27] have suggested a possible mechanisms for to explain why very slow breathing reduces respiratory sinus arrhythmia. During very slow breathing, expiration is of very long duration and vagal cardiac motoneurons may become relatively unresponsive to baroreceptor stimulation.

The importance of ventilation in modulating interaction between sympathetic drive and cardiovascular variability is examined by Van de Borne et al [30]. Chemoreflex stimulation elicts both hyperventilation and sympathetic activation, each of which may have different influences on oscillatory characteristics of cardiovascular variability. During sympathetic activation in healthy humans there is an increase in the low-frequency (LF) normalized powers of RR interval and muscle sympathetic nerve activity (MSNA) variability [31]. This relationship may be lost in pathological conditions such as heart failure, where increased sympathetic drive and tachycardia are associated with attenuated LF power [32].

Ventilation inhibits both cardiac vagal activity [33-35] and sympathetic nerve traffic [36-41].

In their study, Van de Borne et al [30] have performed spectral analysis of RR and MSNA variability during controlled breathing, maximal end-expiratory apnea, isocapnic voluntary hyperventilation and hypercapnia-induced hyperventilation. The authors have observed that LF normalized of RR and MSNA increased during controlled-breathing and decreased during isocapnic and hypercapnic hyperventilation. They have concluded that marked respiratory oscillations in autonomic drive induced by hyperventilation may induce dissociation between RR, MSNA and neurocircolatory variability, perhaps by suppressing central genesis and/or inhibiting transmission of LF cardiovascular rhythms.

Heart rate variability (HRV) is the variation of beat-to-beat intervals. A healthy heart has a large HRV, while decreased or absent variability may indicate cardiac disease. The clinical relevance of HRV was first appreciated in 1965 when Hon and Lee [42] noted that fetal distress was preceded by alterations in interbeat intervals before any appreciable change occurred in heart rate itself. The variations in heart rate may be evaluated by a number of methods. Perhaps the simplest to perform are the time domain measures. . In a continuous ECG record, each QRS complex is detected, and the so-called normal-to-normal (NN) intervals (that is, all intervals between adjacent QRS complexes resulting from sinus node depolarizations) or the instantaneous heart rate is determined. Simple time domain variables that can be calculated include the mean NN interval, the mean heart rate, the difference between the longest and shortest NN interval, the difference between night and day heart rate, and so forth. Other time domain measurements that can be used are variations in instantaneous heart rate secondary to respiration, tilt, Valsalva manoeuvre, or phenylephrine infusion. From a series of instantaneous heart rates or cycle intervals, particularly those recorded over longer periods, traditionally 24 hours, more complex statistical time domain measures can be calculated. These may be divided into two classes: measures derived from direct measurements of the NN intervals or instantaneous heart rate and measures derived from the differences between NN intervals. These variables may be derived from analysis of the total ECG recording or may be calculated using smaller segments of the recording period.

The simplest variable to calculate is the standard deviation of the NN intervals (SDNN), that is, the square root of variance. Since variance is mathematically equal to total power of spectral analysis, SDNN reflects all the cyclic components responsible for variability in the period of recording.

The most commonly used measures derived from interval differences include RMSSD, the square root of the mean squared differences of successive NN intervals, that is expression of parasympathetic component.

Various spectral methods [43] for the analysis of the tachogram have been applied since the late 1960s. Power spectral density (PSD) analysis provides the basic information of how power (variance) distributes as a function of frequency. Fast Fourier Transform is used to obtain power spectral estimates of HRV, total power in the frequency range (0 – 0,40 Hz) is divided into: very low frequency (VLF: < 0,40 Hz), low frequency (LF: 0,04 – 0,15 Hz modulated by both sympathetic and parasympathetic components), and high frequency (HF: 0,15 – 0,40 Hz mediated by parasympathetic system), the integrals underlying respective power density function were measured and expressed in absolute units (ms^2/Hz). Each spectral component was also presented in normalized units (n.u.) by dividing LF or HF by total power minus the very low frequency (LF or HF/total power – VLF). The LF/HF ratio is considered as the principal index of cardiac sympathetic/parasympathetic tone balance.

Heart rate variability analysis is of increasing clinical interest because low variability has been demonstrated to be associated with a poor long-term prognosis after myocardial infarction or heart failure [44, 45]. However, heart rate variability can be influenced also by non-pathological conditions such as activity [46] or different pattern of breathing [18,19]. There is evidence that slow breaths can generate slow fluctuations that can be confused with those related to sympathetic activity. Indeed, slow breaths might increase low frequency (LF) power in RR interval and falsely mimic sympathetic activation.

A study [47] has evaluated the effect of free talking and reading, silently and aloud, on respiration, RR and blood pressure. The data have shown that reading silently increases the speed of breathing, decreases RR variability and increases blood pressure. Reading aloud, free talking and mental arithmetic aloud shift the respiratory frequency into the LF band. The authors conclude that simple mental and verbal activities markedly affect HRV through changes in respiratory frequency. Mental activity appears invariably associated with some degree of sympathetic activation, whose degree likely depends upon the amount of stress involved in performing mental task. These manoeuvres cause different levels of sympathetic activation, as increase in heart rate and blood pressure and decrease in heart rate variability [47-50]. This is evident only for the conditions in which interference with the breathing pattern could be excluded.

The slowing of breathing observed with mental stress has the effect of generating an increase in LF components in RR intervals, regardless of the amount of stress involved in the mental task performed. This is evident by the similarity in RR spectra observed during simple free talking, reading aloud and mental arithmetic, all characterized by a predominance in LF, despite differences in mean heart rate. The slowing of breathing per se thus generates a confounding effect on the RR spectrum, by bringing respiratory sinus arrhythmia into the non-respiratory LF, thought to reflect sympathetic activity.

These data have a very important practical relevance in the interpretation of HRV analysis. Respiration is not normally measured or analysed nor is account taken of such activities, which have been shown to affect the RR spectra. Therefore, caution should be used in the interpretation of short sequences of data. An increase in LF not should be attributed to sympathetic activation during mental tasks might be masked or accentuated by unequal or slow respiratory patterns.

The effects of different breathing patterns on HRV and other cardiovascular parameters, as baroreflex sensitivity and blood pressure, have been investigated by studies conducted on various forms of meditation.

Meditation in its various forms is a traditional exercise with a potential benefit on well-being and health. On a psychosomatic level, these exercises seem to improve the salutogenic potential in men. Especially, the cardiorespiratory interaction seems to play an important role since most meditation techniques make use special low frequency breathing patterns regardless of wether they result from a deliberate guidance of breathing or other mechanisms, for example, the recitation of specific verses.

As described in the first time by Mayer [51], mammals show rhythmic fluctuations in blood pressure and heart rate as a result of autonomic control systems that are influenced by respiration, arousal, and activity. Reciting the Ave Maria prayer and yoga mantras enhances and synchronises inherent cardiovascular rhythms because it slows respiration to almost exactly six respirations per minute, which is essentially the same timing as that of endogenous circulatory rhythms.

These rhythms can be analysed by spectral analysis of cardiovascular fluctuations. A respiratory rate of 6 cycles/minute has shown benefical effects on cardiovascular and respiratory function and increases respiratory sinus arrhythmia, the arterial baroreflex [52], oxygenation of the blood and exercise tolerance [53]. Moreover a slow respiration appears to improve the effects of myocardial ischaemia and to increase calmness and wellbeing [54].

These favourable effects result from synchronization of respiratory and cardiovascular central rhythms and from a modulation of autonomic activity at both central and baroreflex levels.

An experiment [55] was conducted to evaluate and compare the different effects of recitation of the rosary prayer or of mantra during spontaneous and metronome controlled breathing on breathing rate and spontaneous oscillations in RR interval and on blood pressure and cerebral circulation.

The rosary is a repetition 50 times of the Ave Maria, the whole 50 repeated three times. Each cycle, recited half by the priest and half by congregation is, in originally Latin, normally completed within a single slow respiration.

Both the Ave Maria and the yoga mantra had similar effects, slowing respiration to around 6/min and thus having a marked effect on synchronisation and also increased variability in all cardiovascular rhythms. The breathing was more regular during slow breathing, the Ave Maria, and the mantra, whereas it was less regular during free talking than during spontaneous breathing. The spectral peaks of respiration and of all cardiovascular signals were synchronised during the Ave Maria and the mantra sequences. This increased modulation in cardiovascular rhythms influenced the cardiovascular control mechanisms. The arterial baroreflex sensitivity increased on change from spontaneous breathing to controlled slow breathing at 6/min and from free talking to the Ave Maria and the mantra. The authors conclude that recitation of the prayer or mantra has a similar effect to that of slow breathing, increasing the arterial baroreflex, which is a favourable prognostic factor in long term studies in cardiac patients [56-57].

The "complete yoga breathing" is one of the several yogic breathing techniques and it is characterized by a slow, deep breathing at a frequency of about 6 breaths/rminute [58]. Spicuzza et al [59] have tested wheter the practice of yoga can affect hypoxic and hypercapnic chemoreflex sensitivity. To assess the importance of the breathing frequency and of controlled breathing alone, the authors have performed the rebreathing manoeuvres during spontaneous breathing and during controlled breathing at a slow frequency and also at controlled frequency close to spontaneous one.

During spontaneous breathing, the ventilatory responses to both hypoxia and hypercapnia were markedly lower in yoga trainees compared to control group. The study has shown that the practice of yoga breathing decreases the chemoreflex hypoxic and hypercapnic responses. This is due to slower breathing frequency that subjects exhibite spontaneously because similar reduction in ventilatory responses was found also in control subjects when they breath at 6 breaths/minute. If the slower breathing frequency has been the only factor reducing the chemoreflex, then chemoreflex sensitivity should has been increased to values similar to control group when tested at a higher breathing frequency (15 breaths/minute). Because it was found only a minor increase, authors conclude that the long-term effects of yoga mat also independently reduce chemoreflex sensitivity.

Cysarz et al [60] have investigate the effect of poetry recitation on the synchronization between low-frequency breathing patterns and respiratory sinus arrhythmia of hear rate. Healthy volunteers subjects have performed different types of exercises, as recitation of hexameter verse from ancient Greek literature, controlled breathing and spontaneous breathing. Respiratory traces and electrocardiogram has been recorded to evaluate breathing patterns and respiratory sinus arrhythmia. To evaluate the synchronization has been used the index γ, adopted from the analysis of weakly coupled chaotic oscillatiors.

During poetry recitation, it has been observed a prominent cardiorespiratory synchronization, expressed by a high index γ, while, during spontaneous breathing, cardiorespiratory synchronization has been minimal. Controlled breathing exercise has shown cardiorespiratory synchronization to a lesser extent. These results appear to be due to the extent of a low-frequency component in the breathing oscillations that emerge from the hexameter recitation. Therefore, recitation of hexameter verses have a strong influence on respiratory sinus arrhythmia by a prominent low-frequency component in the breathing pattern, generating a strong cardiorespiratory synchronization.

Lehrer et al [61] have examined the effects of "tanden breathing", that involve slow breathing into the lower abdomen, on cardiac variability. Some Zen practitioners have been recruited. During "tanden breathing" heart rate, respiration rate and heart rate variability have been measured. The results of this study have shown an increase of heart rate variability within the frequency range of 0.05 to 0.15 Hz during tanden breathing. This reflects a shift of respiratory sinus arrhythmia from high frequency to slower waves. These data confirm that increased oscillation amplitude during slow breathing is caused by resonance between cardiac variability caused by respiration and that produced by physiological processes underlying slower rhythms.

The impact of meditation on cardiorespiratory synchronization with respect to breathing oscillations and the modulations of heart rate induced by respiration (respiratory sinus arrhythmia, RSA) was investigated in an other study by Cysarz and Bussing [62]. Subjects mainly without any experience in meditation were recruited. Spontaneous breathing, mental task, Zen meditation and Kinhin meditation were performed by these subjects. An electrocardiogram and a respiratory trace were recorded simultaneously and a technique adopted from the analysis of weakly coupled chaotic oscillators was used to quantify the cardiorespiratory synchronization.

Both Zazen and Kinhin meditation have shown a high degree of synchronization, whereas heartbeat and respiration were hardly synchronized during spontaneous breathing.

During the mental task exercise the extent of synchronization has been higher than during spontaneous breathing.

These results are determined by the breathing frequency because the two types of meditation are characterized by a low breathing frequency, which led to a pronounced and in-phase respiratory sinus arrhythmia. The high degree of cardiorespiratory synchronization during meditation in unexperienced meditators suggests that the physiological implications of meditation does not require prior experience in meditation.

We want to evaluate the effects of Zazen and Kinhin meditation on the autonomic function by heart rate variability during exercises of meditation and during night time, when zen practitioners were sleeping.

Nine Zen practitioners with five years of experience took part in the study (4 female, 3 smokers average age 43±7 years). None of the subjects had any history of cardiovascular diseases, especially no hypo- or hypertension or antiarrhythmical therapy or receiving medicaments.

The procedure was divided into six different exercises.

Subsequently, the first session of sitting meditation (Zazen) was practiced (duration: approximately 10 min). This kind of meditation is characterized by breathing at a rate and depth which is only determined by the breathing process itself, that is, any outward intention is minimized and the intention towards the breathing process is increased. The volunteers were advised to avoid any thinking, and to keep the awareness on 'just-sitting' and 'just-breathing'. To facilitate this kind of mediation the subjects sat upright on a cushion and the hands were held together in front of the navel. During the next 7 min, a walking meditation (Kinhin) was practiced. With respect to the breathing modalities it is comparable to Zazen meditation (MZ). In contrast, this exercise is a distinct form of walking with short steps: during inspiration the foot is lifted and the whole body is slightly erected against the gravity. Then, during expiration, the foot is put down a half-step in front and gravity may slightly pull down the body again. Thus, the walking speed is unambiguously intertwined with the breathing process. The hands were held together in front of the sternum and were slightly pressed against each other and the sternum during expiration. For the next 7 min Zazen meditation is practiced again to get back to rest in a calm fashion. At the end of the procedure, the subject sat in a chair again, similar to the beginning of the experiment (duration 10 min).

Autonomic nervous system function was evaluated by heart rate variability (HRV) analysis during 24-hours ECG recording. All Holter recording were performed using a three-channel recorder. Cardiovascular variability was analysed following the recommendations of the Task Force of the European Society of Cardiology and the North American Society of Pacing and Electrophysiology [63].

Spectral estimates of R-R intervals were obtained from stationary regions free of ectopic beats and technical artifacts. The standard deviation of normal-to-normal RR intervals [SDNN (ms) correlated to total autonomic activity] and the square root of the average of the sum of the squares of the differences between adjacent NN intervals [RMS-SD (ms) correlated to parasympathectic system] were calculated and were divided in two periods: night (0 a.m. – 6 a.m.) and day (7 a.m. – 9 p.m.). Fast Fourier Transform was used to obtain power spectral estimates of HRV, total power in the frequency range (0 – 0,40 Hz) was divided into: very low frequency (VLF: < 0,40 Hz), low frequency (LF: 0,04 – 0,15 Hz modulated by sympathetic system), and high frequency (HF: 0,15 – 0,40 Hz mediated by parasympathetic system), the integrals underlying respective power density function were measured and expressed in absolute units (ms^2/Hz). Each spectral component was also presented in normalized units (n.u.) by dividing LF or HF by total power minus the very low frequency (LF or HF / total power – VLF). The LF/HF ratio, considered an index of cardiac sympathetic/parasympathetic tone balance, was also calculated.

Patients were analysed for 24-hours at 10 minutes intervals. Artificial data and arrhythmic events were excluded. The 24-hour period was divided into four parts: night (0 a.m. – 6 a.m.), morning (7 a.m. – 12 a.m.), afternoon (1 p.m. – 6 p.m.) and evening (7 p.m. – 11 p.m.). Data analyses were performed with Del Mar Avionics Accuplus 363, Irvine California, USA. In ten healthy subjects reproducibility was evaluated by means of the interclass correlation coefficient (ICC) comparing baseline values with the results obtained at the fourth week. ICC was > 0,7 for HRV.

Respiratory rate in zen practitioners was 6 breaths/minute.

The analysis of heart rate variability has shown an increase in LF n.u. values and a reduction in HF n.u. components during the practice of zen (respectively LF during zen practice 66.36 ± 4.77 vs LF before zen practice 54.21 ± 6.17, $p < 0.001$ and HF during zen practice 19.57 ± 5.47 vs HF before zen practice 38.64 ± 6.22, $p < 0.001$). Similar results were reported during night time in zen practitioners versus control subjects (LF n.u.: 60.58 ± 6.46 vs 34.80 ± 3.71, $p < 0.001$; HF: 24.06 ± 10.30 vs 59.12 ± 5.16, $p < 0.001$).

Conclusion

The results of this small observational study confirm that ZaZen breathing falls within the range of low frequency heart rate spectral bands.

Repeated training to slow down breathing reduces the spontaneous breathing rate with a long term effects on the cardiovascular control mechanisms. Indeed, when respiration slows to about 6 cycles/min, as in zen practitioners, in the frequency range of the spontaneous LF oscillation, the cardiovascular fluctuations become maximal. The changes in the breathing rate might modify the chemoreflex and the continuous practice in slow breathing can reduce chemoreflex. This change in the automonic control of respiration can persist with a resetting of endogenous circulatory rhythms. Indeed, during night time, when zen practitioners were sleeping, is present an increase of LF components at difference of control subjects.

References

[1] Koepchen, HP, Abel HH, Klussendorf D, Lown B, Malliani A, Prosdocimi M, Editors, *Neural Mechanisms and Cardiovascular Disease* vol. 5, Liviana Press, Padua, (1986).

[2] Shykoff, BE, Naqvi SSJ, Menon AS, Slutsky AS. Respiratory sinus arrhythmia in dogs. *J. Clin. Invest.* 1991; 97:1621-1627.

[3] Malliani, A, Pagani M, Lombardi F, Cerutti S. Cardiovascular neural regulation explored in the frequency domain. *Circulation* 1991; 84: 482-492.

[4] De Boer RW, Karemaker JW, Strackee J. Hemodynamic fluctuations and baroreflex sensitivity in humans: a beat-to-beat model. *Am. J. Physiol.* 1987; 253: 680-689.

[5] Bernardi L, Leuzzi, S, Radaelli A, Passino C, Johnston JA, Sleight P. Low-frequency spontaneous fluctuations of RR interval and blood pressure in conscious humans: a baroreceptor or central phenomenon?. *Clin. Sci.* 1994; 87: 649-654.

[6] Kitney RI, Rompelman O. *The Beat-To-Beat Investigation of Cardiovascular Function*, Clarendon Press, Oxford, (1987).

[7] Bernardi L, Valle F, Coco M, Calciati A, Sleight P. Physical activity influences heart rate variability and very-low frequency components in Holter electrocardiograms. *Cardiovasc. Res.* 1996; 32:234-237.

[8] Eckberg DL. Human sinus arrhythmia is an index of vagal cardiac outflow. *J Appl. Physiol.* 1983; 54:961-966.

[9] Bernardi L, Ricordi L, Lazzari, P, Solda PL, Calciati A, Ferrari MR, Vandea I, Finardi G, Fratino P. Impaired circadian modulation of sympathovagal activity in diabetes. *Circulation* 1992; 86:1443-1452.

[10] Sleight P, La Rovere MT, Mortara A, Pinna G, Maestri R, Leuzzi S, Bianchini B, Bernardi L. Physiology and pathophysiology of heart rate and blood pressure variability in humans: is power spectral analysis largely an index of baroreflex gain?. *Clin. Sci.* 1995; 88:103-109.

[11] Eckberg DL, Sleight P. *Human Baroreflexes in Health and Disease*, Oxford Univ. Press, Oxford, (1992).

[12] Dornhorst AC, Howard P, Leathart GL. Respiratory variations in blood pressure. *Circulation* 1952; 6:553-558.

[13] Melcher A. Respiratory sinus arrhythmia in man. A study in heart rate regulating mechanisms. *Acta Physiol. Scand.* 1976; 435:1-31.

[14] Bernardi L, Porta C, Gabutti A, Spicuzza L, Sleight P. Modulatory effects of respiration. *Auton. Neurosci.* 2001; 90: 47-56.

[15] Bernardi L, Bianchini B, Spadacini G, Leuzzi S, Valle F, Marchesi E, Passino C, Calciati A, Vigano' M, Rinaldi M, Martinelli L, Finardi G, Sleight P. Demonstrable cardiac reinnervation after human heart transplantation by carotid baroreflex modulation of RR interval. *Circulation* 1995; 92:2895-2903.

[16] Bernardi L, Valenti C, Wdowczyck-Szulc J, Frey AW, Rinaldi M, Spadacini G, Passino C, Martinelli L, Vigano' M, Finardi G.Influence of type of surgery on the occurrence of vagal reinnervation after cardiac transplantation. *Circulation* 1998; 97:1368-1374.

[17] Bernardi L, Spadacini G, Bellwon J, Hajiric R, Roskamm H, Frey AW. Effect of breathing rate on oxygen saturation and exercise performance in chronic heart failure. *Lancet* 1998; 351:1308-1311.

[18] Sanderson JE, Yeung LY,Yeung DT, Kay RL, Tomlinson B, Critchley JA, Woo KS, Bernardi L. Impact of changes in respiratory frequency and posture on power spectral analysis of heart rate and systolic blood pressure variability in patients with heart failure and normals. *Clin. Sci.* 1996; 91:35-43.

[19] Bernardi L, Keller F, Sanders M, Reddy PS, Griffith B, Meno F, Pinsky MR. Respiratory sinus arrhythmia in the denervated human heart. *J. Appl. Physiol.* 1989; 67:1447-1455.

[20] Fouad, FM, Tarazi RC, Ferrario CM, Fighaly S, and Alicandri C. Assessment of parasympathetic control of heart rate by a noninvasive method. *Am. J. Physiol. Heart* Circ *Physiol.* 1984; 246: H838-H842.

[21] Katona, PG, and Jih F. Respiratory sinus arrhythmia: noninvasive measure of parasympathetic cardiac control. *J. Appl. Physiol.* 1975;39: 801-805.

[22] Raczkowska, M, Eckberg DL, and Ebert TJ. Muscarinic cholinergic receptors modulate vagal cardiac responses in man. *J. Auton. Nerv. Syst.* 1983;7: 271-278.

[23] Kollai, M, and Koizumi K. Reciprocal and non-reciprocal action of the vagal and sympathetic nerves innervating the heart. *J. Auton. Nerv. Syst.* 1979; 1: 33-52.

[24] Kollai, M, and Mizsei G. Respiratory sinus arrhythmia is a limited measure of cardiac parasympathetic control in man. *J. Physiol.* (Lond) 1990; 424: 329-342.

[25] Berger, R, Saul JP, and RJ C. Transfer function analysis of autonomic regulation. I. Canine atrial rate response. *Am. J. Physiol. Heart Circ. Physiol.* 1989; 256: H142-H152.

[26] Hill-Smith, I, and Purves RD. Synaptic delay in the heart: an ionophoretic study. *J. Physiol.* (Lond). 1978; 279: 31-54.

[27] Taylor JA, Myers CW, Halliwill JR, Seidel H, Eckberg DL. Sympathetic restraint of respiratory sinus arrhythmia: implications for vagal-cardiac tone assessment in humans. *Am. J. Physiol. Heart Circ. Physiol.* 2001; 280(6):H2804-14.

[28] Baskerville, AL, Eckberg DL, and Thompson MA. Arterial pressure and pulse interval responses to repetitive carotid baroreceptor stimuli in man. *J. Physiol.* (Lond). 1979; 297: 61-71.

[29] Eckberg, DL, and Eckberg MJ. Human sinus node responses to repetitive, ramped carotid baroreceptor stimuli. *Am. J. Physiol. Heart Circ. Physiol.* 1982; 242: H638-H644.

[30] Van De Borne P, Montano N, Narkiewicz K, Degaute JP, Malliani A, Pagani M, Somers VK. Importance of ventilation in modulating interaction between sympathetic drive and cardiovascular variability. *Am. J. Physiol. Heart Circ. Physiol.* 2001; 280(2):H722-9.

[31] Pagani, M, Montano N, Porta A, Malliani A, Abboud FM, Birkett C, and Somers VK. The relationship between spectral components of cardiovascular variabilities and direct measures of muscle sympathetic nerve activity in humans. *Circulation.* 1997; 95: 1441-1448.

[32] Van de Borne, P, Montano M, Pagani M, Oren R, and Somers VK. Absence of low-frequency variability of sympathetic nerve activity in severe heart failure. *Circulation.* 1997; 95: 1449-1454.

[33] Anrep, GV, Pascual W, and Rossler R. Respiratory variations of the heart rate. II. The central mechanism of the respiratory arrhythmia and the interrelations between the central and the reflex mechanisms. *Proc. R. Soc. Lond Ser. B.* 1936; 119: 218-230.

[34] De Burgh, DM, Angell-James JE, and Elsner R. Role of carotid-body chemoreceptors and their reflex interactions in bradycardia and cardiac arrest. *Lancet.* 1979; 1: 764-767.

[35] De Burgh, DM, and Scott MJ. An analysis of the primary cardiovascular reflex effects of stimulation of the carotid body chemoreceptors in the dog. *Am. J. Physiol.* 1962; 162: 555-573.

[36] Gootman, PM, Feldman JL, and Cohen MI. Pulmonary afferent influences on respiratory modulation of sympathetic discharge. In: *Central Interaction Between Respiratory and Cardiovascular Control Systems*, edited by Koepchen HP, Hilton SM, and Trzebski A.. Berlin: Springer-Verlag, 1980, p. 172-178.

[37] Hagbarth, KE, and Vallbo AB. Pulse and respiratory grouping of sympathetic impulses in human muscle nerves. *Acta Physiol. Scand.* 1968; 74: 96-108.

[38] Horner, RL, Brooks D, Kozar LF, Gan K, and Phillipson EA. Respiratory-related heart rate variability persists during central apnea in dogs: mechanisms and implications. *J. Appl. Physiol.* 1995; 78: 2003-2013.

[39] Somers, VK, Zavala DC, Mark AL, and Abboud FM. Contrasting effects of hypoxia and hypercapnia on ventilation and sympathetic activity in humans. *J. Appl. Physiol.* 1989; 67: 2101-2106.

[40] Somers, VK, Zavala DC, Mark AL, and Abboud FM. Influence of ventilation and hypocapnia on sympathetic nerve responses to hypoxia in normal humans. *J. Appl. Physiol.* 1989; 67: 2095-2100.

[41] Hon EH, Lee ST. Electronic evaluations of the fetal heart rate patterns preceding fetal death: further observations. *Am. J. Obstet. Gynecol.* 1965;87:814-826.

[42] Kay SM, Marple SL. Spectrum analysis: a modern perspective. *Proc. IEEE.* 1981;69:1380-1419.

[43] Nolan J, Batin PD, Andrews R, Lindsay SJ, Brooksby P, Mullen M, Baig W, Flapan AD, Cowley A, Prescott RJ, Neilson JM, Fox KA. A prospective study of heart rate variability and mortality in chronic heart failure - results of the UK Heart Failure Evaluation and Assessment of Risk Trial (UK-HEART). *Circulation.* 1998;98:1510-1516.

[44] LaRovere M.T., Bigger J.T. Jr, Marcus F.I., et al. "Baroreflex sensitivity and heart rate variability is prediction of total cardiac mortality after myocardial infarction", *Lancet.* 1998; 351:478-484.

[45] Bernardi L, Valle F, Coco M, Calciati A, Sleight P. "Physical activity influences heart rate variability and very low frequency components in Holter electrocardiograms", *Cardiovasc. Res.* 1996; 32:234-237.

[46] Bernardi L, Wdowczyk-Szulc L, Valenti C, Castoldi S, Passino C, Spadacini G, Sleight P. Effects of controlled breathing, mental activity and mental stress with or without verbalization on heart rate variability. *J. Am. Coll. Cardiol.* 2000 May;35(6):1462-9.

[47] Langewitz W., Ruddel H., "Spectral analysis of heart rate variability under mental stress", *J. Hypertens.* 1989; 7 (6):32-33.

[48] Pagani M, Mazzuero G, Ferrari A, Liberati D, Cerutti S, Vaitl D, Tavazzi L, Malliani A.. "Sympathovagal interaction during mental stress. A study using spectral analysis of

heart rate variability in healthy control subjects and patients with a prior myocardial infarction", *Circulation.* 1991; 83 (2) 43-51.

[49] Costoli T, Bartolomucci A, Graiani G, Stilli D, Laviola G, Sgoifo A.. "Assessment of the neural control of the circulation during psychological stress", *J. Aut. Nerv. Sys.* 1991; 35:33-42.

[50] Mayer S. Studien zur Physiologie des Herzens und der Blutgefaesse 6. Abhandlung: ueber spontane Blutdruckschwenkungen. [Studies on the physiology of the heart and the blood vessels 6. *Discourse on fluctuations in blood pressure.*] Sitz Ber Akad Wiss Wien, Mathe-Naturwiss Kl Anat. 1876; 74: 281-307.

[51] Bernardi L, Gabutti A, Porta C, Spicuzza L. Slow breathing reduces chemoreflex response to hypoxia and hypercapnia and increases baroreflex sensitivity. *J. Hypertens.* 2001; 19: 2221-2229.

[52] Bernardi L, Spadacini G, Bellwon J, Hajric R, Roskamm H, Frey AW. Effect of breathing rate on oxygen saturation and exercise performance in chronic heart failure. *Lancet* 1998; 351: 1308-1311.

[53] Friedman EH, Coats AJS. Neurobiology of exaggerated heart rate oscillations during two meditative techniques. *Int. J. Cardiol.* 2000; 73: 199.

[54] Bernardi L, Sleight P, Bandinelli G, Cencetti S, Fattorini L, Wdowczyc-Szulc J, Lagi A. Effect of rosary prayer and yoga mantras on autonomic cardiovascular rhythms: comparative study. *BMJ.* 2001 Dec 22-29;323(7327):1446-9.

[55] La Rovere MT, Bigger Jr JT, Marcus FI, Mortara A, Schwartz PJ. Baroreflex sensitivity and heart-rate variability in prediction of total cardiac mortality after myocardial infarction. *Lancet* 1998; 351: 478-484.

[56] Nolan J, Batin PD, Andrews R, Lindsay SJ, Brooksby P, Mullen M, et al. Prospective study of heart rate variability and mortality in chronic heart failure: results of the United Kingdom heart failure evaluation and assessment of risk trial (UK-heart). *Circulation* 1998; 98: 1510-1516.

[57] Hewitt J. In: *The Yoga of Breathing Posture and Meditation*, Random House, London (1983), pp. 89–91.

[58] Spicuzza L, Gabutti A, Porta C, Montano N, Bernardi L. Yoga and chemoreflex response to hypoxia and hypercapnia. *Lancet.* 2000 Oct 28;356(9240):1495-6.

[59] Cysarz D, von Bonin D, Lackner H, Heusser P, Moser M, Bettermann H. Oscillations of heart rate and respiration synchronize during poetry recitation. *Am. J. Physiol. Heart Circ. Physiol.* 2004 Aug;287(2):H579-87. Epub 2004 Apr 8.

[60] Lehrer P, Sasaki Y, Saito Y. Zazen and cardiac variability. *Psychosom. Med.* 1999 Nov-Dec;61(6):812-21.

[61] Cysarz D, Bussing A. Cardiorespiratory synchronization during Zen meditation. *Eur. J. Appl. Physiol.* 2005 Sep;95(1):88-95. Epub 2005 Jun 7.

[62] Heart rate variability: standards of measurement, physiological interpretation and clinical use. Task Force of the European Society of Cardiology and the North American Society of Pacing and Electrophysiology. *Circulation.* 1996 Mar 1;93(5):1043-65.

In: Circadian Rhythms and Health Research Trends
Editor: Alfredo V. Lignelli, pp. 159-168
ISBN: 978-1-60021-875-0

Chapter IX

Circadian Rhythms in Eye Parameters

Paolo Fogagnolo[1], Federica Capizzi[2], Luca Rossetti[2] and Nicola Orzalesi[2]

[1]G.B.Bietti Foundation for the Study and Research in Ophthalmology, IRCCS (Istituto di Ricovero e Cura a Carattere Scientifico), Rome, Italy
[2]Eye Clinic, San Paolo Hospital; Department of Medicine, Surgery and Odontoiatry, University of Milan, Italy

Abstract

As for many human tissues and organs, the presence of circadian rhythms on eye parameters has been largely demonstrated over the last years. The interest on the role of 24-hour fluctuations in eye parameters interestingly grew up as basic and clinical research progressively put lights on the pathogenetic features of many ocular diseases..

Up-to-now, the main fields of interest is represented by the differences in the 24-hour intraocular pressure (IOP) curve between normal and glaucomatous patients with and without treatment. In fact, a different circadian effect has been shown when comparing comparing medical, laser and surgical treatment of glaucoma; anti-glaucomatous eyedrops are also characterised by peculiar mechanisms of action which reflect on the different effects on 24-hour IOP curve flattening power. Moreover, IOP assessment is influenced by corneal thickness, which is also influenced by circadian fluctuations. Finally, data on 24-hour changes in ocular perfusion and in melatonin shading have been published very recently.

The aims of the present article is to review the published literature in order to ascertain the role of circadian rhythms, both in healthy and pathological eyes, and to investigate their pathogenetic and clinical impact in eye diseases.

The presence of circadian rhythms on eye parameters has been largely demonstrated over the last years, with an interest which progressively increased as basic and clinical research progressively put lights on the pathogenetic features of many ocular diseases.

The aims of the present article is to review the published literature in order to quantify the amount of circadian rhythms both in healthy and pathological eyes, to address their involvement in the pathogenesis of eye diseases and to evaluate their impact in the clinical practice.

The following parameters will be considered: intraocular pressure, central corneal thickness, ocular perfusion, melatonin shading.

1. Intraocular Pressure

Intraocular pressure (IOP) is the result of the balance between aqueous production by ciliary body and its outflow outside the eye through the trabeculum (conventional outflow) and the iris (non-conventional outflow). In normal subjects, normal IOP values are 16 ± 5 mmHg and by convention an IOP > 21 mmHg is considered as outside normal limits (ocular hypertension, OHT). Goldmann applanation tonometer is the gold standard for IOP measurement in sitting position; in the supine position, it is usually measured with Perkins tonometer or other devices such as TonoPen.

IOP is the major risk factor for glaucoma and its progression: the prevalence of the disease rises with increasing IOP [1] and patients with high IOP are prone to a fast progression of visual field defects [2]. Since the beginning of its medical treatment, reducing IOP has been the only way of treating glaucoma and it has been shown that an effective IOP reduction is associated to a reduced or slower progression. Hence, IOP represents the most important parameter for glaucoma detection and follow-up.

As for many human parameters, intraocular pressure follows a circadian rhythm. This has been demonstrated both in animal models and in humans; in particular, a very recent study in a mouse model brilliantly demonstrated that the circadian IOP rhythm is generated by clock genes [3].

According to the classical models, IOP rhythms in human eyes would fit four curves: morning-type, diurnal, nocturnal and flat. The more recent clinical trials suggest that the normal curve for healthy subjects has a peak during early morning; [4-6] while information on glaucoma patients are overall very heterogeneous and sometimes conflicting: it seems probable that, at least for many subjects, a precise and consistent circadian patterns for IOP cannot be established.

Therefore, apart from the shape of IOP curve, increased importance has been given to the detection of IOP peak by means of acrophase analysis. Our recent findings seems to suggest that IOP peak occurs at about 6 a.m. for healthy young and aged subjects compared to 9 a.m. in glaucomatous patients; [6-8] these results are in contrast to those of Weinreb and coauthors, who suggested that IOP peak may occur at late night (4 a.m.), without significant differences between normal and glaucoma subjects [4, 5, 9]. The importance of a correct peak detection is linked to the possibility of customising IOP treatment using drugs endowed with different duration, half-life and curve of efficacy, in order to reduce IOP circadian fluctuation [10].

Actually, also the role of circadian IOP fluctuation in the progression of the disease is not completely clear. Asrani et al. suggested that IOP fluctuations are an independent risk factor for glaucoma progression: in their group of 105 glaucoma patients, who self-measured IOP with a non-contact tonometer throughout 24 hours, they found that subjects with IOP fluctuations below the lowest 25th percentile had stable visual fields in 43%, compared to 12% of the highest 25th percentile [11]. This study is limited by the use of an imprecise tonometer. On the other hand, Bengtsson et al., in a study addressing only daytime IOP fluctuations, suggested that the amount of fluctuations is directly proportional to mean IOP, and that IOP fluctuations are therefore not an independent risk factor for glaucoma progression [12].

Differences in the methods to evaluate IOP fluctuations are also present in literature: Weinreb's group reported data on habitual body position (sitting during daytime, supine during nighttime), while our group provided circadian data for both sitting and supine positions. In any case, it is clear that IOP fluctuates less in young than aged people, and less in healthy subjects compared to glaucoma. A recent study suggested that nychoemeral IOP fluctuations are 8.31 ± 2.58 mmHg in POAG compared to 4.83 ± 2.46 in normals. A finding of IOP higher than 21 mmHg with circadian fluctuations of 6 mmHg or more should be therefore considered as a sign of glaucoma, as it is constantly absent in normal subjects [13].

The treatment of glaucoma aims at reducing IOP; a treatment is deemed as effective if it can achieve a *constant* (therefore reducing circadian IOP fluctuations) and *sufficient* (based on the stage of the disease and on the presence of risk factors) IOP reduction. In industrialised countries, treatment generally follows a stepwise fashion: first, medical treatment (in case of a reduction of efficacy over time, switching to another drug or adding a second or third compound may be considered); second, argon laser trabeculoplasty; and third, surgery (i.e.trabeculectomy).

Different eyedrops are currently used to reduce IOP; their effects have been largely investigated in clinical trials and their effect on circadian IOP is well-established. Beta-blockers (timolol, carteolol, betaxolol, levobunolol) are most popular treatment for glaucoma; they must be instilled twice daily (peak after 2 hours, trough after 12 hours). Circadian effects of timolol in glaucoma and OHT were firstly investigated by Orzalesi et al. [6]: a mean IOP reduction of 20% was obtained, with a loss of efficacy at night (at 3 a.m., IOP under timolol was not statistically different to baseline) and an overall reduction of IOP fluctuation to half (about 6 mmHg in glaucoma and OHT compared to 3 mmHg under timolol).

Prostaglandin analogues represent the most effective class of drugs to treat glaucoma: they induce a constant IOP reduction of about 30% throughout 24-hours, with IOP fluctuations of about 3 mmHg. Moreover, thanks to their pharmacokinetics (peak at 10-12 hours, trough at 20-24 hours), they must be administered only once-a-day. A recent study compared the circadian behaviour of the three commercially available prostaglandin analogues (2 prostaglandins, latanoprost and travoprost, and a prostamide, bimatoprost); it confirmed their efficacy in reducing 24-hour IOP and no statistically significant differences were found between different eyedrops, though bimatoprost achieved the lowest IOP values at 7 over 8 timepoints [8].

Topical inhibitors of carbonic anhydrase (dorzolamide, brinzolamide) must be administered three times daily (peak at 2 hours, trough at 8-10 hours); they achieve an IOP reduction of about 15-20%, but they do not provide a substantial reduction of fluctuations, as they are very effective at night (-25%), but not during the day (-10%) [7].

In clinical practice, alpha-agonists (brimonidine) are administered twice daily (though a correct therapeutic regimen should contemplate their use three times daily, as peak is at 2 hours and trough at 10 hours); they achieve an IOP reduction of about 20%, with a partial loss of efficacy at night (3-6 a.m.), which has been resolved by adopting the "full" regimen [14] When used twice daily, brimonidine obtains a reduction of IOP fluctuations of about 30% [7].

One of the first study on the circadian effects of eyedrops on IOP stated the superiority of timolol over pilocarpine in reducing IOP and its fluctuation [15].

Many fixed combination have been recently introduced in clinical practice. Overall, their mean efficacy is equivalent to the unfixed combination. A substantial advantage of these formulation is to reduce the problems of compliance and the washout of the second eyedrop on the first (if instilled without respecting an interval of 5 minutes). An important disadvantage of the fixed combinations is that one of their compound is administered less frequenly: in the combinations of timolol and prostaglandin analogues, timolol is given once-daily; for the fixed combinations of timolol and dorzolamide and of timolol and brimonidine, the latter is administered twice instead of three times daily. This implies that the circadian IOP reduction under these treatments may not be as flat as when the single components are administered separately, as previously demonstrated for the fixed combinations of timolol and dorzolamide [7].

When inspecting data of clinical studies on circadian IOP, caution should be used, as high individual fluctuations should be present also if the mean IOP curve is flat. This was the case of a recent study addressing the 24-hour effect of the fixed combinations of timolol and latanoprost: the fluctuation for mean values at each timepoint was 2.3 mmHg, but greater fluctuations were found when the within patient curves were analysed, with mean variations of 7.3 (range 3–15) mmHg in supine IOP and fluctuations of more than 5 mmHg in about 60% [16].

Argon laser trabeculoplasty is a viable option to obtain a further IOP decrease when medications are not sufficient. It is a safe procedure which provides a mean IOP reduction of 15-20%, but no data on the 24-hour effect are up-to-now available on literature.

Finally a very interesting study conducted by Konstas over 60 patients with advanced glaucoma compared the effects of surgery and maximum-tolerated medical treatment on circadian IOP [17]. The results showed that a well-functioning trabeculectomy provided a statistically lower mean, peak, and range of IOP for the 24-hour day than maximum tolerated medical therapy. In particular, IOP fluctuations were 2.3 ± 0.8 mmHg after surgery compared to 4.8 ± 2.3 mmHg for medical treatment. Moreover, the majority of peak IOPs occurred outside of normal office hours, thus confirming the clinical importance of the nychtoemeral evaluation of IOP in patients affected by glaucoma.

2. Central Corneal Thickness

The cornea is the anterior one-sixth of the eye, degrading in the periphery to the sclera through the limbus. It is the principal refractive element of the eye and, apart from shape, its optical qualities are determined by transparency. Corneal transparency is due its structure (it contains very few cells, no blood vessels and its stroma is organized in parallel layers that do not diffract light) and to the action of its endothelium, which actively transports water from the stromal-facing surface to the aqueous-facing surface by an interrelated series of active and passive ion exchangers. Critical to this energy-driven process is the role of Na^+/K^+ATPase and carbonic anhydrase. Bicarbonate ions formed by the action of carbonic anhydrase are translocated across the cell membrane, allowing water to passively follow. Moreover, endothelium is the source of nutrition for the whole cornea, as it gets glucose and metabolits from the anterior chamber.

Therefore, changes in corneal thickness are strictly dependent from endothelial activity, and in fact any loss of endothelial integrity or function may determinate corneal swelling. Besides, also changes in epithelium may determine modest variations in corneal thickness.

In clinical practice, corneal thickness is usually measured by means of ultrasonic pachymeters, which are the devices with the highest precision and the lower interoperator and intraoperator variability (<1.5%) [18]. The normal corneal thickness in human eyes is about 0.50-0.55 mm centrally and 1 mm in the periphery.

In the last decades, central corneal thickness (CCT) has become a very important parameter for its implications in refractive surgery, IOP assessment and glaucoma.

Applanation tonometry is influenced by corneal rigidity, which is proportional to its thickness. Goldmann tonometry is calibrated for corneas with a CCT of 535 microns and it obtains accurate measurements in the range of 520-550 microns. Anyway, the vast majority of normal subjects have CCT outside these ranges (Rapuano et al. measured 303 normal corneas and obtained a range of 410 to 625 microns); [19] therefore, Goldmann applanation tonometry may not be accurate in these cases. Many correction factors have been proposed (ranging from 0.2 to 0.7 mmHg for each 10-micron change from 535 microns), but up-to-now their systematic use in the clinical practice seems inappropriate [20]. Moreover, a number of cross sectional studies have shown that ocular hypertensive eyes have thicker CCT, whereas low tension glaucoma (LTG) is characterised by lower mean CCT, and primary open-angle glaucoma (POAG) by normal mean CCT; nevertheless, individual corneal thickness largely varies and overlaps between groups.

Although the relationship between IOP and CCT is very important, up to one year ago no studies had evaluated how an eventual variation on CCT could influence IOP reading in a circadian setting. This was the purpose of a study conducted by Fogagnolo and aggregates [16]. We performed a circadian IOP and CCT series of measurement (at 4, 8, 12 a.m. and 4, 8, 12 p.m.) on 30 consecutive glaucomatous patients; all cases were well-controlled with the same therapeutic regimen (non-fixed combination of latanoprost and timolol). The results of the study demonstrated that glaucomatous patients have a circadian rhythm of CCT, and this curve had the same phase to that esteemed in a meta-analysis on normal, non-white subjects [21]. The peak of this curve was at 4 a.m. and the trough at 4 p.m., with a mean difference of 9 microns. The circadian behaviour of CCT may reflect the well documented changes in corneal metabolism occurring during the night, with increased lactate and corneal swelling. Based on our data, the best interval in which to measure CCT is between 10 a.m. and noon.

Overall, the mean circadian CCT change was small (3.2%, 16 microns), with an inter-individual variability ranging from 1.1 to 6.2% (6-31 microns). On the opposite, mean individual variation for IOP was 45 ± 33%. It is unlikely that such a small variation in CCT may influence the changes occurring between IOP measurements at different time-points; in fact, with respect to the assumption of Doughty's meta-analysis that ''for eyes with chronic disease, [...] even modest changes (that is, 10%) in CCT can be expected to have a measurable impact on tonometry measures,'' [21] we found fluctuations of only 3%.

A similar study on CCT circadian variations was conducted by Harper et al. [22] on normal subjects using a less rigorous method (the study was conducted over a smaller group of subjects and measurements were taken every 6 hours at night), and they found a mean CCT variation of 7.2%.

In conclusion, corneal thickness follows a circadian rhythm, whose amount is – probably – negligible from a clinical viewpoint. On the opposite of IOP, only few works on this topic are available at the moment, although raising interest in the corneal biomechanisms in normal and pathological eyes is developing, which will surely improve our knowledge in the next future.

3. Ocular Blood Flow

One of the pathogenetic features of glaucoma is the loss of the balance between the beneficial effect of ocular perfusion and the negative effect of high IOP at the level of the nerve structures (retinal ganglion cells and optic nerve head).

While normal subjects have a vascular autoregulation which allows a constancy of blood flow when changes in blood pressure occurs (i.e. in recumbent position at night), it is probable that glaucomatous eyes have a vascular autoregulatory defect, which determines a fall in the ophthalmic artery flow velocities [23]. As a confirmation, higher than normal fluctuation between mean ocular perfusion (defined as mean blood pressure minus IOP) has been found in aged, health subjects [24] and in patients affected by low-tension glaucoma[25] when compared to young, healthy subjects.

Epidemiologic studies [26, 27] have shown that low diastolic blood pressure (DBP) and diastolic ocular perfusion pressure (DOPP) are major risk factors for the development and progression of glaucoma. All these studies confirmed that a low DOPP, by convention minor than 55 mmHg, was associated with an increased prevalence and incidence of the disease. It thus seems reasonable that an ideal treatment for glaucoma should not only reduce IOP but also increase DOPP (or at least not affect it) in order to further reduce the rate of progression of the disease.

All drugs used in glaucoma, as previously described, can induce a significant decrease in IOP, but they influence the circadian IOP curve and blood pressure profile in different ways, as recently investigated by Quaranta et al. [28]. In this study, dorzolamide and latanoprost significantly increased DOPP at all time points in comparison with baseline. Timolol induce an increase in DOPP during daytime (from 4 a.m. to 4 p.m.), but not during night time.

Brimonidine induce a significant decrease in mean calculated 24-hour DOPP when compared with baseline; this topical drug also induce a significant decrease in both systolic and diastolic blood pressure. The most likely mechanism for this effect may be referable to a further central reduction in sympathetic activity with reduced catecholamins, leading to reduction in cardiac output and peripheral vascular resistances.

4. Melatonin Shading

Apart from the previously discussed parameters, it is also possible that circadian clocks regulate the biochemical and physiological processes of the retina. Melatonin is one of the most important mediator for retinal activity.

Several investigations on animal models demonstrated that melatonin is synthesized not only by pineal gland but also by retina. According to most studies, the rod and the cone photoreceptors are the most probable candidates for melatonin synthesis in the retina. Since melatonin receptors are present in the retina and retinal melatonin does not contribute to the circulating levels, retinal melatonin probably acts locally as a neuromodulator [29]. It has been shown that melatonin is involved in cone elongation, [30] aggregation of melanin pigments in the pigmented epithelium, [31] suppression of light-adaptive rod outer segment disc shedding, [32] and inhibition of dopamine release [33] Melatonin is considered the major signal of dark adaptation.

Circadian rhythms in the secretion of melatonin has been demonstrated only in isolated retinas of some mammals like rats, mice, turkey. In these models, the nycthoemeral secretion of melatonin is endogenous to the retina itself [34].

Up-to-now, studies on the circadian fluctuation of melatonin in humans are missing. For ethics reasons, it is impossible to directly evaluate retinal secretion of melatonin in humans. Anyway, it is possible to dose melatonin concentration in the saliva, which is directly proportional to the plasmatic concentration of pineal melatonin. Although the relationship between pineal and retinal secretion is uncertain, in both cases the secretion is essentially nocturnal and are suppressed by retinal exposition of the light [29, 35]. Moreover, both circadian rhythms are regulated by light/dark cycle. Therefore, it is reasonable to assume that the two rhythms of melatonin secretion will have the same circadian phase under a normal light/dark cycle [35].

In conclusion, though a strong experimental evidence is still missing, it was suggested that retinal and pineal melatonin would cooperatively regulate the retinal ganglion cell activity in humans; this synergic activity might also be necessary to the normal processing of the visual information [35].

References

[1] Sommer A, Tielsch JM, Katz J, Quigley HA, Gottsch JD, Javitt J, Singh K. Relationship between intraocular pressure and primary open angle glaucoma among

white and black Americans. The Baltimore Eye Survey. *Arch. Ophthalmol.* 1991 Aug;109(8):1090-5.

[2] Davanger M, Ringvold A, Blika S. The probability of having glaucoma at different IOP levels. *Acta Ophthalmol.* (Copenh). 1991 Oct;69(5):565-8.

[3] Maeda A, Tsujiya S, Higashide T, Toida K, Todo T, Ueyama T, Okamura H, Sugiyama K. Circadian intraocular pressure rhythm is generated by clock genes. *Invest. Ophthalmol. Vis. Sci.* 2006 Sep;47(9):4050-2.

[4] Liu JH, Kripke DF, Twa MD, Hoffman RE, Mansberger SL, Rex KM, Girkin CA, Weinreb RN. Twenty-four-hour pattern of intraocular pressure in the aging population. *Invest. Ophthalmol. Vis. Sci.* 1999 Nov;40(12):2912-7.

[5] Liu JH, Kripke DF, Hoffman RE, Twa MD, Loving RT, Rex KM, Gupta N, Weinreb RN. Nocturnal elevation of intraocular pressure in young adults. *Invest. Ophthalmol Vis. Sci.* 1998 Dec;39(13):2707-12.

[6] Orzalesi N, Rossetti L, Invernizzi T, Bottoli A, Autelitano A. Effect of timolol, latanoprost, and dorzolamide on circadian IOP in glaucoma or ocular hypertension. *Invest. Ophthalmol. Vis. Sci.* 2000 Aug;41(9):2566-73.

[7] Orzalesi N, Rossetti L, Bottoli A, Fumagalli E, Fogagnolo P. The effect of latanoprost, brimonidine, and a fixed combination of timolol and dorzolamide on circadian intraocular pressure in patients with glaucoma or ocular hypertension. *Arch. Ophthalmol.* 2003 Apr;121(4):453-7.

[8] Orzalesi N, Rossetti L, Bottoli A, Fogagnolo P. Comparison of the effects of latanoprost, travoprost, and bimatoprost on circadian intraocular pressure in patients with glaucoma or ocular hypertension. *Ophthalmology.* 2006 Feb;113(2):239-46.

[9] Liu JH, Zhang X, Kripke DF, Weinreb RN. Twenty-four-hour intraocular pressure pattern associated with early glaucomatous changes. *Invest. Ophthalmol Vis. Sci.* 2003 Apr;44(4):1586-90.

[10] Blaise P, Guillaume S. Circadian variations in intraocular pressure and their clinical implications. *J. Fr. Ophtalmol.* 2005 Mar;28(3):317-25.

[11] Asrani S, Zeimer R, Wilensky J, Gieser D, Vitale S, Lindenmuth K. Large diurnal fluctuations in intraocular pressure are an independent risk factor in patients with glaucoma. *J. Glaucoma.* 2000 Apr;9(2):134-42.

[12] Bengtsson B, Heijl A. Diurnal IOP fluctuation: not an independent risk factor for glaucomatous visual field loss in high-risk ocular hypertension. *Graefes Arch. Clin. Exp. Ophthalmol.* 2005 Jun;243(6):513-8. Epub 2005 Mar 9.

[13] Sihota R, Saxena R, Gogoi M, Sood A, Gulati V, Pandey RM. A comparison of the circadian rhythm of intraocular pressure in primary chronic angle closure glaucoma, primary open angle glaucoma and normal eyes. *Indian J. Ophthalmol.* 2005 Dec;53(4):243-7.

[14] Konstas AG, Stewart WC, Topouzis F, Tersis I, Holmes KT, Stangos NT. Brimonidine 0.2% given two or three times daily versus timolol maleate 0.5% in primary open-angle glaucoma. *Am. J. Ophthalmol.* 2001 Jun;131(6):729-33.

[15] Hass I, Drance SM. Comparison between pilocarpine and timolol on diurnal pressures in open-angle glaucoma. *Arch. Ophthalmol.* 1980 Mar;98(3):480-1.

[16] Fogagnolo P, Rossetti L, Mazzolani F, Orzalesi N. Circadian variations in central corneal thickness and intraocular pressure in patients with glaucoma. *Br. J. Ophthalmol.* 2006 Jan;90(1):24-8.

[17] Konstas AG, Topouzis F, Leliopoulou O, Pappas T, Georgiadis N, Jenkins JN, Stewart WC. 24-hour intraocular pressure control with maximum medical therapy compared with surgery in patients with advanced open-angle glaucoma. *Ophthalmology.* 2006 May;113(5):761-5.e1.

[18] Stucchi CA, Gennari G, Aimino G, di Carlo I, Bauchiero L. Systematic error in computerized pachymetry. *Ophthalmologica.* 1993;207(4):208-14.

[19] Rapuano CJ, Fishbaugh JA, Strike DJ. Nine point corneal thickness measurements and keratometry readings in normal corneas using ultrasound pachymetry. *Insight.* 1993 Dec;18(4):16-22.

[20] Herndon LW. Measuring intraocular pressure-adjustments for corneal thickness and new technologies. *Curr. Opin. Ophthalmol.* 2006 Apr;17(2):115-9.

[21] Doughty MJ, Zaman ML. Human corneal thickness and its impact on intraocular pressure measures: a review and meta-analysis approach. *Surv. Ophthalmol.* 2000 Mar-Apr;44(5):367-408.

[22] Harper CL, Boulton ME, Bennett D, Marcyniuk B, Jarvis-Evans JH, Tullo AB, Ridgway AE. Diurnal variations in human corneal thickness. *Br. J. Ophthalmol.* 1996 Dec;80(12):1068-72.

[23] Chung HS, Harris A, Evans DW, Kagemann L, Garzozi HJ, Martin B. Vascular aspects in the pathophysiology of glaucomatous optic neuropathy. *Surv. Ophthalmol.* 1999 Jun;43 Suppl 1:S43-50.

[24] Liu JH, Gokhale PA, Loving RT, Kripke DF, Weinreb RN. Laboratory assessment of diurnal and nocturnal ocular perfusion pressures in humans. *J. Ocul. Pharmacol. Ther.* 2003 Aug;19(4):291-7.

[25] Choi J, Jeong J, Cho HS, Kook MS. Effect of nocturnal blood pressure reduction on circadian fluctuation of mean ocular perfusion pressure: a risk factor for normal tension glaucoma. *Invest. Ophthalmol. Vis. Sci.* 2006 Mar;47(3):831-6.

[26] Tielsch JM, Katz J, Sommer A, Quigley HA, Javitt JC. Hypertension, perfusion pressure, and primary open-angle glaucoma. A population-based assessment. *Arch. Ophthalmol.* 1995 Feb;113(2):216-21.

[27] Bonomi L, Marchini G, Marraffa M, Bernardi P, Morbio R, Varotto A. Vascular risk factors for primary open angle glaucoma: the Egna-Neumarkt Study. *Ophthalmology.* 2000 Jul;107(7):1287-93.

[28] Quaranta L, Gandolfo F, Turano R, Rovida F, Pizzolante T, Musig A, Gandolfo E. Effects of topical hypotensive drugs on circadian IOP, blood pressure, and calculated diastolic ocular perfusion pressure in patients with glaucoma. *Invest. Ophthalmol Vis. Sci.* 2006 Jul;47(7):2917-23.

[29] Tosini G. Melatonin circadian rhythm in the retina of mammals. *Chronobiol Int.* 2000; 17:599-612.

[30] Pierce ME, Besharse JC. Melatonin and rhythmic photoreceptor metabolism: melatonin-induced cone elongation is blocked at high light intensity. *Brain Res.* 1987 Mar 10;405(2):400-4.

[31] Kemali M, Milici N, Kemali D. Melatonin and LSD induce similar retinal changes in the frog. Biol. Psychiatry. 1986;21:981-985.

[32] Beherns UD, Douglas RH, Sugden D, Davies DJ, Wagner H-J. Effect of melatonin agonists and antagonists on horizontal cell spinule formation and dopamine release in a fish retina. *Cell tissue Res.* 2000;299:299-316.

[33] Boatright JH, Rubim NM, Iuvone M. Regulation of endogenous dopamine release in amphibian retina by melatonin: the role of GABA. *Vis Neurosci.* 1994; 11:1013-1018.

[34] Dubocovich ML. Role of melatonin in retina. *Prog. Retinal. Eye Res.* 1988;8:129-151.

[35] Rufiange M, Dumont M, Lachapelle P. Correlating retinal function with melatonin secretion in subjects with an early or late circadian phase. *Invest. Ophthalmol. Vis. Sci.* 2002; 43:2491-2499.

In: Circadian Rhythms and Health Research Trends
Editor: Alfredo V. Lignelli, pp. 169-185
ISBN: 978-1-60021-875-0

Chapter X

Circannual Changes in Photoperiod and Environment Affects Fertility in the Pig

Olli Peltoniemi*

Dept Clinical Veterinary Sciences, Faculty of Veterinary Medicine,
University of Helsinki, Pohjoinen pikatie 800,
04920 Saarentaus; Finland

Abstract

The effects of circannual changes in photoperiod and environment on reproductive performance in the pig are reviewed. The ancestral pig is a short day seasonal breeder and therefore seasonal effects on fertility of the pig may be regarded as physiological by nature. In principle, the breeding season in the pig is though to be controlled by the central nervous system, synchronized by the ambient photoperiod and further modified by other environmental cues such as availability of food and ambient temperature. Like in other short day seasonal breeders, breeding season in the pig is favoured in the mid winter in order to provide the offspring with best chances to survive in the following spring four months later. Under hard circumstances when food is difficult to find and temperatures fall too far below the thermal comfort zone, the breeding season of the pig may be delayed. In the modern pig industry, several parameters of reproductive performance indicate seasonal fluctuation in reproductive efficiency corresponding well to the seasonality of breeding seen in the wild for this species. Within the industry, this phenomenon is known as seasonal infertility and is characterised by lower pregnancy rates, lower proportion of females in oestrus when expected and delayed puberty when the pigs are kept in long days. Seasonal infertility is seen as an indicator period of the year, revealing the level of reproductive management and stress in a herd. Low pregnancy rates are likely due to a seasonal disruption of pregnancy, which is economically important and may be caused by a hormonal mechanism. Central effects of the pineal melatonin may suppress LH pulsatility and thereby reduce the pituitary support for CL, retarding the embryonic development and interfering with the physiological

* olli.peltoniemi@helsinki.fi; www.vetmed.helsinki.fi/saari

mechanism for establishment of pregnancy. Detrimental seasonal effects on reproduction may be avoided by adequate lighting programs involving alternation of short and long days, by providing female pigs with abundant feed, by using breed records for genetic selection, by using of cooling systems where the upper thermal comfort zone is exceeded and by providing the pig with adequate shelter, bedding material and freedom to move under cold conditions.

Introduction

Over the 10 000 years of domestication, the pig breeds have been selected for weight gain, reproductive performance and other characteristics associated with meat production. Meat has been the essential outcome this species has been utilised and domesticated for by humans. During the process of domestication, little attention has been paid to seasonality of reproduction. The ancestral pig, similarly to the present European wild boar, used to be a distinct seasonal breeder with a breeding season in November and December in the Northern hemisphere. The fertile individuals in the female population of these distinctly seasonal pig breeds give birth to a litter in spring followed by lactation in summer. During the autumn, the whole female population undergoes complete seasonal anoestrus while preparing the body reserves for the next breeding season and pregnancy (Mauget, 1982).

In most mammals, information regarding the day length and ambient light circumstances is thought to be transferred through a circadian pacemaker in the subrachiasmatic nuclei (SCN) region of the brain (Lincoln et al., 2005). This pacemaker controls the pineal activity and melatonin secretion, which provides the host with an endocrine message indicating the duration of the night and gradual changes in the photoperiod. This message triggers corresponding endocrine responses in the animal, making it possible for the animal to physiologically and behaviourally adapt to the circannual changes in the environment (Tournier et al., 2007).

One of the major physiological functions to be seasonally synchronized for is reproduction. In short day seasonal breeders such as the sheep and wild pig, breeding season occurs in mid winter which enables the offspring to be born in a favourable season from the survival point of view. Parturition at spring or early summer is advantageous for the newborn not only because of suitable ambient temperatures but food may also be readily available (Mauget, 1982).

In the European wild boar, breeding season is synchronized through photoperiod (Tast, 2001). Decreasing daylight in early winter is registered by the circadian-based photoperiodic timer and a corresponding response from the pituitary gonadal axis is triggered (Lincoln, 2006; Lincoln, 2006; Peltoniemi, 2005). Interestingly, the start of the breeding season may be delayed in rough years when there is lack of food in the environment in this species (Mauget, 1982). On the other hand, the breeding season may start earlier than usual in favourable years. However, the cessation of breeding season appears to occur independently of availability of feed (Mauget, 1982).

Seasonal effects on fertility in the domestic pig have been recognised as a topic for research activity for about 25 years (Love, 1978). By definition, seasonal infertility in the female pig includes delayed puberty of gilts (Figure 1), early disruption of pregnancy and, as a result, reduction in farrowing rate (Figure 2). In addition, if sows are kept in individual stalls after mating, weaning to oestrus interval may be prolonged beyond day 10 after weaning (Table 2, Peltoniemi et al., 1999). In the Northern Europe, the period of seasonal infertility may last from early August till early November (Peltoniemi, 1999).

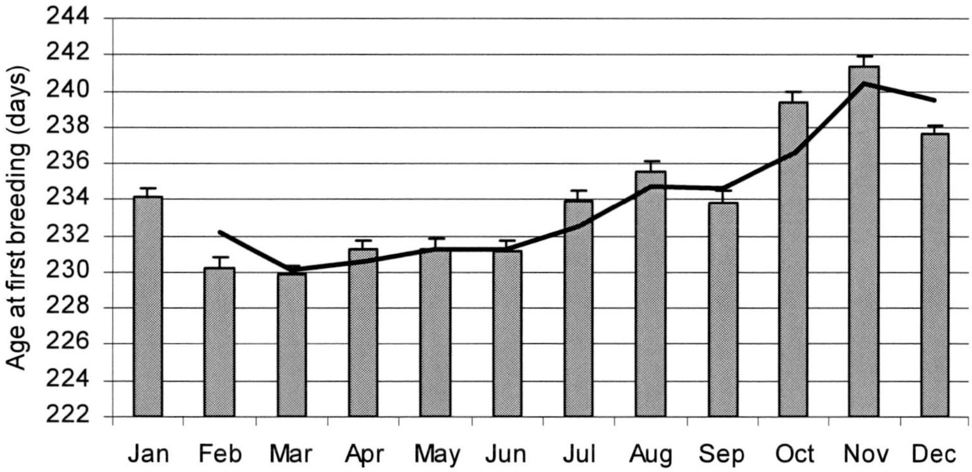

Figure 1. Effect of season on age at first breeding of gilts (data from Peltoniemi et al., 1999).

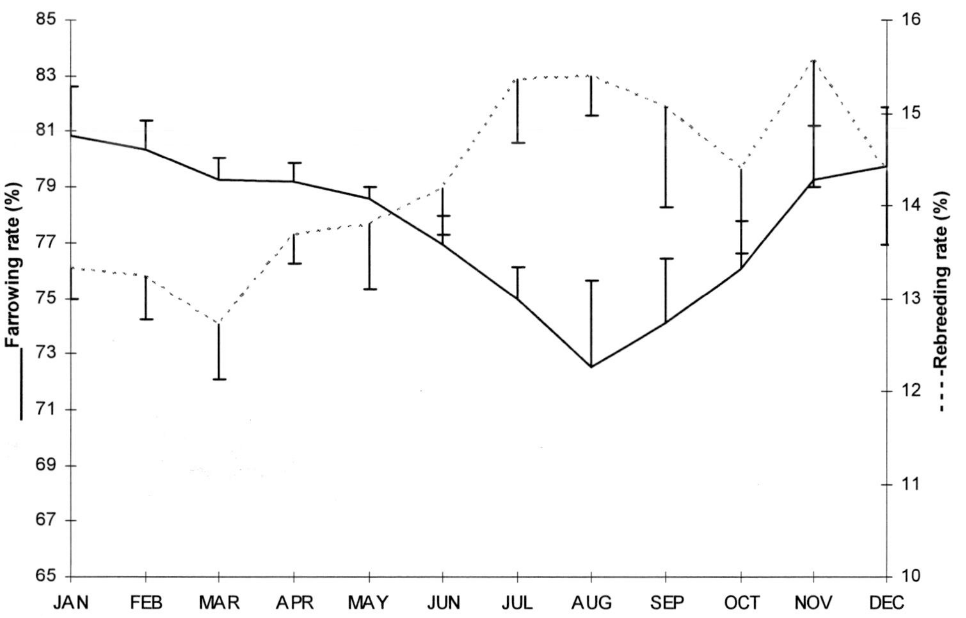

Figure 2. Effect of season on farrowing rate and rebreeding rate in Finland (data from Peltoniemi et al., 1999).

Table 1. Incidence of early disruption of pregnancy on a farm suffering from seasonal infertility (data from Tast et al., 2002)

Season	N of animals inseminated	Farrowing rate %	N of early disruptions
Winter-Spring	52	74	1
Summer-Autumn	80	60	9

Table 2. Proportion of sows becoming to oestrus by Day 10 after weaning and the risk (odds) of a prolonged weaning to oestrus interval. January is regarded as the baseline. Data from Peltoniemi et al., 1999

Month	Percentage of sows oestrus before	becoming to Day 10 (n)	Odds Ratio* (95% C.I.)
	Parity 1	Parity >1	
Jan	91.4 (734)	95.6 (2463)	1
Feb	89.7 (729)	95.1 (2276)	1.14 (0.92-1.42)
Mar	87.7 (892)	94.3 (2645)	1.33 (1.09-1.63)
Apr	87.2 (1005)	93.5 (2635)	1.50 (1.23-1.82)
May	89.4 (1068)	94.8 (2541)	1.18 (0.96-1.44)
Jun	87.8 (960)	94.8 (2578)	1.29 (1.05-1.57)
Jul	83.9 (983)	94.2 (2755)	1.57 (1.29-1.90)
Aug	85.4 (848)	92.8 (2702)	1.70 (1.40-2.06)
Sep	85.0 (827)	92.9 (2833)	1.72 (1.42-2.08)
Oct	83.9 (929)	92.9 (2968)	1.77 (1.47-2.14)
Nov	87.3 (865)	95.4 (2744)	1.22 (0.99-1.49)
Dec	91.0 (1079)	95.8 (3353)	0.98 (0.90-1.20)

* Based on a logistic regression model where herd, breed, parity and lactation length were included as factors.

Effect of Photoperiod on Reproduction in the Pig – How Does It Work?

A great deal of the present understanding about seasonal breeding activities in the wild and domestic pig is still based on the extensive research work carried out with the model short day seasonal breeder, the sheep. In the sheep, both photoperiodism and circannual rhythm generation participate in regulation of the breeding season (Lincoln et a., 2003; Lincoln et al., 2006, Malpaux et al., 1999, Malpaux et al., 2002). Photoperiodism essentially depends on transduction of photoperiod into a nocturnal melatonin signal by the pineal gland that reflects the length of night. Like in most mammals, photoperiodic stimuli every 24 h act to entrain the circadian clockwork of the SCN. In the sheep, ambient photoperiod may dictate whether or not the long term seasonal activity in reproduction is expressed but in some other mammals generation of the circannual rhythm is considered intrinsic and may function quite independently (Woodfill et al., 1994; Lincoln et al., 2003). Calendar cells appear to control seasonal changes in the gonadotrophin / gonadal axis, however, present evidence implies that these cells are mainly located in the mediobasal area of the hypothalamus (Malpaux et al., 1995, 1998; Lincoln et al, 2002; Lincoln et al., 2003).

There is growing evidence to demonstrate that pigs are similar to sheep as far as the mechanism by which season effects reproductive function is concerned. Earlier reports created some confusion since obscure melatonin profiles in the pig were described (Diekman et al., 1997; Bollinger et al., 1997). More recent evidence, however, shows that pigs have a clear circadian pattern of melatonin secretion and therefore the basis of the circannual endocrine control mechanism appears to be the same in wild as well as the domestic pig (Klupiec et al., 1997; Andersson et al., 1998; Tast et al., 2001a, b;Peltoniemi et al., 2005). Recently, we conducted a study to explore whether the seasonally delayed puberty in the domestic gilt would depend on melatonin variation between individual animals. Although delay of puberty turned out not to be associated with melatonin concentration, we were able to, once again, demonstrate a consistent nocturnal rise in melatonin concentration in each experimental animal sampled (Peltoniemi et al., 2005).

It is therefore suggested that the pig is similar to its fellow short day breeder ungulate sheep in terms of regulating mechanisms controlling seasonality of reproduction. The intrinsic circannual rhythm, which affects the GnRH pulse generator is likely to exist in the pig and is subject to seasonal changes in photoperiod. Both the intrinsic circannual rhythm generation and photoperiodism are likely to affect the degree of seasonality of reproductive function that is expressed in the pig.

Factors Affecting Seasonal Infertility

Light Conditions

As discussed above, the essence of lighting – melatonin transduction in the pig appears to be similar to that of other mammals (Tast, 2002). However, the cellular level mechanism by which melatonin exerts its effect on GnRH pulse generation is yet to be confirmed. Some characteristics, however, about the melatonin response to varying photoperiod related factors have been studied in the pig. For instance, light intensity of 40 lx triggered a similar nocturnal melatonin response as a 10 000 lx light environment during the day time (Tast, 2001), which may suggest that light intensity does not play a major role in the mechanism by which seasonal breeding activities are controlled. In addition, change from one photoperiod to another entrained a corresponding change in the melatonin profile within two weeks in the domestic pig (Tast, 2001). Similarly to the lighting programs developed to enhance the fertility of the other short day seasonal breeder, the sheep, it was therefore concluded that it should be possible to develop a lighting program supporting fertility of the pig, (Karsch et al., 1984; Tast, 2002).

So far, there is not enough evidence to give recommendations for different types of production environments so that the biology of the pig as a seasonal breeder would be utilised as much as possible. However, the best recommendation that we can come up with is that the producer would pick up one of the “reasonable” lighting programs, i.e. 16 hours of light and 8 hours of darkness throughout the production cycle and stick to that program (Tast et al., 2005). The lession to be learned is that a producer should not repeatedly change from one program to another, let alone be completely ignorant of the light circumstances that his sows are subjected to. The best lighting program is yet to be recovered.

Effect of Feeding

We have found a positive effect of an abundant post-mating feeding strategy on fertility of the sow in summer-autumn (Peltoniemi OAT, 2000;Love, 1995;Virolainen, 2004). A hypothesis regarding the endocrinology behind this somewhat controversial effect is discussed below.

The positive effects of abundant post mating feeding strategy are shown in tables 3 and 4. The gilts and sows in these studies (Virolainen et al., 2004; 2005) were fed twice a day a total of 2 kg (27 MJ/day) (LLL) or 4 kg (54 MJ/day) (HHH) until the end of the study or 2 kg (27 MJ/day) for 10 days, 4 kg (54 MJ/day) for 7 days and 2 kg (27 MJ/day) until the end of study (LHL).

Tables 3 and 4. Day of slaughter, crown-to-rump length, number of corpora lutea (CL), progesterone concentration in allantoic fluid (al) and in plasma of the jugular vein (jug), number of foetuses and embryo survival rate (means ± SD) in the high (HHH), low (LLL) and modified (LHL) feeding groups on day of slaughter. Table 3 presents information on gilts (I) and Table 4 information on sows (II)

Table 3.

GILTS	HHH mean	± SD	LLL mean	± SD	LHL mean	± SD
Day of slaughter	33.9	0.4	34.3	0.5	33.9	0.4
Crown-to-rump length (cm)	3.8	0.2	3.8	0.2	3.9	0.1
No. of CL	16.8	2.9	26.5	0.7	19.3	9.5
Progesterone (ng/ml), jug	8.2	1.7	12.7	3.1	10.4	1.6
No. of viable foetuses	11.3	4.1	20.0	0	13.3	5.0
Total no. of foetuses	11.8	4.3	24.0	2.8	13.3	5.0
Embryo survival rate (%)	67	18	75	2	69	14
Pregnancy rate (%)	100		25		37.5	

Table 4.

MULTIPAROUS SOWS	HHH mean	± SD	LLL mean	± SD	LHL mean	± SD
Day of slaughter	35.0	0	35.2	1.0	34.8	1.5
Crown-to-rump length (cm)	3.6	0.3	3.8	0.2	3.9	0.4
No. of CL	26.2	5.3	22.6	3.4	18.7	2.5
Progesterone (ng/ml), al	2.2	0.7	1.8	1.0	1.4	1.0
Progesterone (ng/ml), jug	10.1	3.2	13.5	2.9	13.5	3.3
No. of viable foetuses	11	5.1	15.6	3.1	10.3	2.9
Total no. of foetuses	13	7	16.8	2.9	13	1
Embryo survival rate (%)	45	25	69	8	55	10
Pregnancy rate (%)	83		83		60	

In the modified group (LHL), the feeding level was increased within one day but decreased in 0.5 kg/day increments until the low level was reached. A sudden decrease in feeding level was assumed to be more stressful for early pregnant pigs than a sudden increase and therefore the procedure described above was used when changing feeding levels. The results show that in terms of pregnancy rates, gilts do benefit from a high level of feeding after mating whereas this advantage does not appear to be there for the older sows (Virolainen et al., 2004;2005).

Effects of Group Housing and Age

It has also been reported that housing is an important factor in seasonal infertility (Peltoniemi OAT, 2000). Under individual housing, weaning to oestrus interval may be prolonged in summer-autumn, whereas a reduction in farrowing rate is more commonly seen in loose housing systems.

Compared to sows farrowed more than twice, low parity sows are known to be more susceptible to the seasonal reduction in farrowing rate (Love, 1993). Moreover, the highest repeat breeding rate has been seen in low parity sows and gilts. Young sows also showed a seasonal pattern: repeat breeding rate was significantly increased from July to September (Peltoniemi, 1999). The strategy concerning the age structure of herds nowadays is such that sows may not be kept for years. For instance, an average of 3.4 farrowings / sow before removal is the standard practise in Finland. This preference for younger sows is not likely to be changed. Thus, seasonal patterns seen in fertility of the sow may play an ever increasing role in the management of breeding in the future.

As far as the stocking density, new recommendations for stocking density of pregnant sows have been given in Finland at the end of 2005. These recommendations have it that the space allowance per sow should exceed 2,25 square meters, excluding the feeding stalls. This is the baseline for any new piggery to be established in Finland. Regarding seasonal infertility, herds having the least problems with seasonal infertility allow at least 2,5 square meters per sow. However, if group size is high (more than 40 sows / group), one may reduce the space allowance by a little bit without serious consequences.

Regarding the group size, it has been found that, in terms of seasonal infertility, the sow is best off with 4-6 sows as an "optimal" group size (Love et al., 1993). A group size of about 20 appears to be the hardest one to manage whereas in a large group of >40 sows the animals find more space to go to in case of aggressive behaviour being present.As a rule of thumb, we recommend homogenous groups, both in terms of size and in terms of age as the basic planning strategy for breeding units in sow herds (Eliasson-Selling et al., 2000).

Recently, we also made economic simulations for slatted floor vs. deep pit (straw based) as a basis of environment for newly mated sows in breeding units (Peltoniemi et al., unpublished data). For a herd with 300 sows, we found that the building costs were 85 € / sow / year more expensive in a deep pit than in a partly slatted floor system. However, when the good health and reproduction related effects of the deep pit were accounted for together with the running cost simulations, the running costs were 55 € / sow / year cheaper in the deep pit based system as compared with the partly slatted floor system. In the deep pit system with straw, seasonal infertility appeared to be clearly less of a problem as compared with other systems without bedding material (Peltoniemi et al., 1999). We therefore recommend deep pit systems or systems with abundant bedding material for sow producers in order to keep production efficiency high throughout the year.

Boar Effect

The importance of the presence of a boar for attainment of puberty was clearly demonstrated in a study where 80 % of gilts with a daily contact of 30 minutes with a boar reached puberty by 225 days of age whereas only 30 % of the control gilts with no contact with a boar reached puberty (Paterson, 1991). During long days, as low a proportion as 13 % of the gilts without a boar contact reached puberty by 225 days of age.

Seasonal Disruption of Pregnancy

In those sows that do return to oestrus in summer-autumn, pregnancy is established and maintained for a short period, after which pregnancy is disrupted and the whole litter is lost. This seasonal disruption of pregnancy has been recognised as an economically important manifestation of seasonal infertility(Love, 1993). Up to 10 % of established pregnancies may be disrupted due to a seasonal effect with the low parity sows and gilts being most at risk. Fertilization appears to take place like in other seasons, embryos may be transported and the pregnancy is apparently normally established, but quite early on, the embryos die and the sow returns to oestrus after a mating to oestrus interval of 25-35 days, see Table 1. (Love, 1993;Tast, 2002).

As from the biological point of view, this process appears as a wasteful one and there is no clear cut explanation for it. Firstly, from the seasonal point of view, it is feasible that no litter would be born in the middle of winter, when the environment is disadvantageous for the newborn offspring. However, there appears to be no logic for the establishment of pregnancy, but it would be reasonable for the pig to stay anoestrous instead. As far as other mammals are concerned, there does not appear to be a similar phenomenon, other than that seen in a tropical rodent (cloud forest mice)(Love, 1993).

Epidemiological evidence does back up the theory about seasonal disruption of pregnancy. A reasonable number of studies have reported a substantial increase in rebreeding rate during autumn (Xue, 1994; Tantasuparuk, 2000; Tummaruk, 2000; Peltoniemi OAT, 2000). In addition, looking closely at the rebreedings, the oestrus to oestrus interval is clearly prolonged in the autumn months. Concerning the biological evidence for seasonal infertility, it has been shown in at least two studies that in those sows that loose their pregnancy, the pregnancy is first established. This was evidenced by the presence of oestrogen metabolites of embryonic origin (Mattioli, 1987), progesterone profiles as well as by means of real time ultrasonography (Tast, 2002). However, quite early during the course of pregnancy, as evidenced by the oestrogen metabolites, the pregnancy is interrupted. Trying to relate this interruption to the physiology of establishment of pregnancy in the pig, it appears that the first wave of estradiol is present as it should be around day 13 of pregnancy. However, the second wave of embryonic estrogens for establishment of pregnancy at day 19 seems to be missing in these sows. Interestingly enough, studies using estradiol administration as a way to mimic embryonic oestrogen in unmated sows have shown that in case the second phase of higher estradiol concentration is missing, the oestrus to oestrus interval is prolonged to 25-35 days. A similar oestrus to oestrus period usually seen in the 'problem sows' in autumn (Peltoniemi, 2000).

The present hypothesis about the seasonal disruption of pregnancy thus is that in the sows undergoing seasonal disruption of pregnancy, LH secretion is reduced in autumn by a central mechanism involving melatonin. The reduction in LH secretion has a negative effect on progesterone production, which in turn retards the development of embryos to the extent the estradiol secreted by them is inadequate at around day 19 to support the mechanism of the maternal recognition of pregnancy. CL are destroyed as a result within 10 days and the sow returns to oestrus at an oestrus to oestrus interval of 25-35 days (Peltoniemi et al., 2000).

Suppression of LH Secretion during Early Pregnancy

As we have been interested in the mechanism of seasonal disruption of pregnancy, our group has tried to further explore the role of LH secretion in maintenance of early pregnancy in the pig. Earlier work by others had shown that transaction of the hypophysial stalk or hypophysectomy resulted in disruption of early pregnancy, indicating that there were a substance of hypophyseal origin to support CL and maintenance of early pregnancy (Anderson, 1967). These results were interpreted as LH being the factor supporting the CL.

Our group used three models to explore this mechanism. We first implanted pregnant gilts with GnRH agonist implants to down regulate GnRH receptors and thereby suppressed LH pulsatility (Peltoniemi et al., 1995). We then used active and passive immunisation against GnRH to settle the LH pulsatility down in the early pregnant gilt (Tast et al., 2000). Thirdly, we also used a GnRH antagonist to directly down regulate LH pulses (Virolainen, 2004a). By each of these models, regression of CL and disruption of pregnancy in the gilt was induced. As a conclusion, there is enough evidence to state that LH secretion / pulsatility is important to support CL and maintenance of early pregnancy in the pig.

Recently, we examined the effects of feeding strategies on LH secretion during early pregnancy in the gilt and sow (Virolainen et al., 2004b; Virolainen et al., 2005b). Three feeding regimens from mating through until day 34 of pregnancy, high, low and modified (low until day 14, high until day 21 and low thereafter), were applied to explore the above-effects of feeding on LH secretion and fertility. Modified feeding, which was thought to combine the benefits of restricted feeding after ovulation and abundant feeding during implantation, were first tested on three groups of eight gilts in autumn. The pregnancy rate was higher in the high group (100%) than in the low (25%) or modified (38%) groups, although the high group had significantly lower progesterone concentrations on days 9 and 12. The basal and mean LH levels were higher in the high group than in the modified group. The LH amplitude tended to be higher for gilts in the high group. The modified feeding strategy failed to provide the benefits anticipated. Instead, the high feeding strategy provided a distinct advantage in pregnancy rate.

The same regimens were tested during early pregnancy in the multiparous sow (Virolainen et al., 2005b). In this experiment, dietary treatment did not significantly affect LH characteristics measured (mean, amplitude, frequency and basal level). However, progesterone concentration tended to be lower in the high group than in the low group. In the modified group, progesterone concentration seemed to be associated with the level of feeding. The embryonic recovery rate was 69% in the low, 45% in the high and 55% in the modified group. Neither high feeding nor modified feeding provided any benefits for reproductive performance in multiparous sows. Altogether, these studies indicate that both season and feeding level have a significant impact on LH secretion during early pregnancy.

Progesterone Production during Early Pregnancy

Regarding the hormonal factors maintaining pregnancy in the pig, progesterone is generally considered as the most important one. Without progesterone, continuation of pregnancy at any stage of pregnancy is impossible in the pig. A lot of work has been conducted to show the relationship between the peripheral concentrations of progesterone and embryonic survival. This relationship appears a positive one: the higher the progesterone concentration, the higher the embryonic survival. It was also suggested that a low plane of nutrition would be associated with higher progesterone concentration and higher embryonic survival (Jindal R, 1996).

On the other hand, field observations as well as a number of recent experimental studies have put this relationship between nutrition and progesterone / embryonic survival under quite some doubt. It appears that a low plane of nutrition (usually recommended 1.5 X maintenance) may not result in as high pregnancy rates as a more liberal feeding policy.

Our latest findings suggest that progesterone is secreted in a pulsatile manner in the pig, very much in the same way as that seen with LH during early pregnancy (Virolainen, 2005). Most interestingly, it appears that LH pulses from the pituitary are temporally linked to progesterone pulses from the ovary. This finding warrants further investigation and may eventually change the whole concept of how factors like nutrition and season affect the developing embryos.

Stress and Early Pregnancy

It appears that season and stress interact to cause problems in some farms. On the farm level, there may be significant variation in the way stressors have their impact in fertility in the sow. The variation could explain at least some of the pen to pen variation seen in the presence of a seasonal infertility problem on a farm. With 'a good level' of stress, a farm may escape seasonal infertility, whereas the problem is usually seen as a serious one on farms with a lot of stress factors, such as high stocking density, a lot of variation in size and age of females in a group and absence of a boar.

The most recent studies have shown that the early pregnant sow stands well with short lasting stressors, for instance for a period of 48 hours (Turner et al., 2005). Stressors used were food deprivation and ACTH administration (Razdan et al., 2002; Turner et al., 2005). However, many times in practical circumstances stressors are present day after day, even for weeks exerting a potential negative effect on reproduction in the sow. These long acting stressors would involve restricted feeding, social stress caused by pen mates and possibly, poor handling of the pigs by the care taking personnel.

What Can Be Done to Avoid Seasonal Infertility?

Light Programs

We recently investigated whether a simplified artificial lighting programme can improve sow fertility (Tast et al., 2005). Sows in a commercial piggery were allocated either to a long day (16 hours light and eight hours darkness) or a short day (eight hours light and 16 hours darkness), and production data were collected from March to June and July to October. Both regimens effectively eliminated seasonal infertility, and there was no significant difference between the groups in the farrowing rate (90 per cent) or the number of piglets produced. Both regimens had some positive effects on the weaning to oestrus interval, and the culling rate for fertility problems was lower in the short-day group. This study implied that any type of a controlled light program is better in terms of fertility than an uncontrolled lightning in commercial pig production.

Feeding

An abundant post-mating feeding strategy appears to alleviate seasonal infertility and therefore an advice for higher feeding level for the first month of pregnancy may be justified on farms suffering from seasonal infertility (Virolainen et al., 2004b). Furthermore, as far as seasonal infertility is concerned, we have seen a clear advantage of providing sows with hey or straw as an additional component of feed (Figure 3).

Stocking Density

Consultants in the pig industry often tell producers to increase animal inventory before summer-autumn, in order to be prepared for the reduced fertility due to season. While this appears as a reasonable strategy, care should be taken to reduce, not increase the stocking density in summer-autumn. This statement stems on the fact that sows are more susceptible to stress effects of stocking on fertility in summer-autumn than other times of the year.

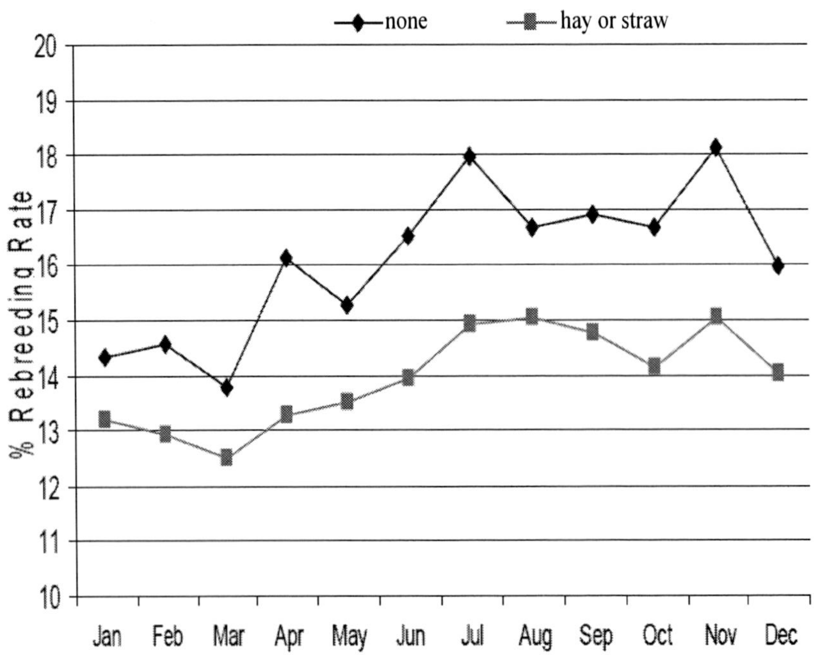

Figure 3. Effect of roughage feed (hay or straw) on seasonally increased rebreeding rate. Data from Peltoniemi et al., 1999.

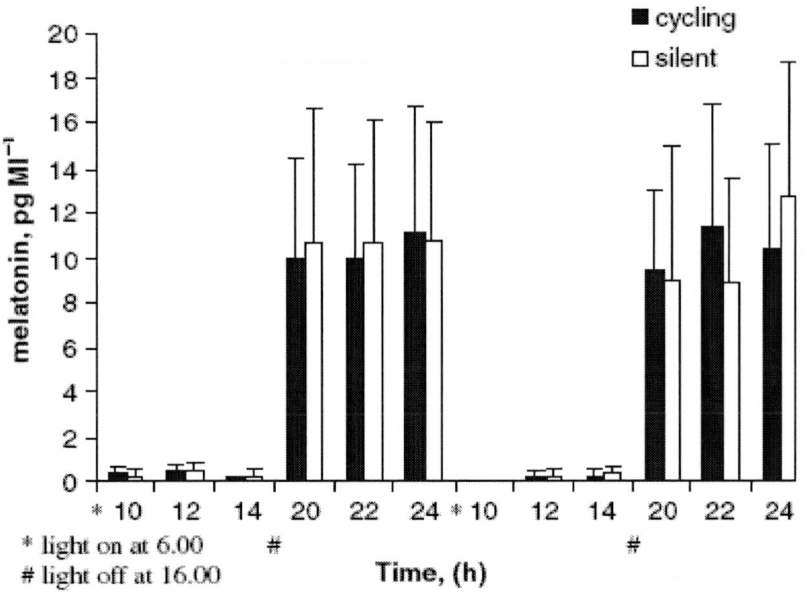

Figure 4. Melatonin concentrations (+ S.D.) of gilts reaching puberty by 7 months of age (CYCLING, n=9) and gilts not reaching puberty by that age (SILENT, n=9). Day time (lights on) samples were taken between 10.00-14.00 and night time samples (lights off) between 20.00-24.00. Data from Peltoniemi et al., 2005.

Hormone Administration

Seasonal anoestrus is considered to be similar to the prepubertal period with regard to the responsiveness of the female pituitary-gonadal axis to endogenous gonadotrophin analogues. Therefore, it is not surprising that gonadotrophin analogues appear to be somewhat effective in induction of oestrus in seasonally anoestrous gilts and sows. However, although oestrus may be induced, fertility figures may not be promising and, on the other hand, consumers wish for no hormonal treatements to be used on a routine basis. Therefore, this approach may not provide with a sustainable key to avoid seasonal effects on reproduction in the domestic pig.

Conclusion

The ancestral pig is a short day seasonal breeder and therefore manifestations of seasonal infertility may be regarded as physiological by nature. Deep down, the domestic pig is still a short day seasonal breeder with breeding season occurring at Christmas time in Europe. Farrowing rate, weaning to oestrus interval and the rate of gilts reaching puberty by the expected time are reduced in summer-autumn. Furthermore, littersize in summer may be reduced, however data on this parameter are contradictory. Seasonal infertility is seen as an indicator period of the year, revealing the level of reproductive management and stress in a herd. Seasonal disruption of pregnancy is economically important and may be caused by a hormonal mechanism. Central effects of the pineal melatonin may suppress LH pulsatility and thereby reduce the pituitary support for CL, retarding the embryonic development and interfering with the physiological mechanism for establishment of pregnancy. Seasonal infertility is best avoided by providing newly mated sow groups with abundant feeding and roughage feed to ad libitum, a reasonable lighting program throughout the year with the light and dark phase alternating for example in a 16 hours of light and 8 hours of dark photoperiodic manner. Groups of sows should be homogenous in size and age and boar should be utilised in heat stimulation and detection. Where the thermal comfort zone of the sow in ambient temperature is exceeded (20-22 degrees of Celcius), cooling systems should be applied.

References

Anderson, L.L., Dyck, G.W., Mori, H., Henricks, D.M., Melampy, R.M., 1967. Ovarian function in pigs following hypophysial stalk transection or hypophysectomy. *American Journal of Physiology*. 212, 1188-1194.

Andersson, H., Wallgren, M., Rydhmer, L., Lundstrom, K., Andersson, K., Forsberg, M., 1998. Photoperiodic effects on pubertal maturation of spermatogenesis, pituitary responsiveness to exogenous GnRH, and expression of boar taint in crossbred boars. *Animal Reproduction Science*. 54, 121-137.

Bollinger, A.L., Wilson, M.E., Pusateri, A.E., Green, M.L., Martin, T.G., Diekman, MA., Melatonin, Gilts, Photoperiod, Puberty., 1997. Lack of a nocturnal rise in serum concentrations of melatonin as gilts attain puberty. *Journal of Animal Science.* 75, 1885-1892.

Diekman, M.A., Arthington, J.A., Clapper, J.A., Green, M.L., Pig-puberty, Melatonin, Puberty., 1997. Failure of melatonin implants to alter onset of puberty in gilts. *Animal Reproduction Science.* 46, 283-288.

Eliasson-Selling L, Hofmo PO and Narum M (2000) The art of oestrus control in group-housed sows – a video presentation Proceedings of the 16th IPVS congress, Melbourne, Australia p 395.

Jindal R, Cosgrove JR, Aherne FX, Foxcroft GR, 1996. Effect of nutrition on embryonal mortality in gilts: association with progesterone. *Journal of Animal Science.* 74, 620-624.

Karsch, F.J., Bittman, E.L., Foster, D.L., Goodman, R.L., Legan, S.J., Robinson, J.E., 1984. Neuroendocrine basis of seasonal reproduction. *Recent Progress in Hormone Research.* 40, 185-225.

Klupiec, C., Evans, G., Love, R.J., Kennaway, D.J., 1997. Clarifying plasma melatonin profiles in domestic pigs: A critical and comparative evaluation of two radioimmunoassay systems. *Journal of Pineal Research.* 22, 65-74.

Lincoln, G., Messager, S., Andersson, H., Hazlerigg, D., 2002. Temporal expression of seven clock genes in the suprachiasmatic nucleus and the pars tuberalis of the sheep: Evidence for an internal coincidence timer. Proceedings of the National Academy of Sciences of the United States of America. 99, 13890-13895.

Lincoln, G.A., 2006. Melatonin entrainment of circannual rhythms. *Chronobiol. Int.* 23, 301-306.

Lincoln, G.A., Clarke, I.J., Hut, R.A., Hazlerigg, D.G., 2006. Characterizing a mammalian circannual pacemaker. *Science.* 314, 1941-1944.

Lincoln, G.A., Andersson, H., Hazlerigg, D., 2003. Clock genes and the long-term regulation of prolactin secretion: Evidence for a photoperiod/circannual timer in the pars tuberalis. *Journal of Neuroendocrinology.* 15, 390-397.

Love, R.J., 1978. Definition of a seasonal infertility problem in pigs. *Veterinary Record.* 103, 443-446.

Love, R.J., Evans, G., Klupiec, C., 1993. Seasonal effects on fertility in gilts and sows. *Journal of Reproduction and Fertility - Supplement.* 48, 191-206.

Love, R.J., Klupiec C, Thornton EJ, Evans G, 1995. An interaction between feeding rate and season affects fertility of sows. *Animal Reproduction Science.* 39, 275-284.

Malpaux, B., Daveau, A., Maurice-Mandon, F., Duarte, G., Chemineau, P., 1998. Evidence that melatonin acts in the premammillary hypothalamic area to control reproduction in the ewe: presence of binding sites and stimulation of luteinizing hormone secretion by in situ microimplant delivery. *Endocrinology.* 139, 1508-1516.

Malpaux, B., Viguie, C., Skinner, D.C., Thiery, J.C., Chemineau, P., 1997. Control of the circannual rhythm of reproduction by melatonin in the ewe. *Brain Research Bulletin* 44, 431-438.

Malpaux, B., Thiery, J.C., Chemineau, P., 1999. Melatonin and the seasonal control of reproduction. *Reproduction and Nutrition Development* 39, 355-366.

Malpaux, B., Tricoire, H., Mailliet, F., Daveau, A., Migaud, M., Skinner, D.C., Pelletier, J., Chemineau, P., 2002. Melatonin and seasonal reproduction: understanding the neuroendocrine mechanisms using the sheep as a model. *Reproduction* Suppl. 59, 167-179.

Mattioli, M., Prandi, A., Camporesi, A., Simoni, A., Seren, E., 1987. Investigation on swine summer infertility in Italy. In *Definition of the summer infertility problem in the pig*, Comission of the European Communities. 33-34.

Mauget, R., 1982. Seasonality of reproduction in the wild boar. In: Foxcroft, J. and Cole, R. (Editors), *Control of pig reproduction.*

Paterson, A.M., Pearce, G.P., D'Antuono, M.F., 1991. Seasonal variation in attainment of puberty in isolated and boar- exposed domestic gilts. *Animal Reproduction Science.* 24, 323-333.

Peltoniemi OAT, Easton BG, Love RJ, Klupiec C, Evans G, 1995. Effect of chronic treatment with a GnRH agonist (Goserelin) on LH secretion and early pregnancy in gilts. *Animal Reproduction Science.* 40, 121-133.

Peltoniemi OAT, Heinonen, M., Leppavuori, A., Love RJ, 1999. Seasonal effects on reproduction in the domestic sow in Finland-A herd record study. *Acta Veterinaria Scandinavica.* 40, 133-144.

Peltoniemi OAT, Tast, A., Love RJ, 2000. Factors effecting reproduction in the pig:seasonal effects and restricted feeding of the pregnant gilt and sow. *Animal Reproduction Science.* 60-61, 173-184.

Peltoniemi, O.A., Tast, A., Virolainen, J.V., Karkamo, V., Heinonen, M., Andersson, M.A., 2005. Night-time melatonin secretion and seasonally delayed puberty in gilts. *Reproduction in Domestic Animals.* 40, 224-227.

Peltoniemi, O.A.T., Love, R.J., Heinonen, M., Tuovinen, V., Saloniemi, H., 1999. Seasonal and management effects on fertility of the sow: a descriptive study. *Animal Reproduction Science.* 55, 47-61.

Peltoniemi, O.A.T., Tast, A., Love, R.J., 2000. Seasonal, environmental and nutritional effects on gonadotrophins - Implications for establishment and maintenance of early pregnancy in the pig. *Reproduction in Domestic Animals.* 96-100.

Razdan, P., Mwanza, A.M., Kindahl, H., Rodriguez-Martinez, H., Hulten, F., Einarsson, S., 2002. Effect of repeated ACTH-stimulation on early embryonic development and hormonal profiles in sows. *Animal Reproduction Science.* 70, 127-137.

Tantasuparuk, W., Lundeheim, N., Dalin, A.M., Kunavongkrit, A., Einarsson, S., 2000. Reproductive performance of purebred Landrace and Yorkshire sows in Thailand with special reference to seasonal influence and parity number. *Theriogenology.* 54, 481-496.

Tast, A., 2002. Endocrinological basis of seasonal infertility. PhD Thesis, University of Helsinki.

Tast, A., Halli, O., Ahlstrom, S., Andersson, H., Love, R.J., Peltoniemi, O.A.T., 2001a. Seasonal alterations in circadian melatonin rhythms of the European wild boar and domestic gilt. *Journal of Pineal Research.* 30, 43-49.

Tast, A., Hälli, O., Virolainen, J.V., Oravainen, J., Heinonen, M., Peltoniemi OAT, 2005. An attempt to develop a simplified artificial lighting programme for commercial piggeries to improve fertility of sows. *Veterinary Record.*

Tast, A., Love RJ, Clarke, I.J., Evans G, 2000. Effects of active and passive gonadotrophin-releasing hormone immunization and establishment of pregnancy in pigs. *Reproduction, Fertility and Development.* 12, 1-6.

Tast, A., Love, R.J., Evans G, Telsfer, S., Giles, R., Nicholls, P., Voultsios, A., Kennaway, D.J., 2001b. The pattern of melatonin secretion is rhythmic in the domestic pig and responds rapidly to changes in daylength. *Journal of Pineal Research.* 31, 294-300.

Tast, A., Love, R.J., Evans, G., Andersson, H., Peltoniemi, O.A., Kennaway, D.J., 2001c. The photophase light intensity does not affect the scotophase melatonin response in the domestic pig. *Animal Reproduction Science.* 65, 283-290.

Tast, A., Peltoniemi, O.A., Virolainen, J.V., Love, R.J., 2002a. Early disruption of pregnancy as a manifestation of seasonal infertility in pigs. *Animal Reproduction Science.* 74, 75-86.

Tast, A., Peltoniemi, O.A.T., Virolainen, J.V., Love, R.J., 2002b. Early disruption of pregnancy as a manifestation of seasonal infertility in pigs. *Animal Reproduction Science.* 74, 75-86.

Tournier, B.B., Dardente, H., Simonneaux, V., Vivien-Roels, B., Pevet, P., Masson-Pevet, M., Vuillez, P., 2007. Seasonal variations of clock gene expression in the suprachiasmatic nuclei and pars tuberalis of the European hamster (Cricetus cricetus). *European Journal of Neuroscience.* 25, 1529-1536.

Tummaruk, P., Lundeheim, N., Einarsson, S., Dalin, A.M., 2000. Reproductive performance of purebred Swedish Landrace and Swedish Yorkshire sows: I. Seasonal variation and parity influence. *Acta Agriculturae Scandinavica Section A-Animal Science.* 50, 205-216.

Turner, A.I., Hemsworth, P.H., Tilbrook, A.J., 2005. Susceptibility of reproduction in female pigs to impairment by stress or elevation of cortisol. *Domestic Animal Endocrinology.* 29, 398-410.

Virolainen, J.V., Love, R.J., Tast, A., Peltoniemi, O.A., 2004a. Effect of a gonadotrophin-releasing hormone antagonist on luteinising hormone secretion and early pregnancy in gilts. *Reproduction, Fertility and Development.* 15, 451-459.

Virolainen, J.V., Love, R.J., Tast, A., Peltoniemi, O.A., 2005a. Plasma progesterone concentration depends on sampling site in pigs. *Animal Reproduction Science.* 86, 305-316.

Virolainen, J.V., Peltoniemi, O.A., Munsterhjelm, C., Tast, A., Einarsson, S., 2005b. Effect of feeding level on progesterone concentration in early pregnant multiparous sows. *Animal Reproduction Science.* 90, 117-126.

Virolainen, J.V., Tast, A., Sorsa, A., Love, R.J., Peltoniemi, O.A., 2004c. Changes in feeding level during early pregnancy affect fertility in gilts. *Animal Reproduction Science.* 80, 341-352.

Virolainen, J.V., Tast, A., Sorsa, A., Love, R.J., Peltoniemi, O.A., 2004b. Changes in feeding level during early pregnancy affect fertility in gilts. *Animal Reproduction Science.* 80, 341-352.

Xue, J.L., Dial, G.D., Marsh, W.E., Davies, P.R., 1994. Multiple manifestations of season on reproductive performance of commercial swine. *Journal of the American Veterinary Medical Association.* 204, 1486-1489.

In: Circadian Rhythms and Health Research Trends
Editor: Alfredo V. Lignelli, pp. 187-198
ISBN: 978-1-60021-875-0

Chapter XI

The Current State of Knowledge about Impact of Working at Night and Working Shifts on Employees' Health

Bartosz Bilski
Poznan University of Medical Sciences, Poland

Through centuries the man's life was dependent on the twenty-four hour cycle of day and night. The factor that seriously limited undertaking of work on a larger scale in the night time was lack of proper lighting. Invention of the bulb by Thomas Alva Edison gradually resulted in usage of the rest of the twenty-four hours for working. For that phenomenon the period of the First World War was crucial, when the considerable needs of the armaments industry contributed to the mass introduction of the night work in Europe. Nowadays we can talk about different problems of the "24-hour society". The part of workers are engaged on the non-standard working hours (near shift and night work - part time work, weekend work, varying working hours, split shifts, seasonal work, on-call work and compressed work week). The shift work is a reality for about 25 percent of the North American working population.

The standard typical working time goes in the iterval starting at 7:00 a.m. and finishing at 8:00 p.m.. T.H. Monk and S. Folkard – authorities in this field, proposed the following definition: "the shift work is making any regular professional activities outside the usual accepted working hours daily, it means 7:00-18:00". Other definitions of the night work are included, among others, in the Convention No 171 and the recommendation No 178 of the International Labour Organization issued in June 1990, the Directive of the European Union 93/104/EC issued in 1993 and the recommendation of the World Health Organization (the Conference in Peking). The effects of the shift work can be divided into biological ones, health ones and sociological ones.

Biological Effects

The biological effects result from disorders of the course of the internal biological rhythms and their only clearly visible effect are sleep disorders. Reactions of the organism to the shift work are very individual. The tests' results indicate that only about 10% of workers do not complain about work in this form. Approximately 20% of them give up quite quickly this kind of work. Whereas 70% of the workers „withstand" such kind of work – however, with appearing symptoms of its intolerance. In the night hours decrease of the working ability has been noticed – it has been found that the lowest level of activity appears about 3 o'clock in the night. At this moment the lowest activity of the following systems is noticed: the cardiovascular system, the respiratory system and the lowest level of metabolism and the lowest body temperature. In the night hours the manual labour requires bigger expenditure of energy. The reason for increased energy demand is worse movement coordination. The shift work is tolerated worse by the elderly people. Decreasing of the shift work tolerance relating to age is connected with decrease of the average melatonin concentration among the elderly people, their earlier waking up, shortening of the sleep time and with increased intake of sleep-inducing drugs and tranquilizers by such persons. It is justified to limit the age of the employed at night for the men above 55 years of age and the women above 50 years of age. The employers often prefer omitting such persons during employing, rather than adapt the work organisation to their abilities.

It should be remembered that the persons who are going to undertake work in such system should be aware (the same for a doctor admitting them to work) of the health contraindications against taking such job. Among such contraindications one should include first of all the sleep disorders, the arterial hypertension of a significant degree, the heart failure in the last year, the insulin-dependent diabetes, some kidney diseases, diseases of the adrenal glands and the thyroid glands, a bronchial asthma, severe gastrointestinal diseases, psychic disorders, epilepsy, cancers and pregnancy. With the exception of pregnancy, these contraindications are usually not strict ones, however, one should remember that such kind of work makes their treatment difficult. Advances in clinical diagnosis and pharmacology offer now better possibilities for treatment of some diseases and therefore they can make possible for some workers to remain on the shift work permanently, if their transfer to daily work would be problematic due to other risk factors or to personal resistance to changing their job. Periodic checks of workers' health status are important tools aimed at detecting the early signs of intolerance or difficulty in adjustment. The periodicity of the medical examination should be set in relation to the various factors related to both working conditions and individual signs. The second health check should take place during the first year of night or shift work and the successive examinations at least every 3 years for workers under 45 years of age and every 2 years for those over 45 years old. Aging decreases the adapting abilities of the worker, in relation to this fact the three shit work is not recommended for the workers over 45 years of age.

It is an interesting fact that one of the measure of functioning of the rhythmical biological processes in the organism is the heart rate/respiratory frequency, at the same time its value over 4 may testify that their course is labile. Such persons tolerate the night work and the shift work worse.

The significant factor having impact upon tolerance of the shift work is the man's chronotype. The persons that tolerate the night work and the shift work much better are the persons of the evening chronotype.

It should be also mentioned that the persons of the low level of physical efficiency tolerate the shift work and the night work worse.

Sociological Effects

The sociological effects of work from the point of view of its impact upon the rhythm of the night work and the shift work are dependent to a high degree on the age, the sex, the marital status, number of children and their age, and also on the worker's personality. Free time in the afternoon and evening hours is very crucial for social contacts, family relations and intimate ones. The shift work and the night work and in the continuous system makes the family relations meaningfully more difficult. Moreover, the conditions that provide the appropriate quantity of sleep for the shift worker in the daily time may limit activity of other members of the household.

Health Effects

The Impact upon Psyche, a Frame of Mind and Sleep

Both the influence of performing work in the night hours and the total number of the worked hours is important for the worker's psyche. The generale rule should be adopted that while performing work for specified amount of time, the number of resting hours should be about twice as big. The work in the night hours is the work that is performed with psychic and intellectual functions that are worse than by day. Mental tasks are peformed best in the morning hours, the tasks connected with processing of data in the late morning hours, whereas the problems that require special concentration – about 8 o'clock in the evening, and the ones requiring quick reaction in the late afternoon hours and the early evening hours. With increase of the working hours per week the increased number of accidents at work is reported. The risk of an accident during the night shift is usually bigger than during working in the day. The maxinum time of reaction to a stimulus occurs between one o'clock and three o'clock. It is the time when the risk of making an error or the risk of an accident at work of the shift worker are the biggest. It is estimated that the risk of an accident for the person that works for the fourth successive night is 1.4 in relation to the risk of such event during work in the first night. Because of such reason, among others, the shifts lasting for 24 hours and longer should be eliminated. For example, in accordance with Landrigan and collaborators, doctors that are employed in the intesive care units during such long turns of duty commit 35.9% more serious errors and even 5.6 times more serious diagnostic errors.

The persons working at night sleep form 5 to 7 hours less per week than the persons that work only in the daytime. This shortening of the sleep time results from shortening of the second stage of sleep and the phase of rapid eye movement (REM). Limitation of the sleep time causes lowering of psychophysical efficiency of the night shift workers. Both the pathologic sleepiness during daytime and also especially during driving back home after the night shift has been found in their case. Working for one night has no significant impact upon the internal mechanisms of the man's organism 24-hour rhythm. Not before having worked for successive nights it become arduous as a result of accumulation of the sleep deficiency. Some test results indicate that long-term (for many years) doing work in the night results in increase of the death rate in the population of such workers. The reason for such phenomenon may be, among other things, shortening of the average sleep time of such person. It has been proved that in the general population the persons that sleep 5.0-5.9 hours per day live shorter. Shortening of the sleep duration, its chronic deficiency and other disorders of its course cause so-called industrial jet lag. In case of long-period shift work and night work chronic tiredness that can manifest itself in the form of so-called the burn out syndrome and also the night shift paralysis appear more frequently. The burn out syndrome may manifest itself through loss of feelings and interest for co-workers and customers, indifference towards their emotional conditions, moreover, there appear negligence at work and excessive reactions to weak stimuli appear. The symptoms of depression resulting from exhaustion are, as follows (according to Pfeifer):

1. Fatigue, lack of enthusiasm, initiative, will and energy.
2. Difficulty in making decisions.
3. Sadness – disposition to crying.
4. Timidity, indisposition towards performing tasks that were carried out with pleasure and with good results in the past.
5. Irritation, anger.
6. Despair, hopelessness.
7. Avoidance of friends, withdrawal from social engagement.
8. Intensified drawsiness or insomnia.
9. Apetite disorder.
10. Various affections such as headache, pains in the muscles, gastric-intestinal troubles, low arterial pressure, dizziness, balance disorders, faintness, that cannot be justified in another way.
11. Problems of the religious nature

It should be mentioned that in the population of shift workers there is the biggest rate of divorces, abuse of medicines and occurring of depression.

Impact upon the Cardiovascular System

So far no increase of the death rate for the shift workers because of diseases of the cardiovascular system has been found. Fragmentation of sleep and its significant shortening may be, however, the reason for occurance of the coronary disease. The relative risk of the ischaemic heart disease in the shift workers is accepted as being around 1.4. In four analysis of five ones that have been carried out so far the relation between occurance of the cardiovascular system and work in this system has been found. The relative risk of the ischaemic heart disease among the shift workers is within the range of 1.3-1.7, when adjusted for other risk factors. It is a interesting fact that the risk of the cardiovascular system diseases increases quite significantly with connection of two risk factors such as smoking cigarettes and the shift work and also when the worker is obese and works under such labour system.

The shift work has an effect on occurance of the arterial hypertension. It may be connected both with the melatonin concentration disorder and also with weakness of the reflex functioning with the baroreceptors of sinus caroticus. The tests results suggest that coexistence of such features as the age over 40 years, the morning chronotype, obesity and value of the heart rate/respiratory frequency > 4 for the persons that work in the night may be conducive to development of the arterial hypertension. For the patients having arterial hypertension shortening of the sleep time causes additional increase of the sleep fragmentation and its significant shortening may be a cause for occurance of the coronary heart disease.

It is important that in case of the cardiovascular system diseases the substantial risk factor that influences their occurance is smoking cigarettes. In the tests that have been performed so far it has been found that among persons that work under the shift system there are more smokers than among such persons that work exclusively in the day time.

Impact on the Digestive System

Irregularity of having meals and their atypical times may have an effect on functioning of the digestive system. Besides, quality of the meals that are consumed in the night is a significant problem. Gastrointestinal and digestive problems such as indigestion, heartburn, stomachache and loss of appetite are more common among the rotating shiftworkers and the night workers than among the day workers. It is less clear if more serious conditions such as peptic ulcers are more common in shiftworkers. Certainly the irregular work, sleep and eating schedules are not helpful for the proper treatment of ulcers. In one of the tests it has been found that the shift workers have lengthened period of low intragastric pH in comparison to the control group. Consumption of a meal results in increase of pH, thus the proper diet is a significant prophylactic factor in this field. One of the tests connects performing of the shift work with frequent occurance of the Helicobacter pylori bacteria, however, only further studies may allow for evaluation whether they are crucial relationships. Another researchers' problem that has been taken into consideration so far is occurance of relationship between performing the shift work and occurance of overweight and obesity among the workers. In two analyses that have been carried out so far no influence of the shift work on increase of the Body Mass Index has been found, however, one thing indicates the relationship between doing such kind of work and the occurance of obesit,y and the next one presents the results indicating that there is small increase of the BMI among such persons. Given the irregularity in type and timing of meals, it is not surprising that the night worker is more likely to have a poorer diet. At night, the loss of appetite often leads to increased snacking on "junk" food rather than eating a full, well-balanced meal. Feelings of fatigue may encourage the consumption of beverages with caffeine to help the worker stay awake.

Impact upon the Endocrine System

The insulin- diabetics are more likely to report the shift work problems, there is no reason why the diabetic on insulin and with a responsible attitude towards their employment should not do shift work. Well-motivated and sensible diabetics should be capable of manipulating their insulin dosage, especially if they monitor their blood glucose levels and use multiple insulin techniques. Diet and medical treatment can minimize the adverse effects of the shift work, but short shift cycles may present difficulties with control.

Glycocorticosteroids – the adrenal cortex hormones have the peak of hormone secretion in the early morning hours. It is of importance in case of some diseases of this organ and appropriate, eventual suplementation of these hormones. Attention must be also turned to the fact that the shift work may influence concentration of these substances among the persons that take steroids for medical purpose – they are popular medicines in reference to several diseases.

The thyroid hormones – as these substances stimulate the metabolic rate it is necessary to verify their dosage in reference to variable metabolic needs connected with the shift work.

Impact upon Procreation

In case of women that perform the shift work the menstruation disorders occur in them statistically more often. The shift work may lead to shortening of the menstrual cycle duration and to abnormal course of its luteal phase. The consequence of the menstrual cycle disorders may be the ovulation disorders – it results in decreasing the chance of getting pregnant by the woman that works under such labour organisation. Increase of the risk of the spontaneous abortion during work in the shift system in comparison with the women that work exclusively in the day or in the night has been found insignificant. Whereas, for example in the population of the nurses working exclusively in the night more than treble bigger increase of risk of the late abortions has been found, however, among those women that worked in the three-shift system the risk of the early abortion has increased twice. In case of a woman being after 23rd week of pregancy performing work in the three-shift system the increase of risk of the premature birth has been noticed. The latest tests indicate relationship between the shift work and 2.1-times bigger risk of low birth weight of a neonate, and the risk of the premature birth has increased twice.

Epilepsy

Insomnia or the night work may increase the risk of occurance of an epiletpic seizure, but many people with well-controlled epilepsy work rotating shift work without problems.

Bronchial Asthma

A small population of asthmatics may have problems with adaptation to quick rotation at the shift work. Sleep deprivation can improve peak expiratory flow rate in the nocturnal asthmatics.

Cancerogenesis

For several last years the almost seventy years old hypothesis suggesting an influence of the melatonin secretion disorder and its level on occurance of some malignant tumours has been verified. The shift work has its impact upon the daily profile of the melatonin concentration fluctuations in the blood and some hormones that are important in the cancerogenesis process in case of some tumours. It is considered that the melatonin concentration fluctuations in the blood are to have an influence specially on development of the mammary gland tumour, colorectal carcinoma and prostatic carcinoma, however, the influence on the cancerogenesis process in the ovarian carcinoma and endometrium is also considered. The influence of melatonin on development of the genital organs tumours may be connected, among others, with the following clinical observations:

- melatonin has the antiestrogenic effect – it decreases secretion of these substances through ovaries,
- it stimulates progestogenes synthesis,
- there is the seasonal variability of occurance, for example, of conditions of hyperplasia of the endometrium (a pre-neoplastic condition) – what can be connected with the levels of the melatonin secretion,
- concentrations of melatonin decrease rapidly in the menopause period – the typical period for occurance of the mammary cancer and the endometrial carcinoma,
- obesity often co-occurs with tumours – among such persons more often occur the daily melatonin level disorders,
- diabetes – that correlates with occurance of some tumours (for example, the endometrial carcinoma), results in decrease of the melatonin secretion by the pineal gland and increases the possibility of calcification of this gland.

Moreover, it is considered that melatonin demonstrates the activity inhibiting vessel growth within the neoplastic tumours. In the experimental researches in rats it has been found that changes of the daily rhythm cause significant disorders of functioning of the immune system (that are characterized, among others, by changes in the blood picture) and they increase the risk of growth of tumours – melatonin was decreasing the risk of such effects. It is also interesting to observe that some cancerogenes, for example, 7,12-dimethylbenz[a]anthracene (DMBA) cause increase of the nightly maximum melatonin level present in the blood in the experimental research, what would testify that there is a physiological mechanism activating the melatonin synthesis in case of significant exposition to a cancerogene.

During the three-year research that has examined relationship between the work rhythm and excretion of the melatonin metabolit - 6-sulphatoximelatonin along with the morning urine and the plasma level of steroid hormones in 80 women before the menopause period it has been found that there is the statistically increased concentration of plasma estradiol among the women employed at the shift work for at least fifteen years in comparison with those who have never worked in such way. It has been found that there is a significant statistically inverse relationship between the number of the worked nights during successive two weeks and the concentration of the melatonin metabolit in urine. Because darkness causes increase of the plasma concentration of melatonin treated as the anticancer substance, the epidemiological research among the population of poeple deprived of the effect of light operation on secretion of melatonin – the blind and the sand-blind, has been carried out. In Swedish research on 1567 completely blind persons and 13292 significantly sand-blind persons the decrease of the risk of occurance of tumours (RR= 0,69; 95% CI = 0,59-0,82) in this population in relation to the general population has been found. The significant statistical relationship between the cases of tumours of the large intestine and decrease of the daily amplitude maximum of the melatonin concentration in the blood has been proved. It has been also proved that the average melatonin concentration in the blood is substantially statistically lower among the patients with the prostate cancer and the breast cancer. In the epidemiological research in the population of 78586 women (nurses) working under the shift system (at least three night shifts within a month) it has been found that in comparison to the women that have never worked rotating shifts, the women that have worked in the shift system up to 14 years the risk of the large intestine cancer has not been increased. Whereas in the analysed population of women after their having worked for over 15 years under such system this risk was definitely increasing – by 35%! on the average. In the same women population the risk of the breast cancer was increasing moderatly in accordance with a burden of the shift work (at least three night shifts within a month). The risk of the breast cancer was increasing in the population with the period of employment from 1 year to 29 years by approximately 8% (in the scope: from 0.99 to 1.18 times in the population having the employment period from 1 year to 14 years and 0.90-1.30 times in the population having the employment period from 15 to 29 years). Not before many years of the shift work (over 30 years) this risk has been significantly statistically increased by 36% on the average. In Seattle among 813 women with the mammary cancer and 793 in the properly selected reference group the customs connected with the night time have been analysed. It has been found that the risk of the berast cancer increased by 14% on the average relates to the persons that often do not sleep in the night time during a period of 10 years prior to making the analysis. The risk of this disease was not increasing in case of the persons whose sleep was often interrupted with articicial lighting up.

Recently a hypothesis has also appeared that relates to the influence of the shift work on activity of the genes mastering the biological rhythms and which are the alarm key moderators of the genes matsering the cell growth cycle and the occurance of apoptosis.

Impact upon Metabolism

The shift work can also influence the medicine effect. In case of many medicines their pharmacocinetics (absorption, metabolism, excretion) and the healing effect are dependent on the daily rhythms. The pharmacocinetic parameters that characterize fluctuation of concentration of most of medicines are usually specified only for the day time. The same dosage of a medicine taken at night demonstrates usually a little bit different power of action than at day. Moreover, the data that relate to the time of administering included in the information on a medicine do not relate in any way to the times of having meals or the sleep time of such a worker. Thus changeable sleep times and waking times can require adaptation of the medicine dosing.

It is also considered that the shift work causes the increase of the risk of intoxication with chemical substances, as it causes desynchronization of the detoxicating mechanisms. It is also a significant problem to evaluate the exposure to chemical substances in the work enviroment and to correct the threshold limit values in relation to the 12-hours shift that is common while working rotating shifts, becasue they are usually related to the 8-hour working day.

Other Problems

The curiosities are the results of some research suggesting that the night shift work can be the factor that protects against development of the Parkinson's disease, whereas little tolerance of the activity of this type may be the early symptom of this disease. Moreover, it is also interesting that the shift work is a factor causing the oxidative stress. Age and obesity are factors that result in bigger sensitivity to such effect of the shift work.

The Ergonomic System of the Shift Work in Relation to the Scientific Data

Some international directives have stressed the need for the adequate organization of the night and shift work and the protection of shift workers' health: in particular, the International Labour Organization Code of Practice on Working Time and Convention No. 171 on night work and, for example, the European Directive No 93/104/EC relating to the aspects of organization of working time. Obviously the best and the simplest solution to the problem of shift work is its elimination, however, in particular branches it is practically impossible.

One can distinguish two levels at which modifications and improvement can be carried out:

- the individual level consisting in helping the workers to limit the stress of shift work,
- the organizational level consisting in proper shaping the shift work model.

The eight-hour work day is usually appropriate for workers' abilities, however, because of the significant work intensity it is justified to shorten the work day. Many data indicate that in some situations the twelve-hour system can be tolerated by workers easier. The time of starting the morning shift has significant meaning for the workers' social functioning. Too early (six o'clock) starting of the morning shift is most often connected with the worker's sleepiness and often with a higher risk of accidents. It is recommended to avoid starting shifts early – at five or six o'clock in the morning. Employers should also take into consideration the workers' requests, social customs, transport availability and safety in the streets. It is the most advisable to start shifts respectively at 7.00 am, 3.00 pm and 9.00 pm. (or at 11.00 pm) for the eight-hour work day. The duration of one shift period should not be longer than 2-3 days, because only quick rotation of shifts makes accumulation of the sleep deficiency impossible. The weekly work time during performing the shift work should be within the limits of 34-38 hours, and the work time for a single shift should not be longer than 9 hours. The system of shift rotation is also an important factor. It is assumed that the most advantageous for tolerance of the shift work is the system in which after having finished the morning shifts a worker goes to the afternoon shifts, and then to the block of the night shifts. This system is advantageous from the point of view of the man's phisiology, because the organism tolerates easier the delay of sleep time than its enforced acceleration. Moreover, owing to such work organization the periods of breaks for rest are getting longer – a worker that works under the shift cycle should have 40% of the free time freely arranged between shifts. Moreover, it is recommended that between the end of the night shifts series and the start of the morning ones there should be at least 24 and in some others' opinion even 48 hours that are free of work. The arrangement of changes in the shift work system should be possibly regular in the longer period of time. The work time under the shift and night work conditions should be inversely proportional to the level of its intensity. If it is possible, the worker should be given the possibility of taking advantage of short sleep between one and three o'clock in the night. It usually allows living with proper vigilance and attention to the end of a work shift. For social reasons it is worth providing free weekends, if possible. The following should be provided for the workers, if possible: proper conditions for resting, proper care of the workers' children, good quality food in the factory bars and buffets, and the possibility of recreation – often in „untypical" times. The worker should maintain constant times of meals with the family members in spite of various working hours. The persons working the afternoon shift should consume the main meal in the middle of a day, and should avoid consuming it during the work shift. However, the persons performing work in the night should consume light meals in the course of the whole shift (it is particularly important to have a warm, light meal between midnight and 2.00 am – it causes delay in development of sleepiness and fatigue), and a light breakfast.

The worker's significant skill is proper use of the free time for naps. The afternoon sleep is especially valuable. It is physiologically connected with increase of concentration of the endogenic melatonin secreted by the digestive track that is filled with food.

The Melatonin Treatment

The sleep-inducing effect of melatonin has been described since the sixties of the twentieth century. The egzogenic melatonin (administered in the form of tablets) shows the sleep-inducing effect when the endogenic melatonin levels are low in the blood, it means in the daytime. The proper supplementation of melatonin has an effect on the subjective evaluation of the sleep quality in the daytime and sleepiness during this time, and it also prolongs duration and improves quality of the afternoon nap. Melatonin in the doses from 0.3-5 mg prolongs the total duration of sleep in the daytime. In several tests the effect of melatonin on sleep in the daytime has been evaluated using polysomnography. In two tests the effect of melatonin on four-hour naps in the daytime has been evaluated. It has been found that melatonin was decreasing the time of falling asleep and it was increasing efficiency of sleep. The changes in the electroencephalografic recording under the influence of oral administration of 5 mg of melatonin about one o'clock, after having restrained oneself from sleep in the nighttime, have a similar nature to the ones that are caused by benzodiazepins. It has been found that the 5 mg dosage accelerates falling asleep in the daytime and increases vigilance during the morning hours. Thus the improvement has been showed in relation to occurence of the states of sleep-vigilance, whereas the influence of melatonin on behaviour and mental abilities requires further analysing. Some trials have been taken in relation to administering melatonin in the morning, after having finished work, expecting easier falling asleep, better rest and, in connection with this, less problems with adaptation to the night work. During one of the tests the limitation of exposition to light after having finished work in the nighttime was also used (dark glasses). In other tests it has not been found that among the persons that are treated with either melatonin or melatonin with limitation of exposition to light there is a minor effect in reference to facilitating falling asleep during the daytime, however, they do not show better adaptation to work in the nighttime. In single tests it has been found that melatonin often has an advantageous impact upon sleep in the daytime after the first worked night, whereas after successive nights the effect of tolerance can appear. The expected effects have not been achieved during the trials of the use of melatonin and phototherapy among the sick having the chronic fatigue syndrome.

Summarizing, the melatonin therapy for sleep disorders that are connected with the shift work or travelling among the time zones does not always give advantageous results.

Bright Light Technology

In order to limit the workers' maladaptation to the shift work the light therapy is also proposed. It should be the bright light (5000-12000 lux) that is used at the beginning of a shift and then a little weaker one (1230-2000 lux).

The Popular Psychostimulants

Apart from long tradition of using coffee in order to improve the night workers' and the shift workers' functioning the modern market also provides with many other products that, beside the caffein content, contain substances that are advertised as either stimulating ones or the diet supplements (L-taurine, glucuronolactone, carbohydrates, vitamins from the B-group vitamins, vitamin C, folic acid, D-Ribose, acethyl l-carnitine, hupercine A, guaranine, teobromine). Biological effects of caffeine on the man's organism are very diverse. They include the effect of this compound on the nervous system, the cardiovascular system, the alimentary system, the respiratory system, the urinary system, and also an effect on metabolic parameters of the organism. Caffeine and its metabolits increase secretion of neuromediators among which serotonine, dopamine, adrenalin, noradrenalin, and GABA and ACTH are mentioned. The effect of these hormones is stimulation of the cerebral cortex work with the consequent facilitation of mental processes and reduction of fatigue. In some studies it has been proved that caffeine has a clear effect on the mood. Particularly significant changes in mood have been observed among the persons that drink a lot of coffee every day. During the research refering to psychiatric patients clear relationship between using caffeine and increase of the feeling of nervousness has been observed. The especially interesting source of caffeine is the *Paulinia guarana* plant that contains up to 2.5 times more caffeine than the coffee grains. Additionally in case of this plant the advantageous effect is influence of the residual substances causing slower absorption of caffeine and its longer half-life period, and also smaller risk of gastro-intestinal disorders charakteristic of consumption of coffee. Moreover, it containts other active caffeine-derivatives – particularly guaranine. It is worth mentioning that some uwanted coffe effects (for example, the lipid balance disorders) are dependent on the method of preparing this beverage. The energizing beverages popular in the market usually contain, apart from caffeine, taurine, glucuronolactone and the addition of vitamins and carbohydrates. Taurine shows an activating effect on the nervous system. It has been observed that using caffeine and taurine together in the form of energizing beverages gives especially positive results. The mechanism of its effect results from numerous interactions with the neurotransmitters, among others with the adrenergic system, the cholinergic one and the GABA one. In many research works the advantageous effect of taurine on concentration and reaction speed has been shown. This amino acid is also the well known mood moderator. It finds its application in stressful situations, and it also has a significant effect on the sleep-wake rhythm, shortening the sleep period.

Apart from caffeine and taurine the energizing beverages pretty often contain glucuronolactone. It is considered that its advantageous effect consists in connecting and accelerating excretion of the anaerobic metabolism products and also in connecting and facilitating transfer of carboxylic acids into the muscle cells, where they are utilized with oxigen under aerobic conditions in the mitochondrial metabolic pathways. Moreover it is the NADPH donator.

The recognised reputable beverages that contain caffeine, taurine and gluconolactone together clearly improve mood, diposition, vitality, some mental activities (the data are partially inconsistent – for example, according to some research they have the effect on the short-term memory, according to others they do not). They effect on the heart efficiency, fluctuation of the arterial blood pressure and they can cause bradycardia resulting from pressure increase. A significant reduction of sleepiness has been noted among the persons that drive cars in the evening or in the night, especially during the period of first 60-90 minutes from consumption. The particularly advantageous effect consists in reduction of sleepiness connected with shortening of sleep during the preceding night.

Summarizing, the popular and present in the market agents that contain the above mentioned substances may really influence many mental activities and increase of physical ability what results in the work efficiency. It must be rememberd, however, that there are also negative effects of overdosage and there is no influence upon many higher brain functions that are important in the advanced mental function.

The daily rhythm disorders and the body functioning disorders resulting from the professional work under the shift and night work system seem to be extremely interesting scientific problems. Moreover, they are significant practical problem of ergonomics and the occupational medicine. As it has been presented above, the shift work and the night work constitute a menace to various systems and organs. Whereas their influence on cancerogenesis and on expression of the biological clock genes is especially alarming and it does require further analyses. The methods of limitation of the shift work negative effect also require further detailed analyses.

In: Circadian Rhythms and Health Research Trends
Editor: Alfredo V. Lignelli, pp. 199-206
ISBN: 978-1-60021-875-0

Chapter XII

Circadian Rhythms of the Cell Proliferation and Cell Migration in the Small Intestine

José Rosa Gomes, Maria Albertina de Miranda Soares and Cristina Lúcia Sant'Ana Costa Ayub
Departamento de Biologia Estrutural, Molecular e Genética,
Universidade Estadual de Ponta Grossa, Paraná, Brasil

Introduction

One of the theories says that the beginning of the life on the earth dates at least 3.5 billion years ago. At this time a great variation of climatic conditions had occurred on the earth caused by the intense sun radiation, electric discharges produced by thunderstorms, volcanic activity and collision with meteors. These climatic conditions of the "primitive" earth allowed the formation of the first organic molecules by prebiotic synthesis and after that, more complex compounds were formed like the ribonucleic acid, which was able to auto replicate and evolved to form the basic genetic structure of all organisms. All those facts culminated with the establishment of life on our planet.

Every day when the sun rises and we get up a conjunct of about 200 different types of cells [1] which form our distinct organs responds to the solar radiation exactly like in that remote time when everything had begun. As confirmed by several researches, the sun radiation is the first and major synchronizer of the innumerous cellular and physiological processes. We can see the role of the solar radiation in the animal physiology when some species get up after hibernation. This is caused by heating of their blood as a consequence of the earth warming by the incidence of radiation from the sun. In animals such as rodents it appears that their metabolism is more functional during the dark period when they are looking for food and reproducing [2].

Although there is still little information about the mechanisms of action of the sunlight on cells, some studies propose that there is a synchronization of cell functions caused by solar radiation at the tissue level. This fact was observed in the process of cell proliferation and migration, gene expression, cell synthesis and secretion and in the activity of a great variety of molecules that work as secondary regulators in many cellular processes [3]. When these cellular processes occur cyclically during the 24 h of the day, there is the stablishement of circadian rhythms which reflect on the activity of the organism during the light or dark period of the day [4].

The majority of the researches on circadian rhythms at the cellular and tissue level were done in mammalian species such as rodents, and there is significant knowledge accumulated about it, especially on the circadian rhythm in the cell proliferation and migration in the small intestinal epithelium that will be subject of this chapter. For this proposal we divided the chapter in the following topics: 1) Circadian variation in cell proliferation and migration in the small intestine of young and adult animals; 2) General aspects of the circadian variation regulation in cellular and tissue level; 3) Role of the circadian variation in the cell proliferation during growth of the small intestine and 4) perspectives.

1. Circadian Variation in Cell Proliferation and migration in the Small Intestine of Young and Adult Animals

1.1. Circadian Variation in Cell Proliferation and Migration

The small intestinal epithelium is organized into two functional compartments. The first one is the crypt where stem cells and Paneth cells are found in the bottom region; in the middle crypt region there are cell populations in constant proliferation and in the crypt-villi junction there are cells in differentiation processes. From this last region the cells can migrate to the crypt as the Paneth cells or go to the second compartment of the small intestine [5]. The second compartment is the villi for where the crypt cells will migrate and execute all functions of the small intestine, i.e., absorption of nutrients or synthesis of important hormones to regulate the digestive process. These functions in the villi are executed during the migration of the crypt cells which takes approximately 7 days and there is a precise regulation between cell proliferation and migration of these cells which follow a circadian rhythm. At the end of this time the cells are lost from the tip villi, and new cells produced in the proliferative compartment replace them. This normal cell renovation occurs in the adult small intestine and this is denominated steady state [5].

The cells constituting the small intestine are phenotypically and genotypically similar but their functionality in terms of cell proliferation and migration is distinct. Then we can find some cell populations with a proliferative behavior independent of other cell groups. In each cell population there is a synchronization of the cell proliferation and migration process during the day. In respect of this, some data are found in the literature about differences in the period of the day where an increase or decrease in the cell proliferation for each cell group is observed. A close relationship between circadian rhythm of cell proliferation of the different cell populations and their position in the crypt reflects spatiotemporal organization of the epithelium [6]. The hierarchic structure of this spatiotemporal organization suggests the existence of several interrelated levels (individual cells, cell subpopulations, and cells with basal and maximum levels of proliferation within subpopulation). Thus each level has its proper temporal and spatial characteristic and their interaction determines spatiotemporal organization of the cell proliferation in the small intestinal crypt epithelium. Most cell populations of the crypt wall present monophasic circadian rhythms, indicating that they have similar but not synchronous rhythms [7].

When the entire cell populations of the crypt are considered, in a steady state condition, it was demonstrated in adult rats distinct cell populations proliferating at 15.00h and 02.00h. Studies in adult mice have demonstrated that the mitotic activity of the whole crypt enterocytes shows a circadian variation, with maximal and minimal values at 12:00 h (zenith) and 20:00 h (nadir), respectively [8].

However, in the small intestine of young rats there is not a steady state between cell proliferation and cell loss on the villi tips which is an acceptable condition because the small intestine of young animals is in a growth process. Thus, the cell lost in the top villi can not be higher than the cell proliferation in the crypt [9].

In rats at the suckling or weaning phase there is not enough information about the circadian variation of the cell proliferation and migration. Data of our group demonstrated that there is already a circadian rhythmicity of the cell proliferation in the small intestine of rats at the weaning phase with a major number of cells cycling at 02.00h [10]. Our suggestion is that young animals acquire a circadian rhythmicity of the cell proliferation in early stages of their development.

1.2. Influence of the Cell proliferation in the Cell Migration

In general the researches show that an increase in the cell migration in the villi occurs a few hours after the cell proliferation had occurred in the crypt.

Earlier investigators assumed that the migration in the gastrointestinal epithelium was a consequence of a force caused by cell proliferation in the proliferative compartment which increases the local pressure forcing the columns of cells upward [11]. However some experiments, where the mitotic activity of the cells was inhibit by radiation, demonstrated that cell migration in the villi was not interrupted, indicating that there are other mechanisms involved in this process [12].

In one experiment where adult rats were submitted to radiation, an increase in the cell migration in the villi was observed at about 18.00 h when the cell proliferation was down [12].

However, in the small intestine of young rats the cell migration rate in the crypt and villus is slower when compared with the rates found in adult animals [13].

All these brief considerations allow us to conclude that there are still few studies available about the regulatory mechanisms of the variation circadian of the cell proliferation and migration in the small intestine of young and adult rats which became this an interesting area for future studies.

2. General Aspects of the Circadian Variation Regulation in Cellular and Tissue Level

2.1. Clock Genes

The mammalian circadian system may be under at least three major regulators: the first one external and more important represented by light/dark cycles and another internal, regulated by the first one, located in the central nervous system (CNS), in the suprachiasmatic nuclei of the hypothalamus, and another external represented possibly by the food. The mammalian circadian system is built up on a multitude of oscillators organized according to a hierarchical model in which the neurons of the suprachiasmatic nuclei may drive the central circadian clock and all the other somatic cells may constitute peripheral clocks. At the molecular level, oscillators are molecules that are interacting to create an internal mechanism which keeps time for the circadian clock by either coordinating or constructing rhythms. The light /dark cycles (synchronizers) can entrain the circadian clockwork through photoreceptors (sensory transducers) relaying the signal to synchronize all the individual oscillators beginning with the neurons of the suprachiamatic nuclei [15].

In recent years, a lot of research has demonstrated an existence of a group of genes involved with the circadian rhythms which were denominated of clock genes. They were first described in *Drosophila melanogaster* mutants which had the circadian clock affected. The genic locus was denominated *period (per)* and from it, other clock genes were identified [16]. For most of the clock genes of *Drosophila melanogaster*, orthologs have been identified in mammals, highlighting a general conservation in the clock circadian mechanism between insects and mammals. In particular, three mammalian *Period* genes (*mPer1, mPer2* and *mPer3*), two Cryptochrome genes (*mCry1* and *mCry2*), as well as *Clock, Bmal1* (ortholog to *cyc*), and *CKI* (ortholog to *dbt*) have been identified [17].

Some of these genes codify transcription factors that directly control the expression of other genes. Other genes codify regulatory proteins which directly control the activity of some proteins through phosphorylation. Some of these proteins are members of the retinoic acid-related orphan nuclear receptors [15].

Actually have been demonstrated in central e periphery tissue of vertebrates, especially in suprachiasmatich nucleus, that there is an intrinsic relationship between the light-dark cycle of the day and the clock gene expression which are directly involved with the circadian rhythm represented by a 24 h. period. Under constant light conditions, the period length of *Per1* mutants is longer than 24 hours, whereas for *Per2* mutants it is shorter and no loss of rhythmicity is observed. This study supports the view that the clock consists of two oscillators acting of opposite manner.

In addition, posttranslational regulation of clock components is essential to prevent accumulation of their proteins in the cell. Accumulation would lead to an equilibrium in transcriptional activation of clock genes and thus to a loss of rhythmicity in gene expression. One gene product that is thought to regulate clock genes posttranslationally is *CKI.* This gene codes for a kinase that is involved in the phosphorylation of *Per* genes. A mutation in the *CKI* gene of hamsters, called *tau* mutation, leads to a very short period length, demonstrating the importance of posttranslational regulation of clock components [18].

2.2. Food Intake

More recently, the role of the food intake was analyzed on the expression of clock genes in peripheral organs. For example, in the liver, although rhythmicity in the central nervous system remained phase-locked to the light-dark cycle, restricted feeding rapidly entrained the liver, shifting its rhythm by 10 hours within 2 days. These results demonstrate that feeding cycles can entrain the liver independently of the CNS and the light cycle, and they suggest the need to reexamine the mammalian circadian hierarchy. They also raise the possibility that peripheral circadian oscillators may be coupled to the CNS primarily through rhythmic behavior, such as feeding [19].

Thus, besides the expression of clock genes in the peripheral organs, it has now been demonstrated that food intake can change the expression standard of these genes in these tissue but not in the central nervous system (CNS). Then the food is a great regulator of the clock rhythms closed with the sun light.

Until now no information was reported on the relationship between the expression of the clock genes and the cell proliferation or cell migration process in the small intestine.

However, our dates have demonstrated that the cell proliferation is significantly reduced in the morning period in food-restricted rats (data not published). These data suggest that the food can also change the circadian rhythm of the cell proliferation in the small intestine of young rats.

From these considerations the conclusion is that the food intake can be one of the regulators of the cell proliferation in the small intestine by changing expression of the clock genes. However, further studies are necessary to confirm this supposition.

3. Role of the Circadian Variation of the Cell Proliferation during Growth of the Small Intestine

The small intestine of younger rats is still growing during the first 2 weeks of the post-natal development, and it was demonstrated that cell proliferation is higher and cell migration is lower when both are compared with the adult small intestine.

In earlier developmental stages, a temporal variation was found in the cell proliferation of mouse small intestinal enterocytes, with a peak of mitotic activity at 12:00h and 04:00h in the duodenum of seven-day-old female and male animals, respectively. In the fourteen-day-old female and male animals the maximal in duodenal mitotic activity appears at 20:00h. A greater rate of mitotic activity was observed in fourteen-day-old animals, which could be explained by maturation of the intestinal crypts in this developmental stage. The mitotic activity of the intestine at either seven-day-old or fourteen-day-old animals is still lower than that found in older mice which could be attributed to the continued renewal of the enterocyte cell population throughout life. The sex-related differences in intestinal mitotic activity could be related to a sexual dimorphism in the hormonal level or growth factor production and/or in the concentration of cellular receptors [6].

The establishment of a circadian rhythm already in the second week of development as demonstrated by works of our group is closed to the transition of the young to adult phase.

The transition from suckling food to solid food, which could to determine the mechanisms that regulate the cell proliferation in the crypt and the cell lost from tip villi should also cause changes in the expression of the kind of digestive enzymes by cell. These processes together allow that the cell proliferation be more pronounced in proximal segments than in distal segments to establish the correct morphology and functions performed by each region of the small intestine as well as to provide the complete steady state conditions in adult life. This transition in the expression of enzymes at brush border of the cells occurs by action of the hormones produced in glands outside the intestinal system and possibly by the neuronal factors of the intrinsic nervous systems of the gastrointestinal tube.

The major hormones that coordinate the small intestinal development are corticoids (cortisol) and thyroid (T_3). The levels of theses hormones reflect the degree of intestinal development and differentiation.

In relation to the effect of those hormones on digestive enzyme expression, studies have demonstrated that the brush border enzyme *lactase-phlorizin hydrolase* (lactase), which is involved in the hydrolysis of lactose (the principal sugar from milk), has its levels decreased at weaning phase, whereas the other brush border enzyme sucrase-isomaltase (sucrase), which hydrolyzes sucrose (the principal sugar found in the adult rat diet), has its levels markedly increased at the same phase. The decrease in *lactase* activity is partially regulated by T_3 while the abrupt raise in the enzymatic activity of the *sucrase* is controlled by cortisol [20]. The effect of T_3 in the decrease of intestinal lactase in adult rats has suggested that the mechanisms of action of this hormone can be generalized to this organ at all stages of the development. The proposal is that this hormone has a critical role in the growth of the intestine [21].

Studies have demonstrated significant morphological changes in the adult rat small intestine caused by T_3 including an increases in villus height which is more pronounced in the proximal small intestine than in the ileum. The increases in villus height coupled with the changes in brush border enzyme expression provide evidence for the linkage between the processes of growth and differentiation on the small intestine [21].

There are studies also showing the role of the diet on digestive enzymes expression. One of these studies shows that the decline of the *lactase* activity can be promoted or delayed by changing the time at which rats decrease the milk consumption and begin the intake of an adult-type diet [22].

Thus, we could conclude that the growth of the small intestine is dependent firstly on the coordination between cell proliferation and cells lost that must be regulated by circadian rhythm and that a set of other factors, among them the hormones, are involved.

We can still conclude that during the growth and differentiation, the small intestine passes by significant morphologic and physiologic alterations where the coordination among the cell proliferation, cell migration and changes on digestive enzymes are directly dependent of the kind of food intake during the first three weeks of the development.

All these observations allow us to suppose that the food intake is the stronger regulator of all these processes.

4. Perpectives

In the normal small intestine of an adult mammalian, for each cell lost in the top villi there is another cell migrating from the crypt to replace it. When this steady state is disrupted, i.e., when the cell production rate into the crypt is bigger than the loss in the top villi there is a metastasis situation or when the loss is bigger than the replacement there is an ulceration condition.

It has been demonstrated that the mitotic activity of the enterocytes from duodenal crypts in EA21a mammary carcinoma-bearing adult mice shows a circadian variation, equivalent to that described in normal mice, but with zenith and nadir values at 00:00 h and 12:00 h, respectively. These results indicate that the mechanism of circadian variation of the mitotic activity is maintained in tumoral enterocytes despite changes in its zenith and nadir values [6]. The authors conclude that these observations must be considered during antitumoral drugs administration which may vary their toxicity according to the circadian rhythm of normal cell proliferation.

Other studies about the circadian rhythmicity of the rectal cell proliferation in patients affected by advanced colon cancer show that there is a fluctuation during 24 hours. The peak of cell proliferation occurs during the morning and afternoon and drops during the night [22]. These results reinforce the idea that the time of drug administration during the day is important to improve its efficacy and decrease its toxicity.

Other studies demonstrating the mechanisms which regulate cell proliferation and migration in normal and abnormal tissues, and principally those which could be showing the time of the day when these processes are more pronounced in both conditions should provide therapeutic strategies to prevent human invasives diseases.

The expression during the G1 phase of the cyclins D_1 and E in non-metastatic adenocarcinoma of the lower rectum in humans showed that these proteins present a similar circadian rhythm, with the highest level obtained at 18:00h and the lowest level obtained at 00:00h. A circadian variation was also observed in the expression of CDKs (Cyclin-Depedent Kinase) inhibitors p16 and p21, with the lowest values obtained at 12:00h and 18:00h for p16 and p21, respectively. These results lead to the conclusion that the expression of these proteic factors involved in the progression from the G1 to S-phase of the cell cycle obeys a circadian variation [23]. Studies demonstrating the role of the circadian variation of the proteic factors involved on the cell cycle regulation consists in a new issue within of the tumor cell biology.

Finally, nowadays the studies in cell proliferation are in evidence due to a great interest in using stem cells to construct organs in vitro to replace one injured by degenerative disease. Then studies corcening cell proliferation and migration in the small intestine *in vivo or in vitro* constitute a great model for the establishment of strategies for organ reconstruction in the future.

References

[1] B. Alberts, D. Bray, J. Lews, M. Ralff, K. Roberts, J. D. Watson. *Artes Médicas. Sul. LTDA*. Porto Alegre (2006).

[2] R. C. Bolles and P. J. Woods. *Animal Behaviour*. 12:427-441 (1964).

[3] L. A. Scheving. *Gastroenterology*. 119:536-549 (2000).

[4] N. Marques and L. Menna-Barreto. Cronobiologia: princípios e aplicações. Edusp, São Paulo, (2003).

[5] H.Quastler and F. G. Sherman. *Exp. Cell Res.*,17:420-38, (1959).

[6] Yu. A. Romanov, A. I. Antokhin, A. Yu. Kozlova, N. A. Zharkova, I. N. Malysheva, and S.S. Filippovich. *Bulletin of Experimental Biology and Medicine*. 136, 599 (2003).

[7] C. G. Barbeito, S. M. Albarenque, J. C. Reyna, M. A. Flamini, P. f. A. Laube and A. F. Badraun. *Cell Biology International,* 26 (1):123-125, (2002).

[8] C. G. Barbeito, N. V. Gonzáles, and A. F. Badrán. *Chronobiology international*. 20, 37 (2003).

[9] A. I. Al-Nafussi and N. A. Wright. *Virchows Arch. 9 (Cell Pathol.)* 40:51-62, (1982).

[10] J. R. Gomes, A.A.M. Pereira, L.Barth, J.S.Silva, M.L.Leite, A.C.M. Wille and M. A. M. Soares. *Cell Prolif.* 38,147:152, (2005).

[11] H. Cheng and C. P. Leblond. *Am. J. Anat.,* 141:461-480, (1974).

[12] P. Kaur and C. S. POTTEN. *Cell Tissue Kinet.,* 19:591-599, (1986).

[13] O. Koldovský, P. Sunshine and N. Kretchemer. *Nature*, 212:1389-1390, (1966).

[14] J. R. Gomes and E. P. Alvares. *Bras. J. Med. Biol. Res.* 31, 281, (1998).

[15] L. Parddini and B. Kaeffer. *Reprod. Nutr.* 5, 463-480, (2006).

[16] R. J. Konopka and S. Benzer *Proc. Nat. Acad. Sci. USA* 68:2112-2116, (1971).

[17] U. Albrecht, Z. S. Sun, G Eichele and C.C. Lee. *Cell* 91:1055-1064, (1997).

[18] H. Kobayashi, K. Oishi and S. Hanai and N. Ishida. *Genes to Cells*, 9: 857-864, (2004).

[19] Duluc I, Galluser M., Raul F. and Freund, J.N. *American Journal of Physiology*. 262, 954 (1992).

[20] R. A. Hodin, S. M. Chamberlain, and M.P. Upton. *Gastroenterology*. 103, 1529 (1992).

[21] D. B. Rhoads, D. H. Rosenbaum, H. Unsal, K. J. Isselbacher, and L. L. Levitsky. *The journal of biological chemistry*. 273, 9510 (1998).

[22] G. Brandi, C. Calabrese, M. A. Pantaleo, A.L. Morselli, G.Di Febo, R. Hakim, G. A. Di Vivo, M.C. Di Marco and G. Biasco. *Câncer Lett.* 28;208(2):193-196 (2004).

[23] J. Griniatsos,O.P. Michail, S. Theocharis, A. A. Velakis, I. Papaconstantinou, E Felikoura, E. Pikoulis, C. Bakoyannis, G. Marinos, J. Branis, O.P. Michail. *World J. Gastroenterol.* 7 (12) 2109-2114 (2006).

Index

A

B

D

E

F

G

H

I

J

K

L

N

O

P

Q

R

S

T

U

V

W

Y

The International Law of the Sea in the 21st Century

State Practice in East Asia

The International Law of the Sea in the 21st Century

State Practice in East Asia

Keyuan Zou
Dalian Maritime University, China
University of Central Lancashire, UK

World Scientific

NEW JERSEY • LONDON • SINGAPORE • BEIJING • SHANGHAI • HONG KONG • TAIPEI • CHENNAI • TOKYO

Published by

World Scientific Publishing Co. Pte. Ltd.
5 Toh Tuck Link, Singapore 596224
USA office: 27 Warren Street, Suite 401-402, Hackensack, NJ 07601
UK office: 57 Shelton Street, Covent Garden, London WC2H 9HE

Library of Congress Cataloging-in-Publication Data
Names: Zou, Keyuan, author.
Title: The international law of the sea in the 21st century : state practice in East Asia / Keyuan Zou, Dalian Maritime University, China University of Central Lancashire, UK.
Description: Singapore ; Hackensack, NJ : World Scientific Publishing Co. Pte. Ltd., [2022] | Includes bibliographical references and index.
Identifiers: LCCN 2021027269 (print) | LCCN 2021027270 (ebook) | ISBN 9789811237058 (hardcover) | ISBN 9789811237065 (ebook) | ISBN 9789811237072 (ebook other)
Subjects: LCSH: Law of the sea--East Asia. | South China Sea--International status. | Japan, Sea of--International status. | East China Sea--International status.
Classification: LCC KZA1693.3 .Z68 2022 (print) | LCC KZA1693.3 (ebook) | DDC 341.4/5091647--dc23
LC record available at https://lccn.loc.gov/2021027269
LC ebook record available at https://lccn.loc.gov/2021027270

British Library Cataloguing-in-Publication Data
A catalogue record for this book is available from the British Library.

For any available supplementary material, please visit
https://www.worldscientific.com/worldscibooks/10.1142/12282#t=suppl

Desk Editors: Balasubramanian Shanmugam/Lixi Dong

Typeset by Stallion Press
Email: enquiries@stallionpress.com

About the Author

ZOU Keyuan is Distinguished Professor, Dalian Maritime University, China. He specializes in international law, in particular law of the sea and international environmental law. He has published over 200 books, book chapters and journal articles in English. His latest books include *Maritime Cooperation in Semi-Enclosed Seas: Asian and European Experiences* (2019), and *The Belt and Road Initiative and the Law of the Sea* (2020). He is member of Editorial Boards of *International Journal of Marine and Coastal Law*, *Ocean Development and International Law*, *Journal of International Wildlife Law and Policy*, *Marine Policy*, *Journal of Territorial and Maritime Studies*, *Chinese Journal of International Law*, and Advisory Boards of *Global Journal of Comparative Law*, *Asia-Pacific Journal of Ocean Law and Policy* and *Korean Journal of International & Comparative Law*.

Contents

Acknowledgments

I gratefully acknowledge the kind permission granted by the publishers Martinus Nijhoff, Oxford University Press, and Taylor & Francis to reproduce the following published articles written by myself.

1. Portions of Chapter 2 were adapted from Zou Keyuan, "Law of the Sea Issues between the United States and East Asian States", *Ocean Development and International Law*, Vol. 39, 2008, 69–93.
2. Portions of Chapter 3 were adapted from Zou Keyuan, "The International Tribunal for the Law of the Sea: Procedures, Practices and Asian States", *Ocean Development and International Law*, Vol. 41, 2010, 131–151.
3. Portions of Chapter 4 were adapted from Zou Keyuan, "Maritime Enforcement of United Nations Security Council Resolutions: Use of Force and Coercive Measures", *International Journal of Marine and Coastal Law*, Vol. 26, 2011, 235–261.
4. Portions of Chapter 7 were adapted from Zou Keyuan, "Sino-Vietnamese Agreement on the Maritime Boundary Delimitation in the Gulf of Tonkin", *Ocean Development and International Law*, Vol. 36, 2005, 13–24.
5. Portions of Chapter 9 were adapted from Zou Keyuan, "How Coastal States Claim Maritime Geographic Features: Legal Clarity or Conundrum?" *Chinese Journal of International Law*, Vol. 11, 2012, 749–765.

List of Abbreviations

ADIZ	Air Defense Identification Zone
ARF	ASEAN Regional Forum
ASEAN	Association of the Southeast Asian Nations
BCN	Biological, Chemical, Nuclear
BOIESA	Bureau of Oceans and International Environmental and Scientific Affairs, United States Department of State
CCG	China Coast Guard
CLCS	Commission on the Limits of the Continental Shelf
CNOOC	China National Offshore Oil Corporation
ECJ	European Court of Justice
EEZ	Exclusive Economic Zone
EU	European Union
FON	Freedom of Navigation
GEF	Global Environmental Facility
IAEA	International Atomic Energy Agency
ICAT	Inter-Agency Coordination Group Against Trafficking in Persons

ICC	International Criminal Court
ICJ	International Court of Justice
ICM	Integrated Coastal Management
ICNT	Informal Composite Negotiating Text
ICSID	International Centre for Settlement of Investment Disputes
ICTR	International Criminal Tribunal for Rwanda
ICTY	International Criminal Tribunal for the Former Yugoslavia
ILC	International Law Commission of the United Nations
IMO	International Maritime Organization
IOM	International Organization for Migration
ISA	International Seabed Authority
ISC	Information Sharing Center
ITLOS	International Tribunal for the Law of the Sea
KOICA	Korean International Cooperation Agency
LOSC	United Nations Convention on the Law of the Sea
MSR	Marine Scientific Research
NLL	Northern Limit Line
nm	Nautical Miles
NOWPAP	Northwest Pacific Action Plan
NPC	National People's Congress
OCS	Outer Continental Shelf
PCA	Permanent Court of Arbitration
PCIJ	Permanent Court of International Justice
PEMSEA	Partnerships in Environmental Management for the Seas of East Asia

PMZ	Provisional Measures Zone
PRC	People's Republic of China
PSI	Proliferation Security Initiative
RACs	Regional Activity Centers
RCU	Regional Coordination Unit
RMSI	Regional Maritime Security Initiative
SDGs	Sustainable Development Goals
SRFC	Sub-Regional Fisheries Commission
UK	United Kingdom
UNCLOS III	Third United Nations Conference on the Law of the Sea
UNDP	United Nations Development Programme
UNEP	United Nations Environmental Program
UNODC	United Nations Office on Drugs and Crime
UNSC	United Nations Security Council
WMD	Weapons of Mass Destruction

Introduction

This is my third single-authored book in the field of the international law of the sea. The previous books — *Law of the Sea in East Asia: Problems and Prospects* (Routledge, 2005) and *China's Marine Legal System and the Law of the Sea* (Martinus Nijhoff, 2005) — were research outcomes resulting from my long-term research project on contemporary issues in the international law of the sea which commenced in Germany where I was an Alexander von Humboldt Fellow associated with the Department of Human Geography of the University of Hannover between 1996 and 1998. This new book can be regarded as a latest reflection of the continued research endeavors in this long-term project.

The international law of the sea is not only the oldest branch in traditional international law but also a fast developing branch in contemporary international law. After the entry into force of the 1982 United Nations Convention on the Law of the Sea in 1994, there have been considerable developments in the field of the law of the sea. Some provisions of the Convention proved to be inadequate, ambiguous or difficult to reach an effective implementation. The Somali piracy is a typical example that relevant clauses in the Convention failed to address this issue adequately. That is why the United Nations Security Council adopted specific resolutions for the purpose of the crackdown on Somali piracy. This new book attempts to reflect the recent developments in the law of the sea field, including the assessment of cases on the law of the sea handled by international judicial organs, legal developments in East Asia and other new emerging issues such as piracy and maritime terrorism.

The book contains three main parts, each of which has three chapters, respectively. While the first chapter will provide an overview of the recent developments in the law of the sea in East Asia when we entered into the 21st century, other chapters will address more specific issues covering, *inter alia*, maritime dispute settlement, maritime boundary delimitation, non-traditional maritime security issues and the impact of maritime powers such as the United States on the development of the law of the sea. While the book takes a holistic approach, it has made a special reference to East Asia, which is the most volatile region in the world today, in terms of economic development as well as maritime disputes. As illustrated in Part III of this book, maritime issues concerning the East and South China Seas are the most complicated ones, thus most difficult to find a solution. Some of the chapters are drawn from my published articles in international journals with acknowledgments.

Part I

The Law of the Sea in the 21st Century and Onward

Chapter 1

Implementing the United Nations Convention on the Law of the Sea in East Asia

East Asian Seas

The East Asian seas, from north to south geographically, include the Sea of Japan, the Yellow Sea, the East China Sea and the South China Sea. All these seas carry the same characteristics as semi-enclosed under the definition provided by the 1982 United Nations Convention on the Law of the Sea (LOSC).[1] The bordering countries include Brunei, Cambodia, China (including Taiwan), Indonesia, Japan, two Koreas, Malaysia, the Philippines, Russia, Singapore, Thailand and Vietnam. Natural resources are abundant in these seas and serve for the peoples around them.

The current marine legal order in the world has been established by and maintained under the LOSC, which is commonly regarded as the constitution of oceans and has incorporated almost all previously existing conventional and customary rules and norms concerning the oceans.[2]

[1]According to Article 122 of the LOSC, "enclosed or semi-enclosed sea" means a gulf, basin or sea surrounded by two or more States and connected to another sea or the ocean by a narrow outlet or consisting entirely or primarily of the territorial seas and EEZs of two or more coastal States. 21 ILM (1982) 1261. The Convention was open for signature on 10 December 1982 and came into effect on 16 November 1994. As of June 2020, there were 164 Contracting Parties to it, including one international organization.

[2]Before the LOSC, the international community had adopted four Geneva Conventions.

Pursuant to the provisions of the LOSC, a coastal State has the right to establish maritime zones under its jurisdiction: internal waters inside the baselines which are used to measure the extent of the territorial sea and other jurisdictional waters, the territorial sea of 12 nautical miles (nm), the exclusive economic zone (EEZ) of 200 nm, and the continental shelf of 200 nm (or up to 350 nm in some cases), outward from the baselines. Within these maritime zones, a coastal State is entitled to enjoy either sovereignty or sovereign rights and to exercise its jurisdiction and enforce its laws and regulations in accordance with international law. All 14 East Asian countries, except Cambodia and North Korea, have acceded to the LOSC (see Table 1.1). In order to implement the LOSC, these countries have adopted relevant domestic laws and regulations for the management of its own maritime zones and maritime activities within their jurisdiction.[3]

The extension of jurisdictional maritime zones of coastal States may cause conflicts and disputes between and among neighboring countries that share the same sea area so that bilateral arrangements are needed to facilitate the maintenance of the marine legal order. These arrangements are concerned with various maritime matters such as fisheries management, maritime boundary delimitation, joint development of marine non-living resources, marine environmental protection and maritime safety. So far there have existed some such bilateral agreements in East Asia (see Table 1.2). Ironically, some bilateral agreements are a source of disruption when they involve and/or encroach on the rights and interests of a third State. A typical example is the Agreement concerning Joint Development of the Southern Part of the Continental Shelf Adjacent to the Two Countries between Japan and South Korea signed in January 1974.[4] China made a strong protest by stating that the agreement had violated China's sovereignty and sovereign rights in the East China Sea.[5] This chapter is designed to assess the legal situation during the post-LOSC era in East Asia by mainly focusing on regional responses at the bilateral and

---

[3]For details on these laws, see Division for Marine Affairs and the Law of the Sea, available at http://www.un.org/Depts/los/LEGISLATIONANDTREATIES/asia.htm.

[4]Text is reprinted in Park Hee Kwon, *The Law of the Sea and Northeast Asia: A Challenge for Cooperation* (The Hague: Kluwer Law International, 2000), pp. 181–198.

[5]For details, see Zhao Lihai, *Studies on the Law of the Sea Issues* (Beijing: Peking University Press, 1996) (in Chinese), pp. 41–57.

Table 1.1: Contracting Parties to the LOSC in East Asia and Their Maritime Claims

States	Date of Ratification (dd/mm/yyyy)	Territorial Sea	Contiguous Zone	EEZ	Continental Shelf
Brunei	05/11/1996	12		200	
Cambodia		12	24	200	200(58)
China	07/06/1996	12	24	200	200/CM
Japan	20/06/1996	3/12	24	200	200/CM
S. Korea	29/01/1996	3/12	24	200	
N. Korea		12		200	
Laos	05/06/1998				
Malaysia	14/10/1996	12		200	200 m/Exp(58)
Mongolia	13/08/1996				
Myanmar	21/05/1996	12	24	200	200/CM
Philippines	08/05/1985			200	Exp
Russia	12/03/1997	12		200	200/CM(58)
Singapore	17/11/1994	3			
Thailand	15/05/2011	12	24	200	200 m/Exp(58)
Vietnam	25/07/1994	12	24	200	200/CM

Notes:

- The date here is the date when the depository the United Nations received the instrument of ratification.
- Taiwan is not a member of the United Nations and not qualified to be a signatory of the LOSC.
- (58): Contracting Party to the 1958 Convention on Continental Shelf.
- CM: continental margin; Exp: exploitability; 200: 200 nm; 200 m: 200 m.
- The Philippine territorial sea is rectangular in accordance with the so-called "Treaty Limits".
- The 3-mile limit, for South Korean, applies to the Korean Strait area, and for Japan, applies to the Soya Strait, the Tsugaru Strait, the eastern and western channels of the Tsushima Strait and the Osumi Straits only.

multilateral levels to the implementation of the LOSC in some critical maritime aspects as addressed in the following sections.

Marine Resources Management

With the depletion of resources and increase in population in the world, marine resources have become ever more important than before. Struggle for marine resources may cause conflicts between the

Table 1.2: Bilateral Maritime Agreements in East Asia

Agreement between the Government of Malaysia and the Government of Indonesia on the delimitation of the continental shelves between the two countries, 27 October 1969.
Treaty between the Republic of Indonesia and Malaysia Relating to the delimitation of the Territorial Seas of the Two Countries in the Strait of Malacca, 17 March 1970.
Agreement between the Government of the Kingdom of Thailand and the Government of the Republic of Indonesia relating to the delimitation of a continental shelf boundary between the two countries in the northern part of the Straits of Malacca and in the Andaman Sea, 17 December 1971.
Agreement between the Government of the Republic of Indonesia, The Government of Malaysia and the Government of the Kingdom of Thailand Relating to the Delimitation of the Continental Shelf Boundaries in the Northern Part of the Strait of Malacca, 21 December 1971.
Agreement Stipulating the Territorial Sea Boundary Lines between Indonesia and the Republic of Singapore in the Strait of Singapore, 25 May 1973.
Agreement between the Republic of Korea and Japan concerning the Establishment of Boundary in the Northern Part of the Continental Shelf Adjacent to the Two Countries, 30 January 1974.
Agreement between the Republic of Korea and Japan concerning Joint Development of the Southern Part of Continental Shelf Adjacent to the Two Countries, 30 January 1974.
Agreement between the Government of the Kingdom of Thailand and the Government of the Republic of Indonesia relating to the delimitation of the sea-bed boundary between the two countries in the Andaman Sea, 11 December 1975.
Treaty between the Kingdom of Thailand and Malaysia relating to the delimitation of the territorial seas of the two countries, 24 October 1979.
MOU between the Kingdom of Thailand and Malaysia on the delimitation of the continental shelf boundary between the two countries in the Gulf of Thailand, 24 October 1979.
Agreement between the Government of the Kingdom of Thailand and the Government of the Socialist Republic of the Union of Burma on the delimitation of the maritime boundary between the two countries in the Andaman Sea, 25 July 1980.
Agreement between the Union of the Soviet Socialist Republics and the Democratic People's Republic of Korea on the Delimitation of the Soviet-Korean National Border, 17 April 1985.
Agreement between the Union of Soviet Socialist Republics and the Democratic People's Republic of Korea on the Delimitation of the Economic Zone and the Continental Shelf, 22 January 1986.
Agreement on Fishery Cooperation between the People's Republic of China and the Union of the Soviet Socialist Republics, 4 October 1988.

Table 1.2: (*Continued*)

Agreement between the Government of the Union of Soviet Socialist Republics and the Government of the Democratic People's Republic of Korea concerning the Regime of the Soviet-Korean State Frontier, 3 September 1990.
Agreement on Fisheries between the People's Republic of China and Japan, 11 November 1997.
Agreement on Fisheries between the Republic of Korea and Japan, 28 November 1998.
Agreement on Fisheries between the People's Republic of China and the Republic of Korea, 3 August 2000.
Agreement on Fishery Cooperation in the Gulf of Tonkin between the People's Republic of China and the People's Republic of Socialist Vietnam, 25 December 2000.
Agreement on Maritime Boundary Delimitation in the Gulf of Tonkin between the People's Republic of China and the People's Republic of Socialist Vietnam, 25 December 2000.
Agreement between the Government of the Socialist Republic of Vietnam and the Government of the Republic of Indonesia concerning the Delimitation of the Continental Shelf Boundary, 26 June 2003.
Treaty between the Republic of Indonesia and the Republic of Singapore relating to the Delimitation of the Territorial Seas of the Two Countries in the Western Part of the Strait of Singapore, 10 March 2009.
Agreement between the Government of the Republic of Indonesia and the Government of the Republic of the Philippines concerning the Delimitation of the EEZ Boundary, 23 May 2014.
Treaty between the Republic of Singapore and the Republic of Indonesia relating to the Delimitation of the Territorial Seas of the Two Countries in the Eastern Part of the Strait of Singapore, 3 September 2014.

countries concerned. The military clashes between North and South Koreas, respectively, in 1999 and 2002 were directly caused by the grabbing of the blue crabs in the Yellow Sea around the Northern Limit Line (NLL).[6] Thus, a legal order governing marine fisheries is necessary. While there is no regional institution or organization relating to fisheries

[6]For details, see Jon M. Van Dyke, Mark J. Valencia and Jenny Miller Garmendia, "The North/South Korea Boundary Dispute in the Yellow (West) Sea", *Marine Policy*, Vol. 27(2), 2003, 143–158.

management in East Asia, there are a number of bilateral fishery agreements. For example, China concluded fishery agreements with Japan (1997), South Korea (2000) and Vietnam (2000), respectively, with an aim at sustainable fishery management in the East China Sea, the Yellow Sea and the Gulf of Tonkin. All the three agreements established joint fisheries zones in the relevant sea areas and fishing in these areas should be conducted in accordance with the terms and conditions under these agreements. Some fishery agreements such as the Sino-Japanese Fishery Agreement assume a dual function: to continue the governance of the joint fishery management and to serve as a provisional arrangement for the future delimitation of the EEZs and continental shelves between China and Japan.

On the other hand, it has to be realized that the fishery agreements either between China and Japan, or China and South Korea, or Japan and South Korea are bilateral ones. They have limitations and also do not completely cover the areas in the East China Sea and the Yellow Sea. Second, because of the bilateral nature, they may affect the interests of a third party. For example, South Korea expressed its dissatisfaction with the Sino-Japanese Fishery Agreement by asking China and Japan to explain how they drew the northen limit of their joint fishing area. They should consult South Korea before reaching their agreement.[7] Third, since bilateral agreements only regulate bilateral relations, fishing activities of third parties are outside any regulation. This is particularly true when Taiwan is concerned. Finally, many fishery resources in the East China Sea and the Yellow Sea are migratory species so that they belong to the same marine ecosystem. In that sense, the East Asian seas urgently need a regional and multilateral fishery arrangement which can be more effective to conserve and manage the fishery resources therein, and the existing bilateral fishery agreements could be the basis for such regional cooperation.

Regarding the management of marine non-living resources, the legal concept of joint development has been applied to East Asian seas. "Joint development" refers to "an agreement between two States to develop so as to share jointly in agreed proportions by inter-State cooperation and national measures the offshore oil and gas in a designated zone of the seabed and subsoil of the continental shelf to which both or either of the

[7] See Chi Young Pak, "Resettlement of the Fisheries Order in Northeast Asia Resulting from the New Fisheries Agreements among Korea, Japan and China", *Korea Observer*, Vol. 30(4), 1999, 614–615.

participating States are entitled in international law".[8] It contains several characteristics: (a) it is an arrangement between two countries; (b) it concerns an overlapping boundary maritime area; (c) it is a provisional arrangement pending the settlement of the boundary delimitation disputes between the countries concerned; (d) it is designed to jointly develop the mineral resources in the disputed area. In East Asia, joint development agreements include, *inter alia*, the Japan–South Korean arrangement in the Sea of Japan and the East China Sea in the 1970s, the Malaysia–Thailand joint development area in the Gulf of Thailand and the Australia–Indonesia joint development zone for the Timor Gap.

The Japanese–South Korean joint development arrangement was the first as such in East Asia. It was based on several agreements signed between the two countries. The arrangement is significant in State practice since it represents the first application of the idea of joint development of offshore oil where the parties failed to agree on boundary delimitation.[9] Under the agreement, concessionaires who are authorized by the two respective governments have an undivided interest with respect to each of the nine defined sub-zones, and one operator is chosen from among the concessionaires so authorized for a particular sub-zone.[10] The agreement establishes a Joint Commission as a consultative body to implement the agreement.

What is more significant is the joint arrangement made by three countries — Malaysia, Thailand and Vietnam — in the Gulf of Thailand for their overlapping claimed sea areas, consisting of two separate but associated bilateral agreements either between Malaysia and Thailand or between Malaysia and Vietnam. In 1979, Malaysia and Thailand signed a Memorandum of Understanding (MOU) to establish, on an interim basis of 50 years, a Malaysia–Thailand Joint Authority "for the purpose of the exploration and exploitation of the non-living natural resources of the

[8] British Institute of International and Comparative Law, *Joint Development of Offshore Oil and Gas: A Model Agreement for States for Joint Development with Explanatory Commentary* (London: British Institute of International and Comparative Law, 1989), p. 45.

[9] Masahiro Miyoshi, "The Joint Development of Offshore Oil and Gas in Relation to Maritime Boundary Delimitation", *Maritime Briefing*, Vol. 2(5), 1999, 1.

[10] *Ibid.*, p. 12.

seabed and subsoil in the overlapping area".[11] More than 10 years later, the two countries worked out the Constitution and other matters relating to the establishment of such an authority, which provides details of the operation in the joint zone.[12] There are two striking characteristics in this joint development scheme: a powerful joint authority which decides on the plan of operation and the work program, to permit operations and conclude transactions or contracts, to approve and extend the period of exploration and exploitation, to approve the work program and budgets of the contractor, and inspect and audit the operator's books and accounts;[13] and the introduction of a production sharing system which includes such terms and conditions as the duration of the contract not exceeding 35 years, the payment of 10% of gross production of petroleum by the contractor to the Joint Authority as royalty, 50% of gross production to be applied by the contractor for the recovery of costs, the remainder of gross production to be profit and divided equally between the Joint Authority and the contractor, all costs of operations to be borne by the contractor, and any dispute arising out of the contract to be referred to arbitration unless settled amicably.[14]

In the same vein, Malaysia and Vietnam also signed an MOU in 1992 for joint development in the Gulf of Thailand. Accordingly, Petronas and Petrovietnam were assigned to undertake respectively petroleum exploration and exploitation in the "defined area". The arrangement between the two State-owned oil companies made in August 1993 established an eight-member Coordination Committee to issue policy guidelines for the management of petroleum operations. This is different from the Thai–Malaysia model in which the Joint Authority is appointed directly by the governments. After the conclusion of the commercial arrangement in July 1997, oil has been extracted from the Bunga Kekwa field.[15] Based on the bilateral arrangements, a tripartite mechanism has been gradually evolving for an overlapping maritime area.

[11] Text is reprinted in Jonathan I. Charney and Lewis M. Alexander (eds.), *International Maritime Boundaries* (Dordrecht: Martinus Nijhoff Publishers, 1993), Vol. 1, pp. 1099–1123.

[12] *Ibid.*

[13] See Article 7 of the 1990 Agreement.

[14] See Article 8 of the 1990 Agreement.

[15] See Nguyen Hong Thao, "Vietnam and Joint Development in the Gulf of Thailand", *Asian Yearbook of International Law*, Vol. 8, 2003, 145.

Encouraged or triggered by all these developments in East Asia, China also put forward the idea of joint development in the disputed sea areas. As early as the 1980s, Deng Xiaoping, the former paramount Chinese leader, made the famous statement regarding China's policy toward disputed areas in China's adjacent seas by applying the concept of joint development. Deng regarded "joint development" as one of the two most important peaceful means for international dispute resolution.[16] Since then, China has been pushing very actively to realize the goal of joint development and reiterated on many international occasions its proposal. On 11 November 2003, the China National Offshore Oil Corporation (CNOOC) and the Philippine National Oil Company agreed to jointly explore oil and gas in the South China Sea through the signing of a letter of intent between the two sides. They also agreed to establish a program to "review, assess and evaluate relevant geographical, geophysical and other technical data available to determine the oil and gas potential in the area".[17] In March 2005, CNOOC, the Philippine National Oil Company and the Vietnam Oil and Gas Group (Petrovietnam) signed a tripartite agreement on joint seismic surveying activities in the South China Sea. The size for cooperation is 143,000 km^2.[18] However, this agreement ceased in 2008 after the first stage of joint activities. In addition, China has proposed joint development with Malaysia and Vietnam in the disputed areas in the South China Sea. Recently, China and the Philippines signed a memorandum of agreement on joint oil exploration in the South China Sea during the visit of Chinese President Xi Jinping in November 2018,[19] but there has been little substantial progress so far. Nevertheless, the complicated situation in the South China Sea indicates that joint development may be the only feasible means for regional cooperation for offshore oil and gas development.

[16] The other is "one country, two systems". See *Selected Works of Deng Xiaoping* (Beijing: Foreign Languages Press, 1994), Vol. 3, p. 87.

[17] "Chinese, Philippine firms join forces to look for oil in South China Sea", *Agence France Presse*, 13 November 2003.

[18] See "China, Philippines and Vietnam sign agreement to explore oil in the South China Sea", *Lianhe Zaobao*, 15 March 2005, available at http://www.zaobao.com/gj/yx501_150305.html.

[19] "PH-China joint oil deal to proceed: Palace", 28 October 2019, available at https://www.pna.gov.ph/articles/1084432.

Marine Environmental Protection

Regional cooperation and institutions for marine environmental protection is not new in State practice. Under the sponsorship of the United Nations Environmental Program (UNEP), several regional arrangements have been made for different sea areas throughout the world. They are the Wider Caribbean Region, the Southeast Pacific Region, the Mediterranean Region, the West and Central African Region, the Red Sea and Gulf of Aden Region, the Kuwait Action Plan Region, the Eastern African Region, the South Asian Seas Region, the South Pacific Region, and the Black Sea Region.[20] All these arrangements bear some common characteristics: they were sponsored by UNEP; they have an action plan as well as a treaty. In 1989, the UNEP Governing Council designated the Northwest Pacific as a new area where a regional action plan and later a regional sea treaty should be developed. The geographical coverage includes the marine environment and coastal zones of China, North and South Koreas, Japan and Russia. Interested countries around the Northwest Pacific sent their representatives to attend the initial contact meeting held in Nairobi shortly thereafter.[21] The subsequent developments showed that the establishment of a regional program in East Asia was not as easy as in other regions. The main culprit may be the abnormal political relations among the coastal States in the East Asian region.[22] The tension in the Korean Peninsula and the Taiwan issue are all obstacles to smooth regional co-operation for marine environmental protection. For example, when the first meeting of National Focal Points and experts convened in Vladivostok in October 1991, North Korea did not send its representative.[23]

---

[20] See Patricia W. Birnie and Alan E. Boyle, *International Law and the Environment* (Oxford: Clarendon Press, 1992), pp. 260–263.

[21] BSAP and NOWPAP — Now We Are 12, *The Siren* (UNEP's Oceans and Coastal Area Program), No. 43, 1 (December 1991); as cited in Jon M. van Dyke, "Promoting Maritime Co-operation in the Asia-Pacific Era", in Dalchoog Kim *et al.* (eds.), *Exploring Maritime Co-operation in Northeast Asia* (Seoul: Institute of East and West Studies, Yonsei University, 1993), p. 204.

[22] See Jin-Hyun Paik, "East Asia and the Law of the Sea", in James Crawford and Donald R. Rothwell (eds.), *The Law of the Sea in the Asia-Pacific Region: Developments and Prospects* (Dordrecht: Martinus Nijhoff, 1995), p. 7.

[23] See Sang-Don Lee, "A Framework for Co-operation for the Protection of the Marine Environment in the Yellow Sea: A Korean View", in Kim *et al.* (eds.), *supra* note 21, p. 73.

In 1994 the Action Plan for the Protection, Management and Development of the Marine and Coastal Environment of the Northwest Pacific Region (NOWPAP) was adopted and contained five objectives: monitoring and assessment of the environment condition; creation of an efficient and effective information base; integrated coastal area planning; integrated coastal area management; and establishment of collaborative and cooperative framework.[24] To implement the Action Plan, a number of projects are designed and carried out in parallel by national institutions with the support from relevant regional and international organizations. For multiple projects, the scattered nature of the various activities, as well as the wide scope of input possible from both within and outside the region, a network of participating institutions coordinated by regional activity centers would be established. Besides, a Regional Coordination Unit (RCU) should be established to ensure the integrated and well-managed execution from within the Region of the projects under the Action Plan. Until such time as an RCU is established and functioning effectively, the member governments designate UNEP as the organization responsible for the coordination of the Implementation of the Action Plan.[25] The RCU Toyama and Busan Offices were established in November 2004, and coordinate with the activities of four NOWPAP Regional Activity Centers (RACs)[26] to ensure integrated and well-managed implementation of the Action Plan.

The other regional marine environmental program is the Partnerships in Environmental Management for the Seas of East Asia (PEMSEA) jointly sponsored by the International Maritime Organization (IMO), the United Nations Development Programme (UNDP), the Global Environmental Facility (GEF) and the World Bank. This regional program, implemented in 1994, has such participating countries as Brunei, Cambodia, China, Indonesia, Japan, Malaysia, North Korea, the

[24] See "Action plan for the protection, management and development of the marine and coastal environment of the Northwest Pacific region", available at http://www.nowpap1.org/2-NOWPAP.htm.

[25] *Ibid.*

[26] Four RACs include Special Monitoring & Coastal Environmental Assessment Regional Activity Centre (based in Japan), Data and Information Network Regional Activity Centre (based in China), Marine Environmental Emergency Preparedness and Response Regional Activity Centre (based in South Korea), and Pollution Monitoring Regional Activity Centre (based in Russia).

Philippines, Singapore, South Korea, Thailand, and Vietnam. Therefore, its geographical coverage is broader than the NOWPAP. Its overall objective is to support the efforts of the participating countries in the prevention and management of marine pollution at both the national and sub-regional levels on a long-term and self-reliant basis.[27] In order to institutionalize Integrated Coastal Management (ICM), the Regional Program established two demonstration sites to demonstrate the application of ICM in Xiamen, China and Batangas Bay, the Philippines. Later such sites were set up in Cambodia, Indonesia, Malaysia, North Korea, Thailand and Vietnam. Another kind of Demonstration Sites for sub-regional sea areas and pollution hot spots were also established in China (Bohai Sea), the Philippines (Manila Bay), Malaysia/Singapore (Malacca Straits), and Thailand (Gulf of Thailand).[28] The program launched the Marine Pollution Monitoring and Information Management Network to help build linkages with participating countries on the status of the marine environment in the East Asian seas. The Sustainable Development Strategy for Seas of East Asia (SDS-SEA) was adopted in December 2003 and updated in 2015[29] in response to the UN Sustainable Development Goals (SDGs).

It should be noted that the implementation of the aforementioned two regional programs depends solely on the willingness and good faith of the participating States since there is no signed legal binding document among the States. On the other hand, it is noted that unlike a regional organization like the Association of Southeast Asian Nations (ASEAN), the marine programs involve cooperation and coordination of international organizations as well as States in East Asia.

Safety of Navigation

In terms of safety of navigation, the most important international treaty is the LOSC, though there are relevant treaties in this respect adopted under the auspices of the International Maritime Organization (IMO).[30]

---

[27]"Fighting pollution in the East Asian seas", *IMO News 4/97*, available at http://www.imo.org/imo/news/3&497/drchua2.htm.

[28]For details, see http://www.pemsea.org/.

[29]See "Regional marine strategy", available at http://www.pemsea.org/our-work/regional-marine-strategy.

[30]Zou Keyuan, "Navigation in the South China Sea: Why Still an Issue?" *The International Journal of Marine and Coastal Law*, Vol. 32(2), 2017, 243.

The LOSC today provides the legal framework with respect to navigational rights binding on all State parties and non-parties including the US, be it through ratification, unilateral declaration or simply as reflecting customary international law.[31] The LOSC has made the *legal* arrangements for navigational rights of vessels in accordance with different sea zones — internal waters, territorial sea, straits used for international navigation, archipelagic waters, the EEZ, the continental shelf, and the high seas and the international seabed (the "Area") — established under the Convention. Since internal waters are part of the territory of a coastal State, no freedom of navigation is granted there and any navigational rights are subject to the regulation of the coastal State.

Territorial sea is also part of the territory of the coastal State which owns the full sovereignty over it. However, due to the expeditiousness of navigation, the right of innocent passage is reserved for foreign vessels under the guarantee of international law. In addition, the coastal State should give appropriate publicity to any danger to navigation, of which it has knowledge, within its territorial sea.[32] On the other hand, the coastal State may adopt laws and regulations on, *inter alia*, the safety of navigation and the regulation of maritime traffic and the protection of navigational aids and facilities and other facilities or installations, in conformity with the provisions of the Convention and other rules of international law, relating to innocent passage through the territorial sea. Furthermore, the coastal State may take necessary steps in its territorial sea to prevent passage which is not innocent.[33]

For navigational safety reasons, the coastal State may, where necessary having regard to the safety of navigation, require foreign ships exercising the right of innocent passage through its territorial sea to use such sea lanes and traffic separation schemes as it may designate or prescribe for the regulation of the passage of ships. In particular, tankers, nuclear-powered ships and ships carrying nuclear or other inherently dangerous or noxious substances or materials may be required to confine their passage to such sea lanes. In the designation of sea lanes and the prescription of traffic separation schemes, the coastal State should take into account:

---

[31] Erik Franckx, "American and Chinese Views on Navigational Rights of Warships", *Chinese Journal of International Law*, Vol. 10, 2011, 192–193.

[32] See Article 21 of the LOSC.

[33] See Article 25 of the LOSC.

(a) the recommendations of the competent international organization;
(b) any channels customarily used for international navigation;
(c) the special characteristics of particular ships and channels; and
(d) the density of traffic.[34]

The coastal State should clearly indicate such sea lanes and traffic separation schemes on charts to which due publicity should be given.[35] Foreign nuclear-powered ships and ships carrying nuclear or other inherently dangerous or noxious substances should, when exercising the right of innocent passage through the territorial sea, carry documents and observe special precautionary measures established for such ships by international agreements.

As to the navigation in the EEZ, the LOSC provides a legal regime similar to that in the high seas.[36] However, since it is an area within national jurisdiction, the coastal State may have the right to lay down necessary laws and regulations relating to navigation safety and marine environmental protection. In this respect, the coastal State may well be aware that its laws and regulations should not hamper the smooth navigation of foreign vessels in and through its EEZ. On the other hand, foreign vessels are obliged to have due regard to the rights and duties of the coastal State and should comply with the laws and regulations adopted by the coastal State in accordance with the LOSC and other applicable rules of international law.[37]

Navigation in the straits used for international navigation was hotly debated during the UNCLOS III. Finally, the LOSC adopted the "transit passage" for foreign vessels passing through the straits used for international navigation. The other sea area within national jurisdiction which foreign vessels enjoy navigational rights is the archipelagic waters. As for the high seas, all States enjoy the freedom of navigation.

The piracy problem is a serious threat to the safety of navigation in East Asian seas. Piracy is regarded as *hostis humani generis*, the enemy of the human race, punishable wherever encountered. Piracy existed in the East Asian region as early as the 14th century. During Zheng He's seven

---

[34] See Article 22 of the LOSC.

[35] *Ibid.*

[36] See J. Ashley Roach, "Today's Customary International Law of the Sea", *Ocean Development and International Law*, Vol. 45(3), (2014), 239–252.

[37] See Article 58(3) of the LOSC.

voyages to the Indian Ocean in the Ming Dynasty (1405–1433), he undertook some actions to suppress pirates in the South China Sea and Southeast Asia for the purpose of maintaining peace and bringing order in the region.[38] Currently, piracy revived again in the region with a serious scale to endanger safety of navigation, persons and properties. The issues concerning piracy will be discussed in Chapter 5.

Closely related to the suppression of piracy is the Regional Maritime Security Initiative (RMSI), which was formally put forward in March 2004 by Admiral Thomas B. Fargo, Navy Commander of the US Pacific Command in his testimony regarding US Pacific Command posture before the House of Armed Services Committee of the US House of Representatives.[39] According to Fargo, the RMSI is designed to implement the "President's Proliferation Security Initiative (PSI)[40] and State Department's Malacca Strait Initiative" with the approach that detailed plans are provided "to build and synchronize interagency and international capacity to fight threats that use the maritime space to facilitate their illicit activity". From Fargo's remarks, it is obvious that this Initiative is designed for the East Asian region, particularly focusing on the safe navigation in the Strait of Malacca. On the other hand, Fargo emphasized that in order to conduct effective interdiction in the sea it is necessary to use high-speed vessels equipped with Special Operations Forces or Marines. Yet, some other countries such as Malaysia and Indonesia were doubtful of the RMSI whether it could really play a positive role in curbing piracy and maritime terrorism. They were also suspicious of the American intention whether it would infringe upon their national

---

[38] See Wang Li, "Zheng He's Voyage and Solution of Pirates Issue", in Nanjing Zheng He Study Association (ed.), *Chinese Heading for the Sea* (Beijing: Sea Tide Publishing House, 1996) (in Chinese), pp. 311–319.

[39] The whole text of the Testimony is available at http://www.pacom.mil/speeches/sst2004/040331housearmedsvscomm.shtml.

[40] PSI is an effort to consider possible collective measures among the participating countries, in accordance with national legal authorities and relevant international law and frameworks, in order to prevent the proliferation of weapons of mass destruction, missiles and their related materials that pose threats to the peace and stability of the international community. The PSI is administered by the "core group" countries, which, at present, consist of 15 countries (Japan, US, UK, Italy, the Netherlands, Australia, France, Germany, Spain, Poland, Portugal, Singapore, Canada, Norway and Russia), see "The Proliferation Security Initiative (PSI) Maritime Interdiction Exercise hosted by Japan", 18 October 2004, available at http://www.mofa.go.jp/policy/un/disarmament/arms/psi/exercise-2.html.

sovereignty and territorial integrity. In order to prevent the potential American military intervention in the Strait affairs, Malaysia and Indonesia decided to formulate a joint patrol to protect this international waterway. Singapore later joined. The tripartite patrol consists of 15–20 military vessels and patrol in the Strait all the year round.[41]

Maritime Boundary Delimitation

There are a number of bilateral agreements concerning maritime boundary delimitation in East Asia. A most important one is the agreement between China and Vietnam regarding the Gulf of Tonkin and will be discussed in Chapter 7. Maritime boundary demarcation often becomes an issue which causes tensions between States concerned in East Asia. The incident in the Yellow Sea between the North and South Koreas resulted directly from the unclear and controversial demarcation of their maritime boundary. In June 1999 and June 2002, respectively, two Koreas had armed skirmishes in the disputed sea areas around the so-called NLL which was unilaterally drawn by the United Nations Command after the conclusion of the Armistice Agreement concerning the Korea War in 1953, but its validity was rejected by North Korea.[42] In response, the North Korean military authorities declared after the 1999 incident (on 2 September 1999) that they had set up a "North Korean Military Demarcation Line in the West Sea (Yellow Sea)" which overlaps with the existing NLL. The different lines existing in the same area definitely have caused and will continue to cause maritime conflicts unless a clear boundary line has been negotiated between the two sides. It is, therefore, not easy to reach agreements on maritime boundary delimitation, though relevant negotiations are now under way, for example, between China and Japan, China and South Korea, and Japan and South Korea.

Recently, the claims to extended continental shelves in East China Sea have further complicated the difficulties in maritime boundary delimitation between countries concerned. The relevant submissions by East

---

[41] See "Indonesia and Malaysia jointly oppose American army to patrol in the Strait of Malacca", *Lianhe Zaobao* (in Chinese), 8 May 2004; "Piratical attacks are more fierce in the Strait of Malacca, Indonesia, Malaysia and Singapore will prevent this jointly", 9 October 2004, available at http://www.people.com.cn/GB/guoji/1029/2905373.html.

[42] A Korean scholar, following the South Korean government position, argues that the NLL has become a bilateral customary law. See Park Hee Kwon, *supra* note 4, p. 108.

Table 1.3: Extended Continental Shelf Submissions in East Asia

State	Date	Submission
Brunei	12/05/09	Preliminary info
China	11/05/09	Preliminary info
	14/12/12	Partial submission
Indonesia	16/06/08	Partial submission
	11/04/19	Partial submission
Japan	12/11/08	Full submission
Malaysia	06/05/09	Joint submission
	12/12/19	Partial submission
Myanmar	16/12/08	Full submission
Philippines	08/04/09	Partial submission
South Korea	11/05/09	Preliminary info
	26/12/12	Partial submission
Vietnam	06/05/09	Joint submission
	07/05/09	Partial submission

Asian States to the Commission on the Limits of the Continental Shelf aggravated the territorial and maritime disputes in both the East China Sea and South China Sea (see Table 1.3).

Islands Disputes and Their Settlement

There are several islands disputes in East Asia, including the Kuril Islands (between Russia and Japan), the Dokdo (between Japan and Korea), the Diaoyu Islands (between China and Japan), and the South China Sea Islands (various claimants).

The islands in the South China Sea, particularly the Spratlys are claimed by five countries including China (including Taiwan), Brunei, Malaysia, the Philippines and Vietnam. There were two armed skirmishes which happened in the South China Sea in 1973 and 1988 between China and Vietnam over the disputes of the South China Sea Islands. The 1995 tension between China and the Philippines over the Mischief Reef also invited a high profile exposure in the mass media. Arrests of fishing boats and fishermen are frequent. The claimed States blame one another for incursions in the disputed areas.[43] In June 2002, Vietnam lodged a protest against China's live ammunition exercises in the South China Sea, but China dismissed Vietnam's protest and stated that the drill fully complied with the international law.[44]

There is, however, a positive sign: all the claimants to the Spratly Islands have pledged to resolve their disputes in a peaceful manner and in accordance with the international law including the LOSC. The ASEAN countries together with China have held several rounds of discussion to formulate a Code of Conduct for the South China Sea.[45] On 4 November 2002, China and all the ASEAN member States signed the Declaration on the Conduct of the Parties in the South China Sea (the 2002 Declaration) in Phnom Penh, Cambodia.[46] It is the most remarkable document ever since signed between China and the ASEAN countries. The 2002 Declaration is designed to consolidate and develop the friendship and cooperation existing between China and ASEAN, to promote a peaceful, friendly and harmonious environment in the South China Sea, and to enhance the principles and objectives of the 1997 Joint Statement of the Meeting of the Heads of State/Government of the Member States of ASEAN and President of the People's Republic of China.

Originally, China was very reluctant to negotiate on such kind of document. It is recalled that during the 9th South China Sea Informal Workshop in December 1998, China opposed the proposal to negotiate a code of conduct for the interested parties concerned. The reasons behind included, *inter alia*, (1) China did not want to receive additional burden or pressure resulting from the code of conduct if it was formulated.

---

[43] For example, the Philippines expressed its alarm over increasing Vietnamese military and fishing vessel incursions into the waters off the disputed Spratly Islands in the South China Sea and there were 205 Vietnamese vessels in areas claimed by the Philippines in the first 10 months in 2001. See "Philippine alarmed over increasing Vietnamese incursions in Spratlys", *Agence France-Presse*, 16 November 2001.

[44] See "Vietnam protests China's exercises in 'off-limits zone'", *Voice of Vietnam*, Hanoi, 11 June 2002; and "China dismisses Vietnam's protest over naval exercise", *Xinhua News Agency*, Beijing, 11 June 2002.

[45] The drafts of the Code of Conduct both from ASEAN and China are reprinted in Hainan Research Institute for the South China Sea (ed.), *Collection of Selected Foreign and Chinese Papers on the South China Sea* (Hainan: Haikou, 2002) (in Chinese), pp. 180–183.

[46] The whole text is available at https://asean.org/?static_post=declaration-on-the-conduct-of-parties-in-the-south-china-sea-2.

Though a code of conduct bears no legal force, it is after all a "gentlemen's agreement" and should be complied with *bona fide*. Any violation would be condemned by other parties. (2) China did not consider the Workshop an appropriate forum to negotiate such code of conduct. The workshop was simply informal and the participants were in their personal capacity. The process of formulating a code of conduct would imply that such informal workshop would turn into a formal one and the personal capacity might become official capacity. China did not want to see such a scenario. (3) China preferred bilateral negotiations to resolve the South China Sea issue rather than multilateral negotiations. If a code of conduct was necessary, it should be done through bilateral diplomatic channel, just as manifested by the code of conduct between China and the Philippines in 1995. What annoyed China more was that the Taiwanese participants at the Workshop attempted to expand its diplomatic space by pushing forward such proposal of negotiating on a code of conduct with their involvement.

Nevertheless, China gradually changed its attitude and began to consider the possibility and benefits of negotiating such a code of conduct. The change may result from two forces, either internal or external. Internally, China carried out a pragmatic foreign policy and "stabilizing the neighboring regions" (*wending zhoubian*) is one of its top priorities. Based on this, China had to maintain a stable environment in the South China Sea. Externally, the pressures and efforts made by the ASEAN countries convinced the Chinese that cooperation with ASEAN was inevitable if China desired a stable environment in the South China Sea. As a result, China put forward its own proposal on the code of conduct in 1999 as a response to the ASEAN counterpart one. There were differences between the two proposals so that the two sides undertook negotiations to reach an agreed document. The 2002 Declaration is a biggest fruit from many rounds of negotiations.

The Declaration reaffirms the Parties' commitment to the use of international law, in particular the LOSC, to conduct confidence building and cooperation. The Parties ensure the freedom of navigation in and overflight above the South China Sea. They promise to resolve their territorial and jurisdictional disputes by peaceful means, without resorting to the threat or use of force. They intend to cooperate in the following matters: (a) marine environmental protection; (b) marine scientific research; (c) safety of navigation and communication at sea; (d) search and rescue operation; and (e) combating transnational crime, including but not limited to trafficking in illicit drugs, piracy and armed robbery at sea, and illegal traffic in arms. The Parties will continue their dialogues on the South China Sea and restraint themselves to take any provocative actions in the area. In comparison, the Declaration absorbed many elements from the previous Chinese proposal, including, but not limited to, the cooperative matters. China's signature can be regarded as a good gesture to show China's willingness to resolve the South China Sea issue by peaceful means. On the other hand, the Parties still face a series of tasks to be accomplished in accordance with the 2002 Declaration. The Declaration itself, though being a first step, is not the code of conduct, and the Parties concerned started its negotiation and will have to work on its adoption in the years to come.

While the dispute over the Spratly Islands is still pending, countries in Southeast Asia decided to use international judicial organs in solving their disputes over islands, which are illustrated by two significant cases before the International Court of Justice (ICJ): the case of *Sovereignty over Pulau Litigan and Pulau Sipadan (Malaysia/Indonesia)* (1998–2002) and the case concerning *Sovereignty over Pedra Branca/ Pulau Batu Puteh, Middle Rocks and South Ledge (Malaysia/Singapore)* (2003–2008). In the Judgment on *Sovereignty over Pulau Litigan and Pulau Sipadan*, the ICJ granted the disputed islands to Malaysia. The Court, when addressing the issue of *effectivités*, confirmed the validity of Malaysia's regulations concerning the control of colleting turtle eggs and the establishment of a bird conservation reserve and regarded them as "regulatory and administrative assertions of authority over territory which is specified by name".[47] In July 2003, Malaysia and Singapore submitted their territorial dispute over tiny islets including the Pedra Branca, Middle Rocks and South Ledge to ICJ for settlement based on the special agreement they signed in February 2003.[48] The ICJ rendered its judgment in 2008, awarding sovereignty over Pedra Branca to Singapore, and Middle

[47] See the ICJ judgment, para. 145, available at http://www.icj-cij.org/icjwww/idocket/iinma/iinmajudgment/iinma_ijudgment_20021217.PDF.

[48] See "Malaysia and Singapore jointly submit a dispute concerning sovereignty over Pedra Branca/Pulau Batu Puteh, Middle Rocks and South Ledge to the International Court of Justice", Press Release 2003/22, 24 July 2003, available at http://www.icj-cij.org/icjwww/ipresscom/ipress2003/ipresscom2003–22_ms_20030724.htm.

Rocks to Malaysia, while sovereignty over South Ledge belongs to the country in whose territorial waters it lies.[49]

Conclusion

The maritime situation has become more complicated after the States bordering on the East Asian seas had acceded to the LOSC. Some countries made their claims to islands and/or maritime zones in the South China Sea by using provisions of the LOSC on the 200 nm EEZ regime to argue that since these islands, islets or reefs are located within their claimed EEZ, these features should belong to them. According to the principle of "land dominating the sea", maritime jurisdiction derives from the sovereignty over land territory, not the other way round. That is to say, with the ownership of a piece of land, a State may have the right to extend its jurisdictional waters including EEZ outwards from that land. To claim a piece of land (no matter how small it is) from the sea is simply a misuse or abuse of the LOSC which no doubt disrupts the marine legal order in East Asia.

Lack of political confidence and trust among the East Asian countries is a further negative factor in the effective implementation of the LOSC. Two countries, China and Korea, are still divided internally. The divided authorities distrust and are hostile toward each other. This is why the Korean Peninsula and the Taiwan Strait are the other two flashpoints, in addition to the South China Sea, in East Asia. The Cold War is over in Europe and other places in the world but still exists in East Asia in the sense that there is no reconciliation between the contesting parties concerned after the Cold War. In contrast, they have been strengthening their military build-up. East Asia remains the second-largest regional arms market after the Middle East and North Africa.[50] China continues to strengthen its naval and air capabilities to prevent Taiwan from *de jure* independence while Taiwan has bought and continues to buy advanced weapons from the United States to deter Mainland China's threat. In case there is an armed conflict in the Taiwan Strait, the marine legal order in

[49] See 'Sovereignty over Pedra Branca/Pulau Batu Puteh, Middle Rocks and South Ledge (Malaysia/Singapore)', available at https://www.icj-cij.org/en/case/130.

[50] Anthony Bergin, "East Asian Naval Developments — Sailing into Rough Seas", *Marine Policy*, Vol. 26, 2002, 121.

East Asia could not be maintained. On the other hand, Japan, which possesses the most advanced fleet in the region, has taken a more active approach in its defense affairs[51] and begun to send troops overseas under its recently passed domestic laws. Its military presence in the South China Sea in the name of anti-piracy patrols worries the neighboring countries that remind of Japan's atrocities and invasions during the Second World War.

Nevertheless, there are three main characteristics in the trend of the marine legal order in East Asia: First, from divergence to convergence. Previously, the State practices in East Asia varied greatly, but after the LOSC the marine legal order was established based on this Convention and States enacted their relevant domestic laws in accordance with it as well.

Second, from rule of power to rule of law. This is not only indicated by the fact that most of the East Asian countries acceded to the LOSC but also by the fact that most of the countries have enacted their basic marine laws such as concerning the territorial sea and EEZ and established their jurisdictional maritime zones. Some countries even use the international judiciary in settling their maritime disputes. Although there are various definitions on rule of law, one thing is clear: the rule of law principle at the international level requires States to use international law rather than power, will of individuals, or even force to govern State-to-State relations. It is remarkable that China, for the first time, expressed clearly that rule of law should be implemented in international relations in 2004. In the Four Opinions of China regarding the work of the United Nations, one of the opinions is the rule of law in international relations. According to China, rule of law is fundamentally important in maintaining world peace and security, enhancing development and protecting human rights. The pick-and-choose mentality toward international legal norms is of no help to promote and realize rule of law and justice in the world community.[52]

Third, from bilateralism to multilateralism (regionalism). The most typical example is the Declaration on the Conduct of Parties in the South China Sea as explained previously. China, a key player in East Asia, has realized, though with some reluctance, that regionalization of the

---

[51] *Ibid.*, p. 126.

[52] See "Zhang Yishang put forward four opinions on the work of the United Nations in the General Assembly", 8 October 2004, available at http://www.people.com.cn/GB/guoji/14549/2904212.html.

maritime issues such as the South China Sea dispute is an inevitable trend, which China is unable to prevent but adapt itself to this trend so that it still can play a significant role in resolving regional maritime issues. In a word, vibrant regional cooperation and a regional mechanism for maintaining and improving the marine legal order in East Asian seas are required by international law including the LOSC, which obliges coastal States to cooperate between/among themselves.

Chapter 2

The United States Accession to the UN Convention on the Law of the Sea and Its Impact on East Asia

Introduction

The United Nations Convention on the Law of the Sea (LOSC),[1] concluded in 1982 and came into effect in 1994, has become a universally accepted part of contemporary international law governing the oceans and human activities related to and in the oceans. As of October 2020, there were 168 treaty parties including one international organization to this monumental international convention. Nevertheless, the United States, a sole superpower in our contemporary world, still remains outside this universally accepted international treaty.

In history, the United States contributed significantly to the development of the international law of the sea. The Harvard Law School facilitated the League of Nations to realize its attempt to codify international law including the law of the sea and first such codification conference was held at The Hague in 1930. The Truman Proclamation in 1945, which claimed jurisdiction and control over natural resources of the subsoil and sea bed of the continental shelf beneath the high seas but contiguous to the coasts of the United States,[2] in practice laid down the foundation for the

[1]Text is reprinted in 21 ILM (1982) 1261.

[2]Proclamation No. 2667, 10 Fed. Reg. 12303 (1945).

establishment of the continental shelf regime in international law as later embodied in the 1958 Convention on the Continental Shelf.[3]

During the Third United Nations Conference on the Law of the Sea (UNCLOS III), 1973–1982, the United States took the lead to many discussions on the formulation of the new legal regimes now incorporated into the LOSC. The legal regime regarding the transit passage through straits used for international navigation has been designed and formulated under the influence of or pressure from the United States. However, while the United States was satisfied with most of the legal arrangements embodied in the LOSC, it was dissatisfied with Part XI of the Convention creating the deep seabed mining regime and such dissatisfaction constituted the main reason for the US objection to the LOSC[4] and its decision not to sign the Convention.[5]

In order to seek universal participation in and effective implementation of the Convention, particularly Part XI, the then Secretary-General Perez de Cuellar took the initiative in July 1990 and convened an informal consultation among all interested States to resolve the issues resulting from Part XI of the LOSC by reaching a new agreement. The informal consultation was focused on nine most controversial issues pertaining to production limitation, expenses of contractual parties, financial provisions in the contract, the enterprise, decision-making process, transfer of technology, compensation fund, review conference and environmental protection. After four years of consultations, the 1994 Agreement relating to the Implementation of Part XI of the LOSC (the 1994 Agreement)[6] was adopted. It has changed, to some fundamental extent, the original deep seabed regime created by Part XI of the LOSC, though the principle of the

---

[3] 499 UNTS 311.

[4] For the position of the United States, see James B. Morell, *The Law of the Sea: The 1982 Treaty and Its Rejection by the United States* (Jefferson, North Carolina: McFarland & Company, 1992), in particular Chapters 4 and 5, pp. 96–189.

[5] It is recalled that during the final negotiations of the Convention, the United States raised a number of objections to Part XI and tried to seek solutions to six specific issues. However, the United States objectives had not been met. See John A. Duff, "A Note on the United States and the Law of the Sea: Looking Back and Moving Forward", *Ocean Development and International Law*, Vol. 35, 2004, 197.

[6] Text is available at http://www.un.org/Depts/los/convention_agreements/texts/unclos/closindxAgree.htm.

common heritage of mankind continued to be maintained.[7] Thus, a major obstacle to the United States accession to the LOSC has been removed. The United States also signed the 1994 Agreement.

Afterward, both the Clinton and Bush Administrations supported the accession of the United States to the Convention. In 2004, the Bush Administration urged members of the Senate to approve the US entry into the LOSC as such approval was in the best interest of the United States.[8] Hearings were held in the Senate. However, since the Obama Administration, there has been no substantial progress toward the ratification of the Convention. This chapter attempts to select several issues in the law of the sea field whose resolution to a large extent depends upon the contributions of the United States after its accession to the LOSC in future.

United States Response to "Excessive" Maritime Claims in the Territorial Sea

There are two major remaining yet controversial issues regarding the territorial sea regime. One is about the innocent passage, in particular relating to the passage for warships and the other is regarding the measurement of straight baselines.

Innocent passage

The right of innocent passage is a well-established rule and constitutes a part of the territorial sea regime. However, while there is a uniform State practice in innocent passage for merchant ships, whether the right of innocent passage should be accorded to warships remains a controversial issue. While the majority of the world community expressly or implicitly allows foreign warships to enjoy the right of innocent passage through territorial seas, other countries, including China, still require prior

[7]As provided in Article 311(6) of the LOSC, the principle of common heritage of mankind should not be amended.

[8]See "Bush Administration Urges Senate Approval of Sea Treaty Law of the Sea Convention Serves U.S. Interests, State's Turner Testifies", (Washington D.C.: US Dept. of State, International Information Programs), 24 March 2004 (on file with the author).

approval or notification.[9] The *Commander's Handbook on the Law of Naval Operations* lists 27 States which require prior permission and 12 others require prior notification.[10]

In 1949, the International Law Commission of the United Nations (ILC) began to draft law of the sea conventions. The 1954 ILC draft did not differ substantially from the annex to the 1930 Hague Conference Report, providing that, "save in exceptional circumstances, warships shall have the right of innocent passage through the territorial sea without previous authorization or notification".[11] But in 1955, ILC adopted a new formula which read: "[t]he coastal State may make the passage of warships through the territorial sea subject to previous authorization or notification. Normally it shall grant innocent passage...".[12] This formula was a concession to a certain State practice which required previous authorization or notification, as the Commentary of the Commission stated that because "a number of States do require previous notification or authorization" it could not "dispute the right of the States to take such a measure".[13]At the 1958 Geneva Conference on the Law of the Sea, however, this provision was not adopted, because it did not achieve the necessary two-thirds majority.[14] Therefore, the Geneva Convention on the Territorial Sea and the Contiguous Zone (the Geneva Convention) was silent on the passage of warships.[15] This silence led to contradictory interpretations of the provisions on innocent passage regarding the right for warships. One opinion held that the passage of warships was not regulated by the 1958 Convention and therefore was left to be decided by customary law. The other view maintained that the right of passage for warships was,

[9]According to the statistics collected by the UN Law of the Sea Office, there are 17 States which still have the special requirement for prior authorization or consent: Algeria, Antigua-Barbuda, Bangladesh, Barbados, China, Grenada, Iran, Maldives, Myanmar, Pakistan, Romania, Saint Vincent and the Grenadines, Somalia, Sri Lanka, Syria, United Arab Emirates and Yemen. See UN Doc. A/51/645, 15 October 1996, p. 67.

[10]Cited in Ivan Shearer, "Military Activities in the Exclusive Economic Zone: The Case of Aerial Surveillance", *Ocean Yearbook*, Vol. 17, 2003, 552.

[11]*Yearbook of the International Law Commission*, Vol. 2, 1954, 161.

[12]*Ibid.*, II, p. 276.

[13]*Yearbook of ILC*, 1956, II, p. 277.

[14]See votes at the 20th Plenary Meeting, First U.N. Conference on the Law of the Sea, *Official Records*, II, pp. 66–68.

[15]Tullio Treves, "Navigation", in Rene-Jean Dupuy and Daniel Vignes (eds.), *A Handbook on the New Law of the Sea* (Dordrecht: Martinus Nijhoff, 1991), II, p. 930.

at least implicitly, provided by the 1958 Convention by the following provisions: "[s]hips of all States, whether coastal or not, shall enjoy the right of innocent passage through the territorial sea", and "[i]f any warship does not comply with the regulations of the coastal state concerning passage through the territorial sea and disregards any request for compliance which is made to it, the coastal state may require the warship to leave the territorial sea".[16]

During UNCLOS III the difference on the right of warships came up again. The maritime powers claimed innocent passage for warships, whereas other countries, particularly the developing ones, emphasized that the importance of their coastal security necessitated imposing conditions on the passage of warships through the territorial sea. They would require either prior authorization or at least prior notification of passage. During the conference, developing countries made several attempts to add a provision requiring prior notification or authorization, but it proved impossible to obtain the necessary consensus.[17] The LOSC, like the 1958 Geneva Convention, contains no provision that expressly allows or denies warships a right of innocent passage. It stipulates that ships of all States enjoy the right of innocent passage through the territorial sea, defining "innocent passage" as transit that is not prejudicial to the peace, good order or security of the coastal State.[18]

At UNCLOS III, the Chinese delegate explained in detail the Chinese position on the regime of passage within the territorial sea: the regime of passage in the territorial sea is of vital importance to the sovereignty and security of the coastal State, so that all relevant provisions shall ensure to protect the sovereignty of the coastal State against any encroachment and threat. In the interest of the convenience of international navigation, China agrees in principle to the regime of innocent passage. However, military vessels and merchant ships should be treated differently by setting down different provisions because military vessels and merchant ships are different in their nature. The coastal State shall decide, in accordance with its own laws and regulations, whether or not the facility of innocent passage through the territorial sea to the foreign military vessel shall be granted.

[16]The Convention on the Territorial Sea and the Contiguous Zone, 29 April 1958, 15 UST 1606. TIAS No. 5639, 516 UNTS 205.

[17]See B. Oxman, "The Third United Nations Conference on the Law of the Sea: The Ninth Session", *American Journal of International Law*, Vol. 75, 1981, 235.

[18]Articles 17 and 19(1) of the LOSC.

It is an infringement of the sovereignty of the coastal State and constitutes a threat to its independence and security if a warship does not comply with the laws and regulations of the coastal State requiring prior notification or prior approval from that State in order to pass through its territorial sea. This shall not be allowed.[19] Since the Informal Composite Negotiating Text (ICNT) of the Conference made no difference between warships and merchant ships, the Chinese delegate suggested a revision of Articles 17, 20 and 29 of ICNT.[20] At the 9th Session of UNCLOS III in 1980, the Chinese delegation pointed out again that China had a serious reservation with respect to the regime of innocent passage through the territorial sea as provided in the ICNT. The Chinese representative stated that "the passage of foreign warships through the territorial sea is a matter which bears upon the sovereignty and national security of the coastal State. Naturally, the coastal State has the right to make the necessary regulations in respect of such passage".[21] The relevant provisions in the new Convention on the Law of the Sea should take full account of such reasonable positions and demands.

The Chinese position was supported by a considerable number of participating countries during UNCLOS III, and China and other countries holding a similar position put forward joint proposals on the issue of innocent passage. It was suggested in a joint proposal to add a paragraph to Article 21 of the ICNT to the effect that the coastal State shall, in accordance with its laws and regulations, have the right to require foreign military vessels to render prior notification to or seek prior approval from that State when passing through the territorial sea.[22] This requirement was asserted to be consistent with principles of international law according to the generally recognized principle that only foreign nonmilitary vessels enjoy the right of innocent passage. The passage of warships through the territorial sea concerns the sovereignty and security of the coastal State,

---

[19] See Speech on the Regime of Passage of Warships by An Zhiyuan, Head of the Chinese Delegation, at the unofficial meeting of the Second Committee, UNCLOS III, in *Collected Documents of the Chinese Delegation to Attend Relevant UN Conferences*, January–June 1978 (Beijing: People Publishing House, September 1978) (in Chinese), p. 131.

[20] *Ibid.*, p. 132.

[21] Speech by Ke Zaishuo, at the Plenary Session, UNCLOS III, in *Collected Documents, supra* note 19, January–June 1980 (Beijing: World Knowledge Press, January 1981), p. 97.

[22] Speech by Shen Weiliang, at the Plenary Session, UNCLOS III, *Ibid.*, July–December 1980, p. 278.

and the coastal State shall certainly have the right to make necessary regulations. It is unfortunate for China's position that the aforementioned proposal was not adopted in the Conference due to the opposition of the maritime powers, particularly the United States and the Soviet Union. At the 11th Session of UNCLOS III in 1982, 27 coastal States, including China, put forward a proposal demanding that a single word "security" be inserted into Article 21(1)(h) of the draft Convention, i.e. "the prevention of infringement of the customs, fiscal, immigration, *security* or sanitary laws and regulations of the coastal State".[23] This proposal again was rejected by the maritime powers.[24] The Chairman of the Conference broke through the impasse by reading the following statement at the Conference while the sponsors of the proposal agreed to withdraw it: "Although the sponsors of the amendment … had proposed the amendment with a view to clarifying the text of the Convention, in response to the President's appeal they have agreed not to press it to a vote. They would, however, like to reaffirm that their decision is without prejudice to the rights of the coastal State to safeguard their security interests, in accordance with Articles 19 and 25 of the draft convention".[25] Thus, the provisions regarding the innocent passage in the ICNT, and the eventual convention, remained unaffected. In an explanatory statement, the Chinese delegate expressed that the provisions on innocent passage through the territorial sea in the draft convention shall not hamper the right of the coastal State, in accordance with its laws and regulations, to require foreign warships to provide advance notice or obtain prior permission from the coastal State when passing through the territorial sea.[26]

Although the provisions on innocent passage in the LOSC, in China's view, are better than those contained in the Geneva Convention, they are

---

[23] Doc.A/CONF.62/L.117, 13 April 1982 (presented by Algeria, Bahrain, Benin, Cape Verde, China, Congo, Egypt, Guinea-Bissau, Iran, Libya, Malta, Morocco, North Korea, Oman, Pakistan, Papua New Guinea, the Philippines, Romania, Sao Tome and Principe, Sierra Leone, Somalia, Sudan, Suriname, Syria, Uruguay, Democratic Yemen), UNCLOS III, *Official Records*, XVI, p. 225.

[24] See Shen Weiliang and Xu Guangjian, "The Third UN Conference on the Law of the Sea and the Law of the Sea Convention", *Chinese Yearbook of International Law* (in Chinese), 1983, p. 411.

[25] An Zhiyuan, *supra* note 19, p. 132.

[26] Speech by Shen Weiliang, at the Plenary Session, UNCOLS III, in *Collected Documents*, *supra* note 19, January–June 1982 (Beijing: World Knowledge Press, April 1983), p. 91.

still regarded as deficiencies of the LOSC. The head of the Chinese delegation stated at the final session of UNCLOS III that there were still shortcomings and even serious defects in the provisions of quite a few articles in the Convention. For example, the provisions on innocent passage through the territorial sea did not make clear the regime of passage of warships through the territorial sea. The Convention was not entirely satisfying to China, and the Chinese Government had already expounded its principle and its position in this respect.[27] Since then, China's position has remained unchanged. When China ratified the LOSC in 1996, it made the following statement with a view to innocent passage:

> The provisions of the United Nations Convention on the Law of the Sea concerning innocent passage through the territorial sea shall not prejudice the right of a coastal State to request, in accordance with its laws and regulations, a foreign State to obtain advance approval from or give prior notification to the coastal State for the passage of its warships through the territorial sea of the coastal State.[28]

In terms of domestic legislation, the 1992 Law on the Territorial Sea and the Contiguous Zone (the Territorial Sea Law) contains several important provisions relating to innocent passage.[29] According to it, foreign ships for non-military purposes shall enjoy the right of innocent passage through China's territorial sea in accordance with the law. Foreign ships used for military purposes shall be subject to permission of the Chinese Government before entering the territorial sea (Article 6). Foreign submarines and other underwater vehicles, when passing through the territorial sea, shall navigate on the surface and show their flag (Article 7). Foreign ships passing through China's territorial sea must

---

[27] Han Xu, Statement, 9 December 1982, UNCLOS III, *Official Records*, XVI; Summary Records of Meetings, 1982, p. 102; also reprinted in Shanghai Institute of International Studies (ed.), *Yearbook of International Situation* (in Chinese), 1983, p. 391.

[28] See the Decision to Ratify the United Nations Convention on the Law of the Sea by the Standing Committee of the National People's Congress, 15 May 1996. See *People's Daily* (*Overseas Edition*) (in Chinese), 16 May 1996. China deposited the instrument of ratification on 7 June 1996.

[29] English version of the text may be found in Office of Ocean Affairs, Bureau of Oceans and International Environmental and Scientific Affairs, US Department of State, *Limits in the Seas*, No. 117 (Straight Baselines Claim: China), 9 July 1996, pp. 11–14.

comply with the laws and regulations of China and their passage shall not be prejudicial to the peace, security and good order of China. Foreign nuclear-powered ships and ships carrying nuclear, noxious or other dangerous substances, when passing through the territorial sea, must carry relevant documents and take special precautionary measures. The Chinese Government shall have the right to take all necessary measures to prevent and stop non-innocent passage through its territorial sea. Cases of foreign ships violating the laws and regulations of China shall be handled by the relevant Chinese organs in accordance with the law (Article 8). The Chinese Government may, for maintaining the safety of navigation or for other special needs, request that foreign ships passing through its territorial sea use the designated sea lanes or navigate according to prescribed traffic separation schemes (Article 9). In the case of violation of the Chinese laws or regulations by a foreign ship for military purposes or a foreign government ship for non-commercial purposes when passing through the territorial sea of China, the Chinese competent authorities shall have the right to order it to leave the territorial sea immediately and the flag State shall bear international responsibility for any loss or damage caused by the ship (Article 10).

In practice, reports about the implementation of China's requirement of prior permission are really rare. However, there is at least one media report in this regard: on 17 April 2001, just after two weeks of the American spy airplane incident, Australia sent three warships to go through the Taiwan Strait. The Australian flotilla, having been asked to leave the Taiwan Strait by the Chinese captain who accused them of breaching China's sea boundaries, refused to change their direction and continued through the strait.[30] While the Australian side alleged that this act was undertaken "quite properly and fully in accordance with international law",[31] it was regarded by the Chinese as breach of Chinese laws and intrusion of Chinese waters. Though it is not known where the Australian warships were exactly sailing, it was indicated that they were passing through China's territorial sea in the Taiwan Strait.[32]

---

[30] "Howard seeks to patch up flotilla row", *South China Morning Post*, 30 April 2001, p. 5.

[31] *Ibid.*

[32] As reported, "in a protest on ANZAC Day, a Chinese naval captain accused the HMA ships SUCCESS, ARUNTA and NEWCASTLE of breaching China's 12-nautical mile territorial zone". See Michael Weaver, "Aussie exercise sea right: China lodges protest",

The Australian spokesman for Minister Alexander Downer confirmed that China had raised an official protest with Australia about the transit.[33]

It is interesting to note that in the 1930s the United States took a similar attitude toward the passage of foreign warships through its territorial sea. In a reply to the Questionnaire of the Preparatory Committee of the League of Nations, the United States made it clear that, in its opinion, warships required authorization by the coastal State before entering its territorial waters.[34] During the Hague Codification Conference, the US delegate further stated that the right of innocent passage as a matter of right did not extend to warships. It was ordinarily granted that the right of innocent passage was primarily in favor of commerce, and so far as warships were concerned, the question was wholly one of usage and comity of nations.[35]After World War II, with the growing strength of its navy, however, the United States changed its attitude, allowing foreign warships to pass innocently through its territorial sea, while asserting that its own warships could enjoy the same right in the territorial seas of other countries.

What is until now the most significant in state practice relating to the right of innocent passage for warships is the Joint Statement on Uniform Interpretation of Rules of International Law Governing Innocent Passage issued by the Soviet Union and the United States on 23 September 1989. According to this statement, all ships, including warships, regardless of cargo, armament or means of propulsion, enjoy the right of innocent passage through the territorial sea in accordance with international law, for which neither prior notification nor authorization is required. Further, a ship passing through the territorial sea which does not engage in any of the activities set out in Article 19(2) of the LOSC is in innocent passage.[36]

Navy News (Australian Royal Navy), 14 May 2001, available at http://www.defence.gov.au/news/navynews/editions/05_14_01/story12.htm.

[33]"Aussie naval transit through Taiwan Straits protested", *China Daily*, 29 April 2001.

[34]League of Nations Doc. C.74.M.39.1929.V (L.N. Pub. No.1929.V.2), pp. 66 and 73, cited in Erik Franckx, "Innocent Passage of Warships: Recent Developments in US–Soviet Relations", *Marine Policy*, Vol. 14, 1990, 485.

[35]See 6th Meeting of the Second Committee on Territorial Waters of the Conference for the Codification of International Law, League of Nations Doc. C.351(b).1930.V (L.N. Pub. No.1930.V.16), p. 59.

[36]See "Uniform Interpretation of Rules of International Law Governing Innocent Passage", 28 ILM (1989), 1444.

The Joint Statement indicates the unification of the practice concerning innocent passage of these two biggest maritime powers.

On the other hand, the fact that there are no express provisions in the LOSC regarding the right of innocent passage of warships may lead to different interpretations. If one looks into all relevant clauses in the Convention, one would easily find that a link is missing between the right of innocent passage and warships. This link was historically dropped out during the 1958 Law of the Sea Conference. This is, no doubt, one of the deficiencies in the Convention that needs an appropriate amendment in the future. It is unknown whether the United States can convince countries with different State practice to change their existing attitude and hopefully can reach with China a similar agreement to the 1989 Joint Statement on Uniform Interpretation of Rules of International Law Governing Innocent Passage. It is noted that after this Statement, some of the countries that set down restrictions on the innocent passage of warships including Bulgaria, Russia, Sweden and Turkey have changed their positions.[37]

Straight baselines

Another controversial area lies in the measurement of the straight baselines. According to the LOSC, straight baselines can be used when the coastal lines meet either of the following conditions: (1) in localities where the coastline is deeply indented and cut into, or (2) if there is a fringe of islands along the coast in its immediate vicinity.[38] The Convention further provides that "[t]he drawing of straight baselines must not depart to any appreciable extent from the general direction of the coast, and the sea areas lying within the lines must be sufficiently closely linked to the land domain to be subject to the regime of internal waters" (Article 7). It can be seen that the use of the straight baselines is limited subject to the aforementioned two conditions. In other words, this method is a supplement to the use of normal baselines.

Although there are no specific technical details about how to draw straight baselines, the United States takes the position that in order to

---

[37]See J. Ashley Roach and Robert W. Smith, *United States Responses to Excessive Maritime Claims*, 2nd ed. (The Hague: Martinus Nijhoff, 1996), pp. 276–277.

[38]Article 7(1) of the LOSC.

meet the LOSC criteria, "straight baseline segments must not depart to any appreciable extent from the general direction of the coastline, by reference to general direction which in each locality shall not exceed 60 miles in length; not exceed 24 miles in length; and result in sea areas situated landward of the straight baseline segments that are sufficiently closely linked to the land domain to be subject to the regime of internal waters".[39]

However, in State practice, in particular in East Asia, the use of straight baselines to measure maritime zones is more common than the use of normal baselines. China uses the method of straight baselines to define the limits of its territorial sea. As for the People's Republic of China (PRC), the 1958 Declaration on China's Territorial Sea declared that: (1) the breadth of the territorial sea of China should be 12 nm, which applies to all territories of China, including the Chinese mainland and its coastal islands, as well as all other islands belonging to China; and (2) China's territorial sea would take, as its baseline, the line composed of the straight lines connecting basepoints on the mainland coast and on the outermost of the islands; the water area extending 12 nm outward from the baseline would be China's territorial sea, and the water areas inside the baseline would be China's inland waters, including the Bohai Sea and the Chiungchow Strait.[40] But at that time the PRC Government did not publicize any geographical coordinates. Only in May 1996 was part of such baselines around the mainland and the Xisha Islands (Paracel Islands) publicized.[41] The publicized baselines are divided into two sets: one comprising 49 base points along features on, and adjacent to, its mainland coast and on Hainan Island beginning at point 1 (*Shandong gaojiao*) on the eastern tip of the Shandong peninsula situated to the southeast of Bohai, south to point 49 situated on the west coast of Hainan Island; and the other comprising 28 base points encompassing the Paracel Islands in the northern part of the South China Sea. However, parts of the baselines

---

[39]Roach and Smith, *supra* note 37, pp. 64–65.

[40]English version may be found in Office of Policy, Law and Regulation, State Oceanic Administration (ed.), *Collection of the Sea Laws and Regulations of the People's Republic of China*, 3rd ed. (Beijing: Ocean Press, 2001), p. 197.

[41]Declaration on the Baseline of the Territorial Sea of the People's Republic of China, 15 May 1996, see Office of Policy, Law and Regulation; *Ibid.*, 206–209.

have been criticized for not being consistent with the criteria set forth in the LOSC. The United States takes the view that

> Much of China's coastline does not meet either of the two LOS Convention geographic conditions required for applying straight baselines. And, for the most part, the waters enclosed by the new straight baseline system do not have the close relationship with the land, but rather reflect the characteristics of high seas or territorial sea.[42]

Further analysis gives the details how China's straight baselines deviate from the LOSC criteria. For example, the coastline from the Shandong peninsula to the area of Shanghai (point 1 to point 11) is essentially smooth with no fringing islands and few indentations. Thus, it is argued, the straight baseline method should not apply.[43] The other criticism of China's straight baselines is that the archipelagic straight baselines encircling the Paracel Islands should not be used because the PRC is not a mid-ocean archipelagic State.[44]

Similar to China, other East Asian countries also have controversial baselines. The United States lodged a protest, in an *aide mémoire*, against a system of 10 straight baseline segments declared by Vietnam in November 1982 and took the view that the baselines claimed by Vietnam did not meet the corresponding criteria and "there is no basis in international law" for such system of straight baselines.[45] Further, "[e]ven if the Mekong delta qualifies the coastline in this locality as deeply indented, the islands selected are not appropriate as basepoints".[46]

---

[42] Bureau of Oceans and International Environmental and Scientific Affairs, United States Department of State (BOIESA), *Limits in the Seas: Straight Baselines Claim: China*, No. 117, 9 July 1996, p. 3.

[43] *Ibid.*, p. 5.

[44] See *Ibid.*, p. 8. For further reference, see Hyun-Soo Kim, "The 1992 Chinese Territorial Sea Law in the Light of the UN Convention", *International and Comparative Law Quarterly*, Vol. 43, 1994, 896–899; and Daniel J. Dzurek, "The People's Republic of China Straight Baseline Claim", *IBRU Boundary and Security Bulletin*, Summer 1996, pp. 77–89.

[45] Roach and Smith, *supra* note 37, p. 102.

[46] W. Michael Reisman and Gayl S. Westerman, *Straight Baselines in International Maritime Boundary Delimitation* (New York: St. Martin's Press, 1992), p. 133.

The United States also criticized the baselines claims made by Japan and South Korea. For Japan, its coastline in many locations does not meet the LOSC geographic conditions required for applying straight baselines.[47] As for South Korea, while acknowledging that its coastline is deeply indented in several areas and there are areas off the coast which are fringed with islands, the United States regards the means by which South Korea has drawn particular straight baseline segments in several locations as not meeting the LOSC criteria.[48]

In December 1998, Taiwan publicized part of its baselines, including those encircling the Taiwan Island and Penghu Islands.[49] The Penghu Channel, once used to be an international waterway, has thus become part of internal waters of Taiwan. The straight baseline segments which encircle the whole Penghu Islands are clearly problematic if they are examined under the criteria of the LOSC. The United States officials also criticized these straight baselines drawn by Taiwan as not consistent with international law.[50]

Baselines are critical for the measurement of maritime zones under national jurisdiction as well as for maritime boundary delimitation between neighboring coastal States. Thus, controversial baselines will definitely cause problems in the aforementioned two areas. When a maritime zone is designated by a coastal State with controversial baselines, then it will become debatable under international law and challengeable by other countries. Potential maritime disputes or conflicts would then arise. The United States may play some role in adopting unified norms applying to the drawing of straight baselines. On the other hand, as it is seen in East Asia, many of the coastal States have drawn "excessive" straight baselines in the eyes of the United States.

---

[47] Bureau of Oceans and International Environmental and Scientific Affairs, United States Department of State (BOIESA), *Limits in the Seas: Straight Baseline and Territorial Sea Claims: Japan*, No. 120, 30 April 1998, p. 2.

[48] BOIESA, *Limits in the Seas: Straight Baseline and Territorial Sea Claims: South Korea*, No. 121, 30 September 1998, p. 2.

[49] See *United Daily News* (Taipei) (in Chinese), 1 January 1999.

[50] See J. Ashley Roach and Robert W. Smith, "Straight Baselines: The Need for a Universally Applied Norm", *Ocean Development and International Law*, Vol. 31, 2000, p. 65. For further details, see J. Ashley Roach, "An International Law Analysis of Taiwan's Maritime Claims", *Taiwan International Law Quarterly*, Vol. 2(1), March 2005, 256–268.

There is an interesting phenomenon in East Asia that "bad baselines beget bad baselines".[51] Now all the countries concerned have realized the need to delimit maritime boundaries between them. It is unknown whether they use the existing baselines as point of departure to undertake maritime boundary delimitation or they will reach an agreement to use a unified standard to adjust their baselines before settling the maritime boundary issues. A Chinese scholar once called for the reconsideration of the straight baselines drawn by China and urged China to negotiate with Japan and Korea to reach a unified criterion for redrawing their respective baselines.[52]

New Issues in the Exclusive Economic Zone

Some significant maritime incidents having occurred in East Asia, for example, EP-3E Spy Airplane Incident in 2001,[53] have raised a number

[51] Roach and Smith, "Straight Baselines: The Need for a Universally Applied Norm", *supra* note 50, p. 53.

[52] Li Linghua, "The Need of a Serious Discussion on Determining the Criterion for Straight Baselines" (in Chinese), unpublished paper, 2005 (on file with the author).

[53] On 1 April 2001, while a US EP-3E Aries II airplane was conducting espionage activities near the Chinese coast in the South China Sea, it was intercepted by two Chinese F-8 fighter jets and then collided with one of the jets. The damaged Chinese jet crashed into the water and the pilot died. The damaged American airplane made an emergency landing in China's Hainan Island at Lingshui and all the crewmembers were safe. The incident immediately became a diplomatic issue between China and the United States. China accused the United States of encroaching on China's territorial sovereignty and of violation of international law as well as of relevant Chinese laws, and demanded an apology and compensation from the American side. The United States responded that the reconnaissance airplane operated outside China's territorial waters and that the airplane landed in distress. For that reason, the United States refused to render an apology; and instead demanded China to immediately return the American crew and the airplane. After several rounds of diplomatic contacts, the United States finally sent a letter to China on 11 April 2001, expressing its sincere regret over the Chinese missing pilot and aircraft and used the word "sorry" for their loss. The letter also used the word "sorry" for the American airplane's entering of China's airspace and landing without verbal clearance. On the next day, China allowed all 24 crewmembers to leave China. However, the American damaged spy airplane did not leave China until 3 July 2001 after it had been dismantled and packed. The Chinese side asked for one million US dollars for the costs relating to the aircraft but the Americans only offered the amount of 34,567 US dollars which was refused by

of legal issues in the context of the development of the law of the sea, particularly of the EEZ regime which is relatively new in comparison with the territorial sea regime. Because of this, it is quite understandable that many new issues will arise during its implementation after the entry into force of the LOSC.

Foreign military activities in the EEZ

According to the LOSC, all the seas in the world shall be used peacefully, and any threat or use of force against the territorial integrity or political independence of any State, or in any other manner inconsistent with the principles of international law embodied in the Charter of the United Nations shall be prohibited.[54] From this basic legal principle, military activities with threatening potentials should not be carried out in the EEZs of other countries.

The first question is what military activities are. According to one scholar, military use of oceans consists of two categories: movement rights and operational rights. The former embraces the notion of mobility and includes such legal rights as transit passage through straits used for international navigation, innocent passage in territorial seas and archipelagic waters, and high seas freedom of navigation and overflight, and the latter includes such activities as task force maneuvering, anchoring, intelligence collection and surveillance, military exercises, ordnance testing and firing, and hydrographic and military surveys.[55] For the purpose of this chapter, military activities refer to those activities in the second category as defined earlier, i.e. other than simple navigation or overflight.

As we know, there is a controversy on whether the conduct of military activities in the EEZ of another country is legitimate. Some States may invoke Article 58(1) of the LOSC to justify their military activities in other countries' EEZ. The provision reads: "[i]n the exclusive economic zone, all States, whether coastal or land-locked, enjoy, subject to

the Chinese. For related information, see "State Jurisdiction and Jurisdictional Immunities: Aerial Incident off the Coast of China", 95 AJIL (2001), pp. 631–633.

[54] See Article 301 of the LOSC.

[55] Charles E. Pirtle, "Military Uses of Ocean Space and the Law of the Sea in the New Millennium", *Ocean Development and International Law*, Vol. 31, 2000, 8.

the relevant provisions of this Convention, the freedoms referred to in Article 87 of navigation and overflight and of the laying of submarine cables and pipelines, and other internationally lawful uses of the sea related to these freedoms, such as those associated with the operation of ships, aircraft and submarine cables and pipelines, and compatible with the other provisions of this Convention". Freedoms in the high seas provided in Article 87 are thus applicable to the EEZ as long as they are not contrary to other provisions of the LOSC. According to maritime powers such as the United States, the wording freedoms "associated with the operation of ships, aircraft" implies the legality of naval maneuvers in a foreign EEZ.[56] One view even considers military exercises, aerial reconnaissance and all other activities of military aircraft freedom of high seas if due regard is paid to the rights and interests of third States.[57] As advocated, since the LOSC mainly provides the rights of navigation and overflight, while keeping silent on the rights of military activities, maritime superpower must defend and enforce such rights for its security interests.[58]

Then the question is whether military use constitutes an internationally lawful use of the ocean. The LOSC does not mention the military use so that it becomes a gray area which leads to different interpretations. This no-mention is criticized as one of the major defects in the new LOSC.[59] On the other hand, it is argued that without an express mention in the Convention, military use is hardly regarded as one of such lawful uses. However, such argument may not be convincing. According to a fundamental legal principle, nothing is illegal if there is no law to make it so.[60] Following this, military use is not prohibited since there is no such prohibition in the LOSC. Second, as the LOSC affirms that matters which are not regulated under it should be continually governed by general

---

[56] See Boleslaw Adam Boczek, "Peacetime Military Activities in the Exclusive Economic Zone of Third Countries", *Ocean Development and International Law*, Vol. 19, 1988, 450.

[57] Kay Hailbronner, "Freedom of the Air and the Convention on the Law of the Sea", *American Journal of International Law*, Vol. 77, 1983, 503.

[58] See Pirtle, *supra* note 55, pp. 8–9.

[59] Shao Jin, "Legal Problems concerning Military Use of Exclusive Economic Zones and Continental Shelves", *Chinese Yearbook of International Law* (in Chinese), 1985, p. 183.

[60] For example, "*nullum crimen sine lege*" (no crime without law); and "*nullum crimen nulla poena sine lege*" (no criminal punishment without law).

international law including customary law. If it is traced back to look at history, military activities were consistently allowed under customary international law, though in the implied form. Third, it is admitted that there is a difficulty in inferring that the establishment of the EEZ has limited foreign military operations other than pure navigation and communication from the text and legislative history of Article 58 of the LOSC.[61]

The fact that there is no express provision under the LOSC prohibiting foreign military activities in the EEZ does not mean that they can be conducted in the EEZ without any regulation. It should be borne in mind that the circumstances now are fundamentally different from those in the past. There was and still remains no controversy regarding the military activities conducted in the high seas which was and is open to all. The EEZ is different from the high seas in that it is an area under national jurisdiction. While military activities are allowed there, the factor of national jurisdiction must be taken into account. There should be some kind of check-and-balance mechanism for foreign military activities in the EEZ. It is hard to understand the logic of the argument that while marine scientific research in the EEZ is subject to the consent of the coastal State, military activities can be conducted freely without any check by the coastal State. On the other hand, even if the military use is an internationally lawful use, it can be argued according to the LOSC that it is limited to navigation and overflight, and other rights as provided in Article 87 of the Convention. This can be seen from some domestic EEZ legislations, such as Suriname's.[62]

---

[61] Francesco Francioni, "Peacetime Use of Force, Military Activities, and the New Law of the Sea", *Cornell International Law Journal*, Vol. 18, 1985, 216.

[62] As it provides, all nations, with the observance of the international law, enjoy: ... 4. Freedom to exercise internationally recognized rights in connection with navigation and communication", Article 5 of Law concerning the extension of the territorial sea and the establishment of a contiguous economic zone of 11 June 1978, in Division for Ocean Affairs and the Law of the Sea, Office of Legal Affairs, United Nations, *The Law of the Sea: National Legislation on the Exclusive Economic Zone* (New York: United Nations, 1993), p. 351. Honduras' law contains a similar provision (Article 2 of Decree No. 921 of 13 June 1980 on the Utilization of Marine Natural Resources), see Division for Ocean Affairs and the Law of the Sea, Office of Legal Affairs, United Nations (1993), *Ibid.*, p. 129.

In practice, costal States, including Bangladesh,[63] Brazil, Cape Verde, India,[64] Malaysia,[65] Pakistan[66] and Uruguay[67] explicitly restrict unapproved military exercises or activities in or over their EEZs conducted by other countries. According to the Brazilian law, military exercises or maneuvers, in particular those that imply the use of weapons or explosives, can only be carried out with the consent of the Brazilian

---

[63]"The Government of the People's Republic of Bangladesh understands that the provisions of the Convention do not authorize other States to carry out in the exclusive economic zone and on the continental shelf military exercise or manoeuvres, in particular, those involving the use of weapons or explosives, without the consent of the coastal State". Bangladesh Declaration, available at http://www.un.org/Depts/los/convention_agreements/convention_declarations.htm.

[64]"The Government of the Republic of India understands that the provisions of the Convention do not authorize other States to carry out in the exclusive economic zone and on the continental shelf military exercises or manoeuvres, in particular those involving the use of weapons or explosives without the consent of the coastal State". Indian Declaration, available at http://www.un.org/Depts/los/convention_agreements/convention_declarations.htm.

[65]"The Malaysian Government also understands that the provisions of the Convention do not authorize other States to carry out military exercises or manoeuvres, in particular those involving the use of weapons or explosives in the exclusive economic zone without the consent of the coastal State". Malaysian Declaration, available at http://www.un.org/Depts/los/convention_agreements/convention_declarations.htm.

[66]"It is the understanding of the Government of the Islamic Republic of Pakistan that the provisions of the Convention on the Law of the Sea do not in any way authorize the carrying out in the exclusive economic zone and in the continental shelf of any coastal State military exercises or manoeuvres by other States, in particular where the use of weapons or explosives is involved, without the consent of the coastal State concerned". Pakistani Declaration, available at http://www.un.org/Depts/los/convention_agreements/convention_declarations.htm.

[67]"In the exclusive economic zone, enjoyment of the freedom of international communication in accordance with the way it is defined and in accordance with other relevant provisions of the Convention excludes any non-peaceful use without the consent of the coastal State — for instance, military exercises or other activities which may affect the rights or interests of that State; and it also excludes the threat or use of force against the territorial integrity, political independence, peace or security of the coastal State". Uruguay Declaration, available at http://www.un.org/Depts/los/convention_agreements/convention_declarations.htm.

Government.[68] Brazil is perhaps the most adamant country which strictly regulates foreign military activities in its EEZ. As early as December 1982 when Brazil signed the LOSC, it made a statement of this kind which was reiterated several times afterward. The United States reacted to it on each occasion by protesting against Brazil's restrictions and stating its reservation of military exercises in Brazil's EEZ as internationally lawful uses of the ocean.[69] The other typical country is Iran which also lays down laws restricting foreign military activities in its EEZ by stipulating that "[f]oreign military activities and practices, collection of information and any other activity inconsistent with the rights and interests of the Islamic Republic of Iran in the exclusive economic zone and the

[68]Article 9 of the Act concerning the Territorial Sea, the Contiguous Zone, the Exclusive Economic Zone and the Continental Shelf of Brazil and other provisions: Act No. 8617 of 4 January 1993, in Division for Ocean Affairs and the Law of the Sea, Office of Legal Affairs, United Nations (1993), *supra* note 62, p. 38.

[69]For details, see Roach and Smith, *United States Responses to Excessive Maritime Claims*, *supra* note 37, pp. 409–413. In addition to the United States, other Western countries are also against the restriction of foreign military activities in the EEZ: Germany ("According to the Convention, the coastal State does not enjoy residual rights in the exclusive economic zone. In particular, the rights and jurisdiction of the coastal State in such zone do not include the rights to obtain notification of military exercises or manoeuvres or to authorize them"); Italy ("According to the Convention, the coastal State does not enjoy residual rights in the exclusive economic zone. In particular, the rights and jurisdiction of the coastal State in such zone do not include the right to obtain notification of military exercises or manoeuvres or to authorize them"); Netherlands ("The Convention does not authorize the coastal State to prohibit military exercises in its exclusive economic zone. The rights of the coastal State in its exclusive economic zone are listed in Article 56 of the Convention, and no such authority is given to the coastal State. In the exclusive economic zone all States enjoy the freedoms of navigation and overflight, subject to the relevant provisions of the Convention"); and the United Kingdom ("The United Kingdom considers that declarations and statements not in conformity with Articles 309 and 310 include, *inter alia*, the following: those which are not in conformity with the provisions of the Convention relating to the exclusive economic zone or the continental shelf, including those which claim coastal state jurisdiction over all installations and structures in the exclusive economic zone or on the continental shelf, and those which purport to require consent for exercises or manoeuvres (including weapons exercises) in those areas"); available at http://www.un.org/Depts/los/convention_agreements/convention_declarations.htm.

continental shelf are prohibited".[70] Because of this legal provision, there was a diplomatic row between Iran and the United States. The United States lodged a protest against it by stating that the prohibition of military activities contravenes international law and the United States reserves its rights in this regard. In reply to the United States protest, the Iranian diplomatic note states that due to the multiplicity of economic activities, it is possible that such activities, for which the coastal State enjoys sovereign rights, could be harmed by military practices and maneuvers; accordingly, those practices which affect the economic activities in the EEZ and the continental shelf are thus prohibited.[71] It is interesting to note that the Iranian explanation does not deny the right of foreign military activities in the EEZ and the only reason for their prohibition results in their possible harm to economic activities there.

In comparison, some restrictions are milder, such as the law of Cape Verde which requires foreign entities and ships exercising high seas freedom or conducting other activities to respect its sovereign rights, its laws and regulations and the principle of the peaceful uses of the oceans.[72]

The regulations aforementioned are made under the rationale that military activities are inherently potential threats to peace and good order of the coastal States. While such regulations are understandable, it should be borne in mind that not all military activities are threatening. Contrarily, some military activities, such as the activities undertaken by the UN peacekeeping forces, are indispensable to maintain peace and good order. In the same thinking, some civilian activities may be threatening and this can be illustrated by a severe marine pollution accident caused by a

---

[70] Article 16 of Act on the Marine Areas of the Islamic Republic of Iran in the Persian Gulf and the Oman Sea, 1993, in Division for Ocean Affairs and the Law of the Sea, Office of Legal Affairs, United Nations, *The Law of the Sea: Current Developments in State Practice No. IV* (New York: United Nations, 1995), p. 67.

[71] See Protest from the United States of America, 11 January 1994; and Comments from the Islamic Republic of Iran concerning the viewpoints of the Government of the United States of America regarding the Act on Marine Areas in the Persian Gulf and the Oman Sea, in Division for Ocean Affairs and the Law of the Sea, Office of Legal Affairs, United Nations (1995); *Ibid.*, pp. 147–151.

[72] Articles 14, 15 and 26 of Law No.60/IV/92 delimiting the maritime areas of the Republic of Cape Verde and revoking Decree-Law No. 126/77 and all legal provisions which contravene this law, in Division for Ocean Affairs and the Law of the Sea, Office of Legal Affairs, United Nations (1995); *Ibid.*, pp. 37, 40.

civilian activity or illegal fishing in the EEZ. In such context, what we should look into is not the form of a certain activity, but its nature. If a military activity is threatening in nature and with clear bad intention and/or in a hostile manner, it should be banned in the EEZ. Otherwise, it can be allowed under certain conditions laid down by the coastal State, similar to the marine scientific research regime under the LOSC. There is no reason why the coastal State is prevented from regulating foreign military activities in its EEZ while it is allowed to regulate foreign marine scientific research there.

There is a discrepancy regarding the concept of the EEZ between the legal term and the operational term. The United States navy divides the ocean into two categories: national waters and international waters, for operational and mobility purposes.[73] The EEZ is accordingly categorized as "international waters". However, it must be pointed out that it is only an expression for operational purposes, thus in no way affecting the legal nature of the EEZ as a maritime zone within national jurisdiction under the LOSC.

It is worth mentioning that the East–West Center once organized several workshops on "military and intelligence gathering activities in the EEZ". The launch of this series of workshops was triggered by the EP-3 Incident between China and the United States. The first one was held in Bali, Indonesia in June 2002, which focused on identifying disagreements and contrasting positions as well as on areas of possible mutual understanding and agreement.[74] The Tokyo Meeting in February 2003 acknowledged that with the technology advances in the EEZs, intelligence gathering activities would increase.[75] The Honolulu Meeting in December 2003 went further and some guidelines for military and intelligence gathering activities in the EEZs were drafted, reflecting the disagreement

[73]National waters include internal waters, territorial seas and archipelagic waters, and international waters include contiguous zones, EEZ, high seas and security zones. See Department of the Navy, *The Commander's Handbook on the Law of Naval Operations*, October 1995, pp. 1-4–1-6.

[74]For details, see East-West Center, *Military and Intelligence Gathering Activities in Exclusive Economic Zones: Consensus and Disagreement: A Summary of the Bali Dialogue* (Honolulu: East-West Center, 2002).

[75]For details, see East-West Center, *The Regime of the Exclusive Economic Zones: Issues and Responses: A Report of the Tokyo Meeting* (Honolulu: East-West Center, 2003).

between maritime powers and developing coastal countries.[76] Based on the aforementioned series workshops, the Guidelines for navigation and overflight in the EEZ was finally adopted in September 2005.[77] Through these efforts, it is hoped that some consensus can be reached in the world community regarding military and intelligence gathering activities in the EEZs, in particular in connection to a possible review of the LOSC after its entry into force for 10 years.

Hydrographic surveying and marine scientific research

There is another gray area, which is also related to military activities in the EEZ addressed previously, i.e. hydrographic surveying when it is undertaken for some military purposes. The International Hydrographic Bureau defines a "hydrographic surveying" as "a survey having for its principal purpose the determination of data relating to bodies of water. A hydrographic surveying may consist of the determination of one or several of the following classes of data: depth of water, configuration and nature of the bottom; directions and force of currents; heights and times of tides and water stages; and location of topographic features and fixed objects for survey and navigation purposes".[78] Another definition is offered by Sam Bateman: "A hydrographic surveying is the collection of information in coastal or relatively shallow areas for the purpose of making nautical charts and similar products to support safety of navigation".[79] Although some wordings like "survey activities" and "hydrographic survey" appear in the LOSC, the Convention does not contain any provision specifically governing this kind of marine activity. The gray and ambiguous area is further widened in the context of the relevant provisions of the LOSC regarding marine scientific research (MSR).

The LOSC has one chapter governing MSR, but no definition on it. During the UNCLOS III, the Chairman of the Third Committee included the following definition in the ISNT, Part III: "Marine scientific research

[76] See Hasjim Djalal, Alexander Yankov and Anthony Bergin, "Draft Guidelines for Military and Intelligence Gathering Activities in the EEZ and their Means and Manner of Implementation and Enforcement", *Marine Policy*, Vol. 29(2), 2005, 175–183.

[77] Text is available at https://nippon.zaidan.info/seikabutsu/2005/00816/pdf/0001.pdf.

[78] Cited in Sam Bateman, "Hydrographic Surveying in the EEZ: Differences and Overlaps with Marine Scientific Research", *Marine Policy*, Vol. 29(2), 2005, 167.

[79] *Ibid.*, p. 167.

means any study or related experimental work designed to increase man's knowledge of the marine environment".[80] However, such a definition was not finally included in the adopted Convention. The United States tried to define MSR as "the general term most often used to describe those activities undertaken in the ocean and coastal waters to expand scientific knowledge of the marine environment and its processes".[81] Some scholars have tried to distinguish fundamental MSR from applied MSR in terms of application of the LOSC: whereas the former has to be granted in normal circumstances, the latter is subject to the coastal State's full discretion.[82] However, such a distinction seems not provide any help to clarify the unclear situation, and even the distinction itself causes problems since there is no clear-cut demarcation to define an MSR as fundamental or applied. The ambiguity of the MSR definition may cause problems for researching States, but have little adverse impact on coastal States since they have a great discretionary power to decide whether an MSR is acceptable under Article 246 of the LOSC.

It is acknowledged that MSR and hydrographic surveying overlap to some extent.[83] There is a reference in the LOSC which may help to some extent clarify the gray area existing between MSR and hydrographic surveying. According to Article 19(2)(j), "the carrying out of research or survey activities" is prohibited when a foreign vessel exercises the right of innocent passage through the territorial sea. This legal expression contains the following two meanings: on one hand, it distinguishes "research activity" from "survey activity", but on the other these two activities are given the same treatment. It is called thus that "hydrographic surveying should only be conducted in the EEZ with the permission of the coastal State".[84]

In State practice, while the United States and the United Kingdom take the view that military hydrographic surveying is part of the freedoms

---

[80] See Alfred H.A. Soons, *Marine Scientific Research and the Law of the Sea* (Deventer, Netherlands: Kluwer Law and Taxation Publishers, 1982), p. 123.

[81] Roach and Smith, *United States Responses to Excessive Maritime Claims*, *supra* note 37, p. 425.

[82] Wolf Plesman and Volker Röben, "Marine Scientific Research: State Practice versus Law of the Sea?" in Rüdiger Wolfrum (ed.), *Law of the Sea at the Crossroads: The Continuing Search for a Universally Accepted Regime* (Berlin: Duncker und Humblot, 1991), p. 375.

[83] Bateman, *supra* note 78, p. 172.

[84] *Ibid.*, p. 172.

of high seas related to "international lawful uses of the sea", Australia and Canada "are understood to seek permission of the coastal State before conducting hydrographic surveying in the EEZ of that State".[85] Some coastal countries like China hold the view that hydrographic surveying is part of MSR and have specific laws governing both MSR and hydrographic surveying.

In 2001, USNS *Bowditch*, an American military surveillance vessel, entered into China's EEZ in the Yellow Sea three times to carry out hydrographic surveying and confronted by Chinese Ocean Surveillance vessels.[86] Chinese side sent monitoring vessels to follow the American ship and attempted to disrupt its operations. Even in one time a Chinese warship forced the *Bowditch* to leave the Chinese EEZ.[87] After the *Bowditch* Incident, China has tightened its control of hydrographic surveying in the maritime zones of its national jurisdiction. One move in legislation is the revision of the 1992 Law of Surveying and Mapping which was taken in 2002. The original provision governing foreign surveying activities has been revised as "Surveying and mapping to be conducted within the territory and other sea areas under the jurisdiction of the People's Republic of China by a foreign organization or individual must be approved by the competent department of surveying and mapping administration under the State Council together with the competent department of surveying and mapping administration of the Army, and shall comply with relevant laws and administrative rules and regulations of the People's Republic of China. A foreign organization or individual that conducts surveying and mapping within the territory of the People's Republic of China must adopt the form of joint venture or of cooperation with a relevant department or unit of the People's Republic of China, and shall not involve any State

[85] *Ibid.*, p. 170.

[86] For details, see You Zhiyong and Zhang Youfeng, "Case Analysis of Foreign-Related Marine Law Enforcement in the East China Sea Zone", in Xiamen University Ocean Law Centre (ed.), *Collected Papers of the Symposium in Commemorating the 20th Anniversary of the Adoption of the UN Convention on the Law of the Sea* (Xiamen University, August 2002) (in Chinese), pp. 81–86.

[87] See John Leicester, "Chinese chase US ship; Jet crash part of spy game", *Herald Sun*, 4 April 2001, at 32; cited in Captain George V. Galdorisi and Commander Alan G. Kaufman, "Military Activities in the Exclusive Economic Zone: Preventing Uncertainty and Defusing Conflict", *California Western International Law Journal*, Vol. 32, 2002, 294.

secret and endanger the State security".[88] The new provision contains some new changes in comparison with the old provision: (1) the original Article 19 has been moved ahead to be Article 7; (2) originally, such activity is only subject to the approval of the competent department of surveying and mapping administration under the State Council but now the approval is made by this department together with its counterpart in the Army; (3) originally foreign organization or individual may conduct surveying and mapping alone within China's territory, but now such activity must be conducted in cooperation with the Chinese counterpart. It is noted that while some more restrictions are imposed, surveying and mapping by a foreign organization or individual may still be conducted alone after China's approval. Clearly, in China's eye, hydrographic surveying is not something that freedom of the seas should apply.

The other reason for the revision of the Surveying and Mapping Law is connected to the implementation of the 1996 Regulations on the Management of the Foreign-Related Marine Scientific Research, which came into force from 1 October 1996.[89] The Regulations apply to

---

[88]Article 7 of the 2002 amended Law of Surveying and Mapping, *People's Daily* (in Chinese), 1 September 2002, p. 5. The amended Law was adopted on 29 August 2002 and came into force from 1 December 2002.

In comparison, the original Article 19 reads "Surveying and mapping to be conducted in the territorial air, land and waters, as well as other sea areas under the jurisdiction of the People's Republic of China by a foreign organization or individual alone or in cooperation with the relevant department or unit of the People's Republic of China shall be subject to the approval by the Government of the People's Republic of China or by the department authorized by it.

A foreign organization or individual that with due approval conducts surveying and mapping in the territorial air, land and waters, as well as other sea areas under the jurisdiction of the People's Republic of China either alone or in cooperation with the relevant department or unit of the People's Republic of China, must comply with relevant laws and administrative rules and regulations of the People's Republic of China and must submit two duplicate copies of the complete surveying and mapping results to the competent department of surveying and mapping administration under the State Council".

The English version of the 1992 Law is available in Office of Policy, Law and Regulation, State Oceanic Administration (ed.), *Collection of the Sea Laws and Regulations of the People's Republic of China*, 3rd ed. (Beijing: Ocean Press, 2001), pp. 300–313.

[89]It is reprinted in *Gazette of the State Council of the People's Republic of China* (in Chinese), 1996. English version is available in Office of Law, Policy and Regulation, *ibid.*, pp. 328–333.

the conduct of survey activities by international organizations, organizations and individuals of any foreign country within sea areas under China's jurisdiction. It is recalled that in the *Bowditch* Incident, the Chinese side asked the *Bowditch* to stop its illegal activities without China's approval under the aforementioned Regulations, but the American vessel only replied that it was doing military hydrographic surveying in international waters.[90] When China addressed the issue to the American side through diplomatic channel, the United States responded that military surveying was not MSR so that it was not subject to the LOSC and to the approval of the coastal State.[91] Clearly, China learnt a lesson from this Incident that the relevant laws and regulations governing MSR did not work in practice with the Americans when they carried out military hydrographic surveying in China's jurisdictional waters.

Unlike China, the United States has a unique legal practice. According to its relevant regulations, the United States promised not to exercise its jurisdiction over the foreign MSR within its EEZ, and foreign nationals or vessels thus do not require permission to undertake MSR in the US EEZ.[92] However, the United States applies the consent system to the MSR on its continental shelf and within its territorial sea because it is a party both to the 1958 Convention on the Continental Shelf and the 1958 Convention on the Territorial Sea. It is not clear whether the United States will change its position after it ratifies the LOSC.

Perhaps the United States has hoped that all the other countries in the world can follow its suit. However, in reality, not so many countries possess the capability to do MSR. The open access regime only benefits big maritime powers and it may not be attractive to most of the developing countries. Due to some excessive State practice based on the unilateral judgment of the United States, an American government officer proposed to establish a freedom of MSR program analogous to the Freedom of

---

[90] See Su Tao and Lin Fangzhong, "Discussion on the Law Governing Military Survey Activities of a Third State in the Sea Areas under the Jurisdiction of the Coastal State", in Li Younqi and Luo Shouben (eds.), *Studies on Basic Issues in the Management of Sea Uses* (Qingdao: Qingdao Ocean University Press, 2002) (in Chinese), pp. 208–210.

[91] You Zhiyong and Zhang Youfeng, *supra* note 86, pp. 84–85.

[92] See "Note to the United Nations of 13 October 1988 on United States Practice Relating to Marine Scientific Research", in United Nations, *The Law of the Sea: National Legislation, Regulations and Supplementary Documents on Marine Scientific Research in Areas Under National Jurisdiction* (New York: United Nations, 1989), p. 276.

Navigation Program to challenge the relevant State regulations and practice so as to guarantee the freedom of MSR.[93] There would be more problems if such a suggestion were turned into a practice. It must be borne in mind that freedom of MSR has no longer existed within the jurisdictional waters of a coastal State. Such a program may not be practical and reasonable.

Finally, if hydrographic surveying, whether military or civilian, was not considered part of MSR, then there should be a need to establish a new legal regime to govern it. As long as the gray area exists, problems regarding survey activities in the EEZ between the coastal State and the conducting State will continue to arise. If hydrographic surveying can be regarded as part of MSR, military surveying could be then considered as abuse of maritime rights of the law of the sea. It is interesting to note that two Chinese marine law enforcers revealed in their article that the *Bowditch* had applied to the Chinese authority for hydrographic surveying but due to some reasons its application was not approved.[94]

Defending or Weakening the International Marine Legal Order?

Freedom of Navigation Program (FON): Defending rights under customary law or maintaining maritime hegemony?

The FON Program was established in 1979 and used to challenge State practice which, in the eye of the United States, was inconsistent with international law including the LOSC as well as to assert the American rights, especially the right of innocent passage in other countries' territorial seas. According to a previously confidential but now unclassified document, *National Security Directive 49* (1990), the FON Program "combines diplomatic action and operational assertions of our navigation and overflight rights to encourage modification of, and to demonstrate non-acquiescence in, maritime claims that are inconsistent with the customary rules of international law pertaining to maritime navigation and

---

[93] See J. Ashley Roach, "Marine Scientific Research and the New Law of the Sea", *Ocean Development and International Law*, Vol. 27, 1996, 68.

[94] You Zhiyong and Zhang Youfeng, *supra* note 86, p. 85.

overflight freedoms".[95] The excessive maritime claims which are subject to FON challenge include: (1) historic bay/historic water claims not recognized by the United States; (2) territorial sea baseline claims not drawn in accordance with the LOSC; (3) territorial sea claims which contain, *inter alia*, requirements for advance notification or authorization for innocent passage for warships, or are in excess of 12 nm; (4) other claims to jurisdiction over maritime areas in excess of 12 nm; and (5) archipelagic claims not in conformance with the LOSC.[96] The authority to implement the FON Program is jointly shared by the Department of State (which acts assertively to preserve US navigation and overflight rights under international law, including lodging protests against excessive maritime claims to avoid legal acquiescence) and the Department of Defense (which plans and administers the operational assertion portion of the FON Program).[97]

Some information about the implementation of this Program is contained in the Department of Defense website. In the Fiscal Year 2019, United States challenged the excessive maritime claims of 22 claimants and its list included several East Asian countries: Burma (excessive straight baselines); Cambodia (excessive straight baselines); China (straight baseline claims; restrictions on foreign aircraft flying through an Air Defense Identification Zone (ADIZ) without the intent to enter national airspace; criminalization of survey activity by foreign entities in the exclusive economic zone; jurisdiction over airspace above the exclusive economic zone; security jurisdiction over the contiguous zone; prior permission required for innocent passage of foreign military ships through the territorial sea; territorial sea and airspace around features not so entitled (i.e. low-tide elevations));Thailand (prior consent required for military exercises or other activities which may affect the rights or interests of the coastal State in the exclusive economic zone); Vietnam (prior permission for warship to enter the territorial sea).[98]

---

[95] National Security Directive 49, 12 October 1990, available at http://bushlibrary.tamu.edu/research/nsd/NSD/NSD%2049/0001.pdf.

[96] *Ibid.*

[97] *Ibid.* The paragraphs about the procedures of operations in the document were colored in black.

[98] 'Annual Freedom of Navigation Report Fiscal Year 2019', available at https://policy.defense.gov/Portals/11/Documents/FY19%20DoD%20FON%20Report%20FINAL.pdf?ver=2020-07-14-140514-643&timestamp=1594749943344.

It is interesting to note that among all the East Asian countries China was challenged during that year for various reasons. This may be related to the deteriorating bilateral relations between China and the US, and also the US' more frequent military activities in the South China Sea. It is noted that in 1998 the two military sides of China and the United States concluded an agreement to establish a consultation mechanism to strengthen military maritime safety and "to promote common understanding regarding activities undertaken by their respective maritime and air forces when operating in accordance with international law, including the principles and regimes reflected in the United Nations Convention on the Law of the Sea".[99] However, there is no sign to indicate that China has made concessions because of the challenges of the FON Program since all the relevant Chinese laws and regulations remain unchanged.

On the other hand, the United States has realized that the FON Program may cause problems at sea or in the air with other countries in its actual operational implementation. There was a serious bumping incident in the Black Sea between the American and Soviet warships in the late 1980s.[100] In order to minimize potential negative impacts in practice, the US Pacific Command prepared the so-called Cooperative Engagement Strategy for the Asia-Pacific region and one of the components in this Strategy is "Operational Law Exchanges" which includes the exchange of views on navigational freedoms embodied in the LOSC, US ocean policy and the Freedom of Navigation Program.[101]

While it is asserted that the FON Program may bring some positive impacts on the development of the international law of the sea, in

---

[99] Agreement between the Department of Defense of the United States and the Ministry of National Defense of the People's Republic of China Establishing a Consultation Mechanism to Strengthen Military Maritime Safety, 19 January 1998. Text is reprinted in 37 I.L.M. 530 (1998).

[100] For details, see W.J. Aceves, "Ambiguities in Plurilingual Treaties: A Case Study of Article 22 of the 1982 Law of the Sea Convention", *Ocean Development and International Law*, Vol. 27, 1996, pp. 187ff. Another incident occurred two years ago on 10 March 1986 with the same American naval vessels. For details, see W.E. Butler, "Innocent Passage and the 1982 Convention: The Influence of Soviet Law and Policy", *American Journal of International Law*, Vol. 81, 1987, 343–346.

[101] See Captain S.W. Horton, "U.S. Pacific Command's Cooperative Engagement Strategy and Freedom of Navigation in the East China Sea", in Myron H. Nordquist and John Norton Moore (eds.), *Security Flashpoints: Oil, Islands, Sea Access and Military Confrontation* (The Hague: Martinus Nijhoff, 1998), p. 445.

particular enhancing the harmonization of State practice with the LOSC, it is mainly designed to preserve the US interest and the global mobility of its naval and air forces. The so-called excessive maritime claims are defined by the United States and in accordance with its own criteria. While some of its criteria are based on universally accepted international law, some are obviously controversial. In this sense, it is doubted that the FON Program can really defend and consolidate the marine legal order centered on the LOSC.

Proliferation Security Initiative (PSI): New rules of international law?

PSI is an effort to consider possible collective measures among the participating countries, in accordance with national legal authorities and relevant international law and frameworks, in order to prevent the proliferation of Weapons of Mass Destruction (WMD), missiles and their related materials that pose threats to the peace and stability of the international community. It was first put forward by President Bush on 31 May 2003 during a speech in Krakow, Poland.[102] The PSI is administered by the "core group" countries, which consist of 11 countries (Japan, US, UK, Italy, the Netherlands, Australia, France, Germany, Spain, Poland, Portugal),[103] and later another four countries joined (Canada, Singapore, Russia and Norway). Other 46 countries supported it as well.[104]

---

[102] "Remarks by the president to the people of Poland, Wawel Royal Castle, Krakow, Poland, 31 May 2003", available at http://www.whitehouse.gov/news/releases/2003/05/20030531–3.html ("When weapons of mass destruction or their components are in transit, we must have the means and authority to seize them. So today I announce a new effort to fight proliferation called the Proliferation Security Initiative. The United States and a number of our close allies, including Poland, have begun working on new agreements to search planes and ships carrying suspect cargo and to seize illegal weapons or missile technologies. Over time, we will extend this partnership as broadly as possible to keep the world's most destructive weapons away from our shores and out of the hands of our common enemies".)

[103] See "The Proliferation Security Initiative (PSI) Maritime Interdiction Exercise hosted by Japan", 18 October 2004, available at http://www.mofa.go.jp/policy/un/disarmament/arms/psi/exercise-2.html.

[104] J. Ashley Roach, "Proliferation Security Initiative (PSI): Countering Proliferation by sea", paper presented to the Conference on "*New Developments of the Law of the Sea and China*", Xiamen, China, 9 March 2012, 2005, p. 3. (on file with the author)

In September 2003, the United States published the principles of PSI. Accordingly,

> The PSI is a broad international partnership of countries which, using their own laws and resources, will coordinate their actions to halt shipments of dangerous technologies to and from states and non-state actors of proliferation concern — at sea, in the air, and on land. The PSI will reinforce, not replace, other nonproliferation mechanisms. Cooperative and coordinated efforts by participating countries will give strength and substance to the broad political consensus against proliferation and help address an increasingly important challenge to international security. The United States is encouraged that all participants have agreed in Paris to abide by these Principles. We support the expansion of PSI to all responsible nations willing to accept the Principles, and will seek the involvement in PSI of such countries.[105]

Since the core element in the PSI is interdiction, the interdiction principles are also laid out together with the PSI principles that participating States are to: (1) undertake effective measures, either alone or in concert with other states, for interdicting the transfer or transport of WMD, their delivery systems, and related materials to and from states and non-state actors of proliferation concern; (2) adopt streamlined procedures for rapid exchange of relevant information concerning suspected proliferation activity; (3) review and work to strengthen their relevant national legal authorities where necessary; and (4) take specific actions in support of interdiction efforts regarding cargoes of WMD, their delivery systems, or related materials.[106] Although the PSI does not target any particular country, countries of concern include North Korea, Iran and Syria.

PSI raises some issues in international law. In the maritime domain, the LOSC grants the exclusive jurisdiction and control of vessels to the flag State on the high seas, and any search or visit is subject to the permission of the flag State. Exceptions do exist in the LOSC, but they are limited to the suppression of piracy, transport of slave, illicit traffic in

[105] "Statement by the Press Secretary: Principles for the Proliferation Security Initiative", 4 September 2003, available at http://www.state.gov/t/np/rls/prsrl/23809.htm.

[106] See "Proliferation Security Initiative: Statement of interdiction principles", 4 September 2003, available at http://www.state.gov/t/np/rls/prsrl/23764.htm.

narcotic drugs, or unauthorized broadcasting.[107] Besides, a warship with reasonable ground may visit a ship without nationality.[108] As commented, "Article 110 is explicitly exhaustive of the bases for interdiction on the high seas and thus creates a significant legal stumbling block to any assertion of authority to interdict foreign ships and aircraft under the PSI".[109] Since there is no expression rule in international law regarding the interdiction of ships suspected of carrying WMD, Chinese scholars perceive such interdiction on the high seas as a grave destruction of the international legal system and a great downgrade of the solemnity of international law.[110] Because of the controversial issues existing in PSI, it is called to improve the PSI by institutionalization.[111]

In addition to the efforts at the international level, the United States has tried to conclude bilateral agreements with relevant countries so as to enhance the effectiveness of the implementation of the PSI. For this purpose, the United States signed non-proliferation shipboarding agreements with Liberia (11 February 2004), Panama (12 May 2004) and Marshall Islands (13 August 2004). According to these agreements, if a ship with either party's flag is suspected of carrying proliferation-related cargo, either party can request the other to confirm the nationality of the ship and if needed, to authorize the boarding, search and possible detention of the ship and its cargo. These agreements, together with PSI partners, cover more than 50% of commercial shipping fleet dead weight tonnage, which is subject to rapid action consent procedures for boarding, search and seizure by the United States.[112] It seems that the United States attempted to create a rule of international law through such kind of agreements by

---

[107] See Articles 99–109 of the LOSC.

[108] See Article 110(1)(d) of the LOSC.

[109] Daniel H. Joyner, "The Proliferation Security Initiative: Nonproliferation, Counterproliferation, and International Law", *Yale Journal of International Law*, Vol. 30, 2005, 534.

[110] See Gao Ying and Sun Bo, "The United States' Proliferation Security Initiative", in China Association of Arms Control and Disarmament (ed.), *2004 Report on International Arms Control and Disarmament* (Beijing: World Knowledge Press, 2004) (in Chinese), p. 170.

[111] See Jack I. Garvey, "The International Institutional Imperative for Countering the Spread of Weapons of Mass Destruction: Assessing the Proliferation Security Initiative", *Journal of Conflict and Security Law*, Vol. 10(2), 2005, 125–147.

[112] See "US–RMI Agreement: Maritime Security Initiative Signed", 15 August 2004, available at http://www.yokwe.net.

shifting its foreign policy "toward a more flexible approach to collective action that eschews both *ad hoc* unilateralism and institutionalized multilateralism".[113]

China's feeling is complex toward PSI. On the one hand, China needs the safety of navigation in the Strait of Malacca since most of China's imported oil reaches China through this strait, but China is concerned with the American intervention in the Strait in that when the military presence of the United States in the Strait may block China's oil transportation once their bilateral relations deteriorate due to some disputes, say, concerning the Taiwan issue. For that reason, China has been considering the construction of an oil pipeline from Myanmar to China to reduce its dependence on shipping oil imports through the Strait of Malacca.[114] As stated by a Chinese diplomat, maritime security is of vital importance for the welfare and economic development of the region, and regional cooperation is indispensable for maritime security.[115] However, China has the doubts whether principles embodied in international law including the UN Charter "would be or could be strictly observed in real actions against maritime threats. Extreme care and sensitiveness is needed when it comes to military involvement".[116] In a word, China prefers a regional arrangement of maritime security for the Strait of Malacca.

In addition, China had a bitter experience regarding the interdiction at sea. On 7 July 1993, the Chinese freighter the *Milky Way* (*yinhe*) departed from Tianjin to the Middle East with the destination of port of Kuwait. On 23 July, the United States accused China of sending to Iran chemical materials (thiodiglycol and thionyl chloride) to be used for making chemical weapons and demanded an inspection on board by the American side. From 1 August, several American warships followed the *Milky Way*, monitoring its movement and taking photos so that the Chinese ship could

---

[113] Michael Byers, "Policing the High Seas: The Proliferation Security Initiative", *American Journal of International Law*, Vol. 98, 2004, 543.

[114] See "Chinese scholars propose way to reduce dependence on Malaccan Strait", *Alexander's Oil & Gas Connections*, Vol. 9(15), 4 August 2004, available at http://www.gasandoil.com/goc/news/nts43127.htm.

[115] Zhao Jianhua, "The Straits of Malacca and challenges ahead: China's perspective", paper presented to the Conference on "Straits of Malacca: Building a Comprehensive Security Environment", 11–13 October 2004, Kuala Lumpur (on file with the author).

[116] *Ibid.*, p. 4.

not sail normally. Due to the American pressures, the *Milky Way* was refused to anchor on port and remained on the high seas until late August. Finally, on 29 August, China agreed to check the containers on board together with the Saudi Arabia representative and American experts. The investigation result proved that there were no chemical materials as the United States alleged. On 4 September 1993, the Chinese Foreign Ministry issued a statement, condemning the American hegemonic attitude and its groundless accusation.[117]

Chinese legal scholars accused the United States of having violated international law including the freedom of the high seas and that the United States should bear legal liability to compensate for the economic loss of the Chinese freighter.[118] From this incident, it is well assumed that China is very averse to the PSI when military interdiction at sea is involved. This also can be seen from China's basic policy regarding the South China Sea as China has always reiterated that it will maintain the freedom of navigation and overflight in the South China Sea; and it has never interfered in such free passage and will not do it in future.[119]

As for the issue of PSI, China expressed its position for the first time by its Foreign Ministry spokesperson stating that

> We agree with the objective of PSI. But the Chinese side thinks that relevant measures under PSI should be taken within the reign of international law in accordance with relevant principles of the UN Charter. We hold reservation on the PSI's possibility of taking coercive interception beyond the reign of international law.[120]

---

[117] See "Milky Way event", available at http://www.china.org.cn/chinese/HIAW/109186.htm; and "The United States forced our Yinhe Freighter to stay on the high seas for 33 days", available at http://news.163.com/2004w09/12686/2004w09_1096092272435.html.

[118] For details, see Zhao Lihai, *Studies on the Law of the Sea Issues* (Beijing: Peking University Press, 1996) (in Chinese), pp. 237–244.

[119] See "China's basic position on the South China Sea issue and its policy assertion on the settlement of the South China Sea dispute", *China Ocean News* (in Chinese), 23 April 2004.

[120] Foreign Ministry Spokesperson Zhang Qiyue's Press Conference on 4 November 2004, available at www.fmprc.gov.cn/eng/xwfw/s2510/t169072.htm.

At the same time, China did not close its door to PSI by stating that "we would like to exchange views with relevant countries on this point".[121] On the other hand, for some reasons, China might support military interdiction or naval inspection in future. A Taiwanese scholar once wrote a piece about how China can stop the United States from selling weapons to Taiwan. One of the alternatives, as he proposed, could be the use of the American experiences in blockading Cuba in the 1960s. China can select some American merchant vessels for inspection by applying the August 17 Communiqué and also learn lessons from the *Milky Way* incident on how to negotiate with the Americans.[122] The American-sponsored PSI, once fully implemented in East Asia, may give China the right of inspection and do the same thing toward American vessels more justifiably.

While RMSI mentioned in Chapter 1 has waned, PSI has been gaining more ground in international law as it develops. The first significant document which gives the PSI some kind of legal support is the United Nations Security Council Resolution 1540, adopted on 28 April 2004, which provides that "all States shall refrain from providing any form of support to non-State actors that attempt to develop, acquire, manufacture, possess, transport, transfer or use nuclear, chemical or biological weapons and their means of delivery", and the Security Council "calls upon all States, in accordance with their national legal authorities and legislation and consistent with international law, to take cooperative action to prevent illicit trafficking in nuclear, chemical or biological weapons, their means of delivery, and related materials".[123] However, it is to be noted that the Resolution only mentions the "non-State actors", which it is most likely means terrorist groups, but the PSI seems to target some so-called "rogue States" based on the intention of the United States. Furthermore, it is argued that "the text of Operative Paragraph 10 does not bestow any additional authority upon states to enforce the PSI and does not exempt states from any international legal obligations they otherwise have".[124]

While it is debated whether Resolution 1540 has provided some legal ground for the PSI, the amendments made in October 2005 to the 1988

---

[121] *Ibid.*

[122] See Shih Chi-yu, "Beijing may inspect American merchant vessels", *Lianhe Zaobao* (in Chinese), 17 August 2004.

[123] UN Doc. S/RES/1540 (2004).

[124] Daniel H. Joyner, *supra* note 109, p. 541.

Convention for the Suppression of Unlawful Acts against the Safety of Maritime Navigation (SUA Convention) have clearly shown the legalization process of the PSI. The review conference, sponsored by IMO, held in October 2005, adopted the 2005 Protocol to the SUA Convention.[125] The interdiction principles embodied in the PSI once legalized will have ramifications for the law of the sea developments and affect the freedom of navigation and overflight. In return, there may be some spillover unfavorable effect on the United States-asserted maritime rights in the EEZs of other coastal States.

The United States proposed declarations and understandings

The Senate Advice and Consent resolution before the US Senate contains a list of 26 declarations and understandings concerning various matters with regards the American interests.[126] While most of the declarations and understandings simply reaffirm the United States existing ocean policy and position, a few of them do raise new issues in the law of the sea. The most controversial one, in this author's eye, is regarding the definition of "military activities": "the United States further declares that its consent to accession to the Convention is conditioned upon the understanding that, under Article 298(1)(b), each State Party has the exclusive right to determine whether its activities are or were 'military activities' and that such determinations are not subject to review".[127] This unilateral and arbitrary assertion will obviously invite protests and opposition from other countries. It is a common sense that any term or definition in an international treaty and/or in relation to the implementation of the treaty is subject to the agreement and consensus of all States Parties concerned. Any unilateral assertion cannot change the terms and conditions of the treaty. In this sense, it can be regarded as an excessive claim. On the other hand, it is commented that this declaration would "have the potential to destabilize the legal order at sea if followed by other state parties to the UNCLOS".[128] It would also cause the phenomenon of "excessive claims beget

[125]For relevant details, see Chapter 5.

[126]The list is annexed to Duff, *supra* note 5, pp. 214–219.

[127]*Ibid.*, p. 215.

[128]Yann-huei Song, "Declarations and Statements with Respect to the 1982 UNCLOS: Potential Legal Disputes between the United States and China after U.S. Accession to the Convention", *Ocean Development and International Law*, Vol. 36, 2005, 268.

excessive claims". Finally, any such excessive claim may devalue the right assertions by the United States in conformity with international law. Another area of controversy is the US interpretation of the expression "other internationally lawful uses of the sea related to these freedoms" contained in Article 58(1) of the LOSC. According to this interpretation, such uses include "*inter alia*, military activities, such as anchoring, launching and landing of aircraft and other military devices, intelligence collection, surveillance and reconnaissance activities, exercises, operations, and conducting military surveys".[129] It is to be borne in mind that quite a number of other countries have made their declarations contradictory to the interpretation of the United States' disputes relating to the interpretation and implementation of the LOSC that may inevitably arise after the US accession.

It is acknowledged that the LOSC allows States to make declarations or statements regarding the application of the Convention at the time of signature, ratification or accession, but for the purpose of harmonizing the laws and regulations of the coastal State with the provisions of the Convention, declarations or statements carry no meaning of excluding or modifying the legal effect of the provisions of the Convention in their application to that State.[130] Therefore, in the context of Article 310 of the LOSC, the relevant American declarations and statements, just like those made by other countries including China only constitute an explanation of the governmental position and are hardly regarded as treaty reservations so as to exclude the application of the LOSC to the United States.

Nevertheless, one phenomenon which deserves close attention of scholars of the law of the sea is that there are quite a number of declarations or statements similar to China's by other states when ratifying the LOSC. Although such declarations could not be regarded as reservations, their *de facto* or cumulative effect on the interpretation of the relevant provisions of the LOSC may not be ignored.

Conclusion

It is clear that remaining outside the LOSC in fact has brought benefits to the United States in the sense that the United States is able to pick-and-choose what benefits it receives from the Convention without taking any

[129] Duff, *supra* note 5, p. 216.

[130] Article 310 of the LOSC.

treaty obligations regarding other provisions which in its eye may be detrimental to its interest and rights. The United States established its 12-nm territorial sea in 1988 and its contiguous zone in 1999 based on the LOSC, though it did not express that the LOSC was directly applied. In this context, the United States regards the LOSC "not as an applicable governing document itself, but rather as the mirror that reflects customary international law".[131]

In conclusion, the United States policy so far toward the LOSC is a mixed picture. First, although the United States has not yet acceded to the LOSC *per se*, it has signed the 1994 Agreement and ratified the Agreement for the Implementation of the Provisions of the Convention Relating to the Conservation and Management of Straddling Fish Stocks and Highly Migratory Fish Stocks (in August 1996). In that sense, the United States has already established its substantial link to the legal arrangements within the LOSC framework. Some of the arrangements such as those embodying in the 1994 Agreement were made under the direct influence of the United States or simply initiated to appease it and accommodate its interests so that the United States will be willing to join the LOSC.

Second, as is revealed, one of the considerations for the United States to accede to the LOSC was the concern that the LOSC would be open for amendment after it came into force, i.e. from November 2004 onward; if the United States was not a party, then it would be left behind any review conferences which might discuss proposals of amendments formulating new international law rules unfavorable for the US interests.[132] Thus, as the United States expressed, one of the reasons to accede to the LOSC is to play a leading role in ocean issues and international law-making.[133] In fact, even while remaining outside the LOSC, the United States had played a key role in the development of the LOSC. This is well illustrated by the adoption of the 1994 Agreement which has fundamentally changed the legal regime established in the original Part XI of the Convention in favor of the American interests.

---

[131] Duff, *supra* note 5, p. 199.

[132] See Song, *supra* note 128, p. 264; and Zhang Xiaoli, "The U.S. Accession to the United Nations Convention on the Law of the Sea", *Ocean Development and Management* (in Chinese), Vol. 3, 2005, 28.

[133] See United Nations Convention on the Law of the Sea Hearing, Testimony of John F. Turner, Assistant Secretary, Bureau of Oceans and International Environmental and Scientific Affairs, 21 October 2003, available at http://foreign.senate.gov/testimony/2003/TurnerTestimony031021.pdf.

Finally, although there are various definitions on rule of law, one thing is clear: the rule of law principle at the international level requires States to use international law rather than power, will of individuals, or even force to govern State-to-State relations. The current prevailing unilateralism in world politics may endanger the spirit and letters of the rule of law. The United States has been criticized to disrupt the international rule of law by withdrawing itself unilaterally from several important international conventions including the Nuclear Non-proliferation Treaty, the Comprehensive Test Ban Treaty, the Biological Weapons Convention, the Kyoto Protocol, and the Rome Statute of the International Criminal Court. All this indicates that the United States tends to adopt the rule of power instead of rule of law in dealing with international affairs, disregarding the existence of international legal norms.[134] A number of other controversial American practices, including the so-called preemptive strike and the Guantánamo Bay detention, have reinforced the impression of many other countries that the United States disregards international law. In the context of the LOSC, some of its proposed declarations and understandings mentioned earlier have also reflected such kind of attitude of the United States. If this policy continues, it will definitely disrupt the legal order in the world community.

There are still obstacles in the way of the United States accession to the LOSC. It is pointed out that opposition to the US accession is largely ideological in the American politics, and "Senators on the right wing would vote against it on the ground that UNCLOS III is perceived to have pandered to the New International Economic Order, which is now discredited in the United States and largely abandoned as a Third World imperative".[135] On the other hand, the United States needs to think that once it has acceded to the LOSC, it is obliged to comply fully with the provisions of the Convention, and the previous pick-and-choose opportunity will no longer exist.

---

[134] "U.S. said violating international treaties", *New York Times*, 4 April 2002. The recent report titled "Rule of Power or Rule of Law? An Assessment of U.S. Policies and Actions Regarding Security-Related Treaties", prepared by the Institute for Energy and Environmental Research and the Lawyers' Committee on Nuclear Policy in 2002, available at http://www.ieer.org/reports/treaties/advisory.html.

[135] See Maritime Awards Society of Canada, *Report on Revisiting the Law of the Sea*, May 2003, p. 6 (on file with the author).

Chapter 3

Rule of Law for Global Oceans: Strengthening the Law of the Sea Dispute Settlement Institutions

Introduction

A dispute, as defined in *Black's Law Dictionary*, refers to "a conflict or controversy, esp. one that has given rise to a particular lawsuit".[1] The most authoritative definition comes from the *Mavrommatis Palestine Concessions* case where the term "dispute" was defined as "a disagreement on a point of law or fact, a conflict of legal views or of interest between the parties".[2] This definition has been inherited by the consequent international judicial tribunals, though with some modifications.[3] The International Court of Justice (ICJ) narrowed the description given by the Permanent Court of International Justice (PCIJ) that embraced both dispute avoidance and dispute settlement activities, to a definition which suits only the specific issue of dispute settlement,[4] as the ICJ emphasized in the Nicaragua case that it is only concerned "to establish, first, that the

---

[1] *Black's Law Dictionary*, 7th edn. (St. Paul, Minn.: West Group, 1999), p. 485.

[2] *Mavrommatis Palestine Concession* (Jurisdiction), PCIJ, Ser. A, No. 2 (1924), p. 11.

[3] The modified definition on international legal dispute is "a disagreement on a question of law or fact, a conflict, a clash of legal views or of interests". See "Contentious Jurisdiction", available at http://www.icj-cij.org/jurisdiction/index.php?p1=5&p2=1.

[4] Cesare P.R. Romano, *The Peaceful Settlement of International Environmental Disputes: A Pragmatic Approach* (The Hague: Kluwer Law International, 2000), p. 11.

dispute before it is a legal dispute, in the sense of a dispute capable of being settled by the application of principles and rules of international law, and secondly, that the Court has jurisdiction to deal with it".[5] This is coincided with the definition given in *Black's Law Dictionary* which also treats "disputes" in the legal and litigation sense.

Apart from definitions given by international judiciary, legal scholars try to offer their articulated definitions. Romano gives the following definition on "dispute":

> Any conflict of views or of interests, taking the form of opposing claims between the parties, specific enough and formulated with sufficient clarity to allow a third party, being either a State or an international body, to try to settle it.[6]

This definition is modified from the classic definition given by PCIJ, but added more complexity and even confusion. Obviously, disputes can be solved by the parties concerned themselves and may not be necessary all the time to be submitted for settlement before a third party.

International disputes are not uncommon in the world community. In international law, there are a number of mechanisms for the settlement of such disputes, including such political means as negotiation and consultation, mediation and good offices, conciliation, investigation, and such judicial means as arbitration and international adjudication as listed in the Charter of the United Nations.[7] In addition, international organizations, whether universal or regional, have played an active role in dispute settlement.

Judicial settlement here refers to the settlement made by the international judiciary that mainly comprises the ICJ and the International Tribunal for the Law of the Sea (ITLOS). Some other international tribunals do not deal, directly or even indirectly, with law of the sea disputes. This is the case with the International Criminal Tribunal for the former Yugoslavia (ICTY). The International Criminal Court (ICC), which was established on 1 July 2002 in accordance with the Rome Statute of

---

[5] *Border and Transborder Armed Actions (Nicaragua v. Honduras)*, Jurisdiction and Admissibility, ICJ Reports, 1988, para. 52, p. 91.

[6] Romano, *supra* note 4, p. 12.

[7] Article 33(1) of the UN Charter.

the International Criminal Court adopted on 17 July 1998, deals with international crimes.

The ICJ is defined as "the principal judicial organ of the United Nations".[8] According to the Statute of ICJ, "the jurisdiction of the Court comprises all cases which the parties refer to it and all matters specially provided for in the Charter of the United Nations or in treaties and conventions in force".[9] Any UN member can declare any time that it recognizes the Court's compulsory jurisdiction in all legal disputes concerning "the interpretation of a treaty; any question of international law; the existence of any fact which, if established, would constitute a breach of an international obligation; the nature or extent of the reparation to be made for the breach of an international obligation".[10] In the event of a dispute as to whether the Court has jurisdiction, the matter should be settled by the decision of the Court.[11]

The acceptance of the Court's compulsory jurisdiction can facilitate the adjudication and more effective function of ICJ. However, the comparable number of States which have accepted such jurisdiction has declined significantly with the increase of UN members and they only account for one-thirds of the total 193 current members (see Table 3.1). In East Asia only four countries have done so (see Table 3.2). The reasons behind are complicated and one of them may attribute to the proliferation of international tribunals. Even after having accepted the compulsory jurisdiction of ICJ, an accepting State may change its decision any time as it wishes following the procedure of the Statute of ICJ. For example, Canada withdrew its acceptance of the compulsory jurisdiction twice, first in the case of its 1969 legislation to protect its Arctic environment in areas beyond those recognized by the then applicable international law and secondly to preserve straddling fish stocks off its east coast exclusive economic zone (EEZ).[12] Some countries exclude law of the sea disputes from the Court's

---

[8] Article 92 of the UN Charter, available at http://www.icj-cij.org/icjwww/ibasicdocuments/ibasictext/ibasicunchart.htm#Chapter14.

[9] Article 36 of the ICJ Statute, available at http://www.icj-cij.org/icjwww/ibasicdocuments/ibasictext/ibasicstatute.htm#CHAPTER_II.

[10] *Ibid.*

[11] *Ibid.*

[12] See Comments of Robert Hage, Minister, Permanent Mission of Canada to the European Union, Brussels, at the round table "Comment apprécier la nouveller Cour?" in Lucius Caflisch (ed.), *The Peaceful Settlement of Disputes between States: Universal and*

Table 3.1: Acceptance of PCIJ/ICJ Compulsory Jurisdiction from 1925 to 2020

Year	States Accepting	States Parties to the Statute
1925 (PCIJ)	23	36
1930	29	42
1935	42	49
1940	32	50
1945 (ICJ)	23	51
1950	35	61
1955	32	64
1960	39	85
1965	40	118
1970	46	129
1975	45	147
1980	47	157
1985	46	162
1990	53	162
1995	59	187
2008	65	192
2020	74	193

Source: Adapted from the ICJ website.

Table 3.2: Asian Countries Accepting ICJ Compulsory Jurisdiction

Cambodia (1957)
India (1974)
Japan (2007)
Pakistan (1960)
Philippines (1972)
Timor-Leste (2012)

Source: Compiled by the Author.

European Perspectives (The Hague: Kluwer Law International, 1998), p. 100. It is interesting to note that the Canadian controversial legislation might have become the first case of marine environmental dispute before ICJ if Canada did not withdraw its declaration.

compulsory jurisdiction, such as Australia which excluded disputes concerning natural resources of the seabed and subsoil of its continental shelf and/or within the meaning of its Pearl Fisheries Acts.[13] The Philippines' Declaration excludes disputes "in respect of the natural resources, including living organisms belonging to sedentary species, of the seabed and subsoil of the continental shelf of the Philippines, or its analogue in an archipelago".[14] Perhaps the primary consideration of such exclusion made by the above countries is for the preservation of their sovereign rights over marine resources rather than refusing the dispute settlement. Nonetheless, they have in practice generated some negative effect on the settlement of law of the sea disputes.

The second main international judicial organ handling law of the sea disputes is the Permanent Court of Arbitration (PCA) which was established in 1899 in accordance with the 1899 Convention for the Pacific Settlement of International Disputes concluded during the first Hague Peace Conference and revised in 1907. As of 2020, the PCA had 122 member States and "provides services for the resolution of disputes involving various combinations of states, state entities, intergovernmental organizations, and private parties".[15] Its Secretariat serves as a registry for arbitral tribunals and commissions. While the scope of its involvement in dispute resolution is very broad, the PCA has the competence in dealing with law of the sea disputes. This competence has been reinforced after the entry into force of the LOSC as the latter has created four categories of dispute settlement mechanisms for States parties to choose and arbitration is one of them. Pursuant to Article 287(3) of the LOSC, arbitration under Annex VII is the default means of dispute settlement if a State has not expressed any preference with respect to the means of dispute resolution available under Article 287(1) of the Convention. Since the LOSC came into force in 1994, 15 cases with three pending have been arbitrated under Annex VII of that Convention. As of 2020, the PCA had acted as registry in 14 of those cases including: (1) *Ireland v. United Kingdom* (MOX Plant case), which was instituted in November 2001 and terminated on 8 June 2008 with an Order to formalize the withdrawal of Ireland's claim against the United Kingdom; (2) *Malaysia v. Singapore*, which was instituted in July 2003 and terminated by an award on agreed

[13] Declaration of 6 February 1954, 186 UNTS 77.

[14] Declaration of 18 January 1972, para.e.I, *ICJ Yearbook* 1994–1995, p. 109.

[15] "About us", available at http://www.pca-cpa.org/showpage.asp?pag_id=1027.

terms rendered on 1 September 2005; (3) *Barbados v. Trinidad and Tobago*, which was instituted in February 2004 and decided by a final award rendered on 11 April 2006; and (4) *Guyana v. Suriname*, which was instituted in February 2004 and decided by a final award rendered on 17 September 2007.[16] Meanwhile the cooperative relationship between the PCA and ITLOS has been established through an exchange of letters with respect to relevant legal and administrative matters.

Arbitration as a legal means of dispute settlement is more flexible than adjudication. The parties to a dispute can choose arbiters to form an arbitral tribunal. The PCA actually possesses a list of arbiter candidates for countries to select for their arbitral tribunals. It does not have the same organizational structure as ICJ or ITLOS. The judges sitting in ICJ or ITLOS can also be chosen as arbiters.[17] Though arbitration is one of the legal means for dispute settlement, there is difference between arbitration and adjudication. The LOSC also allows States parties to make recourse to special arbitration for disputes concerning: (a) fisheries; (b) protection and preservation of the marine environment; (c) marine scientific research; or (d) navigation, including pollution from vessels and by dumping.[18]

It is admitted that there are various institutions which can serve as a venue to solve law of the sea disputes. However, the ITLOS is the specialized judicial organ designed only to handle law of the sea disputes. Therefore, this chapter will limit itself mainly to the procedure and practices of it while some comparisons will be made between the above judicial institutions and the ITLOS. Finally, East Asian States' attitudes toward and practices in judicial settlement of their disputes will be examined through a number of important cases.

---

[16] See https://pca-cpa.org/en/services/arbitration-services/unclos/.

[17] Since 2018, ICJ judges are no longer able to accept appointments to sit as arbitrators in either investor–State or commercial arbitral proceedings. ICJ judges remain able to participate in inter-State arbitrations in exceptional cases, but need to obtain prior authorization before accepting any such appointments and are only permitted to participate in one arbitration procedure at a time. "ICJ Judges to Cease Acting as Arbitrators", available at https://www.fiettalaw.com/pil_news/icj-judges-to-cease-acting-as-arbitrators/.

[18] See Annex VIII to the LOSC.

Law of the Sea Disputes and ITLOS

The ITLOS was established in October 1996 in Hamburg, Germany, under the general framework of the LOSC. The LOSC provides a set of comprehensive compulsory procedures for dispute settlement since management (i.e. avoidance, reduction and settlement) of conflict is one of the five specific goals to maintain a favorable marine legal order.[19] If the parties to a dispute fail to reach settlement by peaceful means of their own choice, they are obliged to choose one or more of the following to settle their disputes: ITLOS; ICJ; arbitral tribunal and special arbitral tribunal when they sign, ratify or accede to the Convention.[20] When a State does not make a choice upon its ratification of the Convention, it is deemed to have accepted the compulsory procedure of arbitration.[21] This deemed acceptance does not affect the State concerned to choose other procedures any time afterward including the resort to the ITLOS.

The ITLOS can establish its jurisdiction over any law of the sea dispute that is submitted to it concerning the interpretation or application of the LOSC and the Agreement relating to the Implementation of Part XI of the Convention. In addition, it can adjudicate cases submitted by parties to other international treaties if such treaties allow it to do so.[22] The ITLOS has established the Seabed Disputes Chamber which consists of 11 judges selected from within the ITLOS and under this chamber some *ad hoc* chambers can be established when they are in need. In addition, the Tribunal has established five special chambers including the Chamber of Summary Procedure, which consists of five judges and two alternates; the Chamber for Fisheries Disputes, which consists of seven judges and is available to deal with disputes concerning the conservation and management of marine living resources; the Chamber for Marine Environment Disputes, which also consists of seven judges and is available to deal with

---

[19] See John Warren Kindt, "Dispute Settlement in International Environmental Issues: The Model Provided by the 1982 Convention on the Law of the Sea", *Vanderbilt Journal of Transnational Law*, Vol. 22, 1989, 1107. The other four goals include security, promotion of efficiency and fair access in ocean use, protection of the environment and promotion of ocean knowledge.

[20] See Article 287 of the LOSC.

[21] See Article 298 of the LOSC.

[22] See Articles 21 and 22 of the Statute of the ITLOS.

disputes relating to the protection and preservation of the marine environment. Another important special chamber is the Chamber for Maritime Delimitation Disputes which was created in March 2007, consisting of eight judges. Finally, the Tribunal can create chambers under Article 15, paragraph 2, of its Statute to deal with a particular dispute if the parties so request. In practice, such a special chamber was established in December 2000 to deal with the *Case concerning the Conservation and Sustainable Exploitation of Swordfish Stocks in the South-Eastern Pacific Ocean* (Chile/European Community). This was the first time that a case had been submitted to a special chamber of the Tribunal.[23] Since its establishment in 1996 up to October 2020, ITLOS had received and dealt with 29 cases (with two still pending). Among them, nine cases are concerning the prompt release of vessels and crews while eight cases are concerning the request by States parties for provisional measures from the Tribunal (see Table 3.3). These two areas have constituted the main judicial activities of the Tribunal and worth further scrutiny.

Prompt release of vessels and crews

Prompt release is a special competence granted by the LOSC to a court or tribunal including the ITLOS. In comparison with other international courts or tribunals, only the ITLOS has exercised this jurisdiction to entertain an application for the prompt release of a detained vessel or its crew. According to Article 292 of the LOSC, where the authorities of a State Party have detained a vessel flying the flag of another State Party and it is alleged that the detaining State has not complied with the provisions of the Convention for the prompt release of the vessel or its crew upon the posting of a reasonable bond or other financial security, the question of release from detention may be submitted to the Tribunal if, within 10 days from the time of detention, the parties have not agreed to submit it to another court or tribunal. Article 292(2) provides that the application for release may be made only by or on behalf of the flag State of the vessel. This clause carries two particularly noted meaning: (a) only the flag State or its agent can make such application; and (b) the wording "on behalf" indicates that not only the State but also other entities which are authorized by that State can submit the application. From the above, it can be seen that

[23] See "Chambers", available at http://www.itlos.org/start2_en.html.

Table 3.3: List of Cases at the ITLOS (1996–2020)

- **Case No. 1**
  The M/V "SAIGA" Case (Saint Vincent and the Grenadines v. Guinea), Prompt Release
- **Case No. 2**
  The M/V "SAIGA" (No. 2) Case (Saint Vincent and the Grenadines v. Guinea)
- **Cases Nos. 3 and 4**
  Southern Bluefin Tuna Cases (New Zealand v. Japan; Australia v. Japan), Provisional Measures
- **Case No. 5**
  The "Camouco" Case (Panama v. France), Prompt Release
- **Case No. 6**
  The "Monte Confurco" Case (Seychelles v. France), Prompt Release
- **Case No. 7**
  Case Concerning the Conservation and Sustainable Exploitation of Swordfish Stocks in the South-Eastern Pacific Ocean (Chile/European Community)
- **Case No. 8**
  The "Grand Prince" Case (Belize v. France), Prompt Release
- **Case No. 9**
  The "Chaisiri Reefer 2" Case (Panama v. Yemen), Prompt Release
- **Case No. 10**
  The MOX Plant Case (Ireland v. United Kingdom), Provisional Measures
- **Case No. 11**
  The "Volga" Case (Russian Federation v. Australia), Prompt Release
- **Case No. 12**
  Case Concerning Land Reclamation by Singapore in and around the Straits of Johor (Malaysia v. Singapore), Provisional Measures
- **Case No. 13**
  The "Juno Trader" Case (Saint Vincent and the Grenadines v. Guinea-Bissau), Prompt Release
- **Case No. 14**
  The "Hoshinmaru" Case (Japan v. Russian Federation), Prompt Release
- **Case No. 15**
  The "Tomimaru" Case (Japan v. Russian Federation), Prompt Release
- **Case No. 16**
  Dispute concerning delimitation of the maritime boundary between Bangladesh and Myanmar in the Bay of Bengal (Bangladesh/Myanmar)

(Continued)

Table 3.3: (*Continued*)

- **Case No. 17**
  Responsibilities and obligations of States sponsoring persons and entities with respect to activities in the Area (Request for Advisory Opinion submitted to the Seabed Disputes Chamber)
- **Case No. 18**
  The M/V "Louisa" Case (*Saint Vincent and the Grenadines v. Kingdom of Spain*)
- **Case No. 19**
  The M/V "Virginia G" Case (Panama/Guinea-Bissau)
- **Case No. 20**
  The "ARA Libertad" Case (*Argentina v. Ghana*), Provisional Measures
- **Case No. 21**
  Request for an Advisory Opinion submitted by the Sub-Regional Fisheries Commission (SRFC) (Request for Advisory Opinion submitted to the Tribunal)
- **Case No. 22**
  The "Arctic Sunrise" Case (*Kingdom of the Netherlands v. Russian Federation*), Provisional Measures
- **Case No. 23**
  Dispute concerning delimitation of the maritime boundary between Ghana and Côte d'Ivoire in the Atlantic Ocean (Ghana/Côte d'Ivoire)
- **Case No. 24**
  The "Enrica Lexie" Incident (*Italy v. India*), Provisional Measures
- **Case No. 25**
  The M/V "Norstar" Case (*Panama v. Italy*)
- **Case No. 26**
  Case concerning the detention of three Ukrainian naval vessels (*Ukraine v. Russian Federation*), Provisional Measures
- **Case No. 27**
  The M/T "San Padre Pio" Case (*Switzerland v. Nigeria*), Provisional Measures
- **Case No. 28**
  Dispute concerning delimitation of the maritime boundary between Mauritius and Maldives in the Indian Ocean (Mauritius/Maldives)
- **Case No. 29**
  The M/T "San Padre Pio" (No. 2) Case (Switzerland/Nigeria)

Source: ITLOS website, available at https://www.itlos.org/en/cases/list-of-cases/.

there are three prerequisites for the ITLOS to exercise its jurisdiction over prompt release cases: (a) both flag and detaining States must be States parties to the LOSC; (b) the parties concerned have not agreed to submit the question of release to another court or tribunal within 10 days from the time of detention; and (c) an application for the prompt release must be submitted by the flag State or by a person or entity duly authorized to do so on behalf of the flag State.[24] Regarding the last requirement, Rule 110 of the Rules of the Tribunal provides additional requirements that there must be a clear authorization by the flag State to a named person or entity and the authorization must be issued by an authority of the State that the Tribunal recognizes as able to act in the name of the State.[25] In the *M/V "Saiga" (No. 1)* case, the application of the Saint Vincent and the Grenadines was submitted by a private attorney who was given the authorization by the Commissioner of Maritime Affairs who had been empowered to issue the authorization by the Attorney General of the State.[26] The purpose of these provisions, according to Lagoni, is "to balance the interests of the detaining state in its measures against the flag state with the interests of the flag state in preventing an excessive detention of vessels flying its flag".[27]

Usually there are two circumstances that the provision on prompt release may be invoked: (a) charges relating to illegal fishing in waters under the jurisdiction of the coastal (detaining) State; and (b) charges or claims by public authorities in respect of pollution of waters under their jurisdiction by reason of unauthorized dumping or irregular discharges.[28] The judicial practices so far of the ITLOS have shown that the applications for prompt release were made for the vessels detained in the EEZ of the coastal (detaining) States, suspected of illegal fishing or other commercial activities related to fishing.

---

[24] See Thomas A. Mensah, "The Tribunal and the Prompt Release of Vessels", *International Journal of Marine and Coastal Law*, Vol. 22(3), 2007, 429–430.

[25] *Ibid.*, p. 431.

[26] *Ibid.*, p. 432.

[27] Rainer Lagoni, "The Prompt Release of Vessels and Crews before the International Tribunal for the Law of the Sea: A Preparatory Report", *International Journal of Marine and Coastal Law*, Vol. 11(2), 1996, 148.

[28] See David H. Anderson, "Investigation, Detention and Release of Foreign Vessels under the UN Convention on the Law of the Sea of 1982 and Other International Agreements", *International Journal of Marine and Coastal Law*, Vol. 11(2), 1996, 177. According to Anderson, Article 292 may be invoked as well in the following situation: (a) fishing for straddling or highly migratory stocks beyond the limits of national jurisdiction; and (b) irregularities or deficiencies on board a vessel, giving rise to reasonable fears of unreasonable threats of future pollution of the sea, with the result that release on bail or bonding is not suited to the needs of the situation facing the port's authorities. *Ibid.*

The rationale behind the stipulations of the LOSC is for the interests of shipping companies to possibly reduce their economic loss, for the crews who could avoid further physical sufferings due to the vessel detention and can also avoid a serious safety and environmental hazard if the detained vessel is promptly released on bond. Thus, the provisions contained in the LOSC "accommodate economic and humanitarian as well as safety and environmental concerns".[29] Another consideration behind is the possible delay of releasing a vessel or crew if parties to a dispute prefer the use of arbitration but it takes a considerable amount of time to establish an arbitral tribunal for that purpose.[30] However, the competence of the ITLOS is only confined to decide on the question of release "without prejudice to the merits of any case before the appropriate domestic forum against the vessel, its owner or its crew".[31] Unlike other disputes which usually need the consent of the disputant parties for settlement, in the case of prompt release, the State party to a dispute can unilaterally submit the application to the ITLOS under Article 292 of the LOSC. For those countries that exclude the jurisdiction of the ITLOS from any potential dispute it may be involved, they can be brought before the Tribunal for cases of prompt release if there is no agreement reached between the disputant parties within 10 days from the time of detention.

It is noted that the provision concerning prompt release does not contain the word "dispute". According to Oxman, the absence of the word "dispute" suggests that the nature of Article 292 of the Convention is different from other provisions of Part XV.[32] On the other hand, since this clause is put together with Article 290 regarding provisional measures within a same section, it is read as containing "a special procedure of a somewhat analogous character".[33]

One important issue that the Tribunal has to decide before issuing its judgment on prompt release is regarding the reasonableness of the bond

---

[29] See Bernard H. Oxman, "Observations on Vessel Release under the United Nations Convention on the Law of the Sea", *International Journal of Marine and Coastal Law*, Vol. 10(2), 1995, 208.

[30] See Louis B. Sohn, "Settlement of Law of the Sea Disputes", *International Journal of Marine and Coastal Law*, Vol. 11(2), 1996, 203–204.

[31] Article 292(3) of the LOSC.

[32] Oxman, *supra* note 29, p. 208.

[33] Anderson, *supra* note 28, p. 167.

to be posted.[34] According to former ITLOS President Judge Mensah, there are some general principles gradually formulated from the judicial practice of the Tribunal: (a) respect for the object of the provisions of the LOSC for prompt release, which in the Tribunal's view is "to reconcile the interest of the flag State to have its vessel and its crew released promptly with the interest of the detaining State to secure appearance in its court of the Master and the payment of penalties".[35] (b) Assessment of what is a reasonable bond must be objective and take into account all information provided by the parties in the case; and (c) the criterion of reasonableness encompasses the amount, the nature and the form of the bond.[36] Based on these principles, the Tribunal considers a list of factors such as the gravity of the alleged offences, the penalties to be imposed or imposable under the laws of the detaining State, the value of the detained vessel or cargo seized, and the amount and form of the bond imposed by the detaining State during the process of determining a reasonable bond to be posted.[37]

The cases of prompt release before the ITLOS have showed that the Tribunal seems to have established its consistent and uniform judicial practice in dealing with such cases. This can be seen from the three cases between 2004 and 2007, in which judgments were unanimously adopted.[38] As it is revealed by the former ITLOS President Judge Mensah, since the *Camouco* case in 2000, the finding that the allegation of non-compliance is "well-founded" has been a regular feature of all judgments in which the Tribunal has ordered the release of a vessel or crew.[39] Under this circumstance, handling of such cases becomes a simple routine or formality for the Tribunal. It is doubtful whether the Tribunal still really needs its whole

[34]For general discussions in this regard, see Erik Franckx, "'Reasonable Bond' in the Practice of the International Tribunal for the Law of the Sea", *California Western International Law Journal*, Vol. 32, 2001–2002, 303–342; and Gao Jianjun, "Reasonableness of the Bond under Article 292 of the LOS Convention: Practice of the ITLOS", *Chinese Journal of International Law*, Vol. 7, 2008, 115–142.

[35]See ITLOS, *Monte Confurco* Judgment (*Seychelles v. France*), 18 December 2000, para 71.

[36]See Mensah, *supra* note 24, p. 440.

[37]Mensah, *supra* note 24, pp. 440–441.

[38]They are the "*Juno Trader*" Case (*Saint Vincent and the Grenadines v. Guinea-Bissau*), the "*Hoshinmaru*" Case (*Japan v. Russian Federation*), and the "*Tomimaru*" Case (*Japan v. Russian Federation*).

[39]Mensah, *supra* note 24, p. 436, note 40.

bench to hear such simple cases. It is then strongly suggested that such cases be dealt with by a special chamber of summary procedure so as to reduce unnecessary costs and enhance the efficiency of the Tribunal, with no need of a request from a State party.

Provisional measures

According to Article 290 of the LOSC and relevant provisions of the ITLOS Statute, if a dispute has been duly submitted to the Tribunal and if the Tribunal considers that *prima facie* it has jurisdiction under Part XV or Part XI, Section 5, of the LOSC, the Tribunal may prescribe any provisional measures which it considers appropriate under the circumstances to preserve the respective rights of the parties to the dispute or to prevent serious harm to the marine environment, pending the final decision. Moreover, the Tribunal may also prescribe provisional measures in the case covered by Article 290(5) of the LOSC which provides that pending the constitution of an arbitral tribunal to which a dispute is being submitted and if, within two weeks from the date of a request for provisional measures, the parties do not agree to submit the request to another court or tribunal, the Tribunal may prescribe provisional measures if it considers that *prima facie* the arbitral tribunal to be constituted would have jurisdiction and that the urgency of the situation so requires. It is to be noted that the cases so far submitted to the ITLOS requesting provisional measures are referred to under the second stipulated condition.

The different usage of wording between the ICJ Statute and the LOSC has been noticed. Although the ICJ also has the power to decide provisional measures, its Statute uses the word "indicate" while the word used in the LOSC for the ITLOS is "prescribe". Therefore, it is commented that the different wording is used "to improve the powers of courts and tribunals having jurisdiction over law of the sea disputes" and to make their provisional measures "have a binding effect upon the parties to the dispute".[40] Another difference from the ICJ Statute is that the LOSC allows the Tribunal to prescribe, modify or revoke provisional measures "only" when it is requested by a party to the dispute.[41] The other difference

---

[40] Rüdiger Wolfrum, "Provisional Measures of the International Tribunal for the Law of the Sea", *Indian Journal of International Law*, Vol. 37, 1997, 433.

[41] See Article 290(3) of the LOSC.

is that the ITLOS can prescribe provisional measures not only for the preservation of the respective rights of the dispute but also for the prevention of serious harm to the marine environment. In this sense, while the ITLOS's competence in prescribing provisional measures is constrained to the request from a State on the one hand, it has, on the other hand, expanded competence in doing so regarding the protection of the marine environment which the ICJ Statute does not mention regarding the competence of the ICJ in this judicial field. As required by the LOSC, the parties to the dispute "shall" comply with the provisional measures prescribed by a court or tribunal.[42]

It is acknowledged that provisional measures will have impacts on the adjudication of the merits of a dispute since "there is a close relationship between the *prima facie* decision concerning provisional measures and the jurisdictional choice relating to the merits to be made at a later stage as the judges are in some way expressing their intimate conviction in this respect".[43] This relationship is particularly relevant to the circumstance that the Tribunal itself will be able to hear the merits of the dispute at a later stage.

In practice, the ITLOS prescribed provisional measures in the *Southern Bluefin Tuna Cases (New Zealand v. Japan; Australia v. Japan)*, the *MOX Plant Case (Ireland v. United Kingdom)* and the *Case concerning Land Reclamation by Singapore in and around the Straits of Johor (Malaysia v. Singapore)*. In addition, the Tribunal also prescribed provisional measures in the *M/V Saiga No. 2 Case (St. Vincent and the Grenadines v. Guinea)* in March 1998 before it adjudicated the merits of the case. Different views have been expressed on the performance of the Tribunal in this respect. For example, one particular measure adopted by the Tribunal in the SBT case appeared to exceed the scope of the request made and was regarded as *extra petita* (but not *ultra petita*).[44] In practice, the provisional measures prescribed by the ITLOS in the SBT case were later revoked by the arbitral tribunal which found that it had no jurisdiction over the dispute. In that case, questions arise as to how the ITLOS

[42] See Article 290(6) of the LOSC.

[43] Francisco Orrego Vicuña, "The International Tribunal for the Law of the Sea and Provisional Measures: Settled Issues and Pending Problems", *International Journal of Marine and Coastal Law*, Vol. 22(3), 2007, 454.

[44] Miguel Garcia Garcia-Revillo, *El Tribunal International del Derecho del Mar* (Biblioteca Diplomatica Española, 2005), p. 492; cited in Orrego Vicuña, *supra* note 43, p. 456.

prescribed such measures based on its reasoning to justify the establishment of "*prima facie* jurisdiction"[45] of a court or tribunal in dealing with the merits of the dispute as required by the LOSC. Suppose that the case were submitted to the ITLOS rather than an arbitral tribunal, the outcome regarding the dispute would be totally different. Some jurists see this as a deficiency in the LOSC as "this case shows how anomalous it is for the Law of the Sea Convention to have given the Hamburg Tribunal injunctive powers in respect of cases always intended by the parties to go elsewhere for their merits to be determined".[46]

Nevertheless, provisional measures sometimes do play some positive role in facilitating the final resolution of a dispute. It is noted that in the jurisprudence of the ITLOS, some of the provisional measures it has prescribed contain recommendations. As commented, "Recommendations may provide useful guidance for the conduct of the parties without creating the burden of sanctions under international law should any conduct be viewed askance".[47] In the *Case concerning Land Reclamation by Singapore in and around the Straits of Johor*, the Tribunal recommended to the parties in dispute to set up a group of independent experts to study and prepare an interim report on the subject matter. Following this recommendation, the two sides formulated such a group and based on the report of the group of experts, solved their dispute through negotiations. In this respect, the function of the Tribunal is very facilitative to the dispute settlement.[48]

---

[45] Regarding *prima facie* jurisdiction, the ICJ once stated that "the dispute appears, *prima facie*, to afford a possible basis on which the jurisdiction of the Court might be founded", See *LaGrand* (*Germany v. US*) (merits), 2001, available at http://www.icj-cij.org/docket/files/104/7736.pdf. In the case of Interhandle, Judge Lauterpacht expressed that "the Court may properly act under the terms of Article 41 provided that there is in existence an instrument … which *prima facie* confers jurisdiction upon the Court and which incorporate no reservations obviously excluding its jurisdiction". *Interhandle Case* (*Switzerland v. US*) 1957 ICJ Report, pp. 118–119.

[46] See Rosalyn Higgins, "The ICJ, the ECJ, and the Integrity of International Law", *International and Comparative Law Quarterly*, Vol. 52, 2003, 19.

[47] Natalie Klein, *Dispute Settlement in the UN Convention on the Law of the Sea* (Cambridge: Cambridge University Press, 2005), p. 85.

[48] According to Robin Churchill, "the establishment of the group was due to an imaginative use of provisional measures by the ITLOS". See Robin Churchill, "Dispute Settlement under the UN Convention on the Law of the Sea: Survey for 2005", *International Journal of Marine and Coastal Law*, Vol. 21(4), 2006, 385.

Limits of the ITLOS

Although the ITLOS is the only special international court for law of the sea disputes and can establish its jurisdiction over any law of the sea dispute, there are obviously constraints and limitations which would compromise the competence of the Tribunal and to some serious extent make its function awkward.

The first such constraint comes from the LOSC itself which has created the ITLOS. The LOSC has established the compulsory dispute settlement mechanisms for law of the sea disputes, but according to its relevant provisions, the ITLOS is merely one of these mechanisms — that means the ITLOS does not have exclusive jurisdiction over all the law of the sea disputes and has to share such jurisdiction with other international courts such as the ICJ or arbitral tribunals such as the PCA. Competition between/among existing international courts and arbitral tribunals is inevitable. The ITLOS is a relatively new judicial institution and so is in a relatively disadvantageous position in the competition.

Competition is also from regional courts or tribunals. The most salient example is the judicial organs formulated under the European Union (EU) legal framework. According to the Treaty of the European Union, EU members have to agree to render some of their sovereign rights to the EU and let the organization exercise them on behalf of its members. This arrangement also affects the settlement of certain kinds of law of the sea disputes. When an EU member made a declaration under the LOSC, it usually inserted a paragraph regarding the EU competence. For example, the French declaration made in April 1996 contains the following statement: "France recalls that, as a State member of the European Community, it has transferred competence to the Community in certain matters covered under the Convention. A detailed statement of the nature and scope of the areas of competence transferred to the European Community will be made in due course in accordance with the provisions of Annex IX of the Convention".[49] On the other hand, when the European Community acceded to the LOSC, it also made a declaration, stating that certain competences have been transferred from its member States to the organization, including those in the protection and preservation of the marine environment, commercial and customs policy, conservation and

[49]The full text of the French declaration is available at http://www.un.org/Depts/los/convention_agreements/convention_declarations.htm.

management of sea fishing resources.[50] Because of this special arrangement, courts established within the EU may have exclusive competence in dealing with certain kinds of law of the sea disputes between EU member States.

This is well illustrated by the *MOX Plant Case* (*Ireland v. United Kingdom*), which involved the ITLOS, an Annex VII arbitral tribunal and the European Court of Justice (ECJ). The case was first submitted by Ireland to the ITLOS requesting provisional measures to stop a UK plant making mixed oxide fuel (MOX) at Sellafield on the Irish Sea coast in Cumbria, England from marine pollution while Ireland instituted proceedings against the United Kingdom (UK) before an arbitral tribunal established in accordance with Annex VII of the LOSC in October 2001. On 3 December 2001, the ITLOS issued an order of provisional measures requesting both parties to exchange information, monitor environmental risks, devise measures to prevent pollution and prepare initial reports.[51] While the case was lodged with the Annex VII Arbitral Tribunal, the Commission of the European Communities instituted proceedings against Ireland under the EC Treaty before the ECJ in October 2003, alleging that Ireland breached relevant provisions of the EC Treaty by instituting arbitration against the UK. The ECJ ruled that Ireland breached Article 292 of the EC Treaty which requires EU members not to submit disputes concerning EC law to any judicial body other than the ECJ. The case lodged with the arbitral tribunal was finally terminated in June 2008 upon the request of Ireland. It is rightly pointed out that "the European Court's judgment would seem to make it very unlikely that any EC Member State would in future risk breaching EC law by instituting proceedings against another Member State before a court or tribunal other than the European Court concerning a dispute relating to the Law of the Sea Convention or any other agreement to which both the EC and its Member States were parties", and EU member States would be "more cautious about using the dispute settlement procedures of the Law of the Sea Convention against non-Member States for fear of intruding on matters of EC competence".[52] This is undoubtedly a discouraging fact that would affect the judicial

[50] See EC declarations, available at http://www.un.org/Depts/los/convention_agreements/convention_declarations.htm.

[51] Text is available at http://www.itlos.org/start2_en.html.

[52] Robin Churchill, "Dispute Settlement under the UN Convention on the Law of the Sea: Survey for 2006", *International Journal of Marine and Coastal Law*, Vol. 22(4), 2007, 470.

operations of the ITLOS as well. In addition, it is to be noted that there are other regional courts existing around the world as well.

The related question is about the choice of procedures to be made by States parties in accordance with Article 287 of the LOSC which provides that "When signing, ratifying or acceding to this Convention or at any time thereafter, a State shall be free to choose, by means of a written declaration, one or more of the following means for the settlement of disputes concerning the interpretation or application of this Convention: (a) the International Tribunal for the Law of the Sea established in accordance with Annex VI; (b) the International Court of Justice; (c) an arbitral tribunal constituted in accordance with Annex VII; (d) a special arbitral tribunal constituted in accordance with Annex VIII for one or more of the categories of disputes specified therein". As of October 2020, altogether 50 States parties had made statements on the choice of procedures. From the list prepared by the UN Office for Ocean Affairs and the Law of the Sea, it can be seen that quite a number of States (including Argentina, Austria, Cape Verde, Croatia, Germany, Greece, Hungary, Trinidad and Tobago, Tunisia, United Republic of Tanzania and Uruguay) selected the ITLOS as the first choice for the settlement of law of the sea disputes when arising, while others selected ICJ (including Denmark, Honduras, Nicaragua, Norway, Sweden and United Kingdom), or arbitration as the first choice. Some States (including Australia, Belgium, Estonia, Finland, Italy, Latvia, Lithuania, Mexico, Oman, Portugal and Spain) listed ITLOS and ICJ in parallel without order of preference.[53] This indicates that despite the fact that the ITLOS is designated as the first dispute settlement mechanism under the LOSC, not all the parties to the Convention favor its jurisdiction. In East Asia, no country has ever made such statements and thus it is assumed that they have accepted arbitration in accordance with Article 287(3) of the LOSC (see Table 3.4).

The second constraint is also from the LOSC which allows States parties to exclude certain disputes from the compulsory dispute settlement mechanisms established under the Convention. Article 298(1) of the LOSC provides that a State may declare not to accept the following categories of disputes: (a) disputes concerning the interpretation or application of Articles 15, 74 and 83 relating to sea boundary delimitations, or those involving historic bays or titles; (b) disputes concerning military

[53] For details, see https://treaties.un.org/Pages/ViewDetailsIII.aspx?src=TREATY&mtdsg_no=XXI-6&chapter=21&Temp=mtdsg3&clang=_en.

Table 3.4: East Asian Acceptance of the Compulsory Mechanisms under the LOSC

States	Date of LOS Convention Ratification (dd/mm/yyyy)	Compulsory Mechanism Accepted
Brunei Darussalam	05/11/1996	Arbitration (deemed)
Cambodia	—	—
China	07/06/1996	*Ibid.*
Indonesia	03/02/1986	*Ibid.*
Japan	20/06/1996	*Ibid.*
Korea (North)	—	—
Korea (South)	29/01/1996	*Ibid.*
Laos	05/06/1998	*Ibid.*
Malaysia	14/10/1996	*Ibid.*
Mongolia	13/08/1996	*Ibid.*
Myanmar	21/05/1996	*Ibid.*
Philippines	08/05/1984	Not specified*
Russia	12/03/1997	Various choices**
Singapore	17/11/1994	Arbitration (deemed)
Thailand	15/05/2011	*Ibid.*
Vietnam	25/07/1994	*Ibid.*

Notes: *Upon signature (10 December 1982), the Philippines made the following declaration regarding the acceptance of the settlement mechanisms:
"The agreement of the Republic of the Philippines to the submission for peaceful resolution, under any of the procedures provided in the Convention, of disputes under Article 298 shall not be considered as a derogation of Philippines sovereignty". Available at http://www.un.org/Depts/los/convention_agreements/convention_declarations.htm.
**Upon signature (10 December 1982), the former USSR government made the following declaration regarding the acceptance of the settlement mechanisms:
"1. The Union of Soviet Socialist Republics declares that, under Article 287 of the United Nations Convention on the Law of the Sea, it chooses an arbitral tribunal constituted in accordance with Annex VII as the basic means for the settlement of disputes concerning the interpretation or application of the Convention. It opts for a special arbitral tribunal constituted in accordance with Annex VIII for the consideration of matters relating to fisheries, the protection and preservation of the marine environment, marine scientific research, and navigation, including pollution from vessels and dumping. It recognizes the competence of the International Tribunal for the Law of the Sea, as provided for in Article 292, in matters relating to the prompt release of detained vessels and crews", available at http://www.un.org/Depts/los/convention_agreements/convention_declarations.htm.
Source: Based on United Nations Office for Ocean Affairs and the Law of the Sea, "Settlement of Disputes Mechanisms", available at http://www.un.org/Depts/los/settlement_of_disputes/choice_procedure.htm.

activities, including military activities by government vessels and aircraft engaged in non-commercial service, and disputes concerning law enforcement activities in regard to the exercise of sovereign rights or jurisdiction excluded from the jurisdiction of a court or tribunal under Article 297, paragraph 2 or 3; and (c) disputes in respect of which the United Nations Security Council is exercising the functions assigned to it by the Charter of the United Nations, unless the Security Council decides to remove the matter from its agenda or calls upon the parties to settle it by the means provided for in the Convention. As of August 2020, there were 38 States parties that made such declarations (see Table 3.5). Among them, some exclude all listed in the aforementioned provision (e.g. China and Republic of Korea) and some exclude one or two of them (e.g. UK and Italy). The effect of such exclusion is clear that disputes concerning the above matters are most likely to be resolved by political means such as bilateral negotiation and consultation. Furthermore, it should be noted that the number of the above States include four permanent UN Security Council members, i.e. China, France, Russia and the UK. It is almost certain that the United States once ratifying the LOSC will make the exclusion statement as well. This can be seen from the draft declarations in a resolution passed by the US Senate Foreign Relations Committee for advice and consent of the full Senate, which will exclude all the items allowed by Article 298(1) of the LOSC.[54] The fact that all five permanent members of the UN Security Council exclude maritime disputes from compulsory dispute settlement mechanisms will no doubt discourage other States to use more of such mechanisms, thus in fact detrimental to the rule of law at the international level.

The third constraint is from the reality that deep seabed mining has not yet begun. According to the LOSC, all disputes relating to Part XI and its annexes concerning deep seabed mining beyond the limits of national jurisdiction are subject to a dispute settlement mechanism created by the Convention. The Seabed Disputes Chamber has broad competences in dealing with disputes relating to deep seabed mining between States parties, between a State party and the International Seabed Authority (ISA), between parties to a contract, between ISA and a prospective contractor, or between the ISA and a state enterprise, or even a natural or juridical

[54] Relevant materials are annexed to John A. Duff, "A Note on the United States and the Law of the Sea: Looking Back and Moving Forward", *Ocean Development and International Law*, Vol. 35(3), 2004, 214–215.

Table 3.5: Declarations in Accordance with Article 298 of the LOSC

State	Date (dd/mm/yyyy)	Exclusion Article 298(1)(a)(b)(c)
Argentina	01/12/1995	All three
Australia	22/03/2002	First one
Belarus*	27/07/2001	All three
Canada	07/11/2003	All three
Cape Verde	10/08/1987	Second one
Chile	25/08/1997	All three
China	25/08/2006	All three
Equatorial Guinea	20/02/2002	First one
Denmark**	16/11/2004	All three
France	11/04/1996	All three
Iceland***	21/06/1985	First one
Italy	13/01/1995	First one
Mexico	06/01/2003	First two
Norway**	24/06/1996	All three
Philippines****	08/05/1984	Unclear
Portugal	03/11/1997	All three
Republic of Korea	18/04/2006	All three
Republic of Palau	27/04/2006	Maritime boundaries
Russia	12/03/1997	All three
Slovenia**	11/10/2001	All three
Spain	19/07/2002	First one
Tunisia	24/04/1985	All three
Ukraine	26/07/1999	First two
United Kingdom	07/04/2003	Second and third ones
Uruguay	10/12/1982	Law enforcement activities

Notes: *The declaration made upon its ratification on 30 August 2006 does not mention maritime boundary delimitation.

**For those countries, they do not accept an arbitral tribunal. It is assumed that they could accept the jurisdiction of international adjudicative bodies.

***Conciliation should be used for any interpretation of Article 83 of the Convention.

****The Philippine declaration states that "The agreement of the Republic of the Philippines to the submission for peaceful resolution, under any of the procedures provided in the Convention, of disputes under Article 298 shall not be considered as a derogation of Philippines sovereignty".

Source: Adapted from UN Office for Ocean Affairs and the Law of the Sea, available at http://www.un.org/Depts/los/convention_agreements/convention_declarations.htm.

person sponsored by a State party.[55] In comparison with the ICJ which is only open to nation States, the Seabed Disputes Chamber can accept disputes submitted by natural or juridical persons. In addition, the jurisdiction of the Chamber is compulsory and no declaration relating to the choice of procedure is required.[56] Of course such jurisdiction is not subject to any exclusion by States parties in accordance with Article 298 of the LOSC. It is well perceived that if deep seabed mining commences, there will be a considerable amount of disputes that need to involve the work of the Seabed Disputes Chamber of the ITLOS for settlement. Unfortunately, there is no prospect that deep seabed mining will take place soon. The commercial exploitation of deep seabed polymetallic nodules seems a "remote possibility".[57] In that case, there are certainly no disputes or cases which would invite the involvement of the ITLOS. It is worth mentioning that the Seabed Disputes Chamber handled the case of *Responsibilities and Obligations of States Sponsoring Persons and Entities with Respect to Activities in the Area* (Request for Advisory Opinion submitted to the Seabed Disputes Chamber) in 2010 and rendered its advisory opinion in 2011.

In comparison, the ICJ can establish its jurisdiction over all legal disputes concerning any question of international law according to the ICJ Statute,[58] but the competence of the ITLOS is only limited to law of the sea disputes. In judicial practice, the ICJ has dealt with more cases on law of the sea disputes than the ITLOS since the establishment of the latter (see Tables 3.3 and 3.6). In addition, it is to be noted that it is generally understood that territorial disputes over offshore or mid-ocean islands are not within the ITLOS competence, though there is a different opinion.[59]

---

[55] See Article 187 of the LOSC.

[56] See Article 287(2) of the LOSC.

[57] See Michael W. Lodge, "International Seabed Authority's Regulation on Prospecting and Exploration for Polymetallic Nodules in the Area", *Journal of Energy and Natural Resources Law*, Vol. 20, 2002, 294.

[58] See Article 36 of the ICJ Statute. Text is available at http://www.icj-cij.org/documents/index.php?p1=4&p2=2&p3=0.

[59] In the view of Judge Wolfrum, in a case of maritime boundary delimitation the ITLOS has competence not only in respect of such delimitation but also in respect of associated issues of delimitation over land or islands and sovereignty over territory. See "Statement by Mr. Rudiger Wolfrum, President of the International Tribunal for the Law of the Sea on Agenda Item 71(a) at the Plenary of the 61st Session of the United Nations General Assembly", 8 December 2006, para. 7, available at http://www.itlos.org/start2_en.html.

Table 3.6: Law of the Sea Cases before the ICJ since the Entry into Force of the LOSC in 1994

2017	Application for revision of the Judgment of 23 May 2008 in the case concerning Sovereignty over Pedra Branca/Pulau Batu Puteh, Middle Rocks and South Ledge (Malaysia/Singapore) (*Malaysia v. Singapore*).
	Request for Interpretation of the Judgment of 23 May 2008 in the case concerning Sovereignty over Pedra Branca/Pulau Batu Puteh, Middle Rocks and South Ledge (Malaysia/Singapore) (*Malaysia v. Singapore*).
2014	Maritime Delimitation in the Caribbean Sea and the Pacific Ocean (*Costa Rica v. Nicaragua*).
2013	Obligation to Negotiate Access to the Pacific Ocean (*Bolivia v. Chile*).
2010	Whaling in the Antarctic (*Australia v. Japan*: New Zealand intervening).
2008	Maritime Dispute (*Peru v. Chile*).
2006	Pulp Mills on the River Uruguay (*Argentina v. Uruguay*).*
2005	Dispute regarding Navigational and Related Rights (*Costa Rica v. Nicaragua*).*
2004	Maritime Delimitation in the Black Sea (*Romania v. Ukraine*).
2003	Sovereignty over Pedra Branca/Pulau Batu Puteh, Middle Rocks and South Ledge (Malaysia/Singapore).
2002	Application for Revision of the Judgment of 11 September 1992 in the Case concerning the *Land, Island and Maritime Frontier Dispute* (*El Salvador/ Honduras: Nicaragua intervening*) (*El Salvador v. Honduras*).
	Frontier Dispute (Benin/Niger).*
2001	Territorial and Maritime Dispute (*Nicaragua v. Colombia*).
1999	Territorial and Maritime Dispute between Nicaragua and Honduras in the Caribbean Sea (*Nicaragua v. Honduras*).
1998	Sovereignty over Pulau Ligitan and Pulau Sipadan (Indonesia/Malaysia).
	Request for Interpretation of the Judgment of 11 June 1998 in the Case concerning the *Land and Maritime Boundary between Cameroon and Nigeria (Cameroon v. Nigeria), Preliminary Objections* (*Nigeria v. Cameroon*).
1996	Kasikili/Sedudu Island (Botswana/Namibia).*
1995	Fisheries Jurisdiction (*Spain v. Canada*).
1994	Land and Maritime Boundary between Cameroon and Nigeria (*Cameroon v. Nigeria*: Equatorial Guinea intervening).

Note: *These cases are disputes concerning rivers but relevant to law of the sea disputes.
Source: Prepared by the author based on the ICJ website.

This is one of the reasons why some of such law of the sea disputes have been submitted to the ICJ rather than to the ITLOS, like the case between Malaysia and Singapore concerning the sovereignty over tiny islets named Pedra Branca and Middle Rocks located at the north entrance of the Straits of Malacca and Singapore.

Looking back at the history of the Third United Nations Conference on the Law of the Sea (UNCLOS III) (1973–1982), there were divergent views regarding the creation of a new international court for law of the sea disputes. Most developing countries, however, supported to have a new court based on their consideration that the ICJ largely represented the Western jurisprudence and lacked universal representation for the whole world community. They did not trust the Euro-centric ICJ.[60] Against this background, it is regrettable to see that since the establishment of the ITLOS, developing countries are still reluctant to use the court being established with their support. Except for the cases concerning prompt release of vessels and provisional measures, they retain their reliance on either the ICJ or arbitration for substantive cases such as maritime boundary delimitation or islands disputes as demonstrated in State practice.

It is generally recognized that the ITLOS has failed to become the pre-eminent forum for the adjudication of maritime delimitations despite the fact that there is a special chamber of maritime boundary disputes within it, and its very existence might be regarded as a symptom of the recent boom of international litigation.[61] So far the ITLOS has only played some supporting or supplementary role in the law of the sea disputes settlement. For its competence in prompt release, it provides legal facilitation to the domestic courts concerned of the countries where the vessels are detained, while for its prescription of provisional measures, it provides assistance to arbitral tribunals which hear the merits of the disputes, despite the fact that the ITLOS is designated as a major, if not a most important, judicial organ to deal with law of the sea disputes. So far the ITLOS has only handled a few cases on their merits, such as the case of

[60] See Zhao Lihai, *Studies on Law of the Sea Issues* (Beijing: Peking University Press, 1996) (in Chinese), pp. 121–122.

[61] Charles Claypoole, "Access to International Justice: A Review of the Trust Funds Available for Law of the Sea-Related Disputes", *International Journal of Marine and Coastal Law*, Vol. 23, 2008, 93.

M/V Saiga No. 2 (St. Vincent and the Grenadines v. Guinea), which was accidentally fallen within the jurisdiction of the Tribunal since it originally should be submitted for arbitration based on the intention of the disputant parties and only later transferred to the Tribunal.[62] Because of this, some States may view the ITLOS as "only a court of first instance, useful for an initial hearing of the facts and for seeking provisional measures or prompt release, but not for a final determination of the dispute".[63] In a word, for the last 24 years since its establishment in 1996, the contribution of the ITLOS to the settlement of law of the sea disputes is not very significant and has not yet met the expectations of the international community. Questions have been thus raised as to whether the ITLOS is in need of reform to respond to emerging challenges.[64]

However, having said that, this does not mean that there is no contribution from the ITLOS to the development of international law, particularly the law of the sea. As the then President of the ITLOS stated, "the Tribunal has been successful in helping States of both developed and developing nations to reach a peaceful solution with respect to cases involving, *inter alia*, the freedom of navigation, prompt release of vessels and their crews, protection and preservation of the marine environment, the commissioning of a nuclear facility and the movement of radioactive materials, land reclamation activities, fisheries, nationality of claims, use of force in law enforcement activities, hot pursuit and the question of the genuine link between the vessel and its flag State".[65] It is pointed out that the Tribunal's examination of the right of the flag State to act on behalf of foreigners on board, particularly relating to ship's crews was influential in the work of the International Law Commission on Diplomatic

---

[62] For a detailed survey of the case, see Louise de La Fayette, "ITLOS and the Saga of the Saiga: Peaceful Settlement of a Law of the Sea Dispute", *International Journal of Marine and Coastal Law*, Vol. 15(3), 2000, 355–392.

[63] See Donald R. Rothwell, "Building on the Strengths and Addressing the Challenges: The Role of Law of the Sea institutions", *Ocean Development and International Law*, Vol. 35, 2004, 148.

[64] *Ibid.*, p. 149.

[65] See "Statement of Judge Rüdiger Wolfrum, President of the International Tribunal for the Law of the Sea, on the Occasion of the Ceremony to Commemorate the Tenth Anniversary of the Tribunal", 29 September 2006, p. 5, available at http://www.itlos.org/start2_en.html.

Protection.[66] Yet, it is recalled that Sir Robert Jennings once wrote that "the primary task of a court of justice is not to 'develop' the law, but to dispose, in accordance with the law, of that particular dispute between the particular parties before it".[67] If there is no case for an international court to handle so as to fulfill such primary task, that court may lose its legitimacy of existence.

Having realized that it could be marginalized among international courts and arbitral tribunals, the ITLOS has made several endeavors for marketing itself. One effort is the establishment of the ITLOS Trust Fund available for law of the sea disputes. A resolution adopted by the UN General Assembly approved the establishment of a voluntary trust fund to assist States in the settlement of disputes through the ITLOS.[68] The balance of the fund as on 31 December 2019 was US$114,622.[69] In comparison with the trust fund for the ICJ established in 1989 (with the amount of US$2,402,864 as of 30 June 2007), the contributed amount to the ITLOS was tiny. It is doubtful how the Tribunal, with this amount of money, can really help, or even encourage disputant States to bring their cases to it. The settlement of law of the sea disputes is sometimes very costly. As it is put, the entire amount of over US$2 million dollars at the ICJ Trust Fund as mentioned above would not be enough to finance the total costs of one party to a land and maritime boundary dispute of relative complexity.[70]

The other is the initiation of regional workshops to be held around the world to impress the localities of the existence and availability of the Tribunal. Since 2006 such workshops, with the generous support from the Korean International Cooperation Agency (KOICA), have been run in Dakar, Libreville, Kingston, Singapore, Bahrain and Buenos Aires, with the intention "to provide government experts working in the maritime

---

[66] See Sir Michael Wood, "The International Tribunal for the Law of the Sea and General International Law", *International Journal of Marine and Coastal Law*, Vol. 22, 2007, 363–364.

[67] Robert Jennings, "Role of the International Court of Justice", *British Yearbook of International Law*, Vol. 68, 1997, 41.

[68] See UN Doc A/RES/55/7, 27 February 2001, particularly Annex 1. Text is available at http://daccessdds.un.org/doc/UNDOC/GEN/N00/559/81/PDF/N0055981.pdf?OpenElement.

[69] See Annual report of the International Tribunal for the Law of the Sea for 2019, SPLOS/30/2, 31 March 2020, p. 20, available at https://undocs.org/en/splos/30/2.

[70] See Charles Claypoole, *supra* note 61, p. 87.

field with insight into the Convention's dispute settlement system" and special attention is given to the jurisdiction of the Tribunal and its procedures.[71] It is hoped that these workshops can really promote the image of the ITLOS and encourage States concerned to use it.

Another endeavor from the ITLOS is its call on States parties to request the Tribunal to give advisory opinions. According to the former ITLOS President, these opinions "can be of great benefit in the solution of international disputes" and the "Tribunal's advisory function may guide parties to a mutually satisfactory result". The legal basis of the advisory competence, according to him, is Article 21 of the ITLOS Statute which vests the Tribunal with jurisdiction with respect to "all matters specifically provided for in any other agreement which confers jurisdiction on the Tribunal" and such agreement may vest competence to issue an advisory opinion in the Tribunal.[72] Here the Tribunal has raised an innovative but controversial point regarding its jurisdictional competences. According to the LOSC, only the Seabed Disputes Chamber of the ITLOS is granted the competence to give advisory opinions at the request of the Assembly or the Council of the International Seabed Authority on legal questions arising within the scope of their activities,[73] while the ITLOS itself has no jurisdiction to give such opinions. It is doubtful whether the Tribunal can enjoy such competence as explained by the then ITLOS President, though he holds a good intention to enable the Tribunal to make contributions to the settlement of international disputes through an alternative. In my personal view, if the ITLOS would really enjoy such a competence, there would be certainly an express provision embodied in the LOSC and/or the ITLOS Statute just like those provisions in the ICJ Statute for the ICJ and in the LOSC for the Seabed Disputes Chamber. Despite the aforementioned controversy, the Tribunal as a whole handled a case of advisory proceedings in 2013 — Request for an Advisory Opinion Submitted by the Sub-Regional Fisheries Commission (SRFC) and rendered its advisory opinion in 2015.[74]

---

[71] See Statement by Rüdiger Wolfrum, 16 June 2008, *supra* note 68, pp. 6–7.

[72] *Ibid.*, p. 5.

[73] See Article 191 of the LOSC. In accordance, the Chamber handled the case of *Responsibilities and obligations of States Sponsoring Persons and Entities with Respect to Activities in the Area* (Request for Advisory Opinion submitted to the Seabed Disputes Chamber) in 2010 and rendered its advisory opinion in 2011.

[74] For details of the case, see https://www.itlos.org/en/cases/list-of-cases/case-no-21/.

Finally, it is worthwhile to look at briefly the question related to the fragmentation of international law which could be caused by the proliferation of international judicial organs. Proliferation of international courts/tribunals has triggered a doubt as to the necessity and efficiency of their establishment. For example, the International Criminal Tribunal for Rwanda (ICTR) only completed eight trials in its first eight years of operation. Second, such proliferation may also bring inconsistency in international jurisprudence as well as overlapping jurisdiction. The judicial practices concerning the SBT case indicate there is a suspicion of conflicting international jurisprudence between the ITLOS and the Annex VII Arbitral Tribunal which revoked the provisional measures granted by the former. It is recalled that Judge Gilbert Guillaume, former President of the ICJ once criticized such proliferation and brought his concerns to the United Nations.[75] Concerns in this respect are also from Asian countries. As a Chinese Delegate to the 6th Committee of the UN General Assembly stated, "with the ever increase of international judicial organs, how to ensure uniform application of international law so as to reduce the negative impact of the fragmentation of international law while these organs properly fulfill their judicial functions is a question that deserves the attention of the international community".[76] Therefore, the most efficient and problem-solving way is to maximize the use of existing international courts, instead of establishing more such courts. On the other hand, there is a need of reform in the existing courts so as to meet the expectations of the international community.

---

[75] See "Address by H.E. Judge Gilbert Guillaume, President of the International Court of Justice, to the United Nations General Assembly", 26 October 2000, available at http://www.icj-cij.org/icjwww/ipresscom/SPEECHES/iSpeechPresident_Guillaume_GA55_20001026.htm. (He gave an example of jurisprudential inconsistency made by the ICTY in ruling on the merits in the Tadic case that "the International Criminal Tribunal for the former Yugoslavia recently disregarded case-law formulated by the International Court of Justice in the dispute between Nicaragua and the United States of America. The Court had found that the United States could not be held responsible for acts committed by the contras in Nicaragua unless it had had 'effective control' over them. After criticizing the view taken by the Court, the Tribunal adopted a less strict standard for Yugoslavia's actions in Bosnia and Herzegovina and replaced the notion of 'effective control' with that of 'overall control', thereby broadening the range of circumstances in which a State's responsibility may be engaged on account of its actions on foreign territory".)

[76] Duan Jielong, "Statement on the Rule of Law at the National and International Levels", *Chinese Journal of International Law*, Vol. 6, 2007, 187.

East Asian Attitudes and Practices

As shown previously there are only three countries in East Asia which accepted ICJ's compulsory jurisdiction. East Asian resort to international courts is rare (see Table 3.7). The reluctance of East Asian countries to use international adjudication for dispute settlement may be rooted in Asian cultures and legal traditions. In China, legalism was long replaced by Confucianism which emphasized the rule by virtue rather than the rule by law. For the ordinary people in the past, the use of the law court for dispute settlement was regarded as unfriendly and confrontational, and a loss of face. They tended to seek a solution by negotiations, and/or third-party mediation. Although things have changed dramatically over time, such unfavorable mentality for law courts may still exist and has been reflected in China's attitude toward international adjudication.

However, it seems that there is a tendency of increase in using international courts in East Asia over the years. There are two significant cases submitted by East Asian States to the ICJ including the case of *Sovereignty over Pulau Litigan and Pulau Sipadan* (Malaysia/Indonesia) (1998–2002) and the case on *Sovereignty over Pedra Branca, Middle Rocks and South Ledge* (Malaysia/Singapore) (2003–2008). Both cases are concerning territorial disputes over small islands in the seas. In the first case, the ICJ granted the disputed islands to Malaysia and in the second one, the Court granted one to Malaysia and one to Singapore so as to reach a so-called "win–win" solution.[77]

There are also five cases involving East Asian States before the ITLOS. One is the case of *Southern Bluefin Tuna* (*New Zealand/Australia v. Japan*). Japan was forced to respond to the case submitted by Australia and New Zealand in July 1999. The two Southern Ocean countries requested the Tribunal to render provisional measures (an interim injunction) to stop Japan's unilateral experimental fishing of Southern Bluefin Tuna (SBT) in 1999. They accused Japan of having breached its obligations under Articles 64 and 116–119 of the LOSC in relation to the conservation and management of the SBT stock by failing to adopt necessary conservation measures for its nationals fishing on the high seas so as to maintain or restore the SBT stock to levels which can produce the

[77]For details about the two cases, see the ICJ website: http://www.icj-cij.org/icjwww/idocket/iinma/iinmajudgment/iinma_ijudgment_20021217.PDF and http://www.icj-cij.org/docket/files/130/14492.pdf?PHPSESSID=a4740e3448ea1887e19ce0a89525d037.

Table 3.7: Cases Submitted by Asian Countries to the ICJ

States	Cases
Cambodia	Temple of Preah Vihear (with Thailand) (1959–1962)
	Request for Interpretation of the Judgment of 15 June 1962 in the Case concerning the Temple of Preah Vihear
	(*Cambodia v. Thailand*) (with Thailand) (2011–2013)
India	Right of Passage over Its Territory (with Portugal) 1955–1960
	Appeal Relating to the Jurisdiction of the ICAO Council (with Pakistan) (1971–1972)
	Trial of Pakistani Prisoners of War (with Pakistan) (1973)
	Aerial Incident of 10 August 1999 (with Pakistan) (1999–2000)
	Obligations concerning Negotiations relating to Cessation of the Nuclear Arms Race and to Nuclear Disarmament (*Marshall Islands v. India*) (2014–2016)
	Jadhav (*India v. Pakistan*) (2017–2019)
Indonesia	Sovereignty over Pulau Litigan and Pulau Sipadan (with Malaysia) (1998–2002)
Iran	Anglo-lanai Oil Co. (with United Kingdom) (1951–1952)
	United States Diplomatic and Consular Staff in Tehran (with United States) (1979–1981)
	Aerial Incident of 3 July 1988 (with United States) (1989–1996)
	Oil Platforms (with United States) (1992– 2003)
Japan	Whaling in the Antarctic (*Australia v. Japan*: New Zealand intervening) (2010–2014)
Malaysia	Sovereignty over Pulau Litigan and Pulau Sipadan (with Indonesia) (1998–2002)
	Sovereignty over Pedra Branca, Middle Rocks and South
	Ledge (with Singapore) (2003–2008)
	Application for revision of the Judgment of 23 May 2008 in the case concerning Sovereignty over Pedra Branca/Pulau Batu Puteh, Middle Rocks and South Ledge (Malaysia/Singapore) (*Malaysia v. Singapore*) (2017–2018)
	Request for Interpretation of the Judgment of 23 May 2008 in the case concerning Sovereignty over Pedra Branca/Pulau Batu Puteh, Middle Rocks and South Ledge (Malaysia/Singapore) (*Malaysia v. Singapore*) (2017–2018)

(*Continued*)

Table 3.7: (*Continued*)

States	Cases
Pakistan	Appeal Relating to the Jurisdiction of the ICAO Council (with India) (1971–1972)
	Trial of Pakistani Prisoners of War (with India) (1973)
	Aerial Incident of 10 August 1999 (with India) (1999–2000)
	Obligations concerning Negotiations relating to Cessation of the Nuclear Arms Race and to Nuclear Disarmament (*Marshall Islands v. Pakistan*) (2014–2016)
	Jadhav (*India v. Pakistan*) (2017–2019)
Singapore	Sovereignty over Pedra Branca, Middle Rocks and South
	Ledge (with Malaysia) (2003–2008)
	Application for revision of the Judgment of 23 May 2008 in the case concerning Sovereignty over Pedra Branca/Pulau Batu Puteh, Middle Rocks and South Ledge (Malaysia/Singapore) (*Malaysia v. Singapore*) (2017–2018)
	Request for Interpretation of the Judgment of 23 May 2008 in the case concerning Sovereignty over Pedra Branca/Pulau Batu Puteh, Middle Rocks and South Ledge (Malaysia/Singapore) (*Malaysia v. Singapore*) (2017–2018)
Thailand	Temple of Preah Vihear (with Cambodia) (1959–1962)
	Request for Interpretation of the Judgment of 15 June 1962 in the Case concerning the Temple of Preah Vihear (*Cambodia v. Thailand*) (with Cambodia) (2011–2013)
Timor-Leste	Questions relating to the Seizure and Detention of Certain Documents and Data (*Timor-Leste v. Australia*) (2013–2015)

Source: Compiled by the Author.

maximum sustainable yield; carrying out unilateral experimental fishing in 1998 and 1999 which has or will result in SBT being taken by Japan over and above previously agreed Commission national allocations; and taking unilateral action contrary to the rights and interests of Australia and New Zealand.[78]

[78] See Order on Southern Bluefin Tuna Cases, 27 August 1999, ITLOS, available at http://www.itlos.org/start2_en.html.

In its response Japan asked the Tribunal to deny the provisional measures requested by Australia and New Zealand. In Japan's view, the two countries do not meet the conditions set forth in international law: "First, the Annex VII tribunal must have *prima facie* jurisdiction. This means among other things that the dispute must concern the interpretation or application of UNCLOS and not some other international agreement. Second, Australia and New Zealand must have attempted in good faith to reach a settlement in accordance with the provisions of UNCLOS Part XV, Section 1".[79] Japan did not think that ITLOS had the authority to prescribe any provisional measures in question. However, Japan had no objection to the jurisdiction of ITLOS if the Tribunal thinks it possesses the jurisdiction over the case. Under this circumstance, Japan requested that "the Tribunal grant Japan provisional relief in the form of prescribing that Australia and New Zealand urgently and in good faith recommence negotiations with Japan for a period of six months to reach a consensus on the outstanding issues between them, including a protocol for a continued Experimental Fishing Program (EFP) and the determination of a Total Allowable Catch (TAC) and national allocations for the year 2000".[80] In addition, Japan asked the Tribunal to "require Australia and New Zealand to fulfill their obligations to continue negotiations over this scientific dispute".[81]

In August 1999, the Tribunal rendered an order containing several decisions: first, the three countries concerned should not aggravate or extend the dispute and their annual catches should not exceed the annual national allocations at the levels last agreed by the parties, respectively. They should refrain from conducting an experimental fishing program involving the taking of a catch of SBT, except with the agreement of the other parties or unless the experimental catch is counted against its annual national allocation, and resume negotiations without delay with a view to reaching agreement on measures for the conservation and management of SBT. Second, each party should submit the initial report to ITLOS and further reports and information upon request. Third, the provisional measures prescribed in the Order should be notified by the Registrar through appropriate means to all States Parties to the Convention participating in the fishery for SBT.

---

[79] *Ibid.*

[80] *Ibid.*

[81] *Ibid.*, para. 33.

Unlike the aforementioned *Sovereignty over Pulau Litigan and Pulau Sipadan* case before ICJ, this one directly concerns fishery management. It is perceived that disputes over fisheries will occur and the precedent set forth by ITLOS is insightful and can be used as one of the legal references to the settlement of fishery disputes between/among East Asian countries by themselves.

The second legal implication of this case for East Asia lies in the fact that Japan was unwillingly brought to the case as it was reported that Japan initially preferred mediation.[82] As we know, Japan did not make any statement on which compulsory jurisdiction mechanism set forth in the LOSC should accept upon its signing and ratification of the convention. Therefore, Japan is deemed to have accepted arbitration in accordance with the relevant provision of the LOSC. ITLOS was clearly not Japan's choice. Australia and New Zealand cleverly used Article 290 of the LOSC, which allows ITLOS to prescribe provisional measures pending a constitution of an arbitral tribunal to which a dispute is being submitted,[83] to force Japan to respond before the Tribunal. This precedent indicates that even without acceptance of the jurisdiction of ITLOS, a State may still be possibly brought for a case before the Tribunal. This has important implications since all the East Asian countries did not make the choice of dispute settlement mechanisms and are therefore deemed to have accepted arbitration under the LOSC. They should be readily prepared to deal with some rather tricky legal situation in which they may be entangled.

The third legal implication for East Asia concerns the relationship between a regional agreement and a global treaty, and in this case, between the 1993 Convention for the Conservation of SBT[84] and the LOSC. The 1993 Convention provides two third-party dispute settlement mechanisms: ICJ and arbitration after the failure of the direct negotiation between/among the parties to a dispute. ITLOS is not listed. However, the LOSC also deals with fishery management, any issue resulting from the interpretation and implementation of a regional fishery agreement will naturally come within the legal domain of the LOSC. This also has impact on procedural matters such as dispute settlement mechanism. As one commentator points out, if the SBT legal reasoning is applied *mutatis*

[82] See Romano, *supra* note 4, p. 201.

[83] See Article 290 of the LOSC.

[84] Text is available at http://www.ccsbt.org/docs/pdf/about_the_commission/convention.pdf.

mutandis to regional fishery regimes, it means that disputes arising out of the implementation of regional agreements might be subject to a dispute settlement mechanism provided for in the LOSC.[85] So far there is no such regional fishery agreement in East Asia. Nevertheless, East Asian countries, particularly those involved in distant fishing, often, sometimes obligatorily, join regional fishery regimes, for example, in the North Pacific and South Pacific. They might be brought to a settlement mechanism against their will once disputes would occur.

On the other hand, there are a number of bilateral fishery agreements signed between East Asian States. Such kind of agreements, in most cases, only provides a mechanism of direct negotiation between the parties to resolve any dispute. The question arises as to what kind of third-party mechanism they have to resort to when direct negotiation has failed. Could one of the parties submit the dispute to ITLOS or other third-party settlement organs? The LOSC allows States parties to opt for a compulsory mechanism for disputes that are related to their sovereign rights to the living resources in the EEZ.[86] That means coastal States may have the right to refuse the use of the compulsory mechanism including ITLOS. However, the recent bilateral fishery agreements signed in East Asia have established joint and/or common fishery management zones within their EEZs, which consist of shared zones between the two relevant countries. Even some of such zones remain the *de facto* status of the high seas, such as the Intermediate Zone established in the East China Sea based on the 1997 Sino-Japanese Fishery Agreement.[87] Although it is not clear about whether a party to a dispute over the implementation of a bilateral fishery agreement, particularly concerning the fishery management in the shared zones, is eligible to resort to a compulsory mechanism under the LOSC, the window of possibility of the ITLOS intervention could always remain open.

Another related question is whether an entity outside the framework of the LOSC can resort to these dispute settlement mechanisms under that convention. What if Taiwan, instead of Japan, had been sued by Australia and New Zealand before ITLOS since Taiwan became a member of the

[85] Romano, *supra* note 4, p. 210.

[86] See Article 297(3) of the LOSC.

[87] See Zou Keyuan, "Sino-Japanese Joint Fishery Management in the East China Sea", *Marine Policy*, Vol. 27(2), 2003, 134.

Southern Bluefin Tuna Commission established by the 1993 Convention in the name of "Fishing Entity of Taiwan" in August 2002?

Following the ITLOS order, an arbitral tribunal is correspondingly established to deal with the substance of the dispute over SBT. In August 2000, the tribunal issued its award ruling that it had no jurisdiction over the case.[88] Though the tribunal declined its jurisdiction to deal with the merits of the dispute, the significance of the case lies in the fact that this arbitral tribunal was the first one established in accordance with Annex VII to the LOSC. With its award, the provisional measures ordered by the ITLOS were naturally no longer in force. It is interesting to note that the case was registered with the International Centre for Settlement of Investment Disputes (ICSID) under the World Bank, rather than with the PCA.

Another case is the *Case Concerning Land Reclamation by Singapore in and around the Straits of Johor (Malaysia v. Singapore)*. The case was first submitted to the ITLOS in September 2003 by Malaysia requesting the Tribunal to prescribe provisional measures to stop Singapore's land reclamation activities in the vicinity of the maritime boundary between the two States or of areas claimed as territorial waters by Malaysia pending the decision of the Arbitral Tribunal.[89] In this case, the Tribunal reiterated its statement expressed in the *Case of MOX Plant* that "the duty to cooperate is a fundamental principle in the prevention of pollution of the marine environment under Part XII of the Convention and general international law".[90] The Tribunal issued an Order in October 2003 prescribing that Malaysia and Singapore should cooperate to establish a group of

---

[88]The text of the award is available at http://untreaty.un.org/cod/riaa/cases/vol_XXIII/1-57.pdf. For a comprehensive survey of the case, see Barbara Kwiatkowska, "The *Australia and New Zealand v. Japan Southern Bluefin Tuna* (Jurisdiction and Admissibility) Award of the First Law of the Sea Convention Annex VII Arbitral Tribunal", *International Journal of Marine and Coastal Law*, Vol. 16(2), 2001, 239–293.

[89]Due to its land scarce, Singapore actively carries out land reclamation. According to a statistic, Singapore's land area increased from 581 km2 in 1966 to 695 km2 in 2003. See Kog Yue-Choong, "Environmental Management and Conflict in Southeast Asia: Land Reclamation and Its Political Impact", IDSS Working Paper No. 101 (Institute for Defence and Strategic Studies, Nanyang Technological University, Singapore), January 2006, p. 7.

[90]Order on the *Case Concerning Land Reclamation by Singapore in and around the Straits of Johor* (*Malaysia v. Singapore*): Request for Provisional Measures, 8 October 2003, para 92, available at http://www.itlos.org/start2_en.html.

independent experts to study and prepare an interim report on the subject matter, directing that Singapore should not "conduct its land reclamation in ways that might cause irreparable prejudice to the rights of Malaysia or serious harm to the marine environment, taking especially into account the reports of the group of independent experts" and deciding that "Malaysia and Singapore shall each submit the initial report referred to in Article 95, paragraph 1, of the Rules, not later than 9 January 2004 to this Tribunal and to the Annex VII arbitral tribunal, unless the arbitral tribunal decides otherwise".[91] Following the Order of the ITLOS, the two sides jointly established the Group of Experts which submitted its Final Report to the two sides on 5 November 2004. Based on the Report, the two sides reached an agreement to the resolution of the said dispute on 26 April 2005. According to it, the two sides agreed to terminate the case by several arrangements including the modification by Singapore of its final design of the shoreline of its land reclamation, Singapore's compensation to affected Malaysia fishermen, and the discussion and monitoring of the environmental impacts in the Straits of Johor by the Malaysia–Singapore Joint Committee on the Environment (MSJCE).[92] Finally, it should be noted that the jurisdiction of ITLOS in the aforementioned two cases is only limited to the prescription of provisional measures without touching upon the merits of the disputes. They are thus not formal as the former ones before the ICJ. However, at least in the second case, the Order issued by the ITLOS played a role in promoting the dispute settlement by an agreement finally reached between the two disputing parties.

There are three critical developments that will influence the East Asian States to accept international adjudication. First, the domestic changes in each individual country. In the case of China, during Mao Tse-tung's era (1949–1976), China refused utterly to abide by any such mechanism, whether judicial or arbitral. During Deng Xiaoping's era (1978–1997), China began to consider the practicality and necessity of international judicial means and accepted international arbitration through the ratification of many international treaties, in particular those relating to economic and trade areas. In the post-Deng era (after 1997), China has become even more receptive, particularly after it joined the WTO and adopted the market economy policy. Communist ideology is on longer a

---

[91] *Ibid.*, para 102.

[92] See Case Concerning Land Reclamation by Singapore in and around the Straits of Johor (*Malaysia v. Singapore*): Settlement Agreement (on file with the author).

guiding principle. Though still reluctant to accept international adjudication, China seems quite ready and comfortable with international arbitration.

A typical example is the China–ASEAN Agreement on Dispute Settlement Mechanism of the Framework Agreement on Comprehensive Economic Co-Operation between the Association of Southeast Asian Nations and the People's Republic of China (the Dispute Settlement Mechanism Agreement) on 29 November 2004.[93] It designates three means of dispute settlement in international law as its working mechanisms including consultation, conciliation or mediation and arbitration. If the consultations could not solve the dispute within the prescribed time limit, the complaining party may make a written request to the party complained against to appoint an arbitral tribunal. The signing of this agreement is hailed as having "far-reaching significance" even beyond the China–ASEAN relations as it is "a sign of the changing attitude of the Chinese government towards international dispute settlement methods", and it is the first time that China "had signed such a special and detailed agreement with other states or international organizations".[94] Nevertheless the Agreement still contains limitations. First, it is limited to economically related disputes within the realm of the Framework Agreement. Second, though there is a stipulation about transparency regarding the information flow between the disputing parties, all the sessions and documents submitted are kept confidential, thus making the arbitral tribunal vulnerable to pressures from the parties to a dispute and raising the doubts of the impartiality of the tribunal. Third, though the decision of the tribunal is final, the implementation provisions are not strong enough to force the failing party to implement the recommendations and rulings of the tribunal. The retaliation measures taken by the complaining party are only temporary and should be removed when they are considered inconsistent with the Framework Agreement. In case such temporary measures are not removed, it is likely then that a new dispute could arise between/among

[93] Text is available at https://asean.org/?static_post=agreement-on-dispute-settlement-mechanism-of-the-framework-agreement-on-comprehensive-economic-co-operation-between-the-association-of-southeast-asian-nations-and-the-people-s-republic-of-china-5. It came into force on 1 January 2005.

[94] See Zeng Lingliang, "ASEAN–China Relations: An International Law Perspective", in John Wong, Zou Keyuan and Zeng Huaqun (eds.), China-ASEAN Relations: *Economic and Legal Dimensions* (Singapore: World Scientific, 2006), pp. 54–55.

the parties to the ACFTA Framework Agreement. It is obvious that the relevant provisions regarding the implementation of the arbitral awards need to be improved in future. It remains to be seen whether the dispute settlement mechanisms established in the Agreement can function well in practice, and by judging the attitudes and policies of China and most of the ASEAN countries toward dispute settlement, it is reasonably assumed that most, if not all, of the potential future disputes will be solved through consultations.

Second, the mushrooming of international and regional organizations and institutions will push East Asian countries toward further integration and interdependence. Potential and existing disputes will be curbed and/or resolved via these arrangements. It is recorded that the EU mediation of the Gabčíkovo–Nagymaros Project dispute played a key role in persuading the parties to submit the dispute to the ICJ.[95] The existing regional organizations such as the Association of the Southeast Asian Nations (ASEAN) may play a similar role in facilitating the settlement of law of the sea disputes in East Asia.

Finally, the globalization process brings East Asian countries to the new front of new challenges and opportunities. The concept of sovereignty has been changed from traditional self-protection to modern interdependence. Interdependence, integration and globalization will enhance the establishment of the concerted and deep-rooted East Asian conviction that rule of law must be applied at the bilateral, multilateral, regional and global levels. In this context, law of the sea disputes can be effectively controlled via cooperative channels and third-party intervention mechanisms so as to realize the goal of their avoidance, prevention and settlement.

[95] See Paul R. William, *International Law and the Resolution of Central and East European Transboundary Environmental Disputes* (New York: Macmillan Press, 2000), p. 223.

Part II

Tackling Non-Traditional Maritime Security Issues

Chapter 4

Maritime Enforcement of United Nations Security Council Resolutions

The Legal Nature of UNSC Resolutions

The Charter of the United Nations designates the United Nations Security Council (UNSC) as one of the principal organs of the United Nations, assuming the "primary responsibility for the maintenance of international peace and security".[1] It plays key role in two major areas concerning international peace and security: dispute settlement and peace and security maintenance and restoration. The second area is more connected to the theme of this chapter.

According to the UN Charter, the Security Council has the power to "determine the existence of any threat to the peace, breach of the peace, or act of aggression and shall make recommendations, or decide what measures shall be taken" "to maintain or restore international peace and security".[2] It may decide what non-military measures, such as economic sanctions and the severance of diplomatic relations, to be employed to give effect to its decisions,[3] and take military actions when the non-military measures have proved to be inadequate. "Such action may include demonstrations, blockade, and other operations by air, sea, or land forces of Members of the United Nations".[4]

[1]Article 24(1) of the UN Charter.

[2]Article 39 of the UN Charter.

[3]Article 41 of the UN Charter.

[4]Article 42 of the UN Charter.

Because of the aforementioned stipulations, it is commonly accepted that UNSC resolutions possess legal binding force, though they are not equivalent to treaties. In fact, the UNSC attempts to prepare its resolutions more like a treaty. This can be seen through the wording and formality of its resolutions. Take Resolution 1540 (2004) as an example.[5] It concerns the proliferation of nuclear, chemical and biological weapons. The resolution contains definitions on "means of delivery", "non-State actor", etc., uses the word "shall" — a very strong word usually appearing in treaties and laws, and intends to bind all States (not UN Members alone).[6]

The terminologies used in the UNSC resolutions also reinforce the degree of its compelling binding force. For example, the term "decide" must be stronger than the terms such as "require" or "urge", which are obviously less compelling. For that reason, when deliberations on draft resolutions take place, UNSC members, due to their different interests and positions, usually debate very hotly on which terms should be adopted for the resolution to be passed.[7]

After North Korea's nuclear weapon test, the UNSC passed Resolution 1718 condemning North Korea's irresponsible act and called it to halt.[8] This resolution has been discussed in the international legal circle as it is regarded to contain "several uncommon provisions".[9] "Unlike the two previous resolutions on the DPRK, which had not explicitly referred to Chapter VII, the new resolution contains provisions legally binding the DPRK and UN member States".[10] Thus, Resolution 1718 is

---

[5]UN Doc. S/RES/1540 (2004). It was adapted on 28 April 2004.

[6]Like "all States *shall* refrain from providing any form of support to non-State actors that attempt to develop, acquire, manufacture, possess, transport, transfer or use nuclear, chemical or biological weapons and their means of delivery". (italic added).

[7]China and Japan took very different views on the draft Resolution 1695 regarding the possible sanctions to be imposed on North Korea. When Japan asked to include Chapter VII of the UN Charter, China disagreed. Regarding the prohibition of export of missile and related materials to North Korea, the Japanese draft used the term "decide" while the Chinese used "call on" instead. As a result, the final version of the resolution uses the term "require" as comprise.

[8]UNSC Resolution 1718 (2006), 14 October 2006, reprinted in 45 ILM 1237 (2006).

[9]See Andreas L. Paulus and Jörn Műller, "Security Council Resolution 1718 on North Korea's Nuclear Test", *ASIL Insight*, Vol. 10(29), 3 November 2006, available at http://www.asil.org/insights/2006/11/insights061103.html.

[10]*Ibid.*

"unique in going beyond the detailed obligations imposed on Iraq in Resolution 687 (1991)".[11]

Since the UNSC resolutions are unilaterally adopted to impose sanctions upon a Member State, despite its authorization from the UN Charter, the countries to be punished are usually very averse to these resolutions and respond very negatively by stating that they will never accept them. This can be seen in UNSC Resolution 1747 adopted on 24 March 2007 against Iran.[12] In response, Iran ceased its cooperation with the International Atomic Energy Agency (IAEA) for certain projects and pledged to continue its nuclear program. The Iranian condemnation and defiance of the UNSC resolution no doubt significantly qualifies the validity of the resolution and has brought more difficulties in its implementation and enforcement.

It is mentioned in the scholarly work that the International Court of Justice (ICJ) never answered the request on the validity of the UNSC resolutions: "in the *Aerial Incident at Lockerbie* it declined to question the validity of a Security Council resolution adopted under Chapter VII. In the *Genocide* case it refused a Bosnian request for the indication of provisional measures which would in effect have opposed the application to Bosnia of an arms embargo imposed by the Council in Resolution 713 (1991)".[13] The superficial reason is that the UNSC has the primary responsibility for the maintenance and restoration of peace and security around the world, it is not appropriate for the ICJ to assess, or even adjudicate the validity of the UNSC resolutions. However, is it possible to suggest that the UNSC resolutions in the past have never been really effectively implemented in practice so that any assessment of them would definitely become a hot potato, inviting more controversies than clarification?

Even if the UNSC Resolution 1718 was hailed as a milestone step for the resolution of the North Korean nuclear issue, the substantive achievements were made not at the table of the Security Council, but through the Six-Party Talks, though Resolution 1718 did create some international

---

[11] *Ibid.*

[12] UN Doc. S/RES/1747 (2007).

[13] Christopher Greenwood, "The International Court of Justice and the Use of Force", in Vaughan Lowe and Malgosia Fitzmaurice (eds.), *Fifty Years of the International Court of Justice: Essays in Honour of Sir Robert Jennings* (Cambridge: Cambridge University Press, 1996), p. 376.

pressure upon North Korea. Related is the question concerning the duration of validity of a UNSC resolution. Unlike a treaty, there is no clause in a UNSC resolution which specifies its duration of validity. Resolution 1718 is the response of the UNSC to the nuclear test undertaken by North Korea.

Nevertheless, it is fair to say that the UNSC has contributed tremendously to world peace and security. The ineffectiveness of its resolution is not the fault of the Council, rather a fault resulting from the deficiencies of our contemporary international structure which is centered upon sovereignty of individual nation States. Without an organ above individual States, the Security Council can actually call for, instead of ordering or instructing, States, more often the UN Member States, to implement and enforce its resolutions. In this sense, the Security Council is more like a consultative body, despite the authority granted by the UN Charter.

Requirements for Implementation and Enforcement

The duty of a Member State

Once a country joins the United Nations, it is bound by the UN Charter, so as the decisions made by the Security Council as the UN Charter provides that UN members agree to accept and carry out the decisions of the Security Council in accordance with the Charter.[14] Furthermore, UN Member States have the duty to contribute to the maintenance of international peace and security by supporting materialistically the actions undertaken by the Security Council, with armed forces, assistance, and facilities, including rights of passage.[15] Article 45 of the Charter clearly provides that

> "In order to enable the United Nations to take urgent military measures, Members shall hold immediately available national air-force contingents for combined international enforcement action. The strength and degree of readiness of these contingents and plans for their combined action shall be determined within the limits laid down in the special agreement or agreements referred to in Article 43, by the Security Council with the assistance of the Military Staff Committee".

[14] Article 25 of the UN Charter.

[15] Article 43 of the UN Charter.

Military actions can be carried out by all or some of the UN Member States, as the Security Council determines. However, the history of UN peacekeeping operations indicates that only some of the States, particularly those within the inflicted region, have been involved in military actions authorized by the UNSC.

Following the UNSC call, UN Member States may take measures domestically in compliance with the UNSC request. After the adoption of several resolutions by the UNSC regarding anti-terrorism and the prevention of nuclear proliferation, China revised its Regulations on the Control of Nuclear Export in November 2006.[16] New purposes such as the prevention of nuclear terrorist act have been added to the Regulations.[17] New provisions are also added. The most significant addition is the provision that "the Recipient Government shall guarantee not to produce enriched uranium with more than 20% density with enriched uranium facilities, technology or any facility based on such technology supplied by China, without the consent of the Chinese Government".[18] Meanwhile, China revised the Regulations on the Control of Exporting Dual-use Nuclear Items and Related Technology in January 2007 as a supplement to the revision of the 2006 Regulations.[19]

Some UN Member States tend to make specific laws or regulations so as to enforce the sanction measures imposed by the UNSC on a State which has breached its international obligations. For example, New Zealand introduced the United Nations Sanctions (Iran) Regulations 2007 to implement UNSC Resolution 1737, which took effect from 23 March 2007, prohibiting the export to or import from Iran of items, materials, equipment and goods that could contribute to Iran's enrichment-related, reprocessing or heavy water-related activities, or to the development of nuclear weapon delivery systems.[20]

[16]The Regulations came into effect on 1 December 2006. The text is available in *People's Daily* (in Chinese), 2 December 2006, p. 6.

[17]Article 1 of the 2006 revised Regulations.

[18]Article 5(5) of the 2006 revised Regulations.

[19]It came into force on 26 January 2007. The text is available at http://www.gov.cn/zwgk/2007–02/16/content_529172.htm.

[20]New Zealand Customs Service, "United Nations import and export sanctions against Iran from 23 March 2007", available at http://www.customs.govt.nz/about/News/unsanctiions iran230207.htm.

Regarding the enforcement of UNSC resolutions by domestic courts, there are two cases in the United States on this matter: in the case of *Diggs v. Richardson* (1976), the court held that the UNSC resolution condemning the continued presence of South Africa in Namibia was "not self-executing" and "not self-enforcing so as to grant individually enforceable rights".[21] In the case of *Diggs v. Shultz* (1972), the court "refused to enjoin the importation of metallurgical chromite from Southern Rhodesia in violation of a United Nations Security Council resolution of 1966, in which the United States joined, and which called for a trade embargo of all that country's products".[22]

In comparison with the ambiguity in US courts, British judicial bodies have more clear-cut answer regarding the validity of UNSC resolutions, though such resolutions are not incorporated into British law. There are several cases held by the House of Lords. In the case of *Kuwait Airways Corp. v. Iraqi Airways Co.* (2002), "the Lords invoked unincorporated Security Council resolutions, *jus cogens* norms and customary international law to determine the status of Iraq's actions as breaches of international law".[23] The *Al-Jedda* case (2006) shows that if there is a conflict between a provision of domestic law and a Chapter VII UNSC resolution, "UN Charter 103 required that the domestic right be overridden".[24]

The maritime factor in enforcement

The most important legal document in this respect is the 1982 United Nations Convention on the Law of the Sea (LOSC) which came into force in 1994. The Convention establishes jurisdictions respectively for coastal States, flag States and port States. As for the safety of navigation and maritime law enforcement, flag States continue to assume principal responsibility. The LOSC provides that "every State shall effectively exercise its jurisdiction and control in administrative, technical and social matters

---

[21] Russell G. Donaldson, "United Nations resolution as judicially enforceable in United States domestic courts", 42 America Law Reports Federal 577, available at Westlaw UK.

[22] *Ibid.*

[23] Katherine R. Thomas, "The Changing Status of International Law in English Domestic Law", *Netherlands International Law Review*, Vol. 53, 2006, 373.

[24] *Ibid.*, p. 395.

over ships flying its flag".[25] It provides that "Every State shall take such measures for ships flying its flag as are necessary to ensure safety at sea with regard, *inter alia*, to: (a) the construction, equipment and seaworthiness of ships; (b) the manning of ships, labour conditions and the training of crews, taking into account the applicable international instruments; (c) the use of signals, the maintenance of communications and the prevention of collisions" (Article 94.3). Flag States should ensure compliance by vessels flying their flag or of their registry with applicable international rules and standards and with their laws and regulations and provide for the effective enforcement of such rules, standards, laws and regulations, irrespective of where a violation occurs (Article 217.1). If a vessel commits a violation of rules and standards, the flag State should provide for immediate investigation and where appropriate institute proceedings in respect of the alleged violation irrespective of where the violation occurred (Article 217.4). The jurisdiction of flag State is particularly important for ships navigating on the high seas or in the exclusive economic zones (EEZs).

A coastal State has sovereignty over its internal and territorial seas, and their air space. For archipelagic States, sovereignty can be extended to their archipelagic waters. Subject to the rules governing innocent passage in the territorial sea and/or archipelagic waters, the coastal State has the power to inspect any suspected vessel, particularly under the authorization of a UN Security Council resolution within its territorial waters. Coastal States can adopt rules and regulations designating sea lanes and traffic schemes. According to the LOSC, a coastal State is entitled to establish a sea belt of 12 nm beyond its territorial sea for its law enforcement and jurisdiction in respect of preventing and punishing infringement of its customs, fiscal, immigration and sanitary laws and regulations within its territory and territorial sea (Article 33). Except for the above, in comparison with flag States in respect of jurisdiction over vessels, the regulatory and enforcement power is more limited.[26] For example, coastal

---

[25] Article 94(1) of the LOSC.

[26] See Ashley Roach, "Enforcement of International Rules and Standards of Navigation Safety in the Malacca and Singapore Straits", *Singapore Journal of International & Comparative Law*, Vol. 3, 1999, 332. This limitation is partly on the account of the attraction of port state control and partly as a reaction to the fear over the risk of unilateral action unreasonably taken by certain coastal States. See David Anderson, "The Roles Flag States, Port States, Coastal States and International Organizations in the Enforcement of

States only exercise enforcement powers against ships not entitled to sovereign immunity which cause or threaten to cause major damage to the marine environment of the straits (Article 233).

The authority of enforcement of port State is originally designed for marine environmental protection under the LOSC, but now extends to cover the security matters such as the Container Security Initiative launched in 2002. The first Memorandum on Port State Control adopted in Paris in 1982 (Paris MOU) requires participating shipping authorities in Europe to establish an effective system of port State control with a view to ensuring that these ships meet international safety, security and environmental standards, and that crew members have adequate living and working conditions.[27] This practice was soon followed by the rest of the world, such as the Tokyo MOU adopted in December 1993 for the Asia-Pacific region. In terms of navigation safety, port States "may enforce in three separate ways: through the exercise of port State control, by enforcement of MARPOL violation on the high seas, and by judicial enforcement of national laws proscribing conduct in port".[28] It is reported that the United States successfully prosecuted a major passenger cruise line operating foreign cruise liners out of US ports for making false statements and presenting false records to the US Coast Guard while in US ports.[29] The port State is empowered to enforce international rules "amounts to an internationally agreed form of extra-territorial jurisdiction".[30]

It is to be noted that the powers of enforcement are exercised by warships, military aircraft, or other governmental service or authorized ships and aircraft. In addition to the aforementioned jurisdiction, the LOSC has established universal jurisdiction for all States over certain crimes at sea including piracy, slave trade, unauthorized broadcasting, and illicit traffic in narcotic drugs or psychotropic substances on high seas.

International Rules and Standards Governing the Safety of Navigation and the Prevention of Pollution from Ships under the UN Convention on the Law of the Sea and Other International Agreements", *Singapore Journal of International & Comparative Law*, Vol. 2, 1998, 577.

[27] Text of the MOU is available at http://www.parismou.org/upload/pdf/MOU%20incl.%2029th%20Amendment.pdf.

[28] Roach, *supra* note 26, p. 334.

[29] *Ibid.*, p. 335. For relevant account, see Anderson, *supra* note 26, pp. 570–571.

[30] Anderson, *supra* note 26, p. 569.

Use of Force in Maritime Enforcement: Legality, Necessity and Proportionality

In the context of enforcing UNSC resolutions, there are two types of use of force which may be involved. One is the direct involvement of the UNSC by way of directly sending peacekeeping troops to the inflicted region and the other is the use of force by UN Member States under the authorization of the UNSC.

Use of force and self-defense

After World War II, use of force is generally prohibited in international relations. This is expressly stipulated in the Charter of the United Nations as Article 2(4) provides that "[a]ll members shall refrain in their international relations from the threat or use of force against the territorial integrity and political independence of any state, or in any manner inconsistent with the purposes of the United Nations". The LOSC has the same stipulation in its Article 301 on peaceful uses of the seas. The principle of peaceful use of oceans fundamentally prohibits the use of force at sea.

The Agreement for the Implementation of the Provisions of the United Nations Convention on the Law of the Sea of 10 December 1982 Relating to the Conservation and Management of Straddling Fish Stocks and Highly Migratory Fish Stocks (the 1995 Agreement) applies the principle of restricting the use of force in the arrest of ship at sea by providing that "[t]he inspecting State shall ensure that its duly authorized inspectors: ...(f) avoid the use of force except when and to the degree necessary to ensure the safety of the inspectors and where the inspectors are obstructed in the execution of their duties. The degree of force used shall not exceed that reasonably required in the circumstances".[31]

Following this principle, States shall refrain from the use of force in the sea areas under national jurisdiction. All the relevant Chinese laws and regulations related to the use of the oceans contain no stipulations authorizing the use of force against illegal activities within the waters under its jurisdiction. The measures China uses are hot pursuit, arrest, detention, and legal proceedings in dealing with illegal activities. Even in the bilateral

[31]Article 22(1)(f) of the 1995 Agreement. Text is available at http://www.un.org/Depts/los/convention_agreements/convention_overview_fish_stocks.htm.

agreements, China prefers adding a provision on the non-use of force. For example, the Fishery Agreement signed between China and Vietnam in December 2000 provides that "[e]ach party, if it finds that small fishing boats of the other party conduct fishing activities in its water area in the buffer zone, may send a warning, or take necessary measures to order them to leave that water area, but by so doing that party should restrain its action not to detain or arrest the vessels in question, nor to use force".[32] However, there is one incident which was against the Chinese position on the use of force in ocean uses including the fishery management: On 8 January 2005 Chinese maritime police shot 8 Vietnamese in the Gulf of Tonkin. While the Chinese side alleged that the shot-dead people were "pirates",[33] Vietnam accused China of seriously violating international law by having killed "innocent" Vietnamese fishermen.[34]

Nonetheless, while it is generally accepted in international law that use of force is prohibited, there is an exception in the case of self-defense. According to Article 51 of the UN Charter:

> Nothing in the present Charter shall impair the inherent right of individual or collective self-defense if an armed attack occurs against a Member of the United Nations, until the Security Council has taken measures necessary to maintain international peace and security. Measures taken by Members in the exercise of this right of self-defense shall be immediately reported to the Security Council and shall not in any way affect the authority and responsibility of the Security Council under the present Charter to take at any time such action as it deems

[32]Article 12 of the Sino-Vietnamese Fishery Agreement. For details, see Zou Keyuan, "Sino-Vietnamese Fishery Agreement for the Gulf of Tonkin", *International Journal of Marine and Coastal Law*, Vol. 17(1), 2002, pp. 127–148.

[33]"Vietnamese Armed Pirating Ships Robbed Chinese Fishing Boats in Beibu Gulf and Shot at Chinese Fishing Boats and Marine Police Patrol Ship in the First Place. The Chinese Marine Police are forced to Shoot Back for Self-Defense. Eight Armed Robbers are Shot Dead and 8 are caught. Weapons and Ammunitions and the Pirating Ship are confiscated", 23 January 2005, available at http://www.fmprc.gov.cn/eng/zxxx/t181176.htm.

[34]"Vietnam's reaction to the January 8 shooting by Chinese naval patrol police who killed, injured and captured many Vietnamese fishermen and called them "pirates", available at http://www.mofa.gov.vn/en/tt_baochi/pbnfn/ns050121102026.

necessary in order to maintain or restore international peace and security.

This Article recognizes the inherent right of individual or collective self-defense if an armed attack occurs. However, there are a number of conditions to be fulfilled before a State may resort to the use of force as self-defense. According to Cassese, the following stringent conditions should be imposed:

(i) the necessity for forcible reaction had to be instant, overwhelming, leaving no choice of means, and no moment for deliberation;
(ii) the use of force was to be exclusively directed to repel the armed attack of the aggressor State;
(iii) force had to be proportionate; and
(iv) the use of force had to be terminated as soon as the aggression had come to an end; and the States acting in self-defense had to comply with the fundamental principles of humanitarian law.[35]

In addition, "armed reprisals in response to small-scale use of force short of an 'armed attack' proper, have been regarded as unlawful both against states and against terrorist organizations".[36]

Jurisprudence of the ICJ

The conditions on the use of force in the circumstances of self-defense are also expounded by international courts. There are several cases handled by the International Court of Justice (ICJ) concerning the legality of use of force and self-defense. The first one is the *Corfu Channel* case in 1949. The ICJ rejected the United Kingdom's statement that its mine-sweeping was a method of self-protection or self-help and declared that the action of the British Navy constituted a violation of Albanian sovereignty.[37] On the other hand, the Court held that the readiness of the British ships to use force if attacked was not an unreasonable precaution so that the United

[35] See Antonio Cassese, "Terrorism is Also Disrupting Some Crucial Legal Categories of International Law", *European Journal of International Law*, Vol. 12(5), 2001, 995.
[36] *Ibid.*, p. 996.
[37] ICJ, Corfu Channel Case (Merits), Judgment of 9 April 1949, available at http://www.icj-cij.org/icjwww/idecisions/isummaries/Iccsummary490409.htm.

Kingdom had not violated international law. It indicates that actual recourse to force in self-defense would have been legitimate in the event of an attack.[38]

The ICJ emphasized that the entitlement to resort to self-defense under Article 51 of the UN Charter is subject to the conditions of necessity and proportionality. In the case concerning *Military and Paramilitary Activities in and against Nicaragua (Nicaragua v. United States of America)*, the Court upheld that "there is a specific rule whereby self-defense would warrant only measures which are proportional to the armed attack and necessary to respond to it, a rule well established in customary international law".[39]

In the *Legality of the Threat or Use of Nuclear Weapons* advisory opinion in 1996, the ICJ concludes unanimously that a threat or use of force by means of nuclear weapons that is contrary to Article 2(4) of the United Nations Charter and that fails to meet all the requirements of Article 51, is unlawful; and a threat or use of nuclear weapons should also be compatible with the requirements of the international law applicable in armed conflict particularly those of the principles and rules of international humanitarian law, as well as with specific obligations under treaties and other undertakings which expressly deal with nuclear weapons. However, the views of the judges are sharply divided and the Court finally reached the following decision by the President's casting vote:

> It follows from the above-mentioned requirements that the threat or use of nuclear weapons would generally be contrary to the rules of international law applicable in armed conflict, and in particular the principles and rules of humanitarian law; However, in view of the current state of international law, and of the elements of fact at its disposal, the Court cannot conclude definitively whether the threat or use of nuclear weapons would be lawful or unlawful in an extreme circumstance of self-defense, in which the very survival of a State would be at stake.[40]

---

[38] See Stanimir A. Alexandrov, *Self-Defense Against the Use of Force in International Law* (The Hague: Kluwer Law International, 1996), p. 123.

[39] ICJ Reports 1986, p. 94, para. 176; and also in ICJ, Advisory Opinion of 8 July 1996 on "Legality of the Threat or Use of Nuclear Weapons", available at http://www.icj-cij.org/icjwww/idecisions/isummaries/iunanaummary960708.htm.

[40] *Ibid.*

From the aforementioned decisions, it can be seen that the ICJ is very cautious of justifying the use of force, even in the circumstance of self-defense.

Recent developments on anti-terrorist campaigns around the world may change some conditions on the use of force in self-defense. One reasonable argument is: "where a State is unable or unwilling to assert control over a terrorist organization located in its territory, the State which is a victim of the terrorist attacks would as a last resort, be permitted to act in self-defence against the terrorist organization in the State in which it is located".[41] While it is admitted that self-defense against non-States entities "has gained legal currency",[42] the doctrine of pre-emptive action still remains controversial in international law.[43]

Use of force under UNSC authorization

In addition, a further lawful use of force is envisaged in Article 42, whereby the UN Security Council may take military enforcement measures in conformity with Chapter VII of the Charter as mentioned previously. As military actions include "demonstrations, blockade, and other operations by air, sea or land forces", maritime enforcement which includes necessary military actions is crucial to the success of the enforcement of the UNSC resolutions. In this respect, it can be said that use of force in enforcing measures under UNSC resolutions is not only an issue concerning whether force should be used or prohibited in accordance with Article 2(4) of the UN Charter, but also an issue of "force necessary to compel compliance with law enforcement measures" under the circumstances that force becomes necessary to enforce UNSC resolutions. The authorization from the UNSC in fact lifts the obligation of any relevant enforcing Member State under Article 2(4) of the UN Charter simply because it is requested to act within the mandate of a relevant

[41]"The Chatham House Principles of International Law on the use of Force in Self-Defence", *ICLQ*, Vol. 55, 2006, p. 970.

[42]Natalino Ronzitti, "The Expanding Law of Self-Defence", *Journal of Conflict & Security Law*, Vol. 11(3), 2006, 357.

[43]For example, the UN High Level Panel Report, *A More Secure World: Our Shared Responsibility*, UN Doc. A/59/565, implicitly rejects the concept of pre-emptive action. For relevant comments, see Christine Gray, "A Crisis of Legitimacy for the UN Collective Security System?" *ICLQ*, Vol. 56, 2007, pp. 157–170.

UNSC resolution. However, even in the case that force is allowed under a UNSC resolution, caution and/or restraint should be taken in every step during the enforcement process so as to avoid any abuse of the authorization from the UNSC.

The International Tribunal for the Law of the Sea (ITLOS) touched upon the issue of use of force and self-defense in the *M/V "Saiga" (No. 2) Case (Saint Vincent and the Grenadines v. Guinea)* in 1999 where the ITLOS expressed a very important view that:

> Although the Convention does not contain express provisions on the use of force in the arrest of ships, international law, which is applicable by virtue of article 293 of the Convention, requires that the use of force must be avoided as far as possible and, where force is unavoidable, it must not go beyond what is reasonable and necessary in the circumstances. Considerations of humanity must apply in the law of the sea, as they do in other areas of international law.[44]

Based on the above, the ITLOS finds that "Guinea used excessive force and endangered human life before and after boarding the *Saiga*, and thereby violated the rights of Saint Vincent and the Grenadines under international law".[45]

On the other hand, the Tribunal does not totally prohibit the use of force in maritime law enforcement operations. Accordingly, when all efforts including auditory or visual signal to stop have failed, then firing of shots across the bows of the ship may be permitted. In addition, there are two considerations that must be taken into account when force is used: (1) it should be a last resort; and (2) "all efforts should be made to ensure that life is not endangered".[46] It is to be noted that the aforementioned considerations are applicable in the circumstances that the targeted ship is unarmed. There must be different degrees of using force when an enforcing vessel encounters a pirate ship or a fishing vessel.

The next important question is whether Article 42 of the UN Charter must be expressly cited in a UNSC resolution in order to justify the use of force in maritime enforcement. It is recalled that following Iraq's invasion

---

[44] See para. 155, Judgment of 1 July 1999, available at http://www.itlos.org/start2_en.html.

[45] *Ibid.*, para. 159.

[46] *Ibid.*, see para. 160.

of Kuwait in August 1990 and imposition of economic sanctions according to UNSC Resolution 661, a similar legal issue arose as to whether a prior authorization from the Security Council for blockade should be required after the United States and the United Kingdom had decided to use warships to enforce UNSC sanctions.[47] It is remembered that after the adoption of Resolution 1695, there were divergent views among UNSC permanent members regarding whether the resolution authorizes to take non-military or military measures in case North Korea would launch another missile test. While the United States took the view that Resolution 1695 "has a perfectly identical legal effect with that under Chapter VII", China and Russia emphasized that Chapter VII was not referred to in the resolution. Suppose the resolution expressly mentioned Chapter VII, particularly Article 42, then there should be no controversy over the interpretation of the text of the resolution. Following this line, when the Security Council decides to take military actions against a State endangering international peace and security, it is better, and in fact necessary to expressly invoke Article 42, whereby the UN Member States which enforce the relevant UNSC resolution can unequivocally use necessary means of force without controversy.

A related issue is to what degree and extent force can be used. The UNSC resolutions usually do not provide guidelines in this regard. For practical purposes, maritime law enforcers under a UNSC resolution may follow general principles and guidelines derived from the Rules of Engagement and those used for police. Rules of Engagement "can be defined as the detailed guidance given to commanding officers prescribing the circumstances in which they may engage the enemy", and "indicate what constitute hostility and the maximum amount of force which can be used".[48] It is also reasonably assumed that rules of naval war such as the 1907 Convention Relative to Certain Restrictions with Regard to the Exercise of the Right of Capture in Naval War could be applicable, in particular when there are encounters of armed conflict during the maritime law enforcement.

Since maritime law enforcement of UNSC resolutions, though involving military personnel (in particular when the police powers are exercised by military authorities), is basically a policing activity at sea, guidelines

---

[47] See Nicholas Grief, *Public International Law in the Airspace of the High Seas* (Dordrecht: Martinus Nijhoff, 1994), p. 236.

[48] *Ibid.*, pp. 212–213.

for policing are applicable. The fundamental principle here is the right to life, as enunciated in the International Covenant on Civil and Political Rights.[49] The European Convention on Human Rights, while reiterating the aforesaid principle, emphasizes that in case of the use of force, it is no more than "absolutely necessary".[50] As regards proportionality, the European Commission of Human Rights once held that "regard must be had to the nature of the aim pursued, the dangers to life and limb inherent in the situation, and the degree of the risk that the force employed might result in loss of life".[51]

The United Nations itself has also exerted its efforts in this respect. In December 1979, the UN General Assembly adopted Resolution 34/169 of the Code of Conduct for Law Enforcement Officials, which consists of eight articles. It reiterates the important principles of proportionality (force to be used only to the extent required) and necessity (force to be used only when strictly necessary).[52] In September 1990, the 8th UN Congress on the Prevention of Crime and the Treatment of Offenders adopted the UN Basic Principles on the Use of Force and Firearms by Law Enforcement Officials. Based on the Code of Conduct, this document further elaborates its-embodied 26 principles concerning the use of force by law enforcers. Principle 5 stipulates that whenever the use of force and firearms is unavoidable, law enforcement officials should exercise restraint, and act in proportion to the seriousness of the offence and the legitimate objective to be achieved. They are required to minimize damage and injury, and respect and preserve human life.[53] Though the above two UN documents are only to be seen as "soft law" without legal binding force, they are for sure a source of reference for maritime law enforcement, particularly under UNSC resolutions. Finally, it is worthwhile to

---

[49] Article 6(1) provides that "Every human being has the inherent right to life. This right shall be protected by law. No one shall be arbitrarily deprived of life".

[50] It lists the following circumstances: (1) in defense of any person from unlawful violence; (2) in order to effect a lawful arrest or to prevent the escape of a person lawfully detained; and (3) in action lawfully taken for the purpose of quelling a riot or insurrection.

[51] European Commission of Human Rights, Application No.10044/82: *Kathleen Stewart v. UK* (10 July 1984, DR. 39/162), cited in Ralph Crawshaw, Barry Devlin and Tom Williamson, *Human Rights and Policing: Standards for Good Behaviour and a Strategy for Change* (The Hague: Kluwer Law International, 1998), p. 97.

[52] *Ibid.*, p. 108.

[53] *Ibid.*, p. 112.

note that while it is accepted that maritime law enforcement is similar to policing, it is questionable to argue that "police action" is not considered "uses of force".[54]

There was a significant case in East Asia related to maritime law enforcement and the use of force. On 22 December 2001, an unidentified ship being pursued by Japan's Coast Guard in the East China Sea for infringing Japan's EEZ sank after being fired on by four Japanese coast guard patrol vessels.[55] All 15 crewmembers on board lost their lives. During the six-hour pursuit, the Japanese vessels fired more than 500 rounds. It is reported that Japan fired the first shot.[56] Though the Japanese law provides that coast guard vessels are allowed to fire at, and even sink, ships that do not obey orders in Japan's territorial waters,[57] there are some conditions on the use of force to the effect that the coast guard officers believe that there exists no other means to stop a vessel from proceeding, whose crew members, passengers or other persons do not comply with the repeated orders to stop, and resist the coast guard officers' efforts to execute their duties.[58] In addition, there is no authorization granted to the Coast Guard to use weapons against foreign vessels beyond Japan's territorial sea. Thus, Japan's use of force to sink the unidentified vessel in China's EEZ finds no ground in its own domestic law. In international law, there is no stipulation in the LOSC which grants Japan to use force against suspicious foreign vessels. Furthermore, there is no right for a coastal State to arrest foreign governmental vessels in its territorial sea and EEZ, much less to use force on it. Japan's argument of self-defense is weak and its action of use of force is regarded as excessive under international law.

---

[54] See Douglas Guilfoyle, "Interdicting Vessels to Enforce the Common Interest: Maritime Countermeasures and the Use of Force", *ICLQ*, Vol. 56, 2007, 79.

[55] According to some other reports, Japan sent altogether 25 patrol vessels and 14 aircraft to chase the mysterious boat. See Lu Lude, "Japan Cannot Do As It Pleases in China's EEZ", *China Ocean News* (in Chinese), 8 March 2002.

[56] See Peter Landers, "Conflict shows a gray area in Japan Law — Tokyo weighs revision to boost defense measures", *Asian Wall Street Journal*, 26 December 2001, p. 3.

[57] Kazunori Takada, "Japan: Legal issues hang over Japan firing on mystery ship", *Reuters English News Service*, 27 December 2001.

[58] Article 20 of Law No. 28 of 27 April 1948 as amended through Law No. 102 of 1999 and Law No. 114 of 2001 (on file with the author).

Turn to the issue of North Korea, unlike Resolution 1695, Resolution 1718 mentions Chapter VII of the Charter but limits its application only with Article 41 which authorizes the UNSC to employ sanction measures but not involving the use of force. It imposes weapon and financial sanctions on North Korea and authorizes UN Member States to inspect cargo to and from North Korea. However, due to the opposition of China and Russia, military sanctions are excluded from the sanction list contained in the resolution. Ironically, in case there would be military actions against North Korea under the authorization of the Security Council, North Korea, as a member of the United Nations, could invoke Article 51 of the UN Charter to defend itself.

Due to the complexity of world politics, difficulties in invoking Chapter VII in order to maintain world peace and security always exist when there are disputes in its invocation. The Korean War is a typical example. The military actions led by the United States under the authorization of the UNSC resolutions[59] were viewed by the Chinese as "aggression". Furthermore, China sent its troops to Korea to fight the American-led UN troops.

Other Coercive Measures in Maritime Law Enforcement

Since use of force is regarded as a last resort, other coercive measures such as interdiction, search and arrest, and hot pursuit, without resorting to the use of force, should be first employed in the maritime law enforcement.

Interdiction

In our contemporary world, maritime interdiction is mainly carried out under the scheme called Proliferation Security Initiative (PSI) which is an effort to consider possible collective measures among the participating countries, in accordance with national legal authorities and relevant international law and frameworks, in order to prevent the proliferation of Weapons of Mass Destruction (WMD), missiles and their related

---

[59]Resolutions include S/RES/82, 25 June 1950; S/RES/83, 27 June 1950 and S/RES/84, 7 July 1950.

materials that pose threats to the peace and stability of the international community.[60]

As for interdiction under UNSC Resolution 1718, UN Member States are authorized to enforce the sanctions measures listed in the resolution including inspection of ships entering or leaving North Korea in order to prevent the proliferation of WMD. China and Russia are concerned that cargo inspection could spark naval confrontations with North Korean boats and view the cargo check as stated in the resolution as non-mandatory, while the United States and Japan would like to carry out such inspection under the PSI scheme.

Inspection should be conducted in accordance with relevant international and national laws. No present international norms governing boarding the ship in the international waters are to be changed by Resolution 1718.[61] Hence, it should be remembered without the consent of the flag State, ships in international waters cannot be boarded. However, when the ship with suspected materials is flying the North Korean flag, it is possible to legally conduct the inspection in accordance with sanction measures contained in Resolution 1718. Furthermore, the inspection of North Korean ships can also be taken when they enter the port of another State. What we should notice is the term the resolution uses: it is "inspection" instead of "interdiction" as the latter is more coercive. The second point is that such inspection is much easier to be taken within the territorial sea than on the high seas due to the maritime jurisdiction variety. Against the background of the Six-Party Talks, the countries directly involved in the North Korean issue include China, Japan, Russia, South Korea and the United States. They may have the right to make on-board inspections in accordance with Resolution 1718, though they hold different positions toward North Korea.

After the adoption of the resolution, North Korea was forced to return to the Six-Party Talks table and finally agreed to abandon its nuclear weapon project and pledged again to make the Korean Peninsula as a nuclear weapon free zone. The February 13 agreement freezes North Korea's nuclear program in exchange for 50,000 tons of heavy fuel oil. What is remarkable in terms of institutional arrangements is the establishment of five working groups concerning respectively denuclearization of

---

[60] For details, see Chapter 2.

[61] See Robert Beckman, "North Korea and the UN: Resolution 1718 and Its Legal Dimensions", IDSS Commentaries, 116/2006, 27 October 2006, p. 2.

the Korean Peninsula, Northeast Asia peace and security mechanism, economic and energy cooperation, normalization of DPRK–US relations, and normalization of DPRK–Japan relations. From the Six Round of the Talk, the first session of which took place in Beijing from 19 to 22 March 2007, the parties involved began to implement in details the goals set forth in the document "Initial Actions for the Implementation of the Joint Statement" they signed in the Fifth Round.

Six-Party Talks and the 1953 Armistice Agreement

Although the Six-Party Talks is helpful for maintaining peace in Korean Peninsula, such talks are currently in suspension. It is recalled that the 1953 Korean War Armistice Agreement[62] signed between the Commander-in-Chief of the United Nations Command and the Supreme Commander of the Korean People's Army and the Commander of the Chinese People's Volunteers is only an interim agreement for ceasefire. It is therefore expected that a peace agreement can be reached for Korea. As commented, "[t]he Korean War ended in an armistice. If this separate forum does indeed replace the armistice with a permanent peace treaty, then one of the most contentious chapters of the Cold War will be closed".[63] On the other hand, though the Armistice Agreement exists largely in a symbolic way at present, any side which resorts to forces in and around Korea Peninsula could be regarded as a breach of that Agreement. Another contemporary implication of the Agreement may be from its arrangement to establish the Neutral Nations Supervisory Commission. This mechanism can be borrowed to monitor the maritime law enforcement actions under the UNSC mandate so as to avoid any abusive practice or disproportionate/unnecessary use of force.

Hot pursuit

Right of hot pursuit is provided in Article 111 of the LOSC. Accordingly, when the competent authorities of the coastal State have good reason to

---

[62] Text of the Agreement is available at http://news.findlaw.com/hdocs/docs/korea/kwarmagr072753.html.

[63] John Feffer, "North Korean Nuclear Agreement: Annotated: What it really means", 1 March 2007, *Guerrilla News Network*, available at http://www.gnn.tv/print/2953/North_Korean_Nuclear_Agreement_Annotated.

believe that a foreign ship has violated its laws and regulations, the hot pursuit of that ship may be undertaken. Such pursuit must commence when the foreign ship or one of its boats is within the internal waters, the archipelagic waters, the territorial sea or the contiguous zone of the pursuing State, and may only be continued outside the territorial sea or the contiguous zone if the pursuit has not been interrupted. If the foreign ship is within a contiguous zone, the pursuit may only be undertaken if there has been a violation of the rights for the protection of which the zone was established.

In addition, the right of hot pursuit shall apply *mutatis mutandis* to violations in the EEZ or on the continental shelf, including the safety zones around continental shelf installations, the laws and regulations of the coastal State applicable in accordance with the LOSC to the EEZ or the continental shelf, including such safety zones.

The LOSC in Article 111 sets forth a number of conditions for the right of hot pursuit to be exercised:

(a) The right of hot pursuit ceases as soon as the ship pursued enters the territorial sea of its own State or of a third State.
(b) The pursuit may only be commenced after a visual or auditory signal to stop has been given at a distance which enables it to be seen or heard by the foreign ship.
(c) The right of hot pursuit may be exercised only by warships or military aircraft, or other ships or aircraft clearly marked and identifiable as being on government service and authorized to that effect.
(d) Where a ship has been stopped or arrested outside the territorial sea in circumstances which do not justify the exercise of the right of hot pursuit, it shall be compensated for any loss or damage that may have been thereby sustained.

Although the right of hot pursuit is provided in the section regarding the high seas in the LOSC, there is an interesting phenomenon that in East Asia the provision governing this right is placed in the national laws governing the territorial sea and/or the EEZ.[64] Japan's Law on the Territorial Sea and the Contiguous Zone promulgated in 1996 provides that the execution of official duties by public officials of Japan in relation to hot

[64] For example, it is stipulated in China's Law on the Territorial Sea (Article 14) and Law on the EEZ (Article 12).

pursuit within the internal waters, territorial sea, and the contiguous zone be undertaken in accordance with Article 111 of the LOSC.[65]

As is recalled, in the *M/V "Saiga" (No. 2) Case (Saint Vincent and the Grenadines v. Guinea)*, Saint Vincent and the Grenadines contends that Guinea did not lawfully exercise the right of hot pursuit under Article 111 of the LOSC in arresting the *Saiga*, but Guinea denies that its pursuit was vitiated by any irregularity and maintains that the officers engaged in the pursuit complied with all the requirements set forth in the LOSC.[66] The ITLOS finds in its judgment that "the conditions for the exercise of the right of hot pursuit under Article 111 of the Convention are cumulative; each of them has to be satisfied for the pursuit to be legitimate under the Convention". After having examined the evidence submitted by both sides, the ITLOS finds that some of the conditions were not fulfilled so that the circumstances under which Guinea stopped and arrested the *Saiga* on 28 October 1997 did not justify the exercise of the right of hot pursuit in accordance with the LOSC.[67]

Conclusion

Under UNSC resolutions, the UN Peacekeeping Operations so far have played a positive, though sometimes ineffective, role in maintaining peace and security in the conflicting areas in the world. Since 1945, "UN peacekeepers have undertaken 61 field missions and participated in the implementation of 172 peaceful settlements that have ended regional conflicts".[68] Maritime peace and security is naturally within the mandate of the UN peacekeeping forces. However, this chapter addresses the issue which differs from the general peacekeeping operations and focuses on maritime law enforcement under the authorization of the UN Security

---

[65] Articles 3 and 5 of Japan's Law on the Territorial Sea and the Contiguous Zone. Text is available in Office of Ocean Affairs, Bureau of Ocean and International Environmental and Scientific Affairs, US Department of State, *Limits in the Seas, No. 120: Straight Baseline and Territorial Sea Claims: Japan*, 30 April 1998, pp. 19–21.

[66] See paras. 139–144, Judgment of 1 July 1999, available at http://www.itlos.org/start2_en.html.

[67] *Ibid.*, see paras. 146–152.

[68] UN Department of Peacekeeping Operations, Fact Sheet, available at http://www.un.org/Depts/dpko/factssheet.pdf.

Council through coercive measures with or without resort of the use of force.

From the aforementioned examination, it can be seen that the effectiveness of the maritime law enforcement to large extent depends, on the one hand, on the capacity and determination of the individual enforcing UN Member States, and on the other, on the close international cooperation among the States, particularly those who take the same position under a UNSC resolution.

Recent State practices in military interventions have shown that some States are not satisfied with the existing law on the use of force, but "[o]btaining authorisation by the Security Council before resorting to non-defensive force remains a legal requirement — and not merely a matter of political convenience".[69] Nevertheless, from the developments regarding the maritime law enforcement of UNSC resolutions, we can also see the developments of contemporary international law. Something legally controversial today may become legal tomorrow as law develops. The general trend is: as long as something is beneficial for the interest of the whole mankind it will eventually be accepted by the entire world community.

[69] Tarcisio Gazzini, "The Rules on the Use of Force at the Beginning of the XXI Century", *Journal of Conflict & Security Law*, Vol. 11(3), 2006, 341.

Chapter 5

International Efforts of Crackdown on Piracy

Introduction

Piracy is traditionally regarded as *hostis humani generis*, the enemy of the human race. Pirates commit acts of murder, robbery, plunder, rape or other villainous deeds at sea, cruelly against humanity. Because of such nature of its offense, piracy is punishable wherever encountered.[1] According to the existing literature, early reference to piracy can be found in *Justinian's Digest* in 529 AD and in *King John's Ordinance* of 1201.[2] However, a law specifically governing the issue of piracy is traced back to the 17th century. The English act on piracy enacted in 1698 was likely the first law of piracy at the national level.[3] Other States like Germany and the United States then followed the suit to enact their laws of piracy.

---

[1]As is said, "a pirate, under the law of nations, is an enemy of the human race; being the enemy of all, he is liable to be punished by all". Appendix to *United States v. Smith*, 5 Wheat. (US) 153 (1820), pp. 7–8; quoted in the Harvard Draft Convention, *American Journal of International Law*, Supplement, Vol. 26(4), 1932, 763.

[2]See Patricia W. Birnie, "Piracy: Past, Present and Future", *Marine Policy*, Vol. 11(3), 1987, 163.

[3]Text is reprinted in Alfred Rubin, *The Law of Piracy* (Newport, RI: Naval War College Press, 1988), pp. 362–369. Unlike the Act of 1536 which was also relevant to piracy, this law was specifically dealing with the suppression of piracy.

Since the early 1990s, piracy has resurged and even increased in some places in the world, particularly in the Southeast Asian region. According to the report prepared by the ICC International Maritime Bureau Piracy Reporting Center (IMB-PRC), between 1 January and 30 June 1999, there were 115 piracy incidents around the world. A more recent report from the International Maritime Bureau (IMB) showed that during January–June 2020, the most piracy incidents were in Africa (38), followed by Southeast Asia (33) and the Americas (17).[4] Though the reported quantity of incidents (98) was significantly lower than that in the same period of 2018 (107), it exceeded those of 2017 (87) and 2019 (78), and was equal to that of 2016.[5] The United Nations General Assembly, for the first time in 1998, called for all States, in particular coastal States in affected regions, to take all necessary and appropriate measures to prevent and combat incidents of piracy and armed robbery at sea. After the September 11 terrorist attack, piracy has been allegedly connected to maritime terrorism. The Bush Administration described terrorism just as immoral as piracy, slave trade and genocide.[6] Thus, piracy constitutes a modern threat to world peace and security.

Sea lines in Southeast Asia are of vital interest for the East Asian countries. More than half of the world's merchant fleet capacity sails through the straits of Malacca, Sunda and Lombok and the South China Sea.[7] More than 10,000 vessels of greater than 10,000 dwt move southward through the South China Sea annually, with well over 8,000 proceeding in the opposite direction.[8] In addition, with the fast growth of economy in East Asia, the trend to greater intra-Asian trade (relative to trade with Europe and North America) results in more shipping in the

---

[4]IMB, Piracy and Armed Robbery Against Ships Report for the Period 1 January–30 June 2020, available at https://www.icc-ccs.org/reports/2020_Q2_IMB_Piracy_Report.pdf.

[5]*Ibid.*

[6]"Pentagon policy chief: Terrorism as immoral as genocide", *Dow Jones International News*, 8 May 2002.

[7]See Stanley B. Weeks, "Sea Lines of Communications (SLOC) Security and Access", in Michael Stankiewicz (ed.), *Maritime Shipping in Northeast Asia: Law of the Sea, Sea Lanes, and Security*, IGCC Policy Paper No. 33 (Institute on Global Conflict and Co-operation, University of California, February 1998), p. 55.

[8]Hal Olson, "Marine Traffic in the South China Sea", *Ocean Yearbook*, Vol. 12, 1996, 137.

littoral waters of Southeast Asia and the South China Sea.[9] Thus, the sea routes in the region are usually regarded as economic lifelines for East Asian countries, particularly Japan. However, piracy endangers the safety of navigation most. Some grave cases, including the *Cheung Son* and the *Tenyu*, demonstrated that pirates hijacked merchant vessels, killed crew members and/or robbed valuables on board the ships.[10]

Improving the International Legal Regime

Crackdown on piracy is mainly governed by international law including the 1982 United Nations Convention on the Law of the Sea (LOSC). Accordingly, States are obliged to cooperate in the suppression of piracy,[11] and can use their warships or military aircraft or similar governmentally authorized ships or aircraft to seize a pirate ship or aircraft and arrest pirates.[12] It is to be noted that the aforementioned piracy suppression clauses are also applicable to the exclusive economic zones (EEZs) to the extent that they are in consistency with the relevant provisions of the LOSC regarding the EEZ which is the maritime zone where the coastal State enjoys sovereign rights to its natural resources and certain kinds of jurisdiction over, *inter alia*, environmental protection and marine scientific research. Though the EEZ is a maritime zone within national jurisdiction, some high seas freedoms are preserved there, such as the freedom of navigation. Since piracy endangers navigation and thus greatly affects the exercise of the navigational freedom, the enforcement of the law of piracy in the EEZ may not be viewed as an impingement on any rights reserved to the coastal State.[13]

---

[9] See Stephen J. Meyrick, "Development in Asian Maritime Trade", in Stankiewicz (ed.), *supra* note 7, 21.

[10] Some cases handled in China are illustrated in Zou Keyuan, "Quelling Sea Piracy in East Asia: China's Domestic Endeavour and Regional Cooperation", *Journal of International Maritime Law*, Vol. 10(4), 2004, 327–342.

[11] See Article 100 of the LOSC. Text of the LOSC is reprinted in 21 ILM (1982) 1261.

[12] See Article 105 of the LOSC.

[13] See Thomas A. Clingan Jr., "The Law of Piracy", in Eric Ellen (ed.), *Piracy at Sea* (Paris: ICC Publishing, 1989), p. 170.

Developments in definition

The term "piracy" is usually referred to a broad range of violent acts at sea. The LOSC defines it as:

> Piracy consists of any of the following acts: (a) any illegal acts of violence or detention, or any act of depredation, committed for private ends by the crew or the passengers of a private ship or a private aircraft, and directed to: (i) on the high seas, against another ship or aircraft, or against persons or property on board such ship or aircraft; (ii) against a ship, aircraft, persons or property in a place outside the jurisdiction of any State; (b) any act of voluntary participation in the operation of a ship or of an aircraft with knowledge of facts making it a pirate ship or aircraft; (c) any act of inciting or of intentionally facilitating an act described in subparagraph (a) or (b).[14]

It consists of five elements: (1) the acts complained against must be crimes of violence such as robbery, murder, assault or rape; (2) committed on the high seas beyond the land territory or territorial sea, or other territorial jurisdiction, of any State; (3) by a private ship, or a public ship which through mutiny or otherwise is no longer under the discipline and effective control of the State which owns it; (4) for private ends; and (5) from one ship to another so that two ships at least are involved.[15]

However, the definition provided for in the LOSC has limitations in respect to the contemporary phenomenon of piracy. First, it defines "piracy" as only for "private ends", though it is argued that such wording could be given a wider interpretation. It is therefore that the terrorist acts at sea for political ends are generally excluded. This is why after the *Achille Lauro* Incident in 1988,[16] the world community needed the

---

[14]Article 101 of the LOSC.

[15]L.F.E. Goldie, "Terrorism, Piracy and the Nyon Agreement", in Yoram Dinstein (ed.), *International Law at a Time of Perplexity: Essays in Honour of Shabtai Rosenne* (Dordrecht: Martinus Nijhoff, 1988), p. 227.

[16]On 3 October 1985, a group of Palestinian guerrillas hijacked the Italian cruise ship *Achille Lauro* while it was in Egyptian territorial waters. The hijackers demanded the release of 50 Palestinians held in Israel in return for the release of the passengers. They ordered the ship to sail to Syria, which refused them port entry. The hijackers then on 8 October killed an American passenger. Several days later the four hijackers gave themselves up to the Egyptian authorities.

Convention on the Suppression of Unlawful Acts against the Safety of Maritime Navigation (the SUA Convention). Second, according to the above definition, piracy *juris gentium* presupposes that a criminal act be exercised by passengers or the crew of a ship against another ship or persons or property on its board. The two-vessel requirement is an ingredient of the crime of piracy, unless a criminal act occurs in *terra nullius*.[17] Thus, "internal seizure" within the ship is hardly regarded as "act of piracy" under the definition of the LOSC.[18] Because of the aforementioned limitations and other alleged deficiencies in the definition, some scholars have suggested revising this definition.[19] It is recalled that in 1970 before the Third United Nations Conference on the Law of the Sea (UNCLOS III), the International Law Association suggested a definition on piracy as "unlawful seizure or taking control of a vessel through violence, threats of violence, surprise, fraud or other means",[20] but it was not taken account during UNCLOS III. In addition, since the above definition is only applicable to the acts of piracy in the high seas or places outside jurisdiction of States, it has a geographic limitation and could not cover the whole picture of contemporary piracy.

The IMB-PRC suggested a definition of piracy "as an act of boarding any vessel with the intent to commit theft or any other crime and with the intent or capability to use force in the furtherance of that act",[21] which seems to be accepted by the shipping industry but has not been recognized both in international law and in domestic law. The International Maritime Organization (IMO) has taken the position to divide acts of piracy into two categories by geographical and legal division of maritime zones:

[17]Natalino Ronzitti, "The Law of the Sea and the Use of Force against Terrorist Activities", in N. Ronzitti (ed.), *Maritime Terrorism and International Law* (Dordrecht: Martinus Nijhoff, 1990), p. 1.

[18]A different view holds that internal seizures could be piracy. See Samuel P. Menefee, "Piracy, Terrorism, and the Insurgent Passenger: A Historical and Legal Perspective, *ibid.*, p. 60. In addition, it is acknowledged that even the internal seizure was not piracy in international law, it is still piracy under municipal law of the flag State.

[19]See Alfred P. Rubin, "Is Piracy Illegal?" *American Journal of International Law*, Vol. 70, 1976, 95.

[20]International Law Association, "Piracy: Sea and Air", *Report of the Fifty-fourth Conference*, 23–29 August 1970, p. 708.

[21]IMB-PRC, *Piracy and Armed Robbery Against Ships*, Report for the Period of January 1–June 30, 1998, Kuala Lumpur, July 1998, p. 1.

piracy in the high seas is defined as "piracy" in accordance with the LOSC, while acts of piracy in ports or national waters (internal waters and territorial sea) are defined as "armed robbery against ships".[22] However, we may notice that the shortcoming of such a division is obvious: piracy is not equivalent to armed robbery and it may also include other violent acts such as murder, assault and rape. Recent piratical incidents showed that hijacking and kidnapping of crew for ransom become a new phenomenon as evidenced by piratical acts in Somalia. What is more questionable is its contents describing what constitutes the "armed robbery against ships". By comparing to the definition on piracy in the LOSC, the IMO definition on armed robbery against ships can be actually described as a definition for any piratical act which occurs in a place within the jurisdiction of States over such a crime since the definition does not limit itself only to "armed robbery" as seen above. Therefore, the attempt of the IMO is clear in that it tried to use an alternative term in covering piratical acts within the waters of national jurisdiction (except the EEZ).

The problem concerning different definitions also affects the number of incidents relating to piratical attacks. As some scholars point out, the definition of the piracy under the LOSC may lead to the conclusion that the low incidence of such acts implies that there is no significant problem of piracy today.[23] On the other hand, the reported 115 incidents by IMB-PRC in 1999 were simply based upon its own definition. This means that there was no difference between incidents in the high seas and those happened in the territorial seas of coastal States. Otherwise, the number would have been much smaller. However, as the Piracy Reporting Center acknowledged that the number they tallied was the lowest possible figure and true figures could be much higher due primarily to the under-reporting of incidents related to its definition.

Finally, the definition problem also exists in the comparison of the international one with the domestic one. A significant number of States define piracy as different from the LOSC. For example, Greece defined piracy as "anyone aboard a ship who, by using corporal violence or threat thereof against persons, commits acts of depredation against another ship

[22] See Birgit S. Olsen, "Piracy and the Law", paper presented to the IMO Seminar and Workshop on Piracy and Armed Robbery against Ships, Singapore, 3–5 February 1999, p. 2. (on file with the author)

[23] Jason Abbot and Neil Renwick, "Pirates? Maritime Piracy and Societal Security in Southeast Asia", *Pacifica Review*, Vol. 11(1), 1999, 11.

on the high seas with the intention to take possession of objects so obtained".[24] Similar legislations are also from Italy,[25] the Philippines,[26] Sri Lanka,[27] Suriname,[28] Thailand,[29] Tanzania,[30] Argentina,[31] Denmark,[32] Estonia,[33] Israel,[34] Korea,[35] Russia,[36] and Ukraine.[37] The United States law

---

[24]Letter from Greek, available at un.org/depts/los/LEGISLATIONANDTREATIES/PDFFILES/GRC_penal_code.pdf.

[25]Article 1135 of Italian Maritime Code (Approved with R. D.30 March 1942, n.327, as subsequently modified and integrated until 2002), available at un.org/depts/los/LEGISLATIONANDTREATIES/PDFFILES/ITA_maratime_law.pdf.

[26]Article 122 of the Revised Penal Code of the Philippines (Act No. 3815), available at un.org/depts/los/LEGISLATIONANDTREATIES/PDFFILES/PHL_revised_penal_code.pdf; and Anti-Piracy and Anti-Highway Robbery Law of 1974, Section 2, available at un.org/depts/los/LEGISLATIONANDTREATIES/PDFFILES/PHL_decree_1974.pdf.

[27]Piracy Act of Sri Lanka (No. 9 of 2001), available at un.org/depts/los/LEGISLATIONANDTREATIES/PDFFILES/LKA_national_%20legislation_piracy.pdf.

[28]Shipping and aviation crimes of Suriname, available at un.org/depts/los/LEGISLATIONANDTREATIES/PDFFILES/SUR_national_%20legislation_piracy.pdf.

[29]Thailand's Act on Prevention and Suppression of Piracy B.E. 2534 (1991), available at un.org/depts/los/LEGISLATIONANDTREATIES/PDFFILES/THA_piracy.pdf.

[30]Tanzania Penal code, Chapter 16 of the Laws (revised) (principal legislation) (issued under Cap. 1, s. 18), Article 66, available at un.org/depts/los/LEGISLATIONANDTREATIES/PDFFILES/TZA_penal_code.pdf

[31]Note Verbale from the Embassy of the Republic of Argentina to the United Kingdom, Addressed to the Secretary-General, available at un.org/depts/los/LEGISLATIONANDTREATIES/PDFFILES/ARG_national_legislation.pdf.

[32]Danish Note (17 April 2009), available at un.org/depts/los/LEGISLATIONANDTREATIES/PDFFILES/DNK_national_legislation_piracy.pdf.

[33]Estonian Legislation on Piracy, available at un.org/depts/los/LEGISLATIONANDTREATIES/PDFFILES/EST_legislation_piracy.pdf.

[34]Letter from Permanent Mission of Israel to the United Nations to UNDOALOS, available at un.org/depts/los/LEGISLATIONANDTREATIES/PDFFILES/ISR_anti_piracy.pdf.

[35]Laws and Ordinances on Punishment of Acts of Piracy and Armed Robbery at Sea of Korea, available at un.org/depts/los/LEGISLATIONANDTREATIES/PDFFILES/KOR_legislation_piracy.pdf.

[36]Information concerning National Legislation on Piracy — Russian Federation, available at un.org/depts/los/LEGISLATIONANDTREATIES/PDFFILES/RUS_national_legislation_piracy.pdf.

[37]Criminal Code of Ukraine 2001 (abstracts), available at un.org/depts/los/LEGISLATIONANDTREATIES/PDFFILES/UKR_criminal_code.pdf.

defines "piracy" as "[a]ny act of piracy as defined by international law if the perpetrators are found in the United States; any act of murder, robbery, or hostility against the United States or against a United States citizen on the high seas, by a citizen of the United States; and acts by aliens against the United States or its citizens that are defined as piracy in the treaty between the nation that the individual is a citizen of and the United States".[38] Though making a reference to the definition in international law, it is not clear that the aforesaid definition includes both "private ends" as defined in the LOSC and "political ends" as defined in the SUA Convention.

It is worth noting that the 2004 Regional Cooperation Agreement on Combating Piracy and Armed Robbery against Ships in Asia (ReCAAP) reflects the divisional definition on piracy and armed robbery by following the definition of the LOSC for "piracy" and the definition from the IMO Code of Practice for the Investigation of the Crimes of Piracy and Armed Robbery against Ships for "armed robbery against ships" (Article 1). It is to be noted that, first, the ReCAAP has slightly modified the LOSC definition by omitting the word "aircraft" as a pirate-attacked target[39]; and second, the significance of the ReCAAP concerning the definition on piracy is that it is the first international treaty which has turned the non-legally binding IMO definition into a legal definition.

Improving the legal regime

International law has established an obligation on States to cooperate in suppression of piracy and grants States certain rights to seize pirate ships

[38] *United States Code*, Title 18, pp. 1651–1653; cited in Jack A. Gottschalk *et al.*, *Jolly Roger with an Uzi: The Rise and Treat of Modern Piracy* (Annapolis, MD: Naval Institute Press, 2000), p. 34.

[39] For reference, see Article 1 of the ReCAAP which provides that "(a) any illegal acts of violence or detention, or any act of depredation, committed for private ends by the crew or the passengers of a private ship or a private aircraft, and directed to: (i) on the high seas, against another ship, or against persons or property on board such ship; (ii) against a ship, persons or property in a place outside the jurisdiction of any State; (b) any act of voluntary participation in the operation of a ship or of an aircraft with knowledge of facts making it a pirate ship or aircraft; (c) any act of inciting or of intentionally facilitating an act described in subparagraph (a) or (b)".

and criminals. The LOSC is a major anti-piracy treaty in contemporary era with the following relevant provisions:

- *Obligation to cooperate*: (Article 100 of the LOSC provides that "All States shall cooperate to the fullest possible extent in the repression of piracy on the high seas or in any other place outside the jurisdiction of any State").
- *Right to seize pirate ships and criminals*: (Article 105 provides that "on the high seas, or in any other place outside the jurisdiction of any State, every State may seize a pirate ship or aircraft, or a ship or aircraft taken by piracy and under the control of pirates, and arrest the persons and seize the property on board").
- *Only warships or military aircraft or similar governmentally authorized ships or aircraft have the power to seize a pirate ship or aircraft.*

In addition, there are supplementary anti-piracy treaties including: (1) SUA Convention and its Protocol.

- It covers the unlawful acts no matter whether they are for political ends or for private ends.
- It aims to punish any person who commits an offense by unlawfully and intentionally seizing or exercising control over a ship by force or threat thereof; or performing an act of violence against a person on board a ship if that act is likely to endanger the safe navigation of that ship; or destroying a ship or causing damage to a ship or to its cargo which is likely to endanger the safe navigation of that ship.
- Each Contracting Party should take necessary measures to establish its jurisdiction over the above offenses or extradite the offender or the alleged offender to the other Contracting Party who has the correspondent jurisdiction.
- The Convention applies to the offenses committed in the ship which is navigating or is scheduled to navigate into, through or from the waters beyond the outer limit of the territorial sea of a coastal State or applies when the offender or the alleged offender is found in the territory of a Contracting Party other than the previous case.
- The 1988 Protocol for the Suppression of Unlawful Acts against the Safety of Fixed Platforms Located on the Continental Shelf (the SUA Protocol) contains similar provisions.

The above two conventions can be regarded as complimentary to the anti-piracy legal measures provided in the LOSC. However, the scope of the territorial application between the LOSC and the SUA Convention is different: while the former applies only to the high seas and the EEZ, the latter applies not only the waters beyond but also waters within national jurisdiction as the SUA Convention clearly states that "[t]his Convention applies if the ship is navigating or is scheduled to navigate into, through, or from waters beyond the outer limit of the territorial sea of a single State, or the lateral limits of its territorial sea with adjacent States". Different scope of application may cause problems in implementation of the above conventions.

The expansion of jurisdictional waters by coastal States under the LOSC causes problems in implementing the aforesaid international treaties too. The EEZ is a major problem. The EEZ, originally being part of the high seas, has perplexed the issue of piracy. Since the articles relating to piracy in the LOSC are fully and unchangeably copied from the 1958 Convention, it may query whether these articles are still applicable to the EEZ, as residual rights and/or obligations in question, despite the changed legal status of the latter. The provisions are ambiguous and controversial, particularly in the context of Article 86 (regarding the non-application of the high seas provisions) of the same convention.[40] In spite of such ambiguity, it is meaningful to note that similar residual rights and/or obligations have been retained in the LOSC formerly belonging to the freedom of high seas, such as the freedom of navigation in the EEZ. Article 58 of the LOSC expressly provides that the piracy provisions are applicable to the EEZs in so far as they are not incompatible with the provisions on EEZs of that convention and in compliance with the laws and regulations adopted by the coastal State. Since piracy is closely related to the safety of navigation, States could assume a corresponding duty or right to suppress piracy in the EEZ of other States provided that anti-piracy measures taken by such States are inadequate. The problem is more exacerbated when a certain coastal State is unable to handle effectively acts of piracy occurring within its EEZ. For such reason, it is argued that the piracy provisions in the LOSC should apply to the EEZ in so far as they are not incompatible with the rights of coastal States set forth in the LOSC. "Since enforcement against a pirate, in normal circumstances, could not

[40] See Bjorn Aune, "Piracy and Its Repression under the 1982 Law of the Sea Convention", *Ocean Yearbook*, Vol. 8, 1989, 29, 36–37.

Table 5.1: Contracting Parties to the LOSC and SUA Convention in East Asia

State	**The LOSC (dd/mm/yyyy)**	**The SUA (dd/mm/yyyy)**
Brunei Darussalam	05/11/1996	04/12/2003
Cambodia		18/08/2006
China	07/06/1996	01/03/1992
Indonesia	03/02/1986	
Japan	20/06/1996	23/07/1998
Korea (North)		
Korea (South)	29/01/1996	14/05/2003
Laos	05/06/1998	20/03/2012
Malaysia	14/10/1996	
Mongolia	13/08/1996	22/11/2005
Myanmar	21/05/1996	19/09/2003
Philippines	08/05/1984	06/01/2004
Singapore	17/11/1994	03/02/2004
Thailand	15/05/2011	
Vietnam	25/07/1994	10/10/2002

be viewed as impinging upon any rights reserved to the coastal State, the law of piracy in the EEZ must be viewed as identical to that applying beyond".[41]

It is to be noted that while most of the East Asian countries acceded to the LOSC, not so many such countries have acceded to the other important international convention on the prevention and suppression of piracy and maritime terrorism, i.e. the SUA Convention (see Table 5.1). It is regrettable that two key Malacca Straits States Indonesia and Malaysia still are left outside the SUA legal framework despite the fact that these two countries participated in the Correspondence Group under the leadership of the United States regarding the revision of the SUA Convention within the IMO.[42] The discussion on the Protocol to the SUA Convention

[41] Thomas A. Clingan Jr, "The Law of Piracy", in Eric Ellen (ed.), *Piracy at Sea* (Paris: ICC Publishing SA, 1989), p. 170.

[42] See Brad J. Kieserman, "Preventing and Defeating Terrorism at Sea: Practical Considerations for Implementation of the Draft Protocol to the Convention for the

is mainly focused on the possible incorporation of the shipboarding regime proposed by the United States (as reflected in the bilateral agreements signed between the United States and relevant flag-State countries) into the SUA Convention. A diplomatic conference on the SUA Convention held in October 2005 adopted the Protocol incorporating substantial amendments aimed at strengthening the SUA Convention in order to provide an appropriate response to the increasing risks posed to maritime navigation by piracy as well as maritime terrorism.[43] This endeavor by the world community will further improve the existing legal regime for the crackdown of piracy.

The 2005 Protocol to the SUA Convention has two major revisions/additions concerning the SUA Convention: first, the Protocol expands the coverage of the unlawful acts under Article 3 of the SUA Convention by adding a new provision to cover "uses against or on a ship or discharging from a ship any explosive, radioactive material or BCN (biological, chemical, nuclear) weapon in a manner that causes or is likely to cause death or serious injury or damage; discharges, from a ship, oil, liquefied natural gas, or other hazardous or noxious substance, in such quantity or concentration that causes or is likely to cause death or serious injury or damage; uses a ship in a manner that causes death or serious injury or damage; transports on board a ship any explosive or radioactive material, knowing that it is intended to be used to cause, or in a threat to cause, death or serious injury or damage for the purpose of intimidating a population, or compelling a Government or an international organization to do or to abstain from doing any act; transports on board a ship any BCN weapon, knowing it to be a BCN weapon; any source material, special fissionable material, or equipment or material especially designed or prepared for the processing, use or production of special fissionable material, knowing that it is intended to be used in a nuclear explosive activity or in any other nuclear activity not under safeguards pursuant to an IAEA comprehensive safeguards agreement; and transports on board a ship any equipment, materials or software or related technology that significantly

---

Suppression of Unlawful Acts against the Safety of Maritime Navigation", in Myron H. Nordquist, John N. Moore and Kuen-chen Fu (eds.), *Recent Developments in the Law of the Sea and China* (Leiden: Martinus Nijhoff, 2006), p. 434.

[43] See "Convention for the Suppression of Unlawful Acts against the Safety of Maritime Navigation", available at http://www.imo.org/home.asp.

contributes to the design, manufacture or delivery of a BCN weapon, with the intention that it will be used for such purpose".[44]

Another major development is the shipboarding regime. According to Article 8 bis of the 2005 Protocol, cooperation and procedures are needed if a State Party desires to board a ship flying the flag of a State Party when the requesting Party has reasonable grounds to suspect that the ship or a person on board the ship is, has been, or is about to be involved in, the commission of an offence under the Convention. The authorization and cooperation of the flag State is required before such a boarding. A State Party may notify the IMO Secretary-General that it would allow authorization to board and search a ship flying its flag, its cargo and persons on board if there is no response from the flag State within four hours. A State Party can also notify that it authorizes a requesting Party to board and search the ship, its cargo and persons on board, and to question the persons on board to determine if an offence has been, or is about to be, committed. The Protocol limits the use of force and includes important safeguarding measures when a State Party takes action against a ship.[45]

It is to be noted that the shipboarding regime, though new to the SUA Convention, exists in international law for almost two decades when the Vienna Convention for the Suppression of Illicit Traffic in Narcotic Drugs and Psychotropic Substances was adopted in December 1988.[46] It is the first international treaty containing the "multilateral shipboarding provision"[47] as stipulated in Article 17 of the Convention. The boarding regime is also incorporated in other international treaties including the Protocol against the Smuggling of Migrants by Land, Sea and Air to the 2000 UN Convention against Transnational Organized Crime,[48] and the 1995 United Nations Agreement for the Implementation of the Provisions of the United Nations Convention on the Law of the Sea of

---

[44] "Convention for the suppression of unlawful acts against the safety of maritime navigation, 1988", available at http://www.imo.org/Conventions/mainframe.asp?topic_id=259&doc_id=686#review.

[45] *Ibid.*

[46] Text is reprinted in 28 ILM (1989) 497. It came into force on 11 November 1990.

[47] See Brad J. Kieserman, *supra* note 42, p. 433.

[48] It entered into force on 29 September 2003, available at http://www.uncjin.org/Documents/Conventions/dcatoc/final_documents_2/convention_eng.pdf.

10 December 1982 relating to the Conservation and Management of Straddling Fish Stocks and Highly Migratory Fish Stocks.[49]

The 2005 SUA Protocol came into force in July 2010 and the current contracting parties are 51. Though in comparison with the SUA Convention which has 166 contracting parties, the SUA Protocol provides the shipboarding regime most applicable to piracy and maritime terrorism for the purpose of maintaining maritime security around the world.

UNSC Resolution 1816

Against the aforesaid background, it is recalled that in June 2008 the United Nations Security Council (UNSC) passed a resolution on combating acts of piracy and armed robbery off Somalia's coast (Resolution 1816).[50] While the Security Council expressed its grave concern about piracy and armed robbery against vessels in the waters off the coast of Somalia, it determined that such piratical incidents exacerbated the situation in Somalia "which continues to constitute a threat to international peace and security in the region".[51] Therefore, the Security Council decided to act under Chapter VII of the Charter of the United Nations. Resolution 1816 was adopted in line with the mandate stipulated in the UN Charter. It urges "States whose naval vessels and military aircraft operate on the high seas and airspace off the coast of Somalia to be vigilant to acts of piracy and armed robbery" and "to cooperate with each other, with the IMO and, as appropriate, with the relevant regional organizations in connection with, and share information about, acts of piracy and armed robbery in the territorial waters and on the high seas off the coast of Somalia, and to render assistance to vessels threatened by or under attack by pirates or armed robbers, in accordance with relevant international law".[52] More significantly, the Security Council decided that "for a period of six months from the date of this resolution, States cooperating with the TFG[53] in the fight against piracy and armed robbery at sea

[49] It entered into force on 11 December 2001. Text, available at http://daccessdds.un.org/doc/UNDOC/GEN/N95/274/67/PDF/N9527467.pdf?OpenElement.

[50] UN Doc S/RES/1816 (2008), 2 June 2008, available at http://daccessdds.un.org/doc/UNDOC/GEN/N08/361/77/PDF/N0836177.pdf?OpenElement.

[51] *Ibid.*

[52] See UN Doc S/RES/1816 (2008).

[53] It refers to the Transitional Federal Government of Somalia.

off the coast of Somalia, for which advance notification has been provided by the TFG to the Secretary General, may: (a) enter the territorial waters of Somalia for the purpose of repressing acts of piracy and armed robbery at sea, in a manner consistent with such action permitted on the high seas with respect to piracy under relevant international law; and (b) use, within the territorial waters of Somalia, in a manner consistent with action permitted on the high seas with respect to piracy under relevant international law, all necessary means to repress acts of piracy and armed robbery".[54] To implement this decision, cooperating States are requested to take appropriate steps to ensure that the activities they undertake do not have the practical effect of denying or impairing the right of innocent passage to the ships of any third State. Furthermore, the Security Council called upon "all States, and in particular flag, port and coastal States, States of the nationality of victims and perpetrators or piracy and armed robbery, and other States with relevant jurisdiction under international law and national legislation, to cooperate in determining jurisdiction, and in the investigation and prosecution of persons responsible for acts of piracy and armed robbery off the coast of Somalia, consistent with applicable international law including international human rights law, and to render assistance by, among other actions, providing disposition and logistics assistance with respect to persons under their jurisdiction and control, such victims and witnesses and persons detained as a result of operations conducted under this resolution".[55]

While the Resolution does not mention the EEZ of Somalia, it is understood that for operational purposes in terms of piracy suppression, the term "high seas" includes the EEZ in accordance with the LOSC, despite the legal fact that the EEZ is a maritime zone within national jurisdiction. In this context, it should be noted that the wording "international waters" contained in Resolution 1816 is not accurately consistent with the relevant provisions of the LOSC. It reflects the approach taken by the American Navy for its operations around the world, which divides the ocean into two categories: national waters and international waters and the EEZ is categorized as "international waters".[56] Furthermore, it is hard to understand that while the resolution refers to the LOSC, it does not

[54] See UN Doc S/RES/1816 (2008).

[55] *Ibid.*

[56] National waters include internal waters, territorial seas and archipelagic waters, and international waters include contiguous zones, EEZ, high seas and security zones.

mention another important anti-piracy international treaty: the SUA Convention.

Resolution 1816 carries a number of significant legal and political implications for the development of the international law of the sea, particularly the law of piracy. It is for the first time that the UNSC has placed the issue of piracy on its agenda and treated it as a matter threatening international peace and security. The Security Council has created a historical precedent. Traditionally, the UNSC by and large regarded armed conflicts as threats to international peace and security, and exerted efforts to eliminate such treats and restore peace and security to the affected regions. It is unprecedented for the Security Council to view piracy as equivalent to armed conflicts. Having said that, it is to be noted that the UNSC action should be interpreted in the context of the whole situation facing Somalia, which is a failing State without an effective government.[57] That is one of the reasons why Resolution 1816 was passed under the title of "Situation of Somalia". Although the reasons for the resurgence of sea piracy off the Somali coast are various, one thing that is certain is that such criminal acts have been caused by the paralysis of the central government, as well as the sustained internal armed conflicts in that country.

Unlike resolutions passed by the UN General Assembly, resolutions passed by the UNSC, particularly those adopted under Chapter VII, have legal effect. When the term "decide" is used, it contains the highest degree of binding force. According to Article 25 of the UN Charter, Member States agree to accept and carry out the decisions of the Security Council in accordance with the Charter.[58] Furthermore, Member States have the duty to contribute to the maintenance of international peace and security by providing material support to the Security Council, through the supply of armed forces, assistance and facilities, including rights of passage.[59] These provisions in the UN Charter constitute a legal foundation for the UN Member States to act in accordance with Resolution 1816.

Resolution 1816 authorizes States, particularly UN Member States, to enter the territorial waters of Somalia for the purpose of repressing acts of piracy and armed robbery at sea by all necessary means. While there are

---

See Department of the Navy, *The Commander's Handbook on the Law of Naval Operations*, October 1995, pp. 1–4–1–6.

[57] See further SC Res. 1814 (2008).

[58] Article 25 of the UN Charter.

[59] See Article 43 of the UN Charter.

some conditions imposed upon the entry into the Somali territorial waters and the use of necessary means, this authorization represents an audacious step taken by the Security Council which impacts on the existing legal regime for the territorial sea. Under general international law, the territorial sea is recognized as an integral part of the territory of a State, which enjoys full sovereignty over it. This sovereignty extends to the air space over the territorial sea as well as to its seabed and subsoil.[60] The inviolability of State sovereignty is well recognized in international law.

During the deliberations on the resolution, some States raised their concern about the entry into the Somali territorial waters under the UN authorization. China expressed the view that the facilitation initiated by the UNSC of international assistance in combating piracy should not produce any negative consequences and such assistance must comply with the LOSC. Furthermore, it must not conflict with existing international legislation. Indonesia and Vietnam held a similar view.[61] In response, the resolution contains a paragraph emphasizing that "the authorization provided in this resolution applies only with respect to the situation in Somalia and shall not affect the rights or obligations or responsibilities of Member States under international law, including any rights or obligations under the Convention, with respect to any other situation, and underscores in particular that it shall not be considered as establishing customary international law".[62] Also, the Security Council comforted its members with the above concerns that such authorization was based on the consent of the TFG.

Clearly, the State of Somalia, as a failing country, has shifted its responsibility for dealing with piratical activities to the international community by giving its consent in the name of the TFG to the United Nations. As we know, Somalia signed the LOSC in 1982 and ratified it in 1989, so that it is obliged to implement relevant provisions of the LOSC in the fight against sea piracy. The failure of Somalia to respect its treaty obligations has resulted in the intervention of the international community under the authorization of the UNSC. This has invited concerns from some other countries which do not possess adequate law enforcement

---

[60] Article 2(2) of the LOSC.

[61] "Security Council condemns acts of piracy, armed robbery off Somalia's coast, authorizes for six months 'all necessary means' to repress such acts", SC/9344, 2 June 2008, available at http://www.un.org/News/Press/docs/2008/sc9344.doc.htm.

[62] See UN Doc S/RES/1816 (2008).

facilities for the control of piracy. It is recalled that piracy was very rampant in Indonesian waters in the past years due to the government's slow and ineffective response to piracy. A source in 2002 revealed that Indonesia needed 239 ships and 115 aircraft to patrol its 13,000 islands effectively but had only 115 vessels and 60 aircraft.[63] Fearing that the Somali case could establish a precedent for the intervention of the international community in the countries which lack effective law enforcement mechanisms, the Indonesian representative to the Security Council emphasized that the resolution should only apply to the territorial waters of Somalia based on that country's prior consent, and that it should not become the basis of rules of customary international law.[64]

Related to Resolution 1816 is the concept of the responsibility to protect. It is argued that "State sovereignty implies responsibility, and the primary responsibility for the protection of its people lies with the state itself". Secondly, "where a population is suffering serious harm, as a result of internal war, insurgency, repression or state failure, and the state in question is unwilling or unable to halt or avert it, the principle of non-intervention yields to the international responsibility to protect".[65] The actions taken by Western countries and Russia against sea piracy in Somali waters imply the application of this concept as such States feel the necessity to protect peoples in distress in addition to the protection of the international interest. Resolution 1816 to some extent reflects this concept by emphasizing the importance of the humanitarian aid to the people of Somalia. It points out that food aid operations carried by the World Food Program have been threatened by sea piracy.[66] However, the controversy remains that this concept benefits powers which have the capacity to intervene, and history tells us that only big powers intervened in small and weak States, thus causing concerns from some developing countries as expressed in the deliberations of the Resolution.

Finally, according to Resolution 1816, UNSC authorization lasts for 6 months from the date of its adoption. The IMO urged the UNSC to

---

[63] "Ship owners urged to install anti-piracy tracking devices", *Agence France-Presse*, 8 May 2002.

[64] "Security Council condemns acts of piracy, armed robbery off Somalia's coast, authorizes for six months 'all necessary means' to repress such acts", *supra* note 47, p. 1.

[65] International Commission on Intervention and State Sovereignty, *The Responsibility to Protect*, Report, December 2001, p. xi.

[66] See UN Doc S/RES/1816 (2008).

extend this authorization for another period. The UN Secretary-General has concurred with the IMO call.[67] As a follow-up, the UNSC adopted Resolution 1838 in October 2008. This resolution addresses the same issues as Resolution 1816, but does not mention the territorial waters of Somalia although it upholds the validity of Resolution 1816.[68] It can be seen from the aforementioned endeavors of the UNSC that the piracy issue remains firmly on its agenda.

State Practice in East Asia

The inadequacy of the international legal regime needs a remedy from domestic legal arrangements. Even a sound international legal regime requires effective implementing efforts taken by individual States. Following are the practices of some East Asian countries.

China

China borders three seas, i.e. the Yellow Sea, the East China Sea and the South China Sea. The South China Sea is a place of piracy and its sea area with points in Hong Kong, Luzon Island, Hainan Island is called "the Hainan Triangle", which is a pirate resort.

China signed the LOSC in 1982 and ratified it in 1996. In addition, China has also ratified the two 1988 SUA legal documents against maritime terrorism. China, as a Member State, has participated in activities relating to the piracy issue sponsored by relevant international organizations such as the IMO.

Under the Chinese legal system, there is no law on piracy, even such a term as "piracy" in law. However, piracy is subject to the Chinese law and law enforcement via its criminal code. According to its amended Criminal Law, certain crimes, particularly some crimes of endangering public security, are relevant to piracy so that the piracy can be punishable under its law. This Law also allows China to enforce anti-piracy conventions within the Chinese territory, as it provides that "for the crimes

[67] See "United Nations Secretary General supports anti-piracy mandate extension", 20 October 2008, available at http://www.imo.org/Newsroom/mainframe.asp?topic_id=10286.

[68] See UN Doc S/RES/1838 (2008), 7 October 2008.

defined in international treaties, concluded or acceded to by the People's Republic of China, which are under the jurisdiction of the People's Republic of China within the framework of the treaty obligations, this Law shall apply".

Other relevant laws are Maritime Traffic Safety Law, 1984 (concerning the safety of navigation); Law on the Territorial Sea and the Contiguous Zone, 1992 (establishing the territorial sea and contiguous zone in China and provides the right of hot pursuit to chase and arrest suspected piracy vessels); and Law on the Exclusive Economic Zone and the Continental Shelf, 1998 (establishing China's jurisdiction in its EEZ and on its continental shelf including the right of hot pursuit). In addition, the Opinion on Strengthening the Safety of Navigation and Fishery in the East China Sea adopted in 1993 is particularly relevant to the suppression of piracy within China's jurisdictional waters.

With the rapid growth of its economy, China has been more dependent on energy resources from abroad. Petroleum products transported to China through sea lanes in East Asia, in particular through the Malacca Straits have become more frequent. About 80% of its imported oil has shipped to China through this sea lane. For this reason, China recently expressed its deep concerns with the navigational safety of the Malacca Straits and its strong interests in Straits' affairs. As reported, China expressed its willingness to help strengthen the maritime security in the Straits though the details have not yet been exposed to the public.[69] On the other hand, the Straits littoral States welcome China's assistance.

There are five cases recorded in China's judicial practice for the crackdown of piracy, including the case of the *Tenyu*, the case of the *Cheung Son*, the case of the *Marine Fortuner*, the case of the *Siam Xanxai* and the case of the *MT Global Mars*.[70] These cases have shown that China has gradually formulated a set of judicial procedures to try piracy cases and rendered harsh punishment on pirates including death penalty and

[69] See "Indonesian Foreign Minister: China is willing to help strengthen the safety of the Malacca Straits", *Liaohe Zaobao* (in Chinese), 2 August 2005, at http://www.zaobao.com/gj/yx050802_501.html.

[70] For details, see Zou Keyuan, "New Developments in the International Law of Piracy", *Chinese Journal of International Law*, 8, 2009, 342–343.

long-term imprisonment. In comparison with trials in other countries, the punishment imposed by Chinese courts is the most severe.[71]

Other East Asian countries

Japan ratified the LOSC in 1996 and adopted its own laws on the territorial sea and on EEZ as well. Punishment of piracy is subject to the Japanese criminal code. Recently, Japan has increased its efforts on the piracy issue, particularly after it ratified the SUA Convention in 1998. Japan realized that the frequent occurrence of piratical incidents and the damages they caused became a major threat to Japan's shipping routes as Japan depends on imports for the vast majority of its oil and other energy supplies and has developed its economy as a trading country.[72] It sent delegations to the Philippines, Malaysia, Singapore and Indonesia in September 2000 to make surveys and exchanges of views regarding specific anti-piracy cooperation and supporting measures in each country. In November 2000, Japan's Coast Guard patrol vessel visited India and Malaysia to seek bilateral cooperation. Japan also sponsored and organized a series of regional conferences on combating piracy and armed robbery against ships.

On the other hand, Japan's initiatives to combat piracy were made in the same pace as Japan expanded its military forces outside its own territory. There is a suspicion in the Southeast Asian region whether Japan takes the advantage of anti-piracy and anti-terrorism as a pretext to realize its military expansion which would create concerns among the Asian countries, particularly linking Japan's recent endeavors to its past of aggression. As commented, fighting piracy has paid dividends for Japan and has become a fairly good way to get Japan's security role in the region accepted.[73] Japan continues its joint anti-piracy military drills with other countries including the Philippines, Thailand and Indonesia.

---

[71] For example, an Indonesian court sentenced a band of hijackers to from only two to four years' imprisonment. See "Pirate attacks have tripled in a decade, IMB report finds", available at http://www.iccwbo.org/home/news_archives/2003/stories/piracy-quarter-1.asp.

[72] See Japanese Ministry of Foreign Affairs, "Diplomatic Bluebook 2001", available at http://www.mofa.go.jp/policy/other/bluebook/2001/chap2–4-d.html.

[73] Anthony Bergin, "East Asian naval developments — sailing into rough seas", *Marine Policy*, Vol. 26, 2002, 126.

Japan adopted the Law on Punishment of and Measures against Acts of Piracy in 2009,[74] and handled the case of *M/V Guanabara*.[75]

As for three Malacca Straits States — Indonesia, Malaysia and Singapore, they ratified the LOSC, but only Singapore is a signatory to the SUA Convention. As reported, there is no piratical incident within Singapore's jurisdictional waters in recent years. This indicates the effectiveness of Singapore's law enforcement, though the water area under Singapore's sovereignty is much smaller than those under the sovereignty of Indonesia or Malaysia.

Indonesia has the most serious problem of piracy in the whole Southeast Asian region. It has been long criticized for its slow and ineffective response to piracy. In order to strengthen its fight against piracy, Indonesia has established a number of anti-piracy centers in Medan on Sumatra, on Batam Island, and in Bangka on Sumatra.

Piracy has frequently harassed Malaysian waters as well, in particular the Straits of Malacca. However, the recent strengthened anti-piracy measures achieved good results. For example, in 2000, there were 75 attacks in the waters off Malaysia, but after the Malaysian authorities took some measures, the number was reduced to 17 in 2001.[76] However, the piratical situation in the Malacca Straits is not stable. It is reported that after the Tsunami Disaster in 2004, piracy resurged rampantly in the Straits again.

Due to the severity of the piracy occurrence in the Straits of Malacca, the above three countries have intensified the degree of their trilateral cooperation. In the end of July 2005, a scheme of maritime air surveillance was discussed, aiming to strengthen the crackdown of piracy in this critical international waterway.[77] In August 2005, the above-said three countries agreed to implement the scheme of air patrol over the Malacca

---

[74] See outline of the draft anti-piracy measure law of Japan, available at un.org/depts/los/LEGISLATIONANDTREATIES/PDFFILES/JPN_anti_piracy.pdf.

[75] Akio Morita, "On the criminal jurisdiction over piracy in states other than arresting state", *Supplement of Law Seminar — Quick Report of Cases*, Vol. 13, 30 August 2013, available at www.internationalcrimesdatabase.org/upload/icd/Prof-Akio-Morita-Piracy-Trial-in-Japan-2013.pdf. (in Japanese).

[76] "ICC report sends six warships into battle against pirates", available at http://www.iccwbo.org/home/news_archives/2002/piracy.asp.

[77] See "Singapore and Malaysia discussed air surveillance over the Malacca Straits", *Liaohe Zaobao* (in Chinese), 30 July 2005, available at http://www.zaobao.com/sp/sp050730_503.html.

Straits from September 2005 and also agreed to establish a Tripartite Technical Experts Group on Maritime Security.[78] It is to be noted that all these tripartite cooperative operations, whether at sea or in the air, are called coordinated patrols, instead of joint patrols. In addition, Singapore is a member of the PSI but Malaysia and Indonesia are skeptical about this Initiative and remain outside it.

Thailand expressed its willingness to join the anti-terrorism agreement among Indonesia, Malaysia and the Philippines. Thailand is also considering joining the air surveillance scheme created by Singapore, Malaysia and Indonesia.[79] Finally, it is worth mentioning that the only country in the region which specifically put "piracy" under its Penal Code is the Philippines. According to the Philippine Penal Code, death penalty will be imposed upon those who commit piracy.[80]

Enhancing Regional Cooperation

In the Southeast Asian seas, different sea zones are established with the ratification of the LOSC by the riparian countries. Therefore, the States concerned have the right as well as the treaty obligation to enforce relevant laws and regulations within their respective jurisdictional sea areas, including the suppression and control of acts of piracy therein. Second, as to the sea areas beyond the territorial seas of the coastal States, a regional cooperation is more required than within the territorial sea. According to some statistics, the average distance of the piratical attacks from shore was 11.55 nm in Indonesia, 68 nm in Northeast Asia and 94.4 nm in the South China Sea.[81]

---

[78] See "Singapore, Malaysia and Indonesia carry air patrol over the Malacca Straits from next month", *Liaohe Zaobao* (in Chinese), 3 August 2005, at http://www.zaobao.com/sp/sp050803_501.html.

[79] See "Thailand needs time to consider air patrol over the Malacca Straits", *Liaohe Zaobao* (in Chinese), 3 August 2005, at http://www.zaobao.com/sp/sp050803_509.html.

[80] See Gabriel V. Trinidad (ed.), *The Revised Penal Code*, 1983 Edition (Quezon City: UP Law Center, 1984), p. 52.

[81] Mark C. Farley, "International and regional trends in maritime piracy 1989–1993", Thesis, Naval Postgraduate School, 1993, p. 24; cited in Barry Hart Dubner, "Maritime violence — The problems with modern day piracy", Issue Paper No. 9/95, Maritime Institute of Malaysia, Kuala Lumpur, p. 9.

Effective law enforcement is extremely difficult in Southeast Asian seas. A typical example is the South China Sea which is vast (more than 200 nm wide) and dotted with numerous uninhabited islands to which pirates can easily retreat.[82] It is said that the Natunas and Spratly Islands are pirate havens.[83] Another main obstacle to the effective law enforcement rests with the overlapping territorial claims to some islands. For example, the Spratly Islands are claimed by five adjacent countries, i.e. Brunei, China including Taiwan, Malaysia, the Philippines, and Vietnam, and the territorial dispute has not yet been resolved. Even if the territorial dispute had been solved, there are still boundary delimitation issues in the South China Sea.

Under such complicated circumstances, cooperation is the only way for maintaining regional maritime security. The "track two" conferences on the South China Sea sponsored by Indonesia put on the agenda the safety of navigation and suppression of piracy as one of the areas of possible cooperation among the interested countries bordering the South China Sea.[84] In November 2002, the Joint Declaration of ASEAN and China on Cooperation in the Field of Non-Traditional Security Issues (the Joint Declaration) was adopted, which initiated full cooperation between ASEAN and China in the field of non-traditional security issues and listed the priority and form of cooperation. The priorities at the current stage of cooperation include "combating trafficking in illegal drugs, people-smuggling including trafficking in women and children, sea piracy, terrorism, arms-smuggling, money-laundering, international economic crime and cybercrime".[85] In addition, the 2002 Declaration on the Conduct of the Parties in the South China Sea[86] also mentions the suppression of piracy and armed robbery at sea.

---

[82] See "IMO to organise missions, regional seminars on piracy", IMOFAX 05/98, 8 April 1998, available at http://www.imo.org/imo/briefing/1998/fax05.htm.

[83] Aune, *supra* note 40, p. 21.

[84] See, for example, the First Meeting of the Technical Working Group on Safety of Navigation, Shipping and Communication in the South China Sea, Jakarta, 3–6 October 1995, in "Workshop and Technical Working Group Statements", available at http://www.law.ubc.ca/centres/scsweb/statement.html.

[85] See Joint Declaration of ASEAN and China on Cooperation in the Field of Non-Traditional Security Issues, 6th ASEAN–China Summit, Phnom Penh, Cambodia, 4 November 2002, available at http://www.aseansec.org/13185.htm.

[86] Text is available at http://www.aseansec.org/13163.htm.

When the coastal State has good reason to believe that a foreign ship has violated its laws and regulations, it may exercise the right of hot pursuit. This can be directed to suspected vessels of piracy. Hot pursuit can begin from internal waters, territorial seas and the contiguous zone or even in the EEZs of the coastal States. The Chinese laws on the territorial sea and on EEZ authorize the Chinese warships or governmental vessels to exercise the right of hot pursuit.[87] However, the right of hot pursuit ceases as soon as the ship pursued enters the territorial sea of its own State or of a third State.[88] For this reason, in the case when piratical attacks occur within territorial seas, and then the attackers flee to the territorial sea of another State, it is suggested that such a situation requires a regional cooperative arrangement between the affected States to coordinate their enforcement policies. Another serious matter relating to the suppression of piracy is that many pirates were let go unpunished under domestic law after the hijacked ship was detained. Such circumstances once happened in Southern China.[89] For this reason, in the *Petro Ranger* incident, the Royal Malaysian Police considered to request the Chinese government to extradite the pirates to stand trial in Malaysia.[90] However, extradition requires an agreement between the two countries concerned, or may be executed under an international treaty to which both countries concerned are the parties.

There is a wide speculation that some of the piratical incidents which occurred in East Asia involved organized criminals since the hijacking of a whole ship and resale of its cargo requires huge resources and detailed planning. Individual pirates do not have these resources.[91] This new form of piracy makes the anti-piracy campaign more difficult.

---

[87] See Article 14 of the Law on the Territorial Sea and the Contiguous Zone; and Article 12 of the Law on Exclusive Economic Zone and the Continental Shelf. Texts of these two laws are reprinted in Office of Policy, Law and Regulation, State Oceanic Administration (ed.), *Collection of the Sea Laws and Regulations of the People's Republic of China*, 3rd ed. (Beijing: Ocean Press, 2001), pp. 201–205, 210–215.

[88] Article 111(3) of the LOSC.

[89] Personal interview with the officer in IMB-PRC, 6 August 1998.

[90] IMB-PRC, *supra* note 21, 7. The IMB-PRC regards this as an effective deterrent against piratical attacks.

[91] "Organized crime takes to the high seas, ICC piracy report finds", 4 February 2002, available at http://www.iccwbo.org/home/news_archives/2002/piracy_report.asp.

Nevertheless, there is a tendency of increased cooperation against piracy. The tripartite cooperation among Indonesia, Malaysia and Singapore has already begun. Another basis for cooperation is the IMB-PRC which was established in Kuala Lumpur on 1 October 1992. The Center acts as a focal point for the industry and liaisons with the law enforcement authorities in the region as well as in the world. Furthermore, in May 2002, Indonesia, Malaysia and the Philippines signed the Agreement on Information Exchange and Establishment of Communication Procedures with a view to preventing the utilization by anyone of their land–air–sea territories for the purpose of committing or furthering such activities as terrorism, money laundering, smuggling, piracy, hijacking, intrusion, illegal entry, drug trafficking, theft of marine resources, marine pollution and illicit trafficking of arms.[92] The agreement is open to other ASEAN countries. Following this agreement, the ASEAN ministerial meeting held in Kuala Lumpur in early May 2002 adopted the Work Program to Implement the ASEAN Plan of Action to Combat Transnational Crime, which focuses eight "priority areas" and piracy is one of them. The ASEAN member States agreed to work toward the harmonization of laws among themselves in order to effectively deal with the issues of transnational crime, and also agreed to develop programs for joint exercises and simulations in various areas to enhance cooperation and coordination in law enforcement and intelligence sharing.[93]

The ASEAN Regional Forum (ARF) for the first time adopted the Statement on Cooperation against Piracy and Other Threats to Maritime Security in June 2003, acknowledging that piracy and armed robbery against ships have been a significant problem in the Asia-Pacific region and effective responses require "regional maritime security strategies and multilateral cooperation in their implementation".[94] It is noted that the ARF countries express their commitment to become parties of the SUA Convention if they have not yet done so. In addition to undertaking necessary actions such as exchange of information, consideration and discussion of new IMB proposals on prescribed traffic lanes for large supertankers with coastguard or naval escort, provision of technical assistance and capacity-building infrastructure for countries that need help, they commit to "endorse the ongoing efforts to establish a legal

---

[92] See "Thailand to join SE Asian nations in terrorism agreement", *Asia Pulse*, 9 May 2002.

[93] "ASEAN plan to fight terror", *The Straits Times*, 18 May 2002.

[94] Text is available at http://www.aseansec.org/14838.htm.

framework for regional cooperation to combat piracy and armed-robberies against ships".[95]

ReCAAP

The most significant development for the international law of piracy lies in the adoption of the Regional Cooperation Agreement on Combating Piracy and Armed Robbery against Ships in Asia (ReCAAP).[96] It was signed by 16 Asian countries including Bangladesh, Brunei, Cambodia, China, India, Indonesia, Japan, Laos, Malaysia, Myanmar, the Philippines, Sri Lanka, Singapore, South Korea, Thailand and Vietnam on 11 November 2004. It is reported that Cambodia, Japan, Laos and Singapore were the first that officially ratified the Agreement on 28 April 2005.[97] The Agreement came into force on 4 September 2006 when it received 10 ratifications.[98]

The Agreement obliges Contracting States: (a) to prevent and suppress piracy and armed robbery against ships; (b) to arrest pirates or persons who have committed armed robbery against ships; (c) to seize ships or aircraft used for committing piracy or armed robbery against ships; and (d) to rescue victim ships and victims of piracy or armed robbery against ships.[99] The Contracting States pledge to implement the Agreement including preventing and suppressing piracy and armed robbery against ships "to the fullest extent possible" "in accordance with their respective national laws and regulations and subject to their available resources or capabilities" (Article 2.1).

For the aforementioned purposes, the Contracting States are required to cooperate between/among themselves. The first area required for cooperation is information sharing. Each Contracting Party designates a focal point responsible for its communication with the Information Sharing Center (ISC), and should "ensure the smooth and effective communication between its designated focal point, and other competent national

---

[95] *Ibid.*

[96] The text is available at http://www.recaap.org/about/pdf/ReCAAP%20Agreement.pdf.

[97] See "Singapore, Japan, Laos and Cambodia take the lead to sign the Anti-Piracy Agreement", *Lianhe Zaobao* (Singapore), 29 April 2005.

[98] It is regretted that two major Straits States Indonesia and Malaysia have not yet ratified the Agreement.

[99] Article 3 of the ReCAAP.

authorities including rescue coordination centers, as well as relevant non-governmental organizations". Each Contracting Party should "make every effort to require its ships, ship owners, or ship operators to promptly notify relevant national authorities including focal points, and the Center when appropriate, of incidents of piracy or armed robbery against ships" (Article 9). Contracting Parties are required to give prompt notification of the information about an imminent threat of, or an incident of, piracy or armed robbery against ships and in the event that a Contracting Party receives an alert from the Center as to an imminent threat of piracy or armed robbery against ships, that Contracting Party should promptly disseminate the alert to ships within the area of such an imminent threat (Article 9). The Agreement has created a right of request for all Contracting Parties regarding the information about piracy and armed robbery against ships.[100] The requested Contracting Party has the obligation to implement such request by taking effective and practical measures.[101]

The second area required for cooperation lies in the endeavors to take legal and judicial measures for the prevention and suppression of piracy and armed robbery including extradition and mutual legal assistance. According to the Agreement, a Contracting Party should endeavor to extradite pirates to the other Contracting Party which has jurisdiction over them and to render mutual legal assistance in criminal matters including the submission of evidence related to piracy and armed robbery at the request of another Contracting Party, but all these endeavors are subject to the national laws and regulations of the Contracting Party concerned.[102]

The third area required for cooperation is in the process of capacity building (including technical assistance such as educational and training programs) as the Agreement obliges each Contracting Party to endeavor to cooperate to the fullest possible extent with other Contracting Parties so as to enhance the capacity to prevent and suppress piracy and armed robbery against ships (Article 14.1). Contracting Parties can also make cooperative arrangements such as joint exercises between/among themselves (Article 15).

---

[100] See Article 10 of the ReCAAP.

[101] See Article 11 of the ReCAAP.

[102] See Articles 12–13 of the ReCAAP.

Another important legal arrangement made by the Agreement is the establishment of the ISC which is located in Singapore.[103] The center is composed of the Governing Council (which is the decision-making body comprising one representative from each Contracting Party) and the Secretariat (which is headed by the Executive Director who is responsible for the administrative, operational and financial matters of the Center in accordance with the policies as determined by the Governing Council and the provisions of the Agreement). It is designed "to promote close cooperation among the Contracting Parties in preventing and suppressing piracy and armed robbery against ships" (Article 4.1). As provided for by the Agreement, the ISC carries the following functions:

(a) to manage and maintain the expeditious flow of information relating to incidents of piracy and armed robbery against ships among the Contracting Parties;
(b) to collect, collate and analyze the information transmitted by the Contracting Parties concerning piracy and armed robbery against ships, including other relevant information, if any, relating to individuals and transnational organized criminal groups committing acts of piracy and armed robbery against ships;
(c) to prepare statistics and reports on the basis of the information gathered and analyzed under subparagraph, and to disseminate them to the Contracting Parties;
(d) to provide an appropriate alert, whenever possible, to the Contracting Parties if there is a reasonable ground to believe that a threat of incidents of piracy or armed robbery against ships is imminent;
(e) to circulate requests referred to in Article 10 and relevant information on the measures taken referred to in Article 11 among the Contracting Parties;
(f) to prepare non-classified statistics and reports based on information gathered and analyzed under subparagraph (b) and to disseminate them to the shipping community and the International Maritime Organization; and
(g) to perform such other functions as may be agreed upon by the Governing Council with a view to preventing and suppressing piracy and armed robbery against ships.[104]

---

[103] See Article 4 of the ReCAAP.

[104] Article 7 of the ReCAAP.

In addition, the ISC also plays an important role in proving capacity building assistance.[105] The ISC was officially launched in November 2007. The daily operations of the ISC is funded by the Host State as well as voluntary contributions from other Contracting Parties or even other sources as agreed by the Governing Council (Article 6.1). In addition to Singapore, three Contracting Parties — China, Japan and South Korea also provide operational funds for the ISC.[106]

There are several characteristics regarding the ReCAAP. First, though the original negotiators of the Agreement are 16 Asian States, the accession to the Agreement is not exclusive; any State can join after its entry into force as provided for in the Agreement (Article 18.5). Second, the ReCAAP is the first specific international treaty concerning the prevention and suppression of piracy. Because of this, it becomes a model of law for other regional legal arrangements. It is reported that a similar agreement will be concluded for the Western Indian Ocean.[107] Third, the ISC established under the ReCAAP is a governmental international organization, different from other organizations which operate similar functions such as the Piracy Reporting Centre (situated in Kuala Lumpur) of the IMB under the International Chamber of Commerce. Finally, it contributes to the legal definition on the piracy and armed robbery against ships as mentioned earlier.

Conclusion

Piracy is a common concern for the whole world community. The effective suppression and control of piracy at the regional level can benefit not

[105] See Article 14.2 of the ReCAAP.

[106] As is reported, both China and South Korea agreed to provide the ISC with the respective amount of US$50,000 and US$100,000 annually for the ISC. See Cai Tianchen, "ReCAAP obtained 100,000 US dollars from South Korea", *Lianhe Zaobao* (Singapore), 29 February 2008, available at http://www.zaobao.com.sg/sp/sp080229_540.shtml.

[107] A relevant meeting took place from 14 to 18 April 2008 in Dares Salaam. See UN Division for Ocean Affairs and the Law of the Sea, "Advance and unedited material on 'maritime security and safety' which will later be integrated into the Secretary-General's report on oceans and the law of the sea for the sixth-third session of the General Assembly and for the ninth meeting of the consultative process", 22 February 2008, p. 6, available at http://www.un.org/Depts/los/consultative_process/documents/maritime_security_and_safety.pdf.

only the adjacent countries but also the interest of the mankind. A regional cooperation can play an active as well as effective role in the suppression of piracy, initiate and coordinate the anti-piracy activities to be sponsored by either individual countries or jointly, resolve some potential jurisdictional conflicts arising from law enforcement among coastal States, and to facilitate and safeguard the navigation in the seas. In this respect, Asian countries set a good example by adopting the aforementioned anti-piracy agreement which legalizes the regional efforts in cracking down on piracy and maritime terrorism. However, this is only a good beginning and it remains to be seen whether the implementation of this Agreement can produce effective result as expected by the world community. It is recalled that despite various efforts taken, piracy in Southeast Asia still remains a serious maritime problem.

Apart from strengthening the efforts for regional cooperation, gaps in State practice should be bridged as well. For example, it is imperative for coastal States, particularly in East Asia, to seriously consider how to define piracy as a statutory crime within their respective domestic legal systems. In addition to the Philippines, more countries in East Asia including Japan, South Korea and Thailand recently criminalized piracy under their legal systems. Nevertheless, many countries such as China still do not have law of piracy at the domestic level. Second is related to the implementation of the relevant international treaties at the domestic level. For example, the implementation of the LOSC including its provisions on piracy at the domestic level needs a set of detailed and enforceable regulations enacted by the coastal State Parties to fulfill their obligations under the Convention.

Finally, the piracy definition and relevant piracy provisions in the LOSC need to be improved in the near future. The old definition is obviously unable to cope with the new situation of piracy in our world. Fortunately, there is a window of opportunity since the LOSC is subject to review and possible amendment any time after its entry into force in 1994.[108] States that would like to have the existing piracy provisions to be amended or to add new provisions can have the opportunity to

---

[108]Article 312 of the LOSC (After the expiry of a period of 10 years from the date of entry into force of this Convention, a State Party may, by written communication addressed to the Secretary-General of the United Nations, propose specific amendments to this Convention, other than those relating to activities in the Area, and request the convening of a conference to consider such proposed amendments.)

submit proposals and/or request to convene the review conference. This opportunity on the other hand imposes an imminent task on the world community to accomplish together with its endeavor to review and amend the SUA Convention. It is believed that after the corresponding revisions of the anti-piracy provisions in international law, the goal to seek effectiveness of piracy crackdown will be ultimately achieved.

Chapter 6

Illegal Human Trafficking by Sea and Countering Measures

Introduction

Maritime security has become a serious concern in East Asia, particularly after the September 11 terrorist attack. Different from traditional maritime security issues such as naval mobility, military activities and arms control, illegal human trafficking or people-smuggling via sea routes is within the category of non-traditional maritime security issues, in parallel with maritime terrorism and sea piracy. Although there is a controversy on whether trafficking of humans can be defined as an international crime against humanity,[1] it no doubt becomes a transnational crime subject to relevant regulations under international law.

Before proceeding to the analysis of the issue, some definitional problems need to be clarified first. The general definition of human trafficking may be found in the Protocol to Prevent, Suppress and Punish Trafficking in Persons, Especially Women and Children, supplementing the United Nations Convention against Transnational Organized Crime (the Trafficking Protocol), which provided that "'Trafficking in persons' shall mean the recruitment, transportation, transfer, harboring or receipt of persons, by means of the threat or use of force or other forms of coercion,

[1] Some scholars tend to define "trafficking of human beings" as "a crime against humanity". For example, see Tom Obokata, "Trafficking of Human Beings as a Crime against Humanity: Some Implications for the International Legal System", *International and Comparative Law Quarterly*, Vol. 54, 2005, 445.

of abduction, of fraud, of deception, of the abuse of power or of a position of vulnerability or of the giving or receiving of payments or benefits to achieve the consent of a person having control over another person, for the purpose of exploitation. Exploitation shall include, at a minimum, the exploitation of the prostitution of others or other forms of sexual exploitation, forced labor or services, slavery or practices similar to slavery, servitude or the removal of organs".[2] In addition, there are a number of similar or related terms. According to the Protocol against the Smuggling of Migrants by Land, Sea and Air (Smuggling Protocol) to the Convention against Transnational Organized Crime, the term, smuggling of migrants, means "the procurement, in order to obtain, directly or indirectly, a financial or other material benefit, of the illegal entry of a person into a State Party of which the person is not a national or a permanent resident".[3] The Smuggling Protocol also provides the definition of "illegal entry" which means crossing borders without complying with the necessary requirements for legal entry into the receiving State.[4]

Furthermore, the Model Law against Smuggling of Migrants, published by the United Nation Office on the Drugs and Crime (UNODC) in 2010, also provides the definition of "smuggled migrant", namely:

> "any person who has been the object of conduct criminalized, regardless of whether the perpetrator is identified, apprehended, prosecuted or convicted".[5]

In 2011, the UNODC adopted the International Framework for Action to Implement the Smuggling of Migrants Protocol, in which the term "migrant smuggling" refers to all the elements that constitute migrant smuggling and related crimes.[6]

---

[2]Article 3 of the Protocol to Prevent, Suppress and Punish Trafficking in Persons, Especially Women and Children, supplementing the United Nations Convention against Transnational Organized Crime, 2000, available at https://www.unodc.org/pdf/crime/a_res_55/res5525e.pdf.

[3]Article 3(a) of the Smuggling Protocol.

[4]Article 3(b) of the Smuggling Protocol.

[5]Article 3(i) of the Model Law against the Smuggling of Migrants, available at https://www.unodc.org/documents/legal-tools/Model_Law_Smuggling_of_Migrants.pdf.

[6]See https://www.unodc.org/documents/human-trafficking/Migrant-Smuggling/Framework_for_Action_Smuggling_of_Migrants.pdf.

It is opined that in comparison, the definition "trafficking in persons" in the Trafficking Protocol is a somehow narrow definition.[7] As Alese C. Wooditch comments, human trafficking, also called trafficking in persons, forms modern-day slavery involving the illegal transport of individuals by force or deception for the purpose of labor, sexual exploitation, or activities in which others benefit financially.[8] According to the US Department of State, "human trafficking", "trafficking in persons", and "modern slavery" are umbrella terms used to refer to both sex trafficking and compelled labor.[9] Likewise, the US Department of Justice concludes that human trafficking, also known as trafficking in persons or modern-day slavery, is a crime that involves compelling or coercing a person to provide labor or services, or to engage in commercial sex acts.[10]

International Legal Framework

The international legal framework for maritime human trafficking mainly consists of two essential sets of international law governing maritime human trafficking: one is concerning the fight against transnational crimes including illegal human trafficking and illegal migration; and the other is related to the governance of maritime order and security.

International law for the suppression of organized crimes

For the first set, the earlier international legal documents related to human migration are those governing refugees protection and diplomatic privileges. For example, the 1951 Convention on the Status of Refugees and its 1967 Protocol are mainly concerning how to provide certain legal protection for people who intend to take refuge in foreign countries. However, they seldom touch upon the issue on how to curb and prevent

---

[7] See Elzbieta M. Gozdziak and Elizabeth A. Collett, "Research on Human Trafficking in North America: A Review of Literature", in Frank Laczko and Elzbieta Gozdziak (eds.), *Data and Research on Human Trafficking: A Global Survey*, Offprint of the Special Issue of International Migration Vol. 43(1/2), 2005, pp. 103–107.

[8] Alese C. Wooditch, "Human trafficking", in *Encyclopedia of Race and Crime*, available at https://www.britannica.com/topic/human-trafficking.

[9] Available at https://www.state.gov/policy-issues/human-trafficking/.

[10] Available at https://www.justice.gov/humantrafficking/what-is-human-trafficking.

illegal human trafficking and to render punishment on those who plot, organize and conduct such illegal activities. The only relevant international treaty is the Final Protocol to the Convention for the Suppression of the Traffic in Persons and of the Exploitation of the Prostitution of Others (Lake Success, New York, 21 March 1950),[11] which came into force on 25 July 1951. It, however, has very limited functions, specifically in dealing with sex trafficking.

Only when it came to the late 1990s, the issue of human smuggling was put on the international agenda. The significant move was the adoption of the United Nations Convention against Transnational Organized Crime (the Palermo Convention) in New York on 15 November 2000, together with three protocols, of which two are directly concerning the illegal human trafficking issue.[12] The Convention came into force on 29 September 2003 and as of July 2020, there were 147 countries including all East Asian countries which ratified or acceded to the Convention (see Table 6.1), but some countries such as China made a reservation regarding the jurisdiction of the International Court of Justice should a dispute arise concerning the application and interpretation of the treaty.[13] As admitted by China, the West once dominated the process of the negotiations, but with the deepening of the negotiations, developing countries coordinated their positions through mechanisms such as "Group of 77 + China" and played an important role in the adoption of the Convention which has balanced rights and obligations of the States parties.[14]

The Palermo Convention aims at promoting international cooperation to prevent and combat transnational organized crime, as the first comprehensive multilateral legal instrument in this regard. The Convention standardizes terminology and concepts, creating a common basis for national crime-control frameworks. Such concepts include "organized criminal group", a definition of which was internationally agreed upon for

[11] Text is available at http://untreaty.un.org/English/TreatyEvent2003/index.htm.

[12] Text is available at https://www.unodc.org/unodc/en/organized-crime/intro/UNTOC.html.

[13] For example, the Chinese reservation reads, "The People's Republic of China makes a reservation with regard to Article 35, paragraph 2 of the Convention and is not bound by the provisions of Article 35, paragraph 2". See UN Treaty website: https://treaties.un.org/pages/ViewDetails.aspx?src=TREATY&mtdsg_no=XVIII-12&chapter=18&clang=_en.

[14] See "China's legal work in the field of the UN combating transnational organized crimes", 9 July 2001, available at http//www.fmprc.gov.cn/ebg/wjb/zzjg/tyfl/2626/2630/t15488.htm.

Table 6.1: United Nations Convention against Transnational Organized Crime in East Asia (as of 31 July 2020)

State	Signature	Ratification/Accession
Brunei		25 Mar 2008
Cambodia	11 Nov 2001	12 Dec 2005
China	12 Dec 2000	27 Aug 2003
Indonesia	12 Dec 2000	20 Apr 2009
Japan	12 Dec 2000	11 Jul 2017
Laos		26 Sept 2003
Malaysia	26 Sep 2002	24 Sept 2004
Mongolia		27 Jun 2008
Myanmar		30 Mar 2004
North Korea		17 Jun 2016
Philippines	14 Dec 2000	28 May 2002
Republic of Korea	13 Dec 2000	05 Nov 2015
Singapore	13 Dec 2000	28 Aug 2007
Thailand	13 Dec 2000	17 Oct 2013
Vietnam	13 Dec 2000	08 Jun 2012

Source: https://treaties.un.org/pages/ViewDetails.aspx?src=TREATY&mtdsg_no=XVIII-12&chapter=18&lang=en.

the first time. The Convention establishes four specific crimes (participation in an organized criminal group, money laundering, corruption and obstruction of justice) to combat activities in which organized criminal groups are commonly engaged. Under the Convention, States parties should criminalize these offences in accordance with the provisions of the Convention.

The Convention contains specific provisions for preventing, investigating and prosecuting these offences as well as serious crime, when they are transnational in nature and involve an organized criminal group. States parties to the Convention are obliged to adopt domestic laws and practices that would prevent or suppress organized crime-related activities. To combat money laundering, countries would have to require their banks to keep accurate records and make them available for inspection by domestic law enforcement authorities. It should be noted that bank secrecy cannot be used to shield criminal activities.

States parties to the Convention are also required to take appropriate action to confiscate illicitly acquired assets. In particular, the Convention created an asset-sharing mechanism under which States parties are encouraged to contribute confiscated assets to bodies working for the fight against organized crime. One of the most important international cooperation components of the Convention is its extradition provision. This provision is vital to ensuring that there are "no safe havens" to which offenders can flee. Under the Convention, fiscal matters should not be a sole ground for refusing extradition.

Mutual legal assistance is another important judicial cooperation tool provided for by the Convention. Under the Convention, assistance is to be channeled through central authorities to regulate the process. One of its innovative elements is that the Convention allows for electronic transmission of requests for quicker processing.

The nature of transnational organized crime makes the protection of victims and witnesses a matter of such importance that the Convention also requires States parties to adopt appropriate measures to protect witnesses from potential intimidation or retaliation. This includes physical protection, relocation and, with appropriate legal safeguards, concealment of identities.

The Convention further calls on States to support the efforts of developing countries to fight transnational organized crime and assist them to implement the Convention through technical cooperation as well as financial and material assistance.

As regards the implementation mechanism, the Convention establishes a conference of the parties to improve the capacity of States parties to combat transnational organized crime. The Conference will first meet within the first year of the entry into force of the Convention.

States parties whose domestic law requires involvement of an organized criminal group for purposes of the offences established in accordance with Article 5, paragraph 1(a)(i), of the Convention and States parties whose domestic law requires an act in furtherance of the agreement for purposes of the offences established in accordance with Article 5, paragraph 1(a)(i), of the Convention shall so inform the depositary at the time of their signature or of the deposit of their instrument of ratification, acceptance, approval of or accession to the Convention (Article 5(3)).

States that make extradition conditional on the existence of a treaty shall inform the depositary whether they will take this Convention as the legal basis for cooperation on extradition with other States parties to this Convention at the time of the deposit of their instrument of ratification, acceptance, approval or accession (Article 16(5)).

Each State party shall designate a central authority that shall have the responsibility and power to receive requests for mutual legal assistance and either to execute them or to transmit them to the competent authorities for execution. This designation of authority shall be notified to the depositary at the time each State deposits its instrument of ratification, acceptance, approval or accession (Article 18(13)).

The Protocol against the Smuggling of Migrants by Land, Sea and Air (Smuggling Protocol) is the most relevant international legal document to the theme of this chapter.[15] The Protocol aims at preventing and combating smuggling, promoting cooperation among States Parties and protecting the rights of smuggled migrants. As in the Convention against Transnational Organized Crime and the Protocol on Trafficking in Persons, many provisions are intended to ensure that the approaches taken by Member States under their domestic legislative and law-enforcement regimes are as coordinated as possible to make collective international measures both efficient and effective. As is the case with the Protocol on Trafficking in Persons, provisions of the Convention against Transnational Organized Crime apply *mutatis mutandis* to this Protocol unless otherwise stated in the Protocol. The Protocol applies to the prevention, investigation and prosecution of the smuggling of migrants as well as to the protection of the rights of persons who have been the object of such offences. States Parties to the Protocol are required to criminalize the basic smuggling of migrants and other forms of activity that support such smuggling. The Protocol specifies that migrants should not become liable for having been smuggled. Specific provisions for smuggling by sea are included because of the seriousness and volume of the problem. Under one such provision, States parties are requested to cooperate to prevent smuggling of migrants by sea and to take necessary measures when it is suspected that a vessel is engaged in the smuggling of migrants.[16] States may board and search vessels believed to be of their own registry.[17] States parties to the Protocol are also required to strengthen border measures and oblige

[15] Text is available at https://www.unodc.org/documents/middleeastandnorthafrica/smuggling-migrants/SoM_Protocol_English.pdf.

[16] See Article 7 of the Protocol.

[17] See Article 8.1 of the Protocol.

commercial carriers of passengers to check the travel documents of those passengers.[18] Another important element of the Protocol is the States' cooperation in the field of public information.[19] States are required to cooperate with each other to raise awareness of the dangers of smuggling to the migrants involved and to raise general awareness of the growing involvement of organized criminal groups.[20] The return of smuggled migrants to their countries of origin is foreseen. The State of origin is required to accept repatriation when the migrants in question have a right of residence in that State at the time of the return.[21] As mentioned previously, the Trafficking Protocol establishes the first common international definition of "trafficking in persons".[22] It is intended to prevent and combat such crime and facilitate international cooperation against it. The Protocol also highlights the problems associated with trafficking in persons that often leads to inhuman, degrading and dangerous exploitation of trafficked persons. As is the case with the parent Palermo Convention, the Protocol is expected to standardize terminology, laws and practices of countries in this area of law.

While the Convention provides for basic measures to prevent and combat transnational organized crime, its Protocols provide for specific measures to deal with specific crimes. As such, the Protocols should be interpreted together with the Convention. The provisions of the Convention apply *mutatis mutandis* to each Protocol. The aforesaid Protocol applies to the prevention, investigation and prosecution of trafficking offences, as well as to the protection of the trafficked persons.

The key definition, "trafficking in persons", is intended to include a range of cases where human beings are exploited by organized criminal groups, particularly where there is an element of duress involved and a transnational aspect, such as the movement of people across borders. According to the definition, the consent of the victim is irrelevant where illicit means are established, although criminal law defenses are preserved.[23]

---

[18] See Article 11 of the Protocol.

[19] See Article 15 of the Protocol.

[20] See Article 15 of the Protocol.

[21] See Article 18 of the Protocol.

[22] See Article 3 of the Protocol.

[23] See Article 3 of the Protocol.

The need for an appropriate balance between crime-control measures and measures to support or protect victims of trafficking arises in two primary places in the Protocol: the provisions expressly providing for protection and support; and provisions dealing with the return of persons to their countries of origin.

The Protocol contains a series of general protection and support measures for victims. These include a list of social support benefits such as counseling, housing, education, medical and psychological assistance and an opportunity for victims to obtain legal status allowing them to remain in the country of the receiving State party, either temporarily or permanently.[24]

Law enforcement agencies of countries which ratify the Protocol would be required to cooperate with each other in identifying offenders and trafficked persons; sharing information about the methods of offenders; and training investigators, enforcement and victim-support personnel. States parties would also be required to implement security and border controls to detect and prevent trafficking. This includes strengthening their own border controls; imposing requirements on commercial carriers to check passports and visas; setting standards for the technical quality of passports and other travel documents; and cooperating in establishing the validity of their own documents when used abroad.

Law of the Sea concerning illegal human trafficking

The 1982 United Nations Convention on the Law of the Sea (LOSC) is the key legal document to formulate the world maritime order today. While not directly governing the prevention and suppression of illegal human trafficking by sea, the Convention contains various clauses which have legal implications for the subject matter discussed in this chapter. The Convention divides maritime zones into two categories: zones within and outside national jurisdiction. The coastal State has the full sovereignty over its internal waters and territorial sea, while having sovereign rights and jurisdiction in its exclusive economic zone (EEZ) and continental shelf. The high seas and international seabed area are beyond national jurisdiction of any coastal State. Within the territorial sea, the coastal State exercises the control necessary to prevent and/or punish infringement of

[24] See Articles 6 and 7 of the Protocol.

its customs, fiscal, immigration or sanitary laws, and such power of control can be extended to the contiguous zone to its territorial sea.[25] Foreign ships enjoy the right of innocent passage in the territorial sea, but the loading or unloading of any person contrary to immigration laws of the coastal State is regarded as "non-innocent"[26] and should be prevented.

Criminal jurisdiction on board a foreign ship by a coastal State within its territorial sea is limited to the following four circumstances: if the consequences of the crime extend to the coastal state; if the crime is of a kind to disturb the peace of the country or the good order of the territorial sea; if the assistance of the local authorities has been requested by the master of the ship or by a diplomatic agent or consular officer of the flag State; or if such measures are necessary for the suppression of illicit traffic in narcotic or psychotropic substances.[27] With regard to illegal human trafficking, whether by a foreign ship or a ship of the coastal State, the aforementioned legal norms should apply since the crime, when occurring within the territorial sea, has the consequences affecting the coastal State, disturbing peace and good order, and comparing to illicit traffic in narcotic or psychotropic substances, it is more serious since it involves human life and creates danger and threat to it.

As to the jurisdiction on the high seas, the principle of flag State applies. According to the LOSC, every State should effectively exercise its jurisdiction and control in administrative, technical and social matters over ships flying its flag.[28] However, warships or governmental vessels of any State may exercise the right of visit.

More significantly, the illegal human trafficking by sea has been coming up with the agenda of the informal consultation meetings of the LOSC States parties for many years and is addressed in the UN report on the law of the sea under the heading "crimes at sea". Early in 2018, some LOSC States parties stressed the importance of ensuring maritime safety and security, including criminal activities at sea, and called for collective international actions.[29] As the UN Secretary-General emphasized in 2019, thousands of people continue to take perilous journeys by sea in many

---

[25] Article 33 of the LOSC.

[26] See Article 19(2)(g) of the LOSC.

[27] See Article 27(1) of the LOSC.

[28] Article 94(1) of the LOSC.

[29] See Report of the 28th Meeting of States Parties, United Nations Convention on the Law of the Sea, New York, 11–14 June 2018, SPLOS/324, para. 92.

parts of the world, risking death and injury, smuggling operations are common along many routes.[30] Criminal activities, including transnational organized crimes at sea, need to be addressed by the international community.[31] Given that, many States parties call on to take effective actions, such as enhancing national capacity, security measures, and, of course, the cooperation with the International Maritime Organisation (IMO).[32]

Recent Developments in Asia

Massive expansion of infrastructure across Asia, such as the Asian Highway Network and the Asia Europe High Speed Railway, does favor the cross-border movement of people in the region, but at the same time, migrant smuggling takes advantage of that development in some cases. Historically, illegal migrants from Asia preferred rich areas as destinations, such as Europe, North America. Such kind of flow is completely credible. New trends, however, occur with some interesting and extraordinary characteristics.

East Asia

In East Asia, smuggled migrants are regularly detected in Japan, Hong Kong, Macau and the Republic of Korea.[33] It should be noted that the situation of smuggled migrants in China, has been tremendously reversed. Until the mid-2000s, among the smuggled migrants originating from East Asia, most of them were Chinese and North Koreans.[34] China turns to be a predominantly receiving country from a sending country around the late-2000s.

---

[30] See Oceans and the law of the sea, Report of the Secretary-General, UNGA/A/74/350, 11 September 2019, para. 27.

[31] *Ibid.*, para. 37.

[32] *Ibid.*, paras. 31–44.

[33] See Migrants Smuggling in Asia: Current Trends and Related Challenges, UNODC, April 2015, p. 50.

[34] In the 1990s when migration from Fujian Province was driven by poverty and unemployment, many Fujianese were smuggled to North America, Europe and, in fewer numbers, to Australia. See Migrants Smuggling in Asia: Current Trends and Related Challenges, UNODC, April 2015, p. 77.

But the real challenge exists in Southeast Asia. Within Southeast Asia, illegal migration is mainly driven by economic disparities. Even though borders are porous, the UNODC estimates that over 80% of irregular migrants rely on smugglers. Around half a million migrants mainly from Myanmar (but also from Cambodia and Laos) are estimated to be smuggled to Thailand each year. Fees for smuggling services are low, ranging from a few dollars to a few hundred dollars, meaning it is often cheaper to use smuggling services than regular labor migration systems. Since the early 2000s, Australia has been a significant destination for illegal migrants from Southwest Asia, or via Southeast Asia. However, such irregular maritime arrivals in Australia nearly came to a halt in 2014 and the sea route from Asia to Australia is at present largely dormant.[35]

In addition, another hotspot is the Mekong River region. The opium economy in the Golden Triangle stimulates the illegal human trafficking across the Mekong River. Efforts continue to be made by Mekong countries, as well as UNODC. For instance, ministers and officials from Cambodia, China, Laos, Myanmar, Thailand and Vietnam, and UNODC endorsed a strategy to address regional challenges, smuggled migration also included.[36] Furthermore, in 2015, China, Cambodia, Laos, Myanmar, Thailand and Vietnam established the Lancang–Mekong Environmental Cooperation Strategy, offering a high-level regime to deal with comprehensive governance of the Mekong River region.[37]

West Asia

During the past almost half century, most of smuggled migrants into Europe are from Southwest Asia. Geographical proximity probably stands behind. Recent tensions in West Asia fundamentally changed the situation. Disorders in Afghanistan, Iraq, Syria and other Islamic countries forced their citizens to smuggle into Europe, and large numbers of them travel illegally to Europe through the Mediterranean Sea.

Since the Syrian civil war, the issue of illegal human trafficking by sea in West Asia has been continually witnessed by the international

[35] See Global Study on Smuggling of Migrants 2018, UNODC, p. 122.

[36] Mekong ministers and UNODC announce new action plan to address regional drug challenges, UNODC, 10 May 2017.

[37] See http://www.chinaaseanenv.org/lmzx/zlyjz/lmhzjz/201711/t20171106_425926.html.

community. Hundreds of thousands of refugees escaped from Syria, via the Mediterranean Sea, and pathetically, many lost their lives in the process of illegal human trafficking, most of whom are children, women and the elderly. The total number of such illegal migrants still remain unknown, but, in Italy, between January and June 2016, the number of unaccompanied minors reaching the country by sea was more than doubled compared to the same period in 2015, reaching about 10,000.[38] However, according to the 2019 Secretary-General Report, approximately 116,650 people reportedly traveled to Europe via the Mediterranean Sea, while the number in 2017 was 172,300.[39] Although the decrease in number in 2019 may be attributed to the constant international efforts, particularly the European Union in deterring the illegal movements, the issue is far from being solved.

The Case of China

The People's Republic of China (PRC), as a biggest and most populace country in East Asia, is inevitably haunted by the human trafficking problem. It often becomes a target of accusation for exporting large number of illegal immigrants to Australia, Europe and North America. There is a very rationale that any writing on illegal human trafficking in East Asia cannot ignore the significant influence of China over the emergence as well as the resolution of the issue.

China as a sending country

In the East Asian region, China is the place where most of the illegal immigration cases occur. In the case of smuggling from China to Europe, illegal migrants usually travel by air for the first part of their journey and then continue by land using cars, trucks or trains.[40] Sea routes are not so common in this scenario. One reason may be the slower speed of travel

[38] Italian Department of Public Security, Office for Central Immigration and Border Police, summary of people landed for the period June 2015–June 2016.

[39] See Oceans and the law of the sea, Report of the Secretary-General, UNGA/A/74/350, 11 September 2019, para. 30.

[40] See Migrant smuggling in Asia and the Pacific: Current trends and challenges, Vol. II, UNODC, July 2018, p. 104.

by boat, than by air, especially to the United Kingdom.[41] On the other hand, organized groups of smuggling of Chinese migrants were quite common in the 1990s, for example, trafficking to the United Kingdom,[42] but late in 2011, there was little evidence of any involvement of Chinese crime gangs in smuggling of migrants to Europe.[43]

Most reported cases of smuggling to North America from China appeared in the late 1990s and early mid-2000s.[44] Nowadays, most smuggled Chinese migrants fly, instead of maritime trafficking, as closely as possible to destinations, landing in countries without visa requirements or where entry controls are weak, and then move clandestinely to the destination by car.[45] Such Chinese migrants normally smuggle to North America through Latin American countries.[46] As for the United States, smuggled Chinese migrants enter into Latin American transit countries via commercial air, then link up with local maritime smuggling groups to bring them to the United States via recreational vessels, fishing vessels or other go-fast vessels.[47]

In addition, there are still a number of cases of illegal human trafficking by sea from China to other countries and regions. Such cases mainly relate to smuggling into Hong Kong and Macau.[48] Due to the geographical proximity, illegal entry into Hong Kong from Mainland China requires a minimum level of organization. Statistically, between 2012 and the first

---

[41] See Migrant smuggling in Asia: Current trends and related challenges, UNODC, April 2015, p. 81.

[42] See Harald Haugom Olsen, "The Snake from Fujian Province to Morecambe Bay: An Analysis of the Problem of Human Trafficking in Sweated Labour", *European Journal of Crime, Criminal Law and Criminal Justice*, Vol. 16, 2008, 12–14.

[43] See Europol, "Facilitated illegal immigration intelligence bulletin", *Intelligence Bulletin*, No. 2 (April 2008), pp. 8–9.

[44] See Migrants smuggling in Asia: Current trends and related challenges, UNODC, April 2015, p. 82.

[45] See Transnational organized crime in East Asia and the Pacific: A threat assessment, UNODC, April 2013, p. iv.

[46] See Migrant smuggling in Asia and the Pacific: Current trends and challenges, Vol. II, UNODC, July 2018, p. 105.

[47] *Ibid.*

[48] For details, see Case law database of smuggling of migrants, https://sherloc.unodc.org/cld/v3/som/cldb/search.html?lng=en.

six months of 2017, about 56% of illegal entries recorded by Hong Kong involved travel by sea.[49]

China as a receiving country

In comparison with the above, illegal immigrants from other countries into China used to be a much less serious problem. However, with the open-door policy and rapid economic growth, more and more foreigners have been to China for various reasons. They mainly come from China's neighboring countries, such as Russia, Cambodia and Vietnam and their main purpose is to make money and improve their living conditions. As a result, some of them are usually involved in illegal businesses such as prostitution.

Nevertheless, most illegal immigrants staying in China are from North Korea. The large scale of illegal escapees from North Korea to China began in 1997 due to food shortage and human rights violations in their home country.[50] It is estimated that China may have hosted 300,000 illegal North Korean immigrants.[51] While the majority of the international community tend to treat these escapees as refugees as defined under the 1951 UN Convention on Refugees, China has refused to grant them such treatment by stating that these people are not refugees. As a result, the principle of non-refoulement is not applicable. On the contrary, China tends to repatriate those illegal Korean immigrants, once arrested, back to North Korea as a fulfillment of its treaty obligation under the Agreement on Mutual Extradition of Escapees and Criminals concluded with North Korea in the 1960s.[52]

Some significant cases have revealed that organized criminal organizations are involved in the smuggling of North Korean illegal immigrants to China. A Japanese snakehead Takayuki Noguchi was deported in August 2004 after he had served an 8-month jail sentence for smuggling two North Koreans out of China. He had been paid 5 million yen (US$4,511) to take the two Koreans from Dalian to Nanning and, with assistance from Chinese collaborators, helped them cross the

[49] See Global study on smuggling of migrants 2018, UNODC, pp. 134–135.

[50] See Eric Yong-Joong Lee, "National and International Legal Concerns regarding Recent North Korean Escapees", *International Journal of Refuge Law*, Vol. 13(1–2), 2001, 143.

[51] *Ibid.*

[52] This agreement is not available to the public so that its details are not known.

Chinese border. He was arrested on 13 December 2003 when he made a second attempt to smuggle another two Koreans out.[53]

In the past decade, the place that most of the illegal immigrants entered into China is Guangdong Province, one of the most developed provinces in China, and in the first quarter of 2016, 3,057 illegal immigrants were detected there.[54] Most of these immigrants are from Southeast Asia, Africa and other less developed countries.

Special case: Trafficking to Taiwan

The PRC regards Taiwan as part of China and the relationship across the Taiwan Strait is internal rather than international. The criminal activities in the Taiwan Strait include, *inter alia*, illegal immigration, smuggling and piracy or armed robbery at sea. Prevention and punishment of such crimes are necessary for the security of the coastal areas of the Strait as well as for the safety of navigation. As early as September 1990, the Red Cross Societies of both sides signed the agreement in Jinmen on mutual repatriation of criminals and criminal suspects. The Joint Agreement resulting from the 1993 Wang-Koo Talks in Singapore reached a consensus that "topics that have been set down for discussion this year include repatriation of illegal immigrants, efforts to crackdown on illegal activities at sea, the resolution of fishery disputes, the protection of intellectual property rights, and mutual assistance on judicial issues".[55] Due to the late-on frustration in the cross-Strait relations, the prospective agreement on prevention and suppression of maritime criminals has been blocked.

Mainland China blamed Taiwan for its unfair policy toward the mainlanders by dividing them into three categories and that most of the mainlanders are prohibited to enter into Taiwan. The second category as defined refers to the mainlanders who are allowed to visit Taiwan after they have visited other foreign countries or regions. Since they are outside China, the relevant Chinese authorities are unable to regulate their visit

---

[53] Jiang Zhuqing, "Japanese snakehead deported", *China Daily*, 18 August 2004, p. 2.

[54] "Guangdong police detected 3057 illegal immigrants in the first quarter, *China News Service*, 24 April 2016, available at http://www.chinanews.com/sh/2016/04–24/7846012.shtml.

[55] See Hungdah Chiu, "The Koo-Wang talks and the prospect of building constructive and stable relations across the Taiwan Straits", *Occasional Papers/Reprints Series in Contemporary Asian Studies*, No. 6, 1993, p. 24.

Table 6.2: Chinese Mainlanders Illegally Entered into Taiwan in Historic Perspective (1987–2000)

Year	Number of Detained	Number of Repatriated
1987	762	760
1988	2,260	1,978
1989	3,384	3,664
1990	5,626	5,057
1991	3,998	4,409
1992	5,446	3,445
1993	5,944	5,986
1994	3,216	4,710
1995	2,248	1,427
1996	1,649	2,250
1997	1,177	1,216
1998	1,294	1,121
1999	1,772	1,166
2000	1,527	1,230
Total	40,303	38,419

Source: Centre for China Studies, *Cross-Strait and Chinese Mainland Human Smuggling Problems: Investigation and Studies* (2001), p. 6.

to Taiwan. Thus, a loophole has been created for human smuggling by organized criminal groups.[56]

As revealed by a study in Taiwan, each year there were thousands of mainlanders sneaking, by sea on most occasions, into the soil of Taiwan (see Table 6.2). Due to the geographic proximity, most of the illegal immigrants were from Fujian Province.[57] According to the aforementioned study, snake heads from both sides cooperated to ship the smuggled

[56]"Office of Taiwan Affairs of the State Council: The so-called simplified state name by Chen Shui-bian reveals his nature of Taiwan independence", 15 September 2004, available at http://tw.people.com.cn/GB/14810/14858/2785998.html.

[57]See Centre for China Studies (Gongdang wenti yanjiu zhongxin), *Cross-Strait and Chinese Mainland Human Smuggling Problems: Investigation and Studies* (Taihai ji zhongguo dalu diqu toudu wenti diaocha yanjiu) (in Chinese) (Bureau of Investigation, Ministry of Justice, 2001), p. 236.

people from the Chinese mainland coast to Taiwan.[58] Such kind of collaboration can be defined as an organized crime. Before Chen Shui-bian came to power, there was some cooperation between Taiwan and Mainland China in the repatriation of the smuggled people from Taiwan back to Mainland China.[59] However, after the 2000 election, the Chinese Government ceased its intercourse with the DPP government in Taiwan, thus affecting the normal repatriation process between the two sides across the Taiwan Strait. While mainlanders are getting rich, the issue is no more serious in the cross-Strait relations.

China's domestic legislation

As a transnational crime, illegal human trafficking is subject to the crack-down measures provided in the Palermo Convention. China ratified the Palermo Convention on 27 August 2003 as well as the LOSC in May 1996. As a treaty party, China is obliged to implement these two international conventions within its territory. In China's practice, there are usually three ways to implement international treaties: (1) express incorporation into the domestic laws of the provisions of international treaties and rules; (2) general stipulations on application of international treaties; and (3) correspondent revision and/or amendment of international treaties which China has ratified or acceded to.[60] When there is a discrepancy between domestic laws and international law, China takes the view that international law is superior to domestic law except for those provisions in international treaties that China has made reservations. The Criminal Law, amended in 2017, further provides that "this Law is applicable to the crimes prescribed in the international treaties concluded or acceded to by the People's Republic of China and over which the People's Republic of China has criminal jurisdiction within its obligation in accordance with the treaties".[61] It thus establishes the

---

[58] There is a detailed list of snake heads in Centre for China Studies, *Ibid.*, pp. 202–230.

[59] See Centre for China Studies, *supra* note 57, pp. 276–278.

[60] See Li Shishi, "The Relationship between the Chinese Legislation, Treaties and International Law", *Chinese Yearbook of International Law* (in Chinese), 1993, pp. 264–266.

[61] Article 9 of the Criminal Law.

universal jurisdiction of China over transnational crimes including illegal human trafficking.

It is recalled that on the occasion of the Congress of the International Association of Penal Law held in Beijing in 2004, Wu Bangguo, then Chairman of the NPC expressed that China, as a responsible country in the world, would continue to take and implement the responsibility to combat transnational and internationally organized crimes. He pledged to work closely with other countries to ensure international criminal justice and better protection of human rights.[62] China has also become a member of the IMO since 1973.

In China's legal system, there are three main categories which concern the prevention and suppression of illegal human trafficking by sea: criminal law, border control regulations and relevant marine laws.

Criminal Law

Illegal human trafficking is a statutory crime under the Chinese Criminal Law. It is punished under "crimes of obstructing national border control". Accordingly, anyone who organizes people to secretly cross the national border is to be sentenced to fixed-term imprisonment from two to seven years, and may also be fined. However, if one of the following situations is involved, the sentence of fixed-term imprisonment should not be less than seven years, in addition to a fine or confiscation of property: (a) ringleaders organizing such crossing; (b) arranging such illegal crossing for many times, or for a large number of people; (c) causing serious injury or death to persons to be organized; (d) depriving or restricting the personal freedom of the persons to be organized; (e) refusing inspection by violence or threatening means; (f) gaining a large amount of illegal money. In addition, anyone who commits other crimes such as murdering, injuring, raping or people selling should be imposed combined punishment in accordance with other relevant provisions of the Criminal Law.[63] For those who transport people secretly across the national border are to be sentenced to imprisonment of not more than 5 years, but if they are repeatedly engaged in such transports or transporting a large number of

[62] See "Wu vows to ensure criminal justice", *China Daily*, 14 September 2004; "China said to take and implement the responsibility and obligation to combat international crimes", 14 September 2004, available at http://www.hsm.con.cn.

[63] Article 318 of the Criminal Law.

people; or using vessels, vehicles or other transportation equipment which do not meet safety standards and may cause serious consequences at any time; or obtaining a large amount of illegal money, imprisonment from 5 to 10 years are to be imposed (Article 321). For those who obtain passports, visas or other exit documents by using false plans of labor export, business transactions or other excuses for the purpose of organizing people to secretly cross the national border, or who provides false or altered passports, visas or other exit documents or sell passports or visas, imprisonment ranging from three to ten years will be imposed.[64] As for individuals, the Law provides that anyone who illegally crosses the national border may be sentenced to imprisonment of not more than one year, criminal detention or public surveillance, and may in addition be subject to a fine (Article 322).

In 1994, the National People's Congress (NPC) adopted the Supplementary Rules for Severely Punishing Crimes of Organizing or Transporting Other People to Secretly Cross the National Border, which according to Appendix II to the 1997 Criminal Law, maintains its validity as a supplement to the Criminal Law.[65] These rules contain three characteristics. First, they reinforce the severity of punishment for those who commit "crimes of obstructing national border control". For example, the imprisonment as provided in Article 318 can be moved up to life imprisonment; and anyone who commits crimes such as murdering, injuring, raping or people selling may be sentenced to death.[66] Second, they detail some provisions of the Criminal Law. For example, anyone who illegally crosses the national border may be imposed fines from 1,000 to 5,000 RMB. Finally, these rules have some new provisions, including the punishment of imprisonment for civil servants, such as those who work at passport, border control or customs offices but help illegal crossings.[67] Relevant crimes stipulated in the Criminal Law include the crime of abducting or selling a woman or child since many women are victims of illegal human trafficking.

---

[64] See Articles 319–320 of the Criminal Law.

[65] Text is reprinted in Peng Liming (ed.), *A Complete Collection of Existing Laws of the People's Republic of China* (Beijing: China Constructing Materials Industry Press, 1998) (in Chinese), pp. 470–472.

[66] See Article 1 of the 1994 Supplementary Rules.

[67] See Article 6 of the 1994 Supplementary Rules.

Border control regulations

The second category of laws and regulations which govern the subject matter are those concerning border control over the entry into and exit from the Chinese territory. There were two major laws in this field: the Law on the Management of the Border Entry and Exit of Foreigners and the Law on the Management of the Border Entry and Exit of Chinese Citizens, both promulgated in November 1985.[68] However, these two laws had been claimed as annulled due to the adoption of the Exit and Entry Administration Law which came into force on 1 July 2013. The entry, transit or sojourn of foreigners is subject to the approval of the relevant Chinese authorities and must accept inspection at the border. While their legitimate rights and interests are protected within the Chinese territory, foreigners have to comply with Chinese laws, and shall not endanger China's national security, harm public interests and disrupt social and public order (Article 3). If foreigners are not able to present valid travel documents, border control agency can deny the entry with no reason (Article 25) and shall return them (Article 26). As for Chinese citizens, they must submit documents, such as passports or other travel documents, when they enter or exit, and must accept inspection of the border control agency at the border (Article 11). Likewise, Chinese citizens who travel between the Mainland and Macau, Hong Kong, Taiwan Region, shall apply for exit and/or entry permits in accordance with this Law (Article 10).

In order to effectively implement the aforementioned two laws, relevant detailed regulations have been adopted and they include, *inter alia*, the 2013 Regulations of the PRC on Administration of the Entry and Exit of Foreigners, the 1991 Measures on the Management of Chinese Citizens to or from the Taiwan Region,[69] the 1986 Interim Measures on the Management of Chinese Citizens to or from the Regions of Hong Kong or Macau for Private Affairs, and the 1995 Regulations on Border Control Inspection for Border Entry and Exit. In addition, the 1995 Notice of the Ministry of Public Security on Strengthening Cooperation in Cracking Down on Smuggling was adopted to mandate coordinately the crackdown on smuggling activities that occurred in coastal provinces at that time.

[68] Texts of these laws are reprinted in Peng, *supra* note 65, pp. 1913–1920.

[69] This law had been amended by No. 661 of Decree of the State Council, 14 June 2015.

Marine laws

In the marine law field, those containing stipulations relevant to illegal human trafficking include the 1979 Regulations Governing Supervision and Control of Foreign Vessels, the 1983 Law on Maritime Traffic Safety,[70] the 1992 Law on the Territorial Sea and the Contiguous Zone and the 1998 Law on the EEZ and the Continental Shelf.[71] The expressions such as "security", "safety" in the aforesaid laws are relevant. In particular, the Law on the Territorial Sea requires foreign vessels to comply with Chinese laws and regulations without doing something prejudicial to China's peace, security and good order. China has the right to take all necessary measures to prevent and stop non-innocent passage through its territorial sea. Cases of foreign ships violating Chinese laws or regulations will be handled by relevant Chinese authorities (Article 8). Following the LOSC, China enjoys the right to exercise control in the contiguous zone to prevent and impose penalties for activities infringing the laws or regulations concerning security, or entry and exit control within its land territory, internal waters or territorial sea (Article 13).

As to hot pursuit, Article 14 of the Law on the Territorial Sea and the Contiguous Zone provides that "[t]he competent authorities of the People's Republic of China may, when they have good reasons to believe that a foreign ship has violated the laws or regulations of the People's Republic of China, exercise the right of hot pursuit against the foreign ship. Such pursuit shall be commenced when the foreign ship or one of its boats or other craft engaged in activities by using the ship pursued as a mother ship is within the internal waters, the territorial sea or the contiguous zone of the People's Republic of China. If the foreign ship is within the contiguous zone of the People's Republic of China, the pursuit may be undertaken only when there has been a violation of the rights as provided for in the relevant laws or regulations listed in Article 13 of this Law. The pursuit, if not interrupted, may be continued outside the territorial sea or the contiguous zone until the ship pursued enters the territorial

[70] Amended in 2016.

[71] English texts of the four laws are available in Office of Policy, Law and Regulation, State Oceanic Administration (ed.), *Collection of the Sea Laws and Regulations of the People's Republic of China*, 3rd edn. (Beijing: Ocean Press, 2001), pp. 314–327, 251–263, 201–205, 210–215.

sea of its own country or of a third State". Similar stipulations are also contained in Article of 12 of the Law on the EEZ and the Continental Shelf.[72]

Generally, illegal human trafficking is subject to the jurisdiction of the Ministry of Public Security. After China's massive cabinet reshuffle plan in 2018, the China Coast Guard (CCG)[73] is responsible for the maritime sector. As the Supplementary Provisions of the Criminal Procedure Law amended in 2018 provides, the CCG has the obligation of law enforcement at sea and enjoys the right of investigation over crimes at sea.[74] On 22 June 2018, the NPC issued a decision concerning the authority of the CCG. This decision provides the power of law enforcement of the CCG including: (1) cracking down on illegal and criminal maritime activities, (2) maintaining maritime security and safeguarding, (3) exploitation and utilization of marine resources, (4) marine ecological environment protection, (5) marine fishery management, and (6) maritime suppression of smuggling and other respects. According to the Notice by the Supreme People's Court, the Supreme People's Procuratorate, and the CCG of Issues concerning Jurisdiction of Maritime Criminal Cases on 20 February 2020, for the purpose of punishing maritime crimes according to the law and protecting the national sovereignty, security, maritime rights and interests, and maritime order, the CCG enjoys the right to place on file for investigation and prosecution over crimes at sea.[75] So far, further relevant detailed regulations or rules are still in the process of drafting.

The Regulations on the Management of Coastal Vessels for Border Public Security issued by the Ministry of Public Security in February 2000 could be regarded as a reference in this respect.[76] Three specific clauses should be mentioned: when a vessel is lost, stolen or hijacked, it must be reported immediately to the adjacent department of public

---

[72] It provides that "[t]he People's Republic of China shall have the right to take necessary measures against violations of the laws or regulations of the People's Republic of China in its exclusive economic zone and the continental shelf, and to investigate according to the law those who are liable and may exercise the right of hot pursuit".

[73] CCG is now affiliated to the Chinese People's Armed Police Forces.

[74] Article 308 of the Criminal Procedure Law.

[75] See the Supreme People's Court website, available at https://www.spp.gov.cn/spp/xwfbh/wsfbt/202002/t20200228_455318.shtml.

[76] The Regulations came into effect on 1 May 2000, available at http://www.msa.gov.cn/flfg.nsf/all/04EFEC7FE780942A48256C70003CB274?OpenDocument.

security or the department of public security which issued the sea-going license (Article 16); any vessel or person on board must not intercept, forcedly anchor, collide or steer other vessels without permission (Article 19); and it is strictly prohibited to use a vessel for smuggling, drugs trafficking, smuggling of weapons, or transport of humans across borders, and or other criminal activities (Article 22). In addition, the anti-smuggling department (General Office for Combating Smuggling), which was established in 1998 and is subordinated to the Customs, also plays some role in curbing illegal human trafficking at sea.

International Cooperation

International cooperation comprises three levels of cooperation: bilateral, regional and global.

Global level

At this level, cooperation is basically based on the international legal framework under the auspices of the United Nations and other international organizations. The most important one is the Palermo Convention. As we can see, the East Asian countries have ratified this convention (see Table 6.1).

Within the framework of the United Nations, relevant regimes have been established to promote international cooperation on this issue. The Inter-Agency Coordination Group Against Trafficking in Persons (ICAT) is a policy forum mandated by the UN General Assembly to improve coordination among UN agencies and other relevant international organizations to facilitate a holistic and comprehensive approach to preventing and combating trafficking in persons, including protection and support for victims of trafficking.[77] The UN Office on Drugs and Crime (UNODC) also establishes the Voluntary Trust Fund for Victims of Human Trafficking providing grassroots humanitarian, legal, and financial aid to victims of trafficking through governmental, inter-governmental and civil society organizations.[78]

---

[77] See http://icat.network/.

[78] See https://www.unodc.org/unodc/en/human-trafficking-fund.html.

Notwithstanding, the International Organization for Migration (IOM) initiated the Voluntary Reporting System on Migrant Smuggling and Related Conduct (VRS-MSRC) in 2013.[79] Under this system, parties are encouraged to share information on migrant smuggling with others. There are 39 States parties to this system till now.[80]

Regional level

Non-traditional security issues have been put on the agenda in various regional fora and targeted as top priorities to be dealt with in regional cooperation. Regional cooperation is a necessity to effectively combat illegal human trafficking in the region. In April 1997, China initiated the concept of new security which contains the core elements of mutual trust, mutual benefit, equality and coordination.[81] In May 2002, China submitted to the ASEAN Regional Forum (ARF) Senior Officials' Conference the document concerning China's stand in strengthening cooperation in non-traditional security fields. In November, the Joint Declaration of

[79]For details, see https://www.unodc.org/southeastasiaandpacific/en/vrs-msrc.html.

[80]These countries are: Afghanistan, Australia, Bangladesh, Bhutan, Brunei Darussalam, Cambodia, China, Fiji, India, Indonesia, Iraq, Japan, Jordan, Kiribati, Lao People's Democratic Republic, Malaysia, Mongolia, Myanmar, Maldives, Nauru, Nepal, New Zealand, Pakistan, Palau, Papua New Guinea, Philippines, Republic of Korea, Samoa, Singapore, Solomon Islands, Sri Lanka, Syrian Arab Republic, Thailand, Timor-Leste, Tonga, Turkey, United Arab Emirates, Vanuatu and Vietnam.

[81]*Mutual trust* means that all countries should transcend differences in ideology and social system, discard the mentality of Cold War and power politics and refrain from mutual suspicion and hostility. They should maintain frequent dialogue and mutual briefings on each other's security and defense policies and major operations. *Mutual benefit* means that all countries should meet the objective needs of social development in the era of globalization, respect each other's security interests and create conditions for others' security while ensuring their own security interests with a view to achieving common security. *Equality* means that all countries, big or small, are equal members of the international community and should respect each other, treat each other as equals, refrain from interfering in other countries' internal affairs and promote the democratization of the international relations. *Coordination* means that all countries should seek peaceful settlement of their disputes through negotiation and carry out wide-ranging and deep-going cooperation on security issues of mutual concern so as to remove any potential dangers and prevent the outbreak of wars and conflicts. See China offers new security concept at ASEAN meetings, available at http://english.cri.com.cn/english/2002/Aug/65936.htm.

ASEAN and China on Cooperation in the Field of Non-Traditional Security Issues (the Joint Declaration) was adopted, which initiated full cooperation between ASEAN and China in the field of non-traditional security issues and listed the priority and form of cooperation. The priorities at the current stage of cooperation include "combating trafficking in illegal drugs, people-smuggling including trafficking in women and children, sea piracy, terrorism, arms-smuggling, money-laundering, international economic crime and cybercrime".[82] As to the multilateral and bilateral cooperation, it aims to: (a) strengthen information exchange, (b) strengthen personnel exchange and training and enhance capacity-building, (c) strengthen practical cooperation on non-traditional security issues, (d) strengthen joint research on non-traditional security issues and (e) explore other areas and modalities of cooperation.[83] There was the ASEAN Ministerial Meeting on Transnational Crime with ministerial counterparts from China, Japan and South Korea in Bangkok in October 2003.

In 2006, China and ASEAN jointly hosted an official symposium concerning the cooperation in law enforcement at sea. China and ASEAN are committed to create an effective regime to jointly crackdown maritime crimes, including illegal human trafficking, and maintain maritime security and stability in this region. It is not easy to firmly establish the regime of law enforcement cooperation between China and ASEAN. In some views, the South China Sea issue is the main obstacle of further cooperation.

Bilateral level

In 1993, China agreed to make cooperation with Japan in maintaining the safety in the East China Sea, including the joint crackdown of maritime smuggling.[84] In November 2002, Hong Kong signed an agreement on

---

[82] See Joint Declaration of ASEAN and China on Cooperation in the Field of Non-Traditional Security Issues, 6th ASEAN–China Summit, Phnom Penh, Cambodia, 4 November 2002, available at https://asean.org/?static_post=joint-declaration-of-asean-and-china-on-cooperation-in-the-field-of-non-traditional-security-issues-6th-asean-china-summit-phnom-penh-4-november-2002–2.

[83] *Ibid.*

[84] Ministry of Foreign Affairs *et al.*, "Opinions on Strengthening the Safety of Navigation and Fishery in the East China Sea", 27 May 1993, in Wang Huai'an *et al.* (eds.),

the readmission of persons residing without authorization with the European Union. The agreement came into effect on 1 March 2004 and helps to tackle illegal immigration through bilateral cooperation.[85] In April 2006, China and the United States reached a principle agreement to repatriate illegal Chinese immigrants in the United States and crackdown on smuggling organizations. As reported, there were 39,000 people within this category. Chinese side expressed that "China will accept the return of illegal immigrants once it has been confirmed they are Chinese citizens and left from the Chinese mainland".[86]

China has begun to negotiate with foreign counterparts for treaties on judicial assistance and/or extradition since 1985. As of September 2020, China signed 72 such treaties (see Table 6.3 for East Asian countries).[87] Domestically, China promulgated its Extradition Law in December 2000 which enables China to deal with extradition cases even without the necessary conclusion of extradition treaties.[88]

Conclusion

Illegal human trafficking and illegal human flow in the world is likely to continue in the future despite the efforts made at international, regional and national levels. There are several key factors which have correlation with or have caused illegal human trafficking. First, it is due to the globalization process. Second, the rigid immigration policy endorsed by advanced countries implicitly and in fact has encouraged such illegal acts since many immigrants are unable to enter these countries through a legal channel. As is noted, most governments encourage the mobility of highly skilled people, but do not welcome lower skilled workers even if there is a demand in job market, thus "resulting in increase in illegal migration

---

Compendium of Laws of the People's Republic of China, Supplementary Volume (Jilin: Jinlin People's Publishing House, 1994) (in Chinese), p. 787.

[85] See "China, EU to fight illegal immigration", *China Daily*, 21 September 2004.

[86] Zhu Zhe, "Talks focus on illegal immigration", *China Daily*, 5 April 2006.

[87] See http://treaty.mfa.gov.cn/Treaty/web/list.jsp?nPageIndex_=1&keywords=%E5%8F%B8%E6%B3%95%E5%8D%8F%E5%8A%A9&chnltype_c=all.

[88] Text is available at http://162.105.148.239:8082/newlaw2001/ShowLawContent.asp?LinkLawId=&DatabaseName=chl&Id=32110. For further details, see Hu Qian and Chen Qiang, "China's Extradition Law of 2000", *Chinese Journal of International Law*, Vol. 1(2), 2002, 645–654.

Table 6.3: Bilateral Judicial Assistance/Extradition Treaties between China and Other East Asian Countries (as of June 2020)

State	Type of Treaty	Signing	Entry into force (dd/mm/yy)
Cambodia	Extradition	09/02/99	13/12/00
Indonesia	Civil & criminal	24/07/00	28/07/06
	Extradition	01/07/09	19/01/18
Japan	Criminal	01/12/07	23/11/08
Laos	Civil & criminal	25/01/99	15/12/01
	Extradition	04/02/02	13/08/03
Malaysia	Criminal	23/11/15	19/02/17
Mongolia	Civil & criminal	31/08/89	29/10/90
	Extradition	19/08/97	10/01/99
North Korea	Civil & criminal	19/11/03	21/01/06
Philippines	Criminal	16/10/00	17/11/12
	Extradition	30/10/01	12/03/06
Singapore	Civil & commercial	28/04/97	27/06/99
South Korea	Criminal	12/11/98	24/03/00
	Extradition	18/10/00	12/04/02
	Civil & commercial	07/07/03	27/04/05
Thailand	Civil & commercial	16/03/94	06/07/97
	Extradition	26/08/93	07/03/99
	Criminal	21/06/03	20/02/05
Vietnam	Civil & criminal	19/10/98	25/12/99

Source: http://treaty.mfa.gov.cn/Treaty/web/list.jsp?nPageIndex_=4&keywords=%E5%BC%95%E6%B8%A1&chnltype_c=all.

often organized by labour-recruiters and people-traffickers".[89] Third, disparity between the poor and the rich in the world community has exacerbated the situation of illegal trafficking.

Finally, in terms of some countries which have a long tradition of emigration, going abroad is a natural phenomenon. In China, such a tradition has been rooted in the places where most of the illegal

[89]Thomas Schindlmayr, "Sovereignty, Legal Regimes, and International Migration", *International Migration*, Vol. 41(2), 2003, 118.

emigrants are going abroad today, such as in Zhejiang, Fujian and Guangdong provinces. The resolution of the issue may go beyond all the legal means addressed in this chapter. The United Nations Declaration of Human Rights guarantees every person in the world a right of free movement, but in reality, personal movement is subject to the regulation of international, regional and national laws.

Part III

Challenges and Issues in East Asia

Chapter 7

Maritime Boundary Delimitation Agreement for the Gulf of Tonkin

Introduction

The Gulf of Tonkin (Beibu Gulf in Chinese and Bac Bo Gulf in Vietnamese) is a shared water area between China and Vietnam. The size of the gulf, as agreed by the two countries, is more than 126,000 km^2,[1] with abundant marine living and non-living resources. With the pace of the new law of the sea developments, China and Vietnam realized the importance and necessity to establish a maritime boundary in the Gulf of Tonkin. The whole process of the negotiation has three stages: brief negotiations in 1974 as initiated by Vietnam; negotiations during the period between October 1977 and June 1978; and the negotiations from 1992 to 2000.[2] The first two stages were fruitless since the relationship between the two countries at that time was very poor. Only after the normalization of their bilateral relations in 1991 did the negotiations enter

[1]The figure that Vietnam refers to is 126,250 km^2 while the Chinese figure is 128,000 km^2. See respectively "Interview by Foreign Minister Nguyen Dy Nien about Tonkin Gulf Delimitation Agreement", 1 July 2004, available at http://www.vnagency.com.vn/newsa.asp?LANGUAGE_ID=2&CATEGORY_ID=29&NEWS_ID=106778; and "Introduction to the Sino-Vietnamese Beibu Gulf Delimitation Agreement", 30 July 2004, available at http://www.fmprc.gov.cn/chn/zxxx/t145558.htm.

[2]See "Interview by Foreign Minister Nguyen Dy Nien about Tonkin Gulf Delimitation Agreement".

into a substantial and fruitful stage, which began in 1992. In 1993, the two sides reached a general agreement on the basic principles for settling the disputes relating to the land border and to the delimitation of the Gulf of Tonkin. During the negotiations on the delimitation, China and Vietnam agreed on the principles of "applying the International Law of the Sea and referring international practices to carry out negotiations on the delimitation of the Tonkin Gulf" and "in line with the principle of equality, taking into consideration all circumstances concerned in the Gulf to reach an equal solution".[3] Within the years between 1992 and 2000, the two sides held seven rounds of negotiation at the governmental level, three meetings between the heads of government-level negotiating teams, 18 rounds of negotiation at the working level and 49 meetings of the Joint Working Team and the Mapping Team.[4] On 25 December 2000, the two sides completed the negotiation process and signed the Agreement on the Delimitation of the Territorial Seas, Exclusive Economic Zones (EEZs) and Continental Shelves in the Beibu Gulf (hereinafter refereed to as the Boundary Agreement) and the Agreement on Fishery Cooperation in the Beibu Gulf (hereinafter referred to as the Fishery Agreement).[5] On 15 June 2004, the 11th National Assembly of Vietnam ratified the Boundary Agreement and on 25 of the same month, China's 10th National People's Congress also ratified the Agreement. On 30 June 2004, the two countries exchanged their ratification instruments in Hanoi and through this process, the Agreement came into force.[6] As for the Fishery Agreement, the two sides also approved it in June 2004 and changed their notes on the same day as the ratification instruments. A new marine legal order based on the 1982 United Nations Convention on the Law of the Sea (LOSC) has been, therefore, established in the Gulf of Tonkin.

---

[3] *Ibid.*

[4] *Ibid.*

[5] An unofficial English text of the Fishery Agreement is attached to Zou Keyuan, "Sino-Vietnamese Fishery Agreement for the Gulf of Tonkin", *International Journal of Marine and Coastal Law*, Vol. 17(1), 2002, 127–148.

[6] For detailed assessment of legal issues before the signing of the Boundary Agreement, see Zou Keyuan, "Maritime Boundary Delimitation in the Gulf of Tonkin", *Ocean Development and International Law*, Vol. 30(3), 1999, 235–254.

Main Contents and Implications of the Agreement

The Boundary Agreement is not a complicated legal instrument, but contains 11 clauses. Article 1 defines the area of the Gulf of Tonkin for the purpose of delimitation of the territorial seas, EEZs and continental shelves of the two countries. However, it should be noted that the defined area as the Gulf in the Agreement may differ from other definitions for other purposes and with other technical methods. For example, one literature refers to the size of the Gulf as having an area of 44,238 km^2.[7] In comparison, the area as defined in the Agreement has much been expanded (more than 126,000 km^2). This indicates that the two sides prefer to have the maritime delimitation with a much broader area. The whole length of the delimitation line is about 500 km.

Article 2 of the Agreement uses 21 geographic points to draw the maritime boundary in the Gulf of Tonkin. By using these coordinates, the segment of the line connecting Point 1–Point 9 is the line to divide the territorial seas of the two countries, whereas the rest of the line connecting Point 9–Point 21 is the line to delimit the EEZs and continental shelves of the two countries in the Gulf of Tonkin (Articles 3–4). However, since the delimitation line starts from the Beilun (Bac Luan in Vietnamese) River estuary, it may also divide internal waters adjacent to the estuary between the two countries, though these waters may be too small and insignificant to mention in the Boundary Agreement.

As we know, both China and Vietnam have adopted straight baselines to measure the breadth of their territorial seas and other maritime zones. China publicized part of its straight lines along its mainland coast as well as for Hainan Island in 1996 when it ratified the LOSC. The straight baselines connected by four geographic coordinates from Yingge Zui (Oanh Ca in Vietnamese) to Junbi Jiao along the coast of Hainan Island facing the Gulf have probably affected the delimitation in the Gulf between the two countries. It is recalled that China deliberately left the baselines for the Gulf of Tonkin undefined because of the maritime delimitation with Vietnam. However, part of its baselines along Hainan Island still produce some impact on the delimitation since this sector of the baselines is within the area of the Gulf of Tonkin as defined by the Boundary Agreement. The discrepancy between the factual result of the maritime delimitation and

[7] See "Beibu Gulf", in *Encyclopaedia Sinica: Geography of China* (Beijing: China Encyclopaedia Press, 1993) (in Chinese), p. 23.

China's original consideration thus lies in the fact that the scope of the Gulf has been expanded. It is unknown whether the segments of China's straight baselines within the Gulf of Tonkin were recognized by Vietnam, as it is recalled that Vietnam lodged official protests with the United Nations against China's publicized baselines respectively in 1996[8] and in 1998.

For Vietnam, it has also adopted the method of straight baselines as proclaimed in its 1982 Statement on the Territorial Sea Baseline of Vietnam. But on the other hand, as the aforementioned Statement provides, the Vietnamese part of the Gulf of Tonkin, as delineated by the 1887 Border Treaty signed between France and China, "constitutes the historic waters and is subject to the juridical regime of internal waters" of Vietnam.[9] Under this circumstance, there is no need to use baselines which are applicable to the territorial sea and beyond. However, the Vietnamese historic waters claim was not recognized by China and the Vietnamese agreement to the negotiation on maritime delimitation was in fact an acquiescent abandonment of its former claim. On the other hand, Vietnam declared a 12-nm territorial sea and published a map that included the marking of the territorial sea in the Gulf of Tonkin in September 1964.[10] Therefore, it is likely that Vietnam went presumably back to its 1964 position during the negotiations, though it is unknown whether the delimitation line between China and Vietnam is based on straight baselines or simply on low-water mark baselines. But it is clear that a small part arising from China's straight baseline segment connected by four coordinates for Hainan Island may have some impact on the delimitation line.[11]

---

[8] UN Doc. A/51/645 (1996), p. 12.

[9] See United Nations, Office for Ocean Affairs and the Law of the Sea, *The Law of the Sea: Baselines: National Legislation with Illustrative Maps* (New York: United Nations, 1989), p. 384.

[10] See Zou (1999), *supra* note 6, p. 245.

[11] However, according to a Chinese official from the Ministry of Foreign Affairs, the majority of the islands in the Gulf of Tonkin are within the straight baselines of the two countries and some are used as basepoints. See "Sino-Vietnamese delimitation in the Gulf of Tonkin does not concern sovereignty over islands", 3 August 2004, available at http://www.peacehall.com/news/gb/china/2004/08/200408031243.shtml. Does this information, if it is true, imply that the two countries have already decided their straight baselines in the Gulf? If so, why do they not publicize them as required by the LOSC?

If the aforementioned reasoning can be established, then another puzzle still remains. The delimitation itself is the evidence that the 1887 French–China red line,[12] which Vietnam had insisted as a demarcation line between the two countries, particularly as the basis for its historic waters, has been discarded. Does it imply that the Vietnamese part of the Gulf of Tonkin is no loner recognized by Vietnam as its historic waters? If so, then Vietnam needs to adjust its legal position and amend its relevant domestic legislation.

For the whole delimited area in the Gulf, Vietnam has obtained 53.23% and China 46.77%. By comparison, Vietnam shares the area 6.46% (about 8,000 km^2)[13] more than China. According to the Vietnamese side, Vietnam is entitled to more areas due to the consideration of the following facts that Vietnam's coastline is longer and Vietnam has more islands in the Gulf, particularly the Bach Long Vi (Bai Long Wei in Chinese) Island lying in the center of the Gulf.[14] It is interesting that in the official explanation of the Boundary Agreement, China did not mention specific figures as Vietnam did and only said that the sea areas divided between the two sides are "basically equivalent".[15] It is unknown why China used such an ambiguous expression instead of using exact figures. One explanation may be that China did not want to cause any dissatisfaction or complaint among the Chinese people. It is recalled that during the negotiation period, China demanded a half-half division. Clearly, China's original desire has not been fulfilled by the negotiation result. However, the Chinese Government is also satisfied with the result in the end. As expressed in the Chinese official explanation, China is quite contented with the Boundary Agreement and regarded the delimitation as "a result of equality".[16]

One important arrangement which is not mentioned in the Boundary Agreement but critical to the delimitation is the effect of islands on the

---

[12]For the illustrative map of this line, see Zou (1999), *supra* note 6, p. 237.

[13]Nguyen Hong Thao, "The Settlement of Disputes in Bac Bo (Tonkin) Gulf", *Vietnam Law & Legal Forum*, Vol. 7(77), January 2001, 17.

[14]See "Interview by Foreign Minister Nguyen Dy Nien about Tonkin Gulf Delimitation Agreement", *supra* note 1.

[15]See "Let the Sino-Vietnamese Maritime Border Become a Peaceful, Friendly and Cooperative Tie — Interview by Wang Yi on the Entry into Force of the Sino-Vietnamese Maritime Delimitation Agreement and Fishery Cooperation Agreement in the Gulf of Tonkin", *People's Daily* (in Chinese), 1 July 2004, p. 3.

[16]*Ibid.*

delimitation line. The two parties finally agreed to give the Bach Long Vi Island 25% effect, thus this mid-ocean island enjoys 12 nm of territorial sea and 3 nm of the EEZ and continental shelf. The other small island — Con Co Island about 13 nm off the coast of Vietnam has been given 50% effect in the delimitation of the EEZs and continental shelves of the two countries at the closing line at the mouth of the Gulf.[17] Since these two islands all belong to Vietnam, China would have wished to minimize their effect on the delimitation as little as possible. The final arrangement is obviously a result of compromise.

Fishery Management

The fishery management is the key issue in the entire negotiation for the boundary delimitation. As for China, the delimitation of the Beibu Gulf was the most relevant to the distribution and utilization of fishery resources in the gulf as well as to the direct interests of tens thousands of Chinese fishermen. Thus, as early as the beginning of the negotiation, the Chinese side stated expressly that the delimitation must be linked to the fishery arrangement, and the agreement for delimitation and the agreement for fishery cooperation must be simultaneously signed and entered into force.[18] The delayed ratifications — which took four years from the conclusion of the Boundary Agreement — by both parties to most extent attributed to the negotiation and subsequent resolution of some remaining issues in the fishery management.

Since details on fishery arrangements are provided for in the Fishery Agreement,[19] the Boundary Agreement only contains one clause regarding cooperation between the two sides in respect to rational use and sustainable development of living resources in the Gulf of Tonkin (Article 8). The Fishery Agreement establishes a Common Fishery Zone, buffer zone for small fishing boats and zone for transitional arrangements. The Common Fishery Zone, about 30,000 km^2, covers most of the fishing grounds of high productivity in the Gulf. In comparison with the Joint

[17] See "Interview by Foreign Minister Nguyen Dy Nien about Tonkin Gulf Delimitation Agreement", *supra* note 1.

[18] See "Introduction to the Sino-Vietnamese Beibu Gulf Delimitation Agreement", *supra* note 1.

[19] For a detailed assessment of the Fishery Agreement, see Zou (2002), *supra* note 5.

Fishery Zone established under the Sino-Japanese Fishery Agreement in the East China Sea, this Common Fishery Zone is considerably large. A bigger joint fishery zone is favorable for China since most of the good fishing grounds are located within Vietnam's EEZ. However, unlike the Boundary Agreement, the Fishery Agreement has a time limit for its effect (12 years with three more years of automatic extension from the date of its entry into force). The year of 2016 was the year of expiry. Though it is reported that the two sides continued their joint fishery inspection in the Gulf during 19–23 April that year,[20] it is unpredictable and also doubtful whether the two sides would still maintain such a big common fishery zone in the Gulf of Tonkin after the expiry of the Fishery Agreement. In addition, the Transitional Zone only lasted for four years. This special arrangement is mainly based on the consideration for the Chinese fishermen to have more time to adjust their fishing life under the new and changed conditions.

There were altogether 700,000–800,000 Chinese fishermen whose living heavily depends on fishing in the Gulf of Tonkin. For Hainan Province only, after the delimitation between China and Vietnam, its traditional fishing grounds have decreased more than one-thirds and 12,000 fishermen have to abandon fishing and find other living means.[21] As a result of the entry into force of the Boundary Agreement and the Fishery Agreement, only few fishermen can still be able to operate in the Gulf of Tonkin. China introduced a method of lot-drawing to decide which fishing vessel can fish first. On 27 June 2004, Guangxi organized a lot-drawing meeting, with owners of 1,000 fishing vessels participating, to determine the order of the fishing vessels to enter into the Common Fishery Zone.[22]

But on the other hand, it should be borne in mind that overexploitation of fisheries resources in the Gulf of Tonkin has already occurred.

[20] It was the 11th joint inspection since 2006. See "China–Vietnam Coast Guards launched the 2016 joint fishery inspection in the Common Fishery Zone of the Beibu Gulf", available at http://www.gov.cn/xinwen/2016–04/19/content_5065786.htm.

[21] "Entry into force of the Sino-Vietnamese Fishery Agreement causes 10,000 fishermen to abandon boats and go to the shore in Hainan", 5 July 2004, available at http://dailynews.muzi.com/11/chinese/1318446.shtml.

[22] See "Sino-Vietnamese Agreement on Fishery Cooperation in the Gulf of Tonkin is coming into force and Guangxi fishermen drew lots for operations in the Common Fishery Zone", *China Ocean News* (in Chinese), 29 June 2004.

According to a scientific estimation, the sustainable catch in the Gulf is 600,000 tons per year.[23] However, in recent years, the catch by the two sides exceeded one million tons. If this situation was allowed to continue, the fishery resources in the Gulf would be depleted within a short period. For the purpose of conserving fishery resources, protection measures are necessary. Even without this agreement, China had to consider the necessary reduction of its fishing capacity in the Gulf of Tonkin for the purpose of sustainable use of fishery resources. Furthermore, China has established the closed season system in the South China Sea since 1999, but it has been seriously challenged by Vietnam and the Philippines, denying China's right to do so. By refusing to recognize this, when Chinese fishermen comply with the regulations not to fish within the closed season, the Vietnamese counterparts continue to fish. As a consequence, China's measures have produced little effect for the fishery conservation and protection in the South China Sea. In this sense, the new fishery agreement has facilitated both China and Vietnam to reach the goal of fishery sustainability in the Gulf as well as in the South China Sea.

After the conclusion of the Fishery Agreement, there were still some remaining issues to be discussed and solved by the two sides, in particular regarding the negotiations on the two additional agreements: the Supplementary Protocol and the Management Measures for the Conservation of the Resources in the Common Fishery Zone in the Gulf of Tonkin. The two sides reached agreement in February 2004 and signed the Protocol on 29 April the same year, thus paving the way for the two countries to ratify the Boundary Agreement and to approve the Fishery Agreement. The entry into force of the Boundary Agreement and the Fishery Agreement has signified the fundamental change in the fishery regime for the Gulf of Tonkin, from the traditional freedom of fishing regime to the EEZ regime based on the LOSC.

Relevant measures have been adopted under the Fishery Agreement. For the Chinese fishermen, they must abide by 12 new regulations when they sail to the Gulf of Tonkin for fishing in the Common Fishery Zone or the Transitional Zone. When they want to fish in the Vietnamese part of the Common Fishery Zone or the Transitional Zone, they must apply for and obtain a special permit; fishing boats have to be equipped with

[23]"Sino-Vietnamese delimitation concerns three problems: Sea area, fishery and oil and gas", 5 August 2004, available at http://www.chinanews.com.cn/news/2004year/2004–08–05/26/468344.shtml.

adequate communications facilities; fishing boats and fishermen must bring the fishing permit and other certificates aboard; fishing boats should hoist Chinese flag and marking plates made by the fishery management department; illegal fishing methods such as use of explosives or poison are prohibited; fishing boats should operate as allowed by the permit and write the fishing log; catch of marine mammals or other endangered species is prohibited, and their by-catch should be immediately returned to the sea; fishing boats should abide by the regulations of collision avoidance and should not hamper normal operations of other fishing boats; maritime incidents between Chinese and Vietnamese fishing boats should be submitted to the competent authorities of either country and their settlement will go through consultation by the two sides or through the Joint Fishery Committee; fishermen should operate when inspectors from either side to make inspections; fishing boats should report when they take refuge in the Vietnamese harbor or waters due to bad weather or emergencies; and they should report to the Vietnamese authorities when they depart.[24] It is likely that the Vietnamese side has adopted the similar regulations as they are both based on the Fishery Agreement and subsequent Protocol and Management Measures.

Non-Living Resources Management

Contrary to the fishery management which is very detailed and comprehensive as an essential part of the package deal between the two countries, China and Vietnam seem very cool toward substantive cooperation in the area of non-living resources development in the Gulf of Tonkin. There is only one simple clause in the Boundary Agreement (Article 7) providing that:

> In case any single geophysical structure of oil and gas or other mineral deposits straddles the demarcation line as provided in Article 2 of this Agreement, the Parties shall, through friendly consultation, reach an agreement on develop the structure or deposit in the most effective way as well as on equal sharing of the profits resulting from the development.

[24] See "Fishing boats must comply with 12 new regulations when they enter and operate in the Sino-Vietnamese agreed waters", *People's Daily* (in Chinese), 1 July 2004, p. 3.

It indicates that at least for the time being there is no intention of the two sides to carry forward any type of joint development of petroleum resources in the Gulf and they leave the opportunity for cooperation in the future. This is interesting because the negotiation at the first stage in the 1970s initiated by Vietnam was actually triggered by the prospect for oil and gas in the Gulf when Vietnam intended to grant the exploration right for some sedimentary blocks in the Gulf to an Italian oil company. In the 1980s, Vietnam proposed that the two countries undertake a joint development program in the Gulf of Tonkin.[25]

On the other hand, unilateral exploration and exploitation of petroleum resources has been undertaken for many years by the two sides in and/or adjacent to the Gulf of Tonkin. According to the prediction of the China National Offshore Petroleum Company, the Gulf area is one of the biggest oil and gas concentrations in the world, having an oil deposit of about 2.29 billion tons and natural gas deposits of about 1,444 billion cu m.[26] The South China Sea Offshore Oil Company pumped 14.2 million tons of crude oil in 1997, up nearly 10% to rank first among China's offshore oil producers.[27] China developed several offshore oilfields in the Yingge Hai Basin adjacent to Hainan Island within the Gulf including the Dongfang 1–1 and Yacheng 13–1.[28] With its strong capacity of developing offshore oil and gas, China may be reluctant to cooperate with Vietnam for petroleum resources in the Gulf. However, China is keen to have joint development with Vietnam in the South China Sea, as China has repeatedly called for it for many years. The Vietnamese attitude and position may be just the opposite.

Post-Agreement Developments

After the entry into force of the Boundary Agreement as well as the Fishery Agreement, necessary follow-up actions by both sides need

[25] See Epsey Cooke Farrell, *The Socialist Republic of Vietnam and the Law of the Sea: An Analysis of Vietnamese Behaviour within the Emerging International Oceans Regime* (The Hague: Martinus Nijhoff, 1998), p. 251.

[26] See Sun Shangzhi (ed.), *Study on the Comprehensive and Co-ordinated Development of China's Beibu Gulf Rim* (Beijing: Meteorological Press, 1997) (in Chinese), p. 10.

[27] China Business Information Network, 26 February 1998.

[28] There are five oilfields developed by China in the Gulf of Tonkin. For details, see http://www.cnooc.com.cn/.

to be undertaken. As Vietnam has stated, the Vietnamese authorities, within their competence, "need to amend, supplement and complete related legal documents in order to create the legal basis for the management and exploitation of the natural resources in the Gulf".[29] Also, Vietnam will provide sufficient information about the agreements for its civil servants and citizens so as to properly implement these agreements. In July 2004, Vietnam held a national conference on the implementation of the two agreements, where Deputy Prime Minister Vu Khoan asked relevant ministries, branches and provinces to closely coordinate in the implementation and to cooperate with the Chinese side.[30] In August 2004, Vietnam established a sub-department for the exploitation and protection of aquatic resources in the Gulf of Tonkin under the Fisheries Ministry's Aquatic Resource Exploitation and Protection Department as a positive response to Prime Minister's decision to accelerate the implementation of the Fishery Agreement.[31] Meanwhile, Vietnam will provide its fishermen with necessary training programs relating to the delimitation line and legal limits of different sea areas in the Gulf of Tonkin.[32]

For the implementation purposes, the Chinese side has to fulfill the same commitment under the treaty obligations. Since the matter of topmost priority in the implementation is related to the fishery management, the Ministry of Agriculture issued two normative documents in June 2004 regulating the implementation of the Fishery Agreement as well as orderly fishing operations to be conducted by the Chinese fishermen. The Urgent Notice on the Implementation of the Sino-Vietnamese Fishery Cooperation Agreement for the Gulf of Tonkin,[33] issued on 14 June 2004, requests all the relevant fishery management departments in Guangdong,

---

[29] See "Interview by Foreign Minister Nguyen Dy Nien about Tonkin Gulf Delimitation Agreement", *supra* note 1.

[30] See "Conference on implementation of Tonkin Gulf demarcation and fisheries agreements", 12 July 2004, available at http://www.vnagency.com.vn/newsa.asp?LANGUAGE_ID=2&CATEGORY_ID=29&NEWS_ID=108471.

[31] See "Implementation of Viet Nam-China Agreement on Fisheries Cooperation", 10 August 2004, available at http://www.vnagency.com.vn/newsa.asp?LANGUAGE_ID=2&CATEGORY_ID=30&NEWS_ID=112982.

[32] See "Training on Tonkin Gulf demarcations for fishermen", 6 July 2004, available at http://www.vnagency.com.vn/newsa.asp?LANGUAGE_ID=2&CATEGORY_ID=29&NEWS_ID=107394.

[33] Text is available at http://www.law-lib.com.

Guangxi and Hainan, which are adjacent to the Gulf of Tonkin, to make a better preparation for the implementation of the Fishery Agreement. The fishery management departments in each province adjacent to the Gulf of Tonkin should issue the fishing permits to specific fishing vessels within the quota granted by the Ministry of Agriculture, and make a comprehensive check of the fishing vessels which are allowed to fish in the Fishery Agreement waters. Second, the fishery management departments should launch training programs for fishery management personnel and fishermen, and disseminate relevant documents and sea charts to fishermen. Third, the fishery management departments should be prepared to deliberate, work out and establish emergency response mechanisms, and improve the communication network and reporting system. Finally, law enforcement teams of the fishery management departments should strengthen their administration in the Fishery Agreement waters, in coordination with the Public Security Department and the Navy, so as to prevent violations from occurring.[34]

The second important regulation is the Public Notice on the Implementation of the Sino-Vietnamese Agreement on Fishery Cooperation in the Gulf of Tonkin issued on 15 June 2004. Accordingly, all the fishing boats must comply with the Fishery Agreement, the Management Measures and the Supplementary Protocol when they enter into the Common Fishery Zone and the Transitional Zone. Legal and valid certificates are required for fishing operations; a special permit issued by the South China Sea Fishery Management Bureau is required for fishing operations in the Vietnamese side of the Common Fishery Zone; a transitional arrangement permit is required for fishing operations in the Vietnamese side of the Transitional Zone. No fishing activities should be conducted within the waters of 15 nm around the Bac Long Vi Island. No trespass of fishing boats beyond the Small Fishing Boat Buffer Zone and fishing boats must leave immediately when they mistakenly enter into the Vietnamese waters. Fishing vessels must accept inspections by competent authorities of both China and Vietnam when they operate in the Common Fishery Zone or the Transitional Zone.[35]

On 4 August 2004, the China Marine Surveillance launched a historical inspection cruise in the Gulf of Tonkin. The inspection was carried by two law enforcement vessels and one aircraft. The vessels sailed along the

[34] *Ibid.*

[35] *Ibid.*

newly delimited maritime boundary line. The mission lasted for 36 h. Taking this as a start point, the China Marine Surveillance carried out routine monitoring and inspection cruises for the maritime boundary in the Gulf of Tonkin.[36]

On 8 August 2004, Wang Yi (China) and Vu Dung (Vietnam), heads of the negotiation delegations of the two countries, met in Gunagxi to exchange views on the remaining issues between them and signed a Minutes to implement the bilateral border agreements. The two sides agreed, in accordance with the Boundary Agreement and the Fishery Agreement, not to take drastic action or force against fishing boats in the Gulf of Tonkin.[37] Between 13 and 14 December 2018, China and Vietnam each assigned two naval vessels to participate in the 25th joint patrol in the Beibu Gulf, during which the fleet sailed for nearly 30 h and they made information exchange on hydrological meteorology, sea and air conditions, formation course and speed, enhancing mutual maritime communications and resource sharing.[38]

Unlike the Boundary Agreement, the Fishery Agreement lasted for 15 years and expired on 30 June 2019. As reported, the relevant departments of China and Vietnam reviewed the implementation of the Agreement in the past and decided to extend it to 30 June 2020.[39] In the time of writing, it is unknown whether the two countries have renewed the Agreement or made some other arrangements concerning the fisheries in the Gulf.

Conclusion

The Boundary Agreement has produced the first maritime boundary that China has ever agreed to share with its neighboring countries. As China

---

[36] "First cruise of the China marine surveillance to safeguard the Beibu Gulf border", 9 August 2004, available at http://www.peopledaily.com.cn/GB/shizheng/1026/2694134.html.

[37] "Special meeting of the heads of the negotiation delegations of China and Vietnam held", *People's Daily* (in Chinese), 10 August 2004, p. 4.

[38] "China, Vietnam wrap up 25th joint patrol in Beibu Gulf", *China Military Online* (20 December 2018), available at http://eng.chinamil.com.cn/view/2018–12/20/content_9384324.htm.

[39] See Ramses Amer and Jianwei Li, "Looking ahead for the Gulf of Tonkin fishery", 25 May 2020, available at https://www.chinausfocus.com/peace-security/looking-ahead-for-the-gulf-of-tonkin-fishery.

still has maritime delimitation problems with eight countries, the success of the delimitation in the Gulf of Tonkin can give China more experiences for the future negotiations with other countries. The practice in using one single line to delimit three different maritime zones (territorial sea, EEZ and continental shelf) indicates that China may follow this practice in the future negotiations with other neighboring countries, bearing in mind that China has used the doctrine of natural prolongation in its claim to the continental shelf in the East China Sea, which would create two different maritime boundary lines in the event that China's claim could be accepted by Japan.

For Vietnam, though the Boundary Agreement is the second of such kind of agreements that Vietnam has signed with its neighboring countries (with Thailand in 1997 and Indonesia 2003), Vietnam has admitted that this one is "the first most comprehensive of its kind".[40]

After the resolution of the two border issues (land border and maritime border in the Gulf of Tonkin), the only remaining boundary issue between the two countries is in the South China Sea, which is much more complicated so that it will be more difficult to be resolved. Nevertheless, China and Vietnam established in 1995 a regular consultation mechanism for the goal of realizing the resolution of this remaining issue. In August 2004, they reached an agreement not to take any action which could complicate or enlarge the issue, not to resort to force or threat of force, including the use of force against fishing boats.[41] It is well perceived that the entry into force of the two bilateral agreements governing the Gulf of Tonkin, the Sino-Vietnamese relations will be further developed and strengthened. As the top leaders of the two countries reached the consensus in May 2004,[42] the principles of "long-term stability, future orientation, good neighborly friendship and comprehensive cooperation" have become guiding principles for the maintenance and development of bilateral relations between China and Vietnam.

---

[40] See "Interview by Foreign Minister Nguyen Dy Nien about Tonkin Gulf Delimitation Agreement", *supra* note 1.

[41] "Special meeting of the heads of the negotiation delegations of China and Vietnam held", *supra* note 37.

[42] See "Consensus on border reached with Viet Nam", *China Daily*, 10 August 2004.

Chapter 8

Maritime Issues between China and Japan and the Prospect for Resolution

Introduction

China and Japan is connected through the East China Sea, which is defined in the 1982 United Nations Convention on the Law of the Sea (LOSC) as a semi-enclosed sea.[1] It covers about 480,000 mi^2 (1,243,190 km^2) and is bounded by the islands of Cheju (north), Kyushu (northeast), Ryukyu (east) and Taiwan (south) and by China's mainland (west).[2] It is a marginal sea with a wide continental shelf, and its average depth is 370 m with the maximum of 2,719 m. In recent years, the East China Sea is not quiet not

[1] According to Article 122 of the LOSC, "enclosed or semi-enclosed sea" means a gulf, basin or sea surrounded by two or more States and connected to another sea or the ocean by a narrow outlet or consisting entirely or primarily of the territorial seas and Exclusive Economic Zones of two or more coastal States. Text is reprinted in 21 I.L.M. (1982) 1261. During the debate on the issue of enclosed and semi-enclosed seas at the second session of the Third United Nations Conference on the Law of the Sea in 1974, the East China Sea was referred to within the general classification of enclosed and semi-enclosed seas. See Michael W. Lodge, "The Fisheries Regimes of Enclosed and Semi-Enclosed Seas and High Seas Enclaves", in Ellen Hey (ed.), *Developments in International Fisheries Law* (The Hague: Kluwer Law International, 1999), p. 197.

[2] "East China Sea", in *The New Encyclopaedia Britannica*, 15th edn., Vol. 3 (Encyclopaedia Britannica, Inc., 1984), p. 756.

only in its tidal waves or natural movements but also in political and legal differences particularly between China and Japan.

Both China and Japan ratified the LOSC in 1996. China has taken several legislative moves in response to the implementation of the LOSC. In 1992, China promulgated the Law on the Territorial Sea and the Contiguous Zone[3] which has improved the territorial sea regime established under the 1958 Declaration on the Territorial Sea. China has set its territorial sea at a breadth of 12 nm and the contiguous zone of 24 nm, measuring from the coastal baselines. This law applies to all of China, including Taiwan and the various islands located in the Chinese adjacent seas.

Another Chinese basic marine law is the Law on the Exclusive Economic Zone (EEZ) and the Continental Shelf which was adopted by the National People's Congress in 1998.[4] This law is designed to guarantee China's exercise of sovereign rights and jurisdiction over its EEZ and continental shelf, and to safeguard China's national maritime rights and interests. According to this law, China's EEZ is the area beyond and adjacent to China's territorial sea, extending up to 200 nm from baselines from which the breadth of the territorial sea is measured. As for China's continental shelf, it comprises the sea-bed and subsoil of the submarine areas that extend beyond China's territorial sea throughout the natural prolongation of its land territory to the outer edge of the continental margin, or to a distance of 200 nm from the baselines from which the breadth of the territorial sea is measured where the outer edge of the continental margin does not extend up to that distance.[5] It is interesting to note that although the provision to define the EEZ is just a copy of the relevant provision of the LOSC, the provision regarding the continental shelf has something new with Chinese characteristics, that is, the emphasis on the natural prolongation of China's rights to the continental shelf, which bears strong implications for the delimitation of the continental shelf in the East China Sea.[6] The Law further provides that EEZs and continental shelves

[3]The English version may be found in Office of Ocean Affairs, Bureau of Oceans and International Environmental and Scientific Affairs, US Department of State, *Limits in the Seas*, No. 117 (Straight Baselines Claim: China), July 9, 1996, pp. 11–14.

[4]The Chinese text is reprinted in *People's Daily* (in Chinese), 30 June 1998. An English translation may be found in *Law of the Sea Bulletin*, No. 38, 1998, pp. 28–31.

[5]Article 2 of the Law on EEZ and Continental Shelf.

[6]Zou Keyuan, *China's Marine Legal System and the Law of the Sea* (Leiden: Martinus Nijhoff, 2005), p. 94.

with overlapping claims between China and the countries with opposite or adjacent coasts should be determined by agreement in accordance with the equitable principle on the basis of international law.

In May 1996, when China was ratifying the LOSC, China issued a declaration on its baselines. China uses the method of straight baselines to define the limits of its territorial sea around part of the mainland and the Xisha (Paracel) Islands.[7] Meanwhile, China stated that it would announce the remaining baselines of its territorial sea at another time.

In the Decision on the Ratification of the LOSC, China made a statement that:

1. In accordance with the provisions of the LOSC, the People's Republic of China shall enjoy sovereign rights and jurisdiction over an EEZ of 200 nm and the continental shelf.
2. The People's Republic of China will effect, through consultations, the delimitation of the boundary of the maritime jurisdiction with the States with coasts opposite or adjacent to China, respectively, on the basis of international law and in accordance with the principle of equitability.
3. The People's Republic of China reaffirms its sovereignty over all its archipelagos and islands as listed in Article 2 of the Law of the People's Republic of China on the territorial sea and the contiguous zone, which was promulgated on 25 February 1992.[8]

The aforementioned statement clearly shows the official position of China regarding the maritime boundary delimitation and reiterates its claims to the disputed islands in the East China Sea.

Japan took the similar domestic legislation process. In 1977, it adopted the Law on Territorial Sea and in 1996 the Law on EEZ and Continental Shelf. What it has done more than China is its promulgation of the Basic Ocean Law in 2007, which provides legal guidance for a

[7]Declaration on the Baseline of the Territorial Sea of the People's Republic of China, 15 May 1996, available at http://www.un.org/Depts/los/LEGISLATIONANDTREATIES/PDFFILES/CHN_1996_Declaration.pdf.

[8]The declaration is available at http://www.un.org/Depts/los/convention_agreements/convention_declarations.htm#China%20after%20ratif.

unified and comprehensive ocean policy and administration.[9] Regarding maritime boundary delimitation with neighboring countries, Japan has advocated the application of the median line as a delimitation line for the EEZ and the continental shelf in the absence of an agreed line with the opposite country. This is reflected in its 1996 EEZ Law.[10] Thus, the ratification of the LOSC and the adoption of relevant domestic laws and regulations consist of a legal basis for maritime interactions between China and Japan in the contemporary time.

Bilateral Fishery Relations

In order to avoid fishery conflicts and to maintain a normal fishing order in the East China Sea, China and Japan began to discuss the fishery matters as early as the 1950s. Since there were no official diplomatic ties between the two at that time, the authorized parties for negotiation were non-governmental entities backed by both their governments. In June 1955, a first (non-governmental) fishery agreement was reached between the Japan–China Fisheries Council (originally known as Japan–China Fisheries Enterprise Association) and the China Fisheries Association. From 1955 to 1975 except for the period from June 1958 to November 1963, three non-governmental fishery agreements (1955, 1963 and 1965) were negotiated and implemented, and they played a very critical role in establishing and developing fishery relations between China and Japan in early times.

What is remarkable in the fishery agreement was the establishment of conservation zones in the East China Sea in the 1950s when sustainable development or other environmental principles had not come into being. It is also remarkable that such conservations zones covered areas in the high seas.

Following the establishment of diplomatic ties, the two countries started their consultations on a governmental fishery agreement. The Fishery Agreement between the Government of the People's Republic of China and the Government of Japan was finally signed on 15 August

[9]A Chinese translated version is available at *China Oceans Law Review*, No. 1, 2008, pp. 128–133.

[10]See Article 1(2) and Article 2(2) of the Law on the Exclusive Economic Zone and the Continental Shelf (Law No. 74 of 1996), *Law of the Sea Bulletin*, No. 33, 1997, 94–95.

1975, and came into force on 23 December 1975.[11] Meanwhile, the non-governmental agreement was terminated. The 1975 Agreement had been revised twice in 1978 and 1985.[12] Although the 1975 Agreement introduced more rigid protective measures than the non-governmental agreements, it was largely the same as the non-governmental ones.

The entry into force of the LOSC in 1994 ushered in a new era of fishery relations between China and Japan. Both countries announced their EEZs based on the relevant provisions of the LOSC as well as on their respective domestic legislations. Since the broadest width of the East China Sea is less than 400 nm, the whole sea area becomes EEZs that are shared by China, Japan and Korea. The fishery relationship between the two sides inevitably needed a new adjustment. After several rounds of negotiation, the two sides finally reached an agreement in September 1997 regarding the fishery management in the East China Sea.[13] The new agreement came into force on 1 June 2000.

The agreement contains some significant provisions in response to the changed situation: (a) affirming the principle of fishery resources conservation and protection: Pursuant to the relevant provisions in the LOSC and environmental requirements from Agenda 21 and others, the Agreement contains as one of its purposes the establishment of a new fishery order in accordance with the LOSC, conserving and utilizing rationally marine living resources of common concern, and maintaining the normal operation order at sea. Both sides agree to cooperate to conduct scientific research in fishery and to conserve marine living resources.[14] Each should adopt necessary measures to ensure compliance

---

[11] Text in Fishery Administrative Bureau, Ministry of Agriculture, PRC (ed.), *Sino-Japanese Governmental Fishery Agreements and Non-Governmental Protocols on the Safety of Fishing Operations* (in Chinese), April 1993, pp. 1–19.

[12] They are concerned with the establishment of a horsepower restriction line inside which trawlers and purse seiners of 600 hp or more are prohibited to enter; closed areas or suspension areas which are completely closed during designated periods; and fishing restrictions concerning minimum body length, minimum mesh size, light intensity fish-attracting devices, incidental catch limit. See Mark J. Valencia, *A Maritime Regime for Northeast Asia* (Hong Kong: Oxford University Press, 1996), p. 258.

[13] Fishery Agreement between the People's Republic of China and Japan, 11 November 1997. An unofficial English translation is available in Zou Keyuan, *Law of the Sea in East Asia: Issues and Prospects* (London: Routledge, 2005), pp. 175–180.

[14] Article10 of the Sino-Japanese Fishery Agreement.

by their nationals and fishing boats with the provisions of the Fishery Agreement and the conservation measures and other conditions provided for in the relevant laws and regulations of the other Party when they are engaged in fishery activities in the other's EEZ, and should inform each other of such conservation measures and other conditions provided for in its relevant laws and regulations.[15] (b) Providing reciprocal fishing rights: The Agreement applies to the EEZs of both countries. However, this does not include all the EEZs as the Agreement excludes the EEZ area south of 27°N, and west of 125°30′E in the East China Sea where Taiwan and the disputed Diaoyu/Senkaku Islands are located. Within the EEZ, China and Japan give each other's nationals and fishing boats the right to fish in each other's EEZs pursuant to the principle of reciprocity, the Fishery Agreement and their relevant domestic laws and regulations. The competent authorities of each Party issue fishing permits to nationals and fishing boats of the other Party, and may levy appropriate fees on the issuance of such permits. The issuance of fishing permits to the fishermen of the other side should follow the relevant provisions of the Agreement. How many permits should be issued and how many tons of fish can be caught by the other side in its EEZ is a complicated and difficult matter which both sides need to discuss and reach an agreement. (c) Establishing the Provisional Measures Zone (PMZ): The Agreement creates the PMZ which is located in the middle of the East China Sea, 52 nm from the baselines of the territorial seas for the Chinese mainland coast and for the coast of the Ryukyu Islands with its northern limit on the parallel of 30°40′N and its southern limit on the parallel of 27°N. For conservation and quantity of fishery resources in the PMZ, both sides should adopt, based on decisions made by the Sino-Japanese Fishery Joint Committee, appropriate management measures in order to protect marine living resources from the harm of being overexploited. Each party should take administrative and other necessary measures for its nationals and fishing boats fishing in the PMZ, and should not impose administrative and other measures on nationals and fishing boats of the other Party in this water area. The establishment of a common fishery zone is a typical form of fishery cooperation for shared waters between any two countries. There are many examples in the world. The Sino-Japanese PMZ is just a new addition. What is new is that the PMZ is the first such zone between China and Japan, though there was some kind of fishery cooperation between the two sides in history.

[15]Article 4 of the Sino-Japanese Fishery Agreement.

It indicates that the fishery cooperation between these two countries has entered into a new era. (d) Maintaining the Joint Fishery Committee: In order to implement the fishery agreement as well as to coordinate respective fishery management procedures, both sides decided to establish the Joint Fishery Committee which consists of four members, two of whom are appointed by each Party. The decision-making mechanism is based on the unanimous consent of the committee members. Both sides have to respect the recommendation made by the Joint Fishery Committee, and adopt necessary measures in accordance with the decision made by the Fishery Committee. The Committee may be convened annually either in China or in Japan and temporary meetings, if necessary, may be held.[16]

Cooperation in Handling Maritime Issues in the EEZ

The EEZ is a new maritime zone created by the LOSC and issues inevitably arise regarding activities in that zone. China and Japan had a dispute over Chinese scientific research vessels conducting marine scientific research (MSR) in the East China Sea. Japan accused China of violating Japan's sovereignty and maritime interests. China argued that there was no clear boundary delimitation between the two countries in that sea. As we know, freedom of MSR is a right of high seas. Before the LOSC and the establishment of EEZs by China and Japan, scientific research could be conducted in the East China Sea beyond the territorial seas of the two countries freely. But after the LOSC, the legal status of the sea areas which were high seas has changed and maritime scientific research in the EEZ is subject to the consent of the coastal State as provided for in the LOSC. Since there is no maritime boundary between China and Japan for their respective EEZs, some sea areas in the East China Sea are overlapping in their claims.

In order to defuse the tension resulting from scientific research in overlapping areas in the EEZ, the two sides held talks on 15 September 2000 so as to establish a framework for mutual prior notification of any marine survey activities in the waters around the two countries. As a result, an agreement was reached in 2001 on prior notification of any MSR activities in the disputed waters including the following main

[16]Article 11 of the Sino-Japanese Fishery Agreement.

points: (a) notification is to take place two months in advance of any scheduled MSR investigation to be undertaken in the sea area adjacent to the jurisdictional waters of the other side; and (b) notification is to include the name of the institution and of the vessel involved in the MSR investigation, the purpose and contents of the investigation, and the period and the area of the investigation.[17] The arrangement is a provisional measure pending the maritime boundary settlement between the two states, but it should ease any unnecessary tension resulting from MSR activities.

Another incident in the EEZ which involved the maritime interaction of the two countries happened on 22 December 2001 when an unidentified ship being pursued by Japan's Coast Guard in the East China Sea for infringing Japan's EEZ sank after being fired on by four Japanese coast guard patrol vessels.[18] After the incident, Japan requested to salvage the sunken ship as a necessary step to investigate the whole matter and China gave its consent to Japan's salvage proposal. According to the spokesman of the Chinese Foreign Ministry, the consent was based on the LOSC and relevant Chinese domestic laws and China had the right to monitor the whole salvage.[19] For the negative impact on normal fishery order and the loss of the Chinese fishermen caused by the salvage, Japan agreed to offer a compensation of 150,000,000 yen for such negative impact and fishery loss.[20]

Dispute in Oil and Gas Exploration and Exploitation

However, more serious issues than scientific research or salvage exist in the East China Sea, such as oil and gas exploration and development. After China began to explore the Chunxiao gas field, Japan expressed its discontent and accused China of encroaching upon Japan's rights in the East China Sea. The two sides held their consultation on 25 October 2004. At the sixth round of consultation in July 2006, the two sides reached a consensus to establish a maritime hotline communication mechanism in

---

[17] See *Hong Kong Economic Journal* (in Chinese), 9 February 2001, p. 10.

[18] See Chapter 5.

[19] See "Chinese side monitor under the law Japan's investigation of the sunken ship in the East China Sea", *People's Daily* (in Chinese), 30 April 2002, p. 4.

[20] The two sides reached the agreement on 27 December 2002 after five rounds of consultation. See http://www.fmprc.gov.cn/chn/gxh/zlb/tyfg/t5905.htm.

Table 8.1: Rounds of Negotiation on the 2008 Consensus between China and Japan

Round	Time	Venue
1	25 October 2004	Beijing
2	30–31 May 2005	Beijing
3	30 Sept.–1 Oct. 2005	Tokyo
4	6–7 March 2006	Beijing
5	18 May 2006	Tokyo
6	8–9 July 2006	Beijing
7	29 March 2007	Tokyo
8	25 May 2007	Beijing
9	26 June 2007	Tokyo
10	11 October 2007	Beijing
11	14 November 2007	Tokyo

dealing with unexpected matters.[21] During the following consultation, China put forward two proposals on joint development in the East China Sea but Japan only considered the one in the northern sea area.[22] After 11 rounds of consultation (see Table 8.1), the two countries finally reached a consensus agreement on 17 June 2008. They pledged to cooperate in the East China Sea and turn it into a sea of peace, cooperation and friendship. A joint development scheme has been created and a small patch of joint development zone is identified. In addition, China allows Japanese legal persons to participate in the development of Chunxiao Oil and Gas Field in accordance with Chinese laws.[23]

[21] "China and Japan reached the principle consensus on establishing maritime hotline", available at http://world.people.com.cn/GB/4572649.html.

[22] "Japan consider accepting the proposal on joint development in the northern sea area in the East China Sea", 15 May 2006, available at http://world.people.com.cn/GB/1029/42354/4371017.html.

[23] "China, Japan reach principled consensus on East China Sea issue", 18 June 2008, available at http://www.chinadaily.com.cn/china/2008–06/18/content_6774860.htm. For a brief assessment of this agreement, see Gao Jianjun, 'A Note on the 2008 Cooperation Consensus between China and Japan in the East China Sea', Ocean Development and International Law, Vol. 40, 2009, 291–303.

The conclusion of this agreement is in line with the spirit and provisions of the LOSC which encourages States concerned to work out provisional arrangements including joint development agreement pending the settlement of their maritime boundary disputes. The two sides planned to finalize the details based on the aforementioned agreement in September 2013. However, due to the incident of the collision of a Chinese fishing vessel with a Japanese Coast Guard vessel in the disputed area, China asked for an indefinite postponement of any implementation of the joint development agreement. So far there is no progress in this regard.

Issues Concerning the Outer Limit of Continental Shelf

Article 76 of the LOSC provides the criteria to determine the outer limit of the continental shelf up to 350 nm from the baselines from which the breadth of the territorial sea is measured or 100 nm from the 2,500 m isobath:

(i) a line delineated by reference to the outermost fixed points at each of which the thickness of sedimentary rocks is at least 1% of the shortest distance from such point to the foot of the continental slope; or
(ii) a line delineated by reference to fixed points not more than 60 nm from the foot of the continental slope.

The deadline for any submission from a coastal State, which became the party to the LOSC before 13 May 1999, on its outer continental shelf to the Commission on the Limits of the Continental Shelf (CLCS) was 13 May 2009. The Commission has a function to consider the data and other material submitted by coastal States concerning the outer limits of the continental shelf in areas where those limits extend beyond 200 nm, and to make recommendations in accordance with Article 76 of the LOSC and the Statement of Understanding adopted on 29 August 1980 by the Third United Nations Conference on the Law of the Sea. Accordingly, Japan submitted its claim on 12 November 2008.

On 6 February 2009, China lodged its objection to part of the Japanese submission regarding the Okinotori Shima, which, in China's view, is in fact a rock which does not have any ground to claim continental shelf and requests the Commission not to take any action on the portions extended from that rock.[24] South Korea shared the same view as the Chinese. However, the Commission is not the body which has the mandate to interpret the provisions of the LOSC as it reiterated that it had no role on matters relating to the legal interpretation of Article 121 of the Convention.[25] On the other hand, if there is a dispute regarding a submission between different countries, the Commission has no mandate to review such a submission.

On the other hand, China submitted its Preliminary Information concerning the outer continental shelf reaching the Okinawa Trough in the East China Sea in 2009 and a formal Submission in December 2012 to the CLCS.[26] Of course, it is protested by Japan.

Territorial Dispute over the Diaoyu/Senkaku Islands

The dispute over the Diaoyu/Senkaku Islands between China and Japan has now become an historic issue affecting their bilateral relations. Located in the East China Sea, the Diaoyu/Senkaku has a total land area of 7 km^2 with the Diaoyudao (釣魚島) or Uotsuri Jima (魚釣島) as the largest island among this group of islands.

As we know, the Diaoyu/Senkaku issue was frozen in the fishery agreement by providing that both sides respect the current fishing operations in the part of the East China Sea in the north of the PMZ, pay due consideration to the traditional fishing of the other country and the resource situation of the said area, and not unduly damage the interest of fishing of the other country in the said area, in considering the traditional and cooperative fishing relations.[27] As explained by the Japanese side, the

---

[24] China's Objection is available at http://www.un.org/Depts/los/clcs_new/submissions_files/jpn08/chn_6feb09_e.pdf.

[25] "Statement by the Chairman of the Commission on the Limits of the Continental Shelf on the progress of work in the Commission", CLCS/64, 1 October 2009, available at http://daccess-dds-ny.un.org/doc/UNDOC/GEN/N09/536/21/PDF/N0953621.pdf?OpenElement

[26] See details, see https://www.un.org/Depts/los/clcs_new/submissions_files/preliminary/chn2009preliminaryinformation_english.pdf and https://www.un.org/Depts/los/clcs_new/submissions_files/submission_chn_63_2012.htm.

[27] Agreed Minutes between the Government of the People's Republic of China and the Government of Japan, 11 September 1997.

purpose of the fishery agreement was to establish a fishing order between Japan and the PRC and had no direct relationship with the territorial title of the Senkaku Islands. The territorial waters of the Senkaku Islands are not an area where the agreement applies or would affect its status.[28] At present, the Diaoyu/Senkaku Islands are under Japan's control, but the dispute goes back over a century.[29] On 12 September 2012, China announced its baselines for the Diaoyu Islands[30] as "a powerful counter-measure and deterrent to behaviors that infringe upon China's territorial sovereignty".[31] There is no immediate prospect that the dispute could be solved in the near future.

Maritime Boundary Delimitation

The boundary delimitation of the EEZ and the continental shelf between China and Japan is very complicated. As we know, Japan has advocated the application of the median line as a delimitation line for the EEZ and the continental shelf with three main reasons:

(1) It would be inappropriate if the outer limit of the EEZ remained undecided when delimitation talks did not reach any agreement for a long time.
(2) The traditional position of Japan that delimitation of the EEZ should be made in accordance with the median line principle should be maintained.

---

[28] See Nobukatsu Kanehara and Yutaka Arima, "New Fishing Order — Japan's New Agreements on Fisheries with the Republic of Korea and with the People's Republic of China", *Japanese Annual of International Law*, Vol. 42, 1999, 27–28.

[29] For the official positions of both governments, see "The basic view on the sovereignty over the Senkaku Islands", available at http://www.mofa.go.jp/region/asia-paci/senkaku/senkaku.html; and "The issue of Diaoyu Islands", available at http://www.fmprc.gov.cn/eng/3790.html.

[30] China's Statement is available at https://www.un.org/Depts/los/LEGISLATIONANDTREATIES/PDFFILES/DEPOSIT/chn_mzn89_2012_e.pdf.

[31] "Diaoyu Islands baseline announcement significant: Chinese diplomat", 14 September 2012, available at http://en.people.cn/90883/7948501.html.

(3) It is appropriate to maintain consistency with the Law on the Provisional Measures related to the Fishery Zone of 1977 which adopted the median line principle.[32]

However, China's position is different and does not think that the equidistance method is the only criterion for delimitation. Instead, it has advocated the application of the natural prolongation principle for the delimitation of the continental shelf with Japan.[33] As early as the 1970s, when Japan and South Korea concluded the agreement for joint development in the East China Sea, China sent its strong protest against such a deal on the ground that it encroached upon the sovereignty of China over its legitimate continental shelf.[34] Natural prolongation is quite meaningful for China because, at least in the East China Sea, the continental shelf from Mainland China is very broad and China has used the concept of natural prolongation to support its claim to the continental shelf in the East China Sea. Since the general trend in State practice concerning boundary delimitation of the EEZ/continental shelf is toward a single line to delimit the two different and closely associated maritime zones, it is reasonable to wonder whether natural prolongation could still play a significant role in such delimitation. A single but adjusted line may be more practical.

As to the EEZ delimitation, it seems that China will agree to the median line for the two countries. If China insists on natural prolongation for the delimitation of the continental shelf while agreeing to the median

[32]Toshihisa Takasa, "The Conclusion by Japan of the United Nations Convention on the Law of the Sea (UNCLOS) and the Adjustment of Maritime Legal Regime", *Japanese Annual of International Law*, Vol. 39, 1996, 139.

[33]As is stated, China adheres to the following principles for the delimitation of the continental shelf: (1) application of the concept of natural prolongation; (2) delimitation through negotiation or consultation; and (3) consultation regarding equitable principles taking into account all relevant circumstances. Wang Tieya, "China and the Law of the Sea", in Douglas M. Johnston and Norman G. Letalik (eds.), *The Law of the Sea and Ocean Industry: New Opportunities and Restraints* (Honolulu: Law of the Sea Institute, University of Hawaii, 1984), p. 587.

[34]For reference, see Wei-chin Lee, "Troubles under the Water: Sino-Japanese Conflict of Sovereignty on the Continental Shelf in the East China Sea", *Ocean Development and International Law*, Vol. 18(5), 1987, 585–611.

line as the line of delimitation of the EEZ, then there would be two different delimitation lines in the East China Sea with Japan, thus definitely bringing difficulties in law enforcement and in the exercise of jurisdiction for both countries.

Nevertheless, some agreements relating to maritime boundary delimitation have been reached between China and Japan. The 1997 Fishery Agreement recognized the respective undisputed parts of the EEZs in the East China Sea and established a PMZ in the overlapping sea area there, waiting for a final settlement of the EEZ delimitation. Despite its nature as provisional, the Agreement has narrowed the disputed area in the East China Sea between the two countries. The consensus agreement reached in June 2008 regarding joint development in the East China Sea is also relevant to maritime boundary delimitation.

It has to be noted that there is a very important and even critical factor affecting the maritime boundary delimitation, i.e. the dispute over the Diaoyu/Senkaku Islands. Currently, Japan has maintained a tight control over the islands, while the China Coast Guard sails around the islands regularly to maintain its claim. As is observed, China's assertions of sovereignty over the Diaoyu Islands are not constant and it did not lodge protests until the very late 1960s and early 1970s. This may affect the effectiveness of its claims to the islands.[35] Nevertheless, the determination of whose legal title to the disputed islands has to be based on various factors and reasons. This can be seen from the judicial rulings of the International Court of Justice (ICJ), which awarded islands to the existing occupiers in the cases of the *Sovereignty over Pulau Litigan and Pulau Sipadan* (Malaysia/Indonesia) and the *Sovereignty over Pedra Branca, Middle Rocks and South Ledge* (Malaysia/Singapore) but transferred the legal tile of the Bakassi Peninsula (which occupied by Nigeria) to Cameroon in the case of *Land and Maritime Boundary between Cameroon and Nigeria*.[36]

It is obvious that who owns these islands, can own a large area of maritime zones generating from these islands. Based on the LOSC, a small islet can generate a large area of maritime zone around it (12-nm territorial sea, and in some cases 200-nm EEZ from the baselines from

[35] See Steven Wei Su, "The Territorial Dispute over the Tiaoyu/Senkaku Islands: An Update", *Ocean Development and International Law*, Vol. 36, 2005, 55.

[36] For details of these cases, see the ICJ website: http://www.icj-cij.org/homepage/index.php?lang=en.

which the breadth of the territorial sea is measured). In that case, no country would give up its claims to the disputed islands and has maintained its firm position of asserting sovereignty over them. What is more complicated is the fact that these territorial claims are intricately connected to nationalism. This can be seen from the mass protests in China regarding the Diaoyu Islands against Japan between August and September 2012. Thus, any settlement of these maritime territorial disputes continues to remain as a remote possibility.

Third Party Interests

Maritime issues are usually affecting the interest of third parties. In the East China Sea, China lodged a strong protest against the establishment of a Japan–Korea Joint Development Zone in the East China Sea in the 1970s. Likewise, the bilateral fishery agreement between China and Japan or between Japan and Korea also affected the interests of the third party as any bilateral arrangement may affect other relations and interests since they share the same sea area and use the same pool of marine living resources.

When Japan and Korea concluded their new fishery agreement, China considered that the Japan–Korea Fishery Agreement encroached on China's sovereign rights over its EEZ in the border areas among the three countries and stated that China's rights and interests in the EEZ and its fishery activities should not be subject to the limitation of that agreement. China maintained that the delimitation of the overlapping area among the three countries should be subject to consultations among the parties concerned, and exclusion of any party from the delimitation negotiations would be a violation of international law.[37] On the other hand, South Korea also expressed its dissatisfaction with the Sino-Japanese Fishery Agreement by asking China and Japan to explain how they drew the

[37]"MF Spokesman expresses that the Japan-Korea Fishery Agreement encroaches on China's sovereign rights over its EEZ", *People's Daily* (in Chinese), 23 January 1999. For further reference, see Fan Xiaoli, "Comments on the New Japan–Korea Fishery Agreement", *Ocean Development and Management* (in Chinese), Vol. 17(1), 2000, 68–70.

northern-limit line of their joint fishing area. They should consult South Korea before reaching the agreement.[38]

On the other hand, it has to be realized that the fishery agreements either between China and Japan, or China and South Korea, or Japan and South Korea are bilateral ones. They have limitations and also do not completely cover the areas in the East China Sea and the Yellow Sea. Second, because of the bilateral nature, they may affect the interests of a third party as illustrated earlier in the Japan–Korean Fishery Agreement. Third, since bilateral agreements only regulate bilateral relations, fishing activities of third parties are outside any regulation. Finally, many fishery resources in the East China Sea and the Yellow Sea are migratory species so that they belong to the same marine ecosystem. In that sense, the East Asian seas urgently need a regional and multilateral fishery arrangement which can be more effective to conserve and manage the fishery resources therein, and the newly concluded bilateral fishery agreements including the Sino-Japanese one could be the basis for such regional cooperation.

In addition, in May 2009 China and Republic of Korea both submitted respectively their Preliminary Information regarding their outer continental shelf in the East China Sea to the CLCS in accordance with relevant provisions of the LOSC and other decisions made by the State Parties Conferences on the Law of the Sea.[39] The two countries have met the requirement and the submission of the preliminary information enables them to make their full submissions in a later time. Both China and South Korea sent respectively their formal Submissions claiming extended continental shelf in the East China Sea to the CLCS in December 2012. Since such claims have been extended to the Okinawa Trough, it surely affects the interests of Japan. Furthermore, since they are both in the East China Sea, their claims are conflicting each other as well.

---

[38] See Chi Young Pak, "Resettlement of the Fisheries Order in Northeast Asia resulting from the New Fisheries Agreements among Korea, Japan and China", *Korea Observer*, Vol. 30(4), 1999, 614–615. For other related reference, see Mark J. Valencia and Yong Hee Lee, "The South Korea–Russia–Japan Fisheries Imbroglio", *Marine Policy*, Vol. 26, 2002, 337–343.

[39] For details on China's submission, see http://www.un.org/Depts/los/clcs_new/submissions_files/preliminary/chn2009preliminaryinformation_english.pdf and on Korea's submission, see http://www.un.org/Depts/los/clcs_new/submissions_files/preliminary/kor_2009preliminaryinformation.pdf.

Possible Settlement

As discussed in Chapter 3, there are a number of mechanisms for the settlement of maritime disputes in international law, including such political means as negotiation and consultation, mediation and good offices, conciliation, investigation, and such judicial means as arbitration and international adjudication. China always prefers bilateral negotiation and consultation for dispute settlement between States. In the international arena, China has advocated negotiation as the most practical means of dispute resolution. In practice, China has resolved a number of bilateral disputes between itself and other countries via negotiation and consultation, on border disputes, nationality, etc.

Nevertheless, for the resort of judicial means, China's attitude is rather conservative. During the Sino-India border conflict in 1962, China refused India's proposal to submit the border dispute to arbitration by stating that "the Sino-India border dispute is an important matter concerning the sovereignty of the two countries, and the vast size of more than 100,000 sq km of territories. It is self-evident that it can only be resolved through direct bilateral negotiations. It is never possible to seek a settlement from any form of international arbitration".[40] China's hesitation in this respect can be seen in its attitude toward the dispute settlement mechanism set forth in the 1997 Convention on the Law of the Non-Navigational Uses of International Watercourses. During the negotiations, China expressed its concerns over any arrangement for compulsory mechanism. The Chinese representative stated that "China was not opposed to binding procedures, only to the adoption of such procedures without the consent of the parties".[41] That means China will not accept any compulsory mechanism. China also refused to participate in the South China Sea arbitration case initiated by the Philippines in 2013.

Some reasons may be behind this passivity: first, China, having suffered tremendously from Western colonization and aggression in history,

---

[40] See Gao Yanping, "International Dispute Settlement", in Wang Tieya (ed.), *International Law* (Beijing: Law Press, 1995) (in Chinese), pp. 611–612.

[41] See UN Doc. A/C.6/51/SR.59 (1997), para.3, p. 3; cited in Attila Tanzi, "Recent Trends in International Water Law Dispute Settlement", in International Bureau of the Permanent Court of Arbitration (ed.), *International Investments and Protection of the Environment: The Role of Dispute Resolution Mechanisms* (The Hague: Kluwer Law International, 2001), p. 139.

did not trust the international judiciary centered on the Western legal principles and doctrines; and second, China stood in alignment with other communist countries, holding the perception that the world was in a perpetual state of conflict between capitalist States and socialist States and since the interests and policies of these States were so opposed, it was not possible to resolve any dispute through impartial adjudication.[42]

Some conventions require the contracting States to accept compulsory judicial dispute settlement procedures, such as the LOSC. Upon the ratification of the LOSC, China did not identify which compulsory mechanism it had accepted. Under such circumstances, it was deemed to have accepted the mechanism of arbitration.[43]

Meanwhile, China's attitude toward the role of international courts in dispute settlement is even more passive. China has usually made a reservation about the clause of judicial settlement by ICJ in the treaties to which China is a party. On 5 September 1972, China formally declared that it did not recognize the Statement of the former Chinese Government on Acceptance of the Compulsory Jurisdiction of the ICJ. In fact, China refused to settle any dispute with other countries through ICJ.[44] For this reason, some scholars suggested that China accept international judicial settlement to some degree so as to improve China's international image.[45] On the other hand, as a UN Security Council member, China has nominated judges of Chinese nationality to ICJ as well as to other international courts.

In order to be involved in dealing with the world ocean affairs, including the mechanism for dispute settlement stipulated in the LOSC and to safeguard its own maritime interest, China ratified the LOSC on 15 May 1996, which gave it a right of eligibility to nominate a candidate to be one of the ITLOS judges. It is to be noted that in August 2006 China made a declaration to exclude certain disputes including those concerning

---

[42]See Paul R. William, *International Law and the Resolution of Central and East European Transboundary Environmental Disputes* (New York: Macmillan Press, 2000), p. 199.

[43]Article 287(3) of the LOSC provides that "A State Party, which is a party to a dispute not covered by a declaration in force, shall be deemed to have accepted arbitration in accordance with Annex VII".

[44]Gao, *supra* note 40, p. 612.

[45]See Shen Wei, "Dispute Settlement Mechanism of the UN Convention on the Law of the Sea", *Ocean Development and Management* (in Chinese), Vol. 13(3), 1996, 53–54.

maritime boundary delimitation from compulsory third-party settlement procedures in accordance with Article 298 of the LOSC.[46] It is not a surprise and follows China's long-standing tradition of non-interference from any international judicial body in the disputes in which China is involved. Whenever China accedes to or ratifies a treaty, it makes a reservation on the clause concerning compulsory jurisdiction of the ICJ. However, the settlement mechanism in the LOSC is much trickier. First, a State has to accept one of the four juridical means. This means disputes regarding the application and interpretation of the LOSC which China is involved are subject to arbitration at least, except for those matters to be excluded by the Chinese statement. Second, even if China excludes certain specific disputes, they will still be subject to the third-party conciliation mechanism which can produce tremendous pressure on the States concerned to find a way of solving their disputes.

Therefore, China's attitude toward international judiciary is somewhat ambivalent. On the one hand, as a big country China would like to play a part in that area. On the other hand, it is short of qualified legal experts who could provide adequate services in the international judiciary, which would benefit it for its own interests. Furthermore, due to the fact that the composition of the judges is mainly of the West, developing countries including China are doubtful whether the international judiciary can maintain the necessary impartiality and justice. Only when China has cherished enough confidence in that field can it play its due role. The domestic legal reform in China may provide some help to push China closer to the international rule of law and China's willingness to resort to the international judiciary for dispute settlement.

In comparison, Japan's attitude is more active and positive. It accepted ICJ's compulsory jurisdiction as early as 1958. Accordingly, Japan recognized compulsory jurisdiction "over all disputes". This statement is very generous, but may exclude disputes "which the parties thereto have agreed or shall agree to refer for final and binding decision to arbitration or judicial settlement".[47] As reported, Japan's attitude toward international adjudication has completely changed since the World War II. It adopted the policy of submitting a dispute to adjudication unless

[46]The full text is available at http://www.un.org/Depts/los/convention_agreements/convention_declarations.htm#China%20after%20ratification.

[47]See Japan's Declaration of 15 September 1958, *ICJ Yearbook*, 1994–1995, p. 95.

it is resolved through negotiations.[48] It is recalled that in the 1950s, Japan proposed to submit a dispute with Australia concerning Japan's pearl fishing activities in the Timor and Arafura Seas, situated off the northern coast of Australia which is beyond its territorial waters but within its continental shelf claim.[49] Since Japan has accepted the compulsory jurisdiction of the ICJ, if China likes to use the ICJ in resolving the territorial disputes over the aforementioned islands, it may have some advantages to bring the case to the court.

Although Japan did not choose a compulsory dispute settlement mechanism under the LOSC upon its ratification, it has been involved in three cases before the ITLOS: two concerning prompt release and one requesting provisional measures. The two prompt release cases all involved Japan whose vessels were detained by Russia in its EEZ for alleged illegal fishing. For the *Hoshinmaru* case, the two disputing parties were not in agreement with the amount of the security bond imposed by the Russian authorities. The Tribunal's judgment in August 2007 settled the dispute by adjusting this amount.[50] For the *Tomimaru* case submitted by Japan in 2007, the Tribunal simply ruled that there was no object on which it needs to be called upon for a decision since the targeted Japanese vessel had been already confiscated by the Russian authorities.[51] In the case of *Southern Bluefin Tuna (New Zealand/Australia v. Japan)*, Japan was forced to respond to the case submitted by Australia and New Zealand in July 1999.[52]

As indicated in Chapter 1, some countries in East Asia resorted to third party dispute mechanisms to settle their disputes and the cases include the case on *Sovereignty over Pulau Litigan and Pulau Sipadan* (Malaysia/Indonesia) (1998–2002) and the case on *Sovereignty over Pedra Branca, Middle Rocks and South Ledge* (Malaysia/Singapore) (2003–2008). It is interesting to see whether China and Japan would learn from their East Asian neighbors to settle their sovereign and maritime disputes in the East China Sea.

---

[48] See K. Yokota, "International Adjudication and Japan", *Japanese Annual of International Law*, Vol. 17, 1973, 20.

[49] See P. Macalister-Smith, "Pearl Fisheries", *Encyclopedia of Public International Law*, Vol. 11, p. 257.

[50] For details, see http://www.itlos.org/start2_en.html.

[51] *Ibid.*

[52] See relevant discussions in Chapter 3.

Conclusion

After the aforementioned observations, a couple of concluding remarks can be made. First, as neighboring countries maritime encounters and interactions are inevitable between China and Japan, so as the maritime disputes. Second, it can be seen that while maritime cooperation has been carried out between the two sides, there are a number of maritime disputes to be resolved. The maritime disputes have historical reasons and also been exacerbated by the new maritime legal regime centered on the LOSC. Third, apparently both sides prefer a solution to their disputes by peaceful means and third party judicial settlement is thus recommended. Finally, as both sides exert their sincere efforts toward the direction of the settlement of their maritime and territorial disputes, the East China Sea is expected to become a sea of peace and cooperation, rather than a sea of conflict and tension.

Chapter 9

Maritime Geographic Features: Seeking a Legal Clarity

Introduction

As a general principle, a coastal State claims maritime zones and geographic features in accordance with the principle of "land dominating the sea", which means that all maritime entitlements derive from land. Following this principle, coastal States can extend their sovereignty, sovereign rights and jurisdiction to sea areas adjacent to their coasts. According to the 1982 United Nations Convention on the Law of the Sea (LOSC), a coastal State is entitled to have a 12-nm territorial sea, 200-nm exclusive economic zone (EEZ) and at least 200-nm continental shelf.[1] The entitlement to these maritime zones includes water columns, natural resources and geographical features such as the continental shelf.

However, the claims of coastal States to geographical features are different from claims to maritime zones in general due to the different nature of the geographic features. Without mentioning other geographic features such as coastlines and deep seabed, this chapter is designed to focus on maritime geographic features of islands/islets, reefs, submerged features and continental shelf.

[1]For details on these maritime zones, see the relevant provisions of the LOSC.

Impact of the Truman Proclamation

When we discuss the issue concerning the claims to maritime geographic features, we cannot escape the significant, somewhat fundamental document in history, i.e. the Truman Proclamation issued on 28 September 1945. In that Proclamation, Truman, the then American President proclaimed that:

> Having concern for the urgency of conserving and prudently utilizing its natural resources, the Government of the United States regards the natural resources of the subsoil and sea bed of the continental shelf beneath the high seas but contiguous to the coasts of the United States, subject to jurisdiction and control. In cases where the continental shelf extends to the shores of another State, or is shared with an adjacent State, the boundary shall be determined by the United States and the Sate concerned in accordance with equitable principles. The character as high seas of the waters above the continental shelf and the right to their free and unimpeded navigation are in no way thus affected.[2]

This Proclamation is the first that had claimed the subsoil and seabed of the continental shelf in the name of the utilization and conservation of the natural resources. It has a huge impact on the development of the law of the sea as well as American marine policy. There are subsequently at least two aspects of impact that generated from the Proclamation: one internal and the other external.

As for the internal impact, the United States enacted the Submerged Lands Act in 1953. In the Act, the term "lands beneath navigable waters" means

(1) "all lands within the boundaries of each of the respective States which are covered by nontidal waters that were navigable under the laws of the United States at the time each State became a member of the Union, or acquired sovereignty over such lands and water thereafter, up to the ordinary high water mark as heretofore or hereafter modified by accretion, erosion, and reliction;

[2]Proclamation 2667 of 28 September 1945: Policy of the United States with respect to the natural resources of the subsoil and sea bed of the continental shelf, available at http://www.oceancommission.gov/documents/gov_oceans/truman.pdf.

(2) all lands permanently or periodically covered by tidal waters up to but not above the line of mean high tide and seaward to a line three geographical miles distant from the coast line of each such State and to the boundary line of each such State where in any case such boundary as it existed at the time such State became a member of the Union, or as heretofore approved by Congress, extend seaward (or into the Gulf of Mexico) beyond three geographical miles; and
(3) all filled in, made, or reclaimed lands which formerly lands beneath navigable waters, as hereinabove defined".[3]

The Act is designed to return the title to submerged lands to the states and promote the exploration and development of petroleum deposits in coastal waters.[4]

Following the Truman Proclamation, the US Congress also enacted the other important act — the Outer Continental Shelf Lands Act in the same year, which defines the "outer continental shelf" (OCS) as "all submerged lands lying seaward of state coastal waters (three miles offshore) which are under US jurisdiction".[5] The Secretary of Interior is authorized to promulgate regulations to lease the OCS in an effort to prevent waste and conserve natural resources and to grant leases to the highest responsible qualified bidder as determined by competitive bidding procedures.[6] The OCS is a vital national resource reserve held by the Federal Government for the public.

It is pointed out that the term "OCS" is different from the similar term found in the LOSC as the former only refers to the submerged lands outside of the area of lands beneath navigable waters as belonging to the

---

[3] 43 U.S.C. §§1301–1315 (2002).

[4] *Ibid.*

[5] Outer Continental Shelf Lands Act of 1953 (43 U.S.C. 1331–1356, P.L. 212, Ch. 345, 7 August 1953, 67 Stat. 462) as amended by P.L. 93–627, 3 January 1975, 88 Stat. 2130; P.L. 95–372, 18 September 1978, 92 Stat. 629; and P.L. 98–498, 19 October 1984, 98 Stat. 2296. As the Congress did not include the geological definition of the continental shelf in the Act, "Congress may purposely have left the definition vague so that the Act would adapt to developing international concepts regarding the Continental Shelf doctrine". See Donna Darm, "The Outward Limit of the Department of Interior's Authority over Submerged Lands — The Effect of Customary International Law on the Outer Continental Shelf Lands Act", *Washington Law Review*, Vol. 60, 1984–85, 678.

[6] *Ibid.*

relevant states, i.e. only beyond three geographical miles. It may include the areas beyond 200 nm in the context of the LOSC. As the US has not yet ratified the Convention, it is not eligible for the moment to submit its OCS claims to the Commission on the Limits of Continental Shelf (CLCS) for consideration.

The external impact is reflected in the State practice around the world following the Truman Proclamation. The first bunch of countries were from Latin America: Mexico issued a Declaration on 29 October 1945, Argentina on 9 October 1946, Chile in June 1947, Peru on 1 August 1947, Costa Rica on 27 July 1948, Nicaragua in 1949 and Honduras on 7 March 1950. In addition, Great Britain issued the Bahamas (Alteration of Boundaries) Order in Council and the Jamaica (Alteration of Boundaries) Order in Council on 26 November 1948.[7] In 1949, a series of Middle East governments also issued such proclamations and included Saudi Arabia (28 May), Bahrain (5 June), Qatar (8 June), Abu Dhabi (10 June), Kuwait (12 June), Dubai (14 June), Shayah (16 June), Umm al Oeiwain (20 June) and Ajman (20 June).[8]

All such declarations and proclamations are related to the national claims to submerged lands such as the continental shelf. Some countries made new legislations and/or amended the old ones to legalize their claims. Article 5 of the 1950 Constitution of Nicaragua provides that the "continental platform and the submarine shelves were enumerated, together with the adjacent islands, the subsoil of the territorial waters, the air space and the stratosphere as forming part of national territory".[9]

In Asia, Pakistan issued a declaration on 9 March 1950 stating that "the sea bed along the coasts of Pakistan extending to the one hundred fathom contour into the open sea should, with effect from the date of this declaration, be included in the territories of Pakistan".[10] It is worth mentioning that the Ministry of the Interior of the Republic of China publicized in February 1948 a location map with a U-shaped line to claim

[7]See H. Lauterpacht, "Sovereignty over Submarine Areas", *British Yearbook of International Law*, Vol. 27, 1950, 380–381.

[8]*Ibid.*, p. 381.

[9]*Ibid.*, p. 381. Brazil and El Salvador hold the same position regarding the continental shelf.

[10]*Gazette of Pakistan, Extraordinary*, 14 March 1950; cited in Lauterpacht, *ibid.*, p. 381.

maritime features in the South China Sea.[11] The publication of the map was clearly influenced by the Truman Proclamation. But unlike the countries which issued declarations, proclamations and even decrees, China only published a map to indicate its maritime claim. Suppose China had followed other coastal States to issue a declaration instead of simply publishing a map, the legal effect of China's claim in the South China Sea would be much more reinforced.

As H. Lauterpacht summarizes, the aforementioned instruments have two features in common: "they all disclaim any intention of interfering with the principle of navigation on the high seas"; and "none of them has drawn itself the protest of any state except in cases in which the proclamation of rights over the submarine areas has been used for asserting exorbitant claims lacking any foundation in law and alien to the apparent occasion which prompted them".[12] Despite the fact that some countries such as Argentina, Nicaragua, etc., claimed full sovereignty over the continental shelf, it was understood then that the rights reflected in the Truman Proclamation should be the rights of jurisdiction and control over the submarine areas and their resources, though it was admitted that "the rights of exclusive control and jurisdiction claimed in the submarine areas are rights indistinguishable from sovereignty."[13]

Conventional Framework for the Continental Shelf

The Truman Proclamation also triggered the birth of the Convention on the Continental Shelf. Based on the draft prepared by the International Law Commission, the Geneva Convention on the Continental Shelf was adopted in 1958. The term of continental shelf used in the Convention refers: "(a) to the seabed and subsoil of the submarine areas adjacent to the coast but outside the area of the territorial sea, to a depth of 200 m or, beyond that limit, to where the depth of the superjacent waters admits of the exploitation of the natural resources of the said areas; (b) to the seabed

---

[11]For details about this line, See Zou Keyuan, "The Chinese Traditional Maritime Boundary Line in the South China Sea and Its Legal Consequences for the Resolution of the Dispute over the Spratly Islands", *International Journal of Marine and Coastal Law*, Vol. 14(1), 1999, 27–55.

[12]Lauterpacht, *supra* note 7, p. 383.

[13]*Ibid.*, p. 415.

and subsoil of similar submarine areas adjacent to the coasts of islands".[14] Over the continental shelf, the coastal State exercises sovereign rights for the purpose of exploring it and exploiting its natural resources (Article 2) and "the rights of the coastal State over the continental shelf do not affect the legal status of the superjacent waters as high seas, or that of the air-space above those waters" (Article 3). It is to be noted that the use of the term "sovereign rights" indicate that a coastal State could not treat the continental shelf as its national territory like islands or the territorial sea. Following Article 3, freedom of navigation and overflight is preserved above the continental shelf.

Different from the 1958 Convention, the LOSC defines the continental shelf of a coastal State as: "comprises the seabed and subsoil of the submarine areas that extend beyond its territorial sea throughout the natural prolongation of its land territory to the outer edge of the continental margin, or to a distance of 200 nautical miles from the baselines from which the breadth of the territorial sea is measured where the outer edge of the continental margin does not extend up to that distance".[15] That means a coastal State can claim more extensive continental shelf in accordance with the LOSC. Besides the aforesaid definition, the LOSC also allows coastal States to extend their continental shelves up to 350 nm if recommended by the CLCS.

It is to be noted that the LOSC contains some overlapping provisions in the part of EEZ and the part of the continental shelf. Due to this fact, State practices show that coastal States usually incorporate these two parts into one legislation, such as the 1998 Law on the EEZ and the Continental Shelf of the People's Republic of China.[16] However, they have differences in terms of sovereign rights and national jurisdiction. The EEZ regime is more focused on marine resources rather than maritime geographic features as the LOSC stipulates clearly that the EEZ is "an area beyond and

---

[14]Article 1 of the Convention on the Continental Shelf. Text is available at http://untreaty.un.org/ilc/texts/instruments/english/conventions/8_1_1958_continental_shelf.pdf.

[15]Article 76 of the LOSC. The same provision also defines the continental margin as "comprises the submerged prolongation of the land mass of the coastal State, and consists of the seabed and subsoil of the shelf, the slope and the rise. It does not include the deep ocean floor with its oceanic ridges or the subsoil thereof".

[16]An unofficial English text translated by this author is available in Zou Keyuan, *China's Marine Legal System and the Law of the Sea* (Leiden: Martinus Nijhoff, 2005), pp. 342–345.

adjacent to the territorial sea"[17] and coastal States enjoy "sovereign rights for the purpose of exploring and exploiting, conserving and managing the natural resources, whether living or non-living, of the waters superjacent to the seabed and of the seabed and its subsoil".[18] Therefore, it is difficult, if not totally impossible, for coastal States to claim maritime geographic features by invoking relevant provisions in the LOSC governing the EEZ regime.

Legal Status of Islands

According to the LOSC, an "island" "is a naturally formed area of land, surrounded by water, which is above water at high tide".[19] This definition contains a number of essential factors: a piece of land; naturally formed; surrounded by water; and above the sea surface at high tide. But the Convention does not explain to what extent a piece of land surrounded by water and above water at high tide can be regarded as an island. Once defined as an island, it is treated as a piece of land which can have its own territorial sea, EEZ and continental shelf.

For the purpose of measuring the breadth of the territorial sea, in the case of islands situated on atolls or of islands having fringing reefs, then the baseline is the seaward low-water line of the reef.[20]

In State practice, some countries use the term "island" in a very broad sense. Like Japan, it names "Okinotori" as *shima* (island in Japanese). China uses the term *qundao* (archipelagoes or group of islands in Chinese) to name all the features, even including some permanently submerged features such as Macclesfield Bank in the South China Sea. The Spratly Islands in Chinese is called *Nansha* Archipelago. Even for a same maritime geographic feature, there are different names from different countries. Scarborough Reef in Chinese is *Huangyan* Island and Okinotorishima in Chinese is *Chongzhiniao* Reef.

The LOSC further provides in Article 121 that "Rocks which cannot sustain human habitation or economic life of their own shall have no

[17] Article 55 of the LOSC.

[18] See Article 56 of the LOSC.

[19] See Article 121 of the LOSC.

[20] Article 6 of the LOSC.

exclusive economic zone or continental shelf".[21] As there is a big difference between an island and a rock in terms of maritime zone generation, coastal States attempt to consolidate a tiny piece of maritime land in order to make it get qualified as an island. The most significant example is Okinotorishima (Douglas Reef) of Japan.[22] As suggested by the International Hydrographic Bureau, a rock has an area less than 0.001 square miles (or 2,590 m^2),[23] but there are other different definitions. The problem on how to define an island still remains a controversial issue in international law, which will be discussed further in the following sections.

The South China Sea arbitration (2013–2016) has touched upon the regime of islands and explained in details the differences between an island and a rock. According to the arbitral award issued on 12 July 2016, "the use of the word 'rock' does not limit the provision to features composed of solid rock. The geological and geomorphological characteristics of a high-tide feature are not relevant to its classification pursuant to Article 121(3)".[24] "[T]he status of a feature is to be determined on the basis of its natural capacity, without external additions or modifications intended to increase its capacity to sustain human habitation or an economic life of its own".[25]

As for "human habitation", according to the Tribunal, "the critical factor is the non-transient character of the inhabitation, such that the inhabitants can fairly be said to constitute the natural population of the feature, for whose benefit the resources of the EEZ were seen to merit protection. The term 'human habitation' should be understood to involve the inhabitation of the feature by a stable community of people for whom the feature constitutes a home and on which they can remain".[26]

---

[21] Article 121(3) of the LOSC.

[22] Yann-huei Song, "Okinotorishima: A 'Rock' or an 'Island'? Recent Maritime Boundary Controversy between Japan and Taiwan/China", in Seoung-Yong Hong and Jon M. Van Dyke (eds.), *Maritime Boundary Disputes, Settlement Processes, and the Law of the Sea* (Leiden: Martinus Nijhoff, 2009), p. 148.

[23] Cited in Leticia Diaz, Barry Hart Dubner and Jason Parent, "When is a 'Rock' an 'Island'? — Another Unilateral Declaration Defies 'Norms' of International Law", *Michigan State Journal of International Law*, Vol. 15, 2007, 535.

[24] The South China Sea Arbitration Award of 12 July 2016, available at http://www.pcacases.com/pcadocs/PH-CN%20-%2020160712%20-%20Award.pdf, para. 482.

[25] *Ibid.*, para. 541.

[26] *Ibid.*, para. 542.

As for the term "economic life of their own", it is "linked to the requirement of human habitation, and the two will in most instances go hand in hand".[27] "The Tribunal considers that the 'economic life' in question will ordinarily be the life and livelihoods of the human population inhabiting and making its home on a maritime feature or group of features. Additionally, Article 121(3) makes clear that the economic life in question must pertain to the feature as 'of its own'. Economic life, therefore, must be oriented around the feature itself and not focused solely on the waters or seabed of the surrounding territorial sea. Economic activity that is entirely dependent on external resources or devoted to using a feature as an object for extractive activities without the involvement of a local population would also fall inherently short with respect to this necessary link to the feature itself. Extractive economic activity to harvest the natural resources of a feature for the benefit of a population elsewhere certainly constitutes the exploitation of resources for economic gain, but it cannot reasonably be considered to constitute the economic life of an island as its own".[28]

In the Tribunal's view, the text of Article 121(3) is disjunctive, such that the ability to sustain either human habitation or an economic life of its own would suffice to entitle a high-tide feature to an EEZ and continental shelf. However, as a practical matter, the Tribunal considers that a maritime feature will ordinarily only possess an economic life of its own if it is also inhabited by a stable human community. One exception to that view should be noted for the case of populations sustaining themselves through a network of related maritime features. The Tribunal does not believe that maritime features can or should be considered in an atomized fashion. A population that is able to inhabit an area only by making use of multiple maritime features does not fail to inhabit the feature on the grounds that its habitation is not sustained by a single feature individually. Likewise, a population whose livelihood and economic life extends across a constellation of maritime features is not disabled from recognizing that such features possess an economic life of their own merely because not all of the features are directly inhabited.[29]

The Tribunal also considers the capacity of a maritime feature to sustain human habitation or an economic life of its own, and takes the view

[27] *Ibid.*, para. 543.

[28] *Ibid.*, para. 543.

[29] *Ibid.*, para. 544.

that "the capacity of a feature is necessarily an objective criterion" and has no relation to the question of sovereignty over the feature. According to it, the principal factors that contribute to the natural capacity of a feature include the presence of water, food and shelter in sufficient quantities to enable a group of persons to live on the feature for an indeterminate period of time. "Such factors would also include considerations that would bear on the conditions for inhabiting and developing an economic life on a feature, including the prevailing climate, the proximity of the feature to other inhabited areas and populations, and the potential for livelihoods on and around the feature".[30]

The Tribunal considers that "the capacity of a feature should be assessed with due regard to the potential for a group of small island features to collectively sustain human habitation and economic life. On the one hand, the requirement in Article 121(3) that the feature itself sustain human habitation or economic life clearly excludes a dependence on external supply. A feature that is only capable of sustaining habitation through the continued delivery of supplies from outside does not meet the requirements of Article 121(3). Nor does economic activity that remains entirely dependent on external resources or that is devoted to using a feature as an object for extractive activities, without the involvement of a local population, constitute a feature's 'own' economic life".[31] "If a feature is entirely barren of vegetation and lacks drinkable water and the foodstuffs necessary even for basic survival, it will be apparent that it also lacks the capacity to sustain human habitation. The opposite conclusion could likewise be reached where the physical characteristics of a large feature make it definitively habitable".[32]

The South China Sea arbitral award, though attempting to interpret Article 121 of the LOSC, has raised a number of controversial issues in the law of the sea concerning islands, group of islands, or even archipelagic States and waters. The Tribunal re-defined the definition on island in Article 121: there would be no "island" in the meaning of Article 121 and the high tide geographic features are divided into two categories: fully entitled island and rock. If a high tide geographic feature would become a fully entitled island, it must meet the criteria set forth by the Tribunal, such as permanent residence, natural food and water, etc. Second, the

[30] *Ibid.*, para. 545.

[31] *Ibid.*, para. 547.

[32] *Ibid.*, paras. 540–548.

Tribunal eliminated the meaning of the phrase "economic life of its own" in Article 121(3) by subjecting it to "human habitation". In the eye of the Tribunal, without "human habitation", "economic life" becomes meaningless. Such an explanation raises the question on how to interpret the relevant wordings in this paragraph: does the word "or" mean in parallel one or the other, or mean "and" or "plus"? It seems that there is suspicion of excessive, if not abusive interpretation by the Tribunal of the relevant provisions in the LOSC.

Another related issue is about artificial islands. It is remembered that in the 1950s when the International Law Commission deliberated over whether to incorporate a requirement of natural formation into the definition of islands, one of its members Hersch Lauterpacht expressed his concern that "if artificial islands erected within the territorial sea were to have a territorial sea of their own, then a state could erect a series of small artificial islands just within its territorial sea and a few miles apart", which "might in that way double the extent of its territorial sea".[33]

Legal Status of Low-Tide Elevations

The legal status of low-tide elevations is not very clear in international law. In the 19th century, Great Britain considered low-tide elevations as islands if they were located in the immediate vicinity of the coasts.[34] In the Anglo-Norwegian Fisheries case (1949–1951), the United Kingdom made the following submission:

> An elevation of the sea bed, only above water at low tide, which is situated within the territorial belt of a mainland (or of a permanently dry island) counts as a piece of territory for the purpose of the delimitation of territorial waters. It is sufficient for the purpose of this rule that the elevation is only partially within the territorial belt, in which case the whole of the elevation is so taken into account.[35]

---

[33] International Law Commission 1954, "260th Meeting Summary Record", *Yearbook of the International Law Commission*, Vol. 1, 1954, 94.

[34] Hugo Ignacio Llanos, "Low-Tide Elevations: Reassessing Their Impact on Maritime Delimitation", *Pace International Law Review*, Vol. 14, 2002, 255.

[35] Anglo-Norwegian Fisheries case *(United Kingdom v. Norway)*, ICJ Pleadings, Vol. 2, p. 78; cited in Geoffrey Marston, "Low-Tide Elevations and Straight Baselines", *British Yearbook of International Law*, Vol. 46, 1972–1973, 406.

The LOSC only mentions them in the context of maritime boundary determination. As Article 13 provides:

> A low-tide elevation is a naturally formed area of land which is surrounded by and above water at low tide but submerged at high tide. Where a low-tide elevation is situated wholly or partly at a distance not exceeding the breadth of the territorial sea from the mainland or an island, the low-water line on that elevation may be used as the baseline for measuring the breadth of the territorial sea.
>
> 2. Where a low-tide elevation is wholly situated at a distance exceeding the breadth of the territorial sea from the mainland or an island, it has no territorial sea of its own.

Here the Convention only gives low-tide elevations a limited legal status and it is not allowed to have its own territorial sea, much less other maritime zones.

In its judgment in the Case Concerning Maritime Delimitation and Territorial Questions between Qatar and Bahrain (*Qatar v. Bahrain*), the International Court of Justice discussed the status issue of low-tide elevations and made the following statement that:

> International treaty law is silent on the question whether low-tide elevations can be considered to be 'territory'. Nor is the Court aware of a uniform and widespread State practice which might have given rise to a customary rule which unequivocally permits or excludes appropriation of low-tide elevations. It is only in the context of the law of the sea that a number of permissive rules have been established with regard to low-tide elevations which are situated at a relatively short distance from a coast.[36]

The same Court further states that:

> The few existing rules do not justify a general assumption that low-tide elevations are territory in the same sense as islands. It has never been disputed that islands constitute terra firma, and are subject to the rules

---

[36] Judgment of 16 March 2001, para. 205, available at http://www.icj-cij.org/docket/files/87/7027.pdf.

> and principles of territorial acquisition; the difference in effects which the law of the sea attributes to islands and low-tide elevations is considerable. It is thus not established that in the absence of other rules and legal principles, low-tide elevations can, from the viewpoint of the acquisition of sovereignty, be fully assimilated with islands or other land territory.[37]

Nevertheless, in the maritime delimitation between the said States, the Court gave partial effect to a low-tide elevation, which had an impact on the delimitation line. As is pointed out, it made this the first precedent in international case law to give effect to this type of feature.[38]

In the case of Sovereignty over Pedra Branca/Pulau Batu Puteh, Middle Rocks and South Ledge (Malaysia/Singapore),[39] both the contesting parties asked the Court to decide on which party would own the sovereignty over South Ledge, a low-tide elevation in the disputed area. In the judgment, the Court only decides that sovereignty over South Ledge belongs to the State in the territorial waters of which it is located.

In domestic jurisprudence, a case from the United States is insightful. Concerning Dinkum Sands, Alaska claimed it was an island, while the United States argued that it was only rarely aforementioned the mean high water (an average elevation of high tides taken over a 19-year period) and could not be considered an island.[40] The Court held that there was "no precedent for treating as an island a feature that oscillates above and below mean high water", and thus concluded that Dinkum Sands was not an island and Alaska was not entitled to claim the submerged lands within 3 miles of that feature.[41]

---

[37] Judgment of 16 March 2001, para. 206, available at http://www.icj-cij.org/docket/files/87/7027.pdf.

[38] Llanos, *supra* note 24, p. 255.

[39] For details, see Judgment of 23 May 2008, available at http://www.icj-cij.org/docket/index.php?p1=3&p2=3&k=2b&case=130&code=masi&p3=4.

[40] See David J. Bederman, "US State Boundaries — Submerged Lands Act — 1958 Geneva Convention on the Territorial Sea — Straight Baselines — Fringing Islands — Ice Elevations", *American Journal of International Law*, Vol. 92, 1998, 85.

[41] *United States v. Alaska*. 117 S.Ct. 1888 (1997), cited in Bederman, *ibid.*, p. 86.

Legal Dilemma for Hybrid Dry Features

The term "hybrid dry features" used herewith refers to those geographic features which are above water surface, whether at high tide or low tide, but with artificial constructions and installations. The issue is closely related to the legal status of artificial islands.

While there is no definition under the LOSC, there are attempts in the international law academia to define "artificial islands". The *Encyclopedia of Public International Law* defines artificial island as a temporary or permanent fixed platform made by man surrounded by water and above water at high tide.[42]

The LOSC to some extent defines the legal status of artificial islands. Its definition on islands clearly excludes any artificial island. Article 60 provides that "Artificial islands, installations and structures do not possess the status of islands. They have no territorial sea of their own, and their presence does not affect the delimitation of the territorial sea, the exclusive economic zone or the continental shelf".[43] While artificial islands do not generate any maritime zones, coastal States are allowed to establish safety zones around them. But such safety zone around an artificial island should not exceed a distance of 500 m, "measured from each point of their outer edge, except as authorized by generally accepted international standards or as recommended by the competent international organization".[44]

Taking the South China Sea as an example, geographically speaking, there are hundreds of natural features in that sea area including islands/islets, atolls, reefs and banks. According to a Chinese source, the Spratly Islands consist of 14 islands/islets, 6 banks, 113 submerged reefs, 35 underwater banks and 21 underwater shoals.[45] Another source from a Western scholar tells us that there are about 170 features in the South China Sea among which only about 36 tiny islands are above water at

---

[42] Rudolf Bernhardt, *Encyclopedia of Public International Law*, Law of the Sea Volume, 1989, p. 38.

[43] Article 60 of the LOSC.

[44] *Ibid.*

[45] Zeng Zhaoxuan (eds.), *Islands in the South China Sea* (Guangzhou: Guangdong People's Publishing House, 1986) (in Chinese), p. 3.

high tide.[46] From these two figures, we can see that natural features submerged under water are more than those above water. This special geographical characteristic of the South China Sea decides the importance of artificial islands and installations in political, economic and strategic dimensions of the Spratly Islands.

Artificial islands constructed on natural rocks and reefs with the nature of permanence become a new issue in international law. China (PRC) has occupied several reefs in the Spratlys since 1988 and for the purpose of military stationing or for other purposes, it built artificial structures on these reefs (Johnson South/*Chigua* Reef, Subi/*Zhubi* Reef, Gaven/*Nanxun* Reef, Cuarteron/*Huayang* Reef, Hughes/*Dongmen* Reef, and Mischief/Meiji Reef). Some of them have been later expanded by reclamation to become more than natural islands, such as the one established on Fiery Cross/*Yongshu* Reef and Mischief/*Meiji* Reef. The Swallow Reef (*Terumbu LayangLayang* in Malay) occupied by Malaysia has been massively reclaimed and has a fishing port, 15-room diving resort and a 1.5 km airstrip. In this sense, it is acknowledged that due to artificial installations, for many of the islands, "it has become difficult to distinguish what is the natural feature and what is man-made".[47] When it is safe to say that the relevant provisions of the LOSC are applicable to natural islands and artificial islands, the legal status of the hybrid category is special and difficult to define under the LOSC, and even under the applicable international law as a whole. If it is defined as a natural feature, it is mixed with artificial installations and structures; if it is defined as an artificial island, it is not artificially fixed to the seabed, rather is supported by a natural base such as a reef whether above water at high tide or not. Apparently, there is no regulation in international law governing such kind of natural and artificial combinations. The relevant 1958 Convention on the Continental Shelf contains provisions only concerning the construction of artificial islands on the continental shelf as it provides that the coastal State is entitled to construct and maintain or operate on the continental shelf installations and other devices necessary for its exploration

[46] *Daniel J. Dzurek*, The Spratly Islands Dispute: Who's on First? IBRU (International Boundaries Research Unit) *Maritime Briefing*, Vol. 2(1), 1996, 1.

[47] Robert W. Smith, "Maritime Delimitation in the South China Sea: Potentiality and Challenges", *Ocean Development and International Law*, Vol. 41(3), 2010, 223.

and exploitation of its natural resources, but such artificial installations and devices do not possess the status of islands.[48]

Finally, the issue of artificial islands is related to the application of Article 121(3) of the LOSC concerning rocks. The current situation of artificial construction in the South China Sea or elsewhere has been in fact triggered by this provision since States, in order to extend their maritime spaces, make every effort in turning those “rocks” into “islands” which can fulfill the conditions to sustain human habitation or economic life of their own. The Okinotorishima is a typical example as mentioned previously. The question is whether such artificial construction on the reef has changed the legal status of it. As expressed by China, the construction of artificial facilities on the reef “will not change its legal status”.[49] However, this is only China’s legal position regarding the Okinotorishima and it is contrary to the real intention of Japan. What Japan has done and will do is to make the reef fulfill the conditions prescribed in Article 121(3) of the LOSC so as to enable Japan to claim not only territorial sea but also EEZ and continental shelf from that reef. Due to the opposition of China and South Korea, the CLCS suspended its consideration of Japan’s submission concerning the part at least partially generating from Okinotorishima (Southern Kyushu–Palau Ridge Region) while giving its recommendations for other parts of Japan’s extended continental shelf claims.[50] But this setback does not prevent Japan from unilaterally claiming EEZ as well as continental shelf (though limited to 200 nm) allegedly generating from that reef.

Nevertheless, Japan’s attempt of conversion may lead to two possible legal consequences: one is the change to Japan’s heart that the reef will eventually turn into a natural island fulfilling the relevant conditions set forth in the LOSC; and the other, which might be contrary to Japan’s wish, is the change that the natural reef will eventually be treated by the

---

[48]Article 5 of the 1958 Convention on the Continental Shelf.

[49]See Qin Jize, Li Xiaokun and Cheng Guangjin, “Japan atoll expansion ‘hurts neighbors’”, *China Daily*, 11 February 2010, available at http://www.chinadaily.com.cn/world/2010–02/11/content_9461259.htm.

[50]For details, see UN Doc CLCS/74, 30 April 2012, available at http://daccess-dds-ny.un.org/doc/UNDOC/GEN/N12/326/32/PDF/N1232632.pdf?OpenElement. For the relevant analysis, see Michael Sheng-ti Gau, “Recent Decisions by the Commission on the Limits of the Continental Shelf on Japan’s Submission of Outer Continental Shelf”, *Chinese Journal of International Law*, Vol. 11, 2012, 487–504.

international community as an artificial island, which under the existing international law can only have a safety zone of 500 m. Thus, it would not be allowed to claim EEZ and continental shelf and also lose the entitlement to territorial sea due to the substantial construction of artificial installations and structures.

A different question may arise as well: if a reef with a lighthouse can enjoy the entitlement to extensive maritime zones as argued by some legal scholars,[51] is it fair that the reefs with substantial artificial works in the South China Sea are treated as "artificial islands"? It is interesting to note that in the *Tidelands Case (United States v. California)*, the Supreme Court decided that California was entitled to all mineral rights to lands outside inland waters that were submerged at the date California entered the Union but have since been enclosed or reclaimed by means of *artificial structures* (italic added).[52]

Legal Dilemma for Submerged Geographic Features

Having discussed earlier, we can see that there are relevant conventional and/or customary rules, though far from satisfactory, governing the claims to islands, low-tide elevations, artificial islands and continental shelf. Apparently, there is a missing link in the international law of the sea which governs the claims to the submerged geographic features between dry features above the water surface and the subsoil and seabed of the continental shelf.

However, in State practice, coastal States do claim such submerged maritime features. In East Asia, these features include, *inter alia*,

---

[51] For example, as is observed by two Dutch scholars, an increasing number of scholars subscribe to the view that a lighthouse or other aid to navigation built on an island gives the island "an economic life of its own" due to its value to shipping. Barbara Kwiatkowska and Alfred. H. A. Soons, "Entitlement to Maritime Areas of Rocks Which Cannot Sustain Human Habitation or Economic Life of Their Own", *Netherlands Yearbook of International Law*, Vol. 31, 1990, 167–168; cited in Robert Beckman and Clive Schofield, 'Moving Beyond Disputes over Island Sovereignty: ICJ Decision Sets Stage for Maritime Boundary Delimitation in the Singapore Strait', *Ocean Development and International Law*, Vol. 40, 2009, 10.

[52] *United States v. California*, 381 U.S. 139 (1965), 176; cited in Barry Hart Dubner, Peter hanks and Michael Pryles, "Demarcation of Authority over Coastal Waters and Submerged Lands in the United States and Australia", *Stetson Law Review*, Vol. 10, 1980–81, 240.

James Shoal, Macclesfield Bank, Vanguard Bank and Reed Bank in the South China Sea and Socotra/Ieodo/Suyan Reef in the East China Sea. As there is no express applicable rule in international law which States could invoke to support its claim, the basis for such claims may vary in State practice. China has claimed such features mainly based on the concept of historic rights within its unilaterally drawn U-shaped line around the South China Sea and also on the ground that these submerged features are integral part of the Spratly Islands/*Nansha Qundao* belonging to China, while South Korea claimed the Ieodo based on the EEZ regime provided for in the LOSC. While some coastal States invoke the continental shelf provisions either in the 1958 Convention on the Continental Shelf or the LOSC, the sector principle has been used to claim maritime geographic features within the polar regions.

The question here now is how a coastal State can claim submerged features as it can in respect to other features such as islands or continental shelf. The International Court of Justice once emphasized the geographic correlation between coast and submerged areas off the coast and regarded it as "the basis of the coastal state's legal title" as "the coast of the territory of the state is the decisive factor for title to submarine areas adjacent to it".[53] Here it is obvious that once a coastal State wishes to claim certain submerged features, there should be support from its coast either from the mainland or an island.

It is wondered if a coastal State can claim a continental shelf which is the seabed and subsoil, why it is not allowed to claim a submerged geographic feature. It is recalled when the Truman Proclamation was issued, the claim to the continental shelf included submerged features. That is why the term "submerged lands" was used instead of the modern name "continental shelf". It is interesting to note that on 21 April 2010, Republic of Trinidad and Tobago and Grenada signed a maritime delimitation agreement in which they use the title "delimitation of marine and submarine areas" instead of using EEZ and continental shelf.[54] The wording "submarine areas may contain not only continental shelf, but also

---

[53] Continental Shelf (Tunisia/Libya), 24 February 1982 ICJ 18, ¶¶73–74; cited in Julia Lisztwan, "Stability of Maritime Boundary Agreements", *Yale Journal of International Law*, Vol. 37, 2012, 168.

[54] See the Treaty between the Republic of Trinidad and Tobago and Grenada on the Delimitation of Marine and Submarine Areas, available at http://www.gov.gd/egov/docs/other/delimitation_treaty_trinidad_tobago_grenada.pdf.

submerged features". In this respect, a principle of reasonableness should apply. As Lauterpacht opines, "Its authority and continued usefulness can be safeguarded only if it is acted upon in accordance with its true purpose and the ever valid test of reasonableness. Its true purpose is to ensure freedom of navigation, unhampered by exclusive claims of individual states, and freedom of utilization of the resources of the sea to a degree to which they can be equitably utilized by all".[55] Of course, a coastal State may not enjoy full sovereignty over such features, but only sovereign rights; and such features themselves certainly do not generate any maritime zones.

Currently, climate change has a significant impact upon the change of maritime geographic features. Global warming causes sea-level rise, leading to a threat not only to maritime features but also to island countries. In the coming years, islands could become low-tide elevations, and low-tide elevations become submerged features. In order to preserve the status of the islands and other features, coastal States have taken various measures. Japan has adopted the Law on Preservation of Low-Water Line in 2010. It regulates certain acts that may damage low-water lines from which those maritime zones are measured in the "Low-Water Line Protection Areas" to be designated in accordance with the law and provides for matters relating to the specified remote port facilities to be built on the remote islands.[56]

It is admitted that climate change would cause a lot of legal problems in international law. For example, does a state still exist when all the islands it is based on have disappeared from the water surface? Does an island change its legal status when it has become a low-elevation or even submerged feature? What about the maritime boundary delimitation after such changes? Some legal scholars have suggested expanding the LOSC to allow technological installations to replace the lost territory of an inundated island state, ultimately in the form of a legal framework to allow the nationals of that State to maintain the State's sovereign rights.[57] The legal

[55] Lauterpacht, *supra* note 7, p. 407.

[56] See Hiroko Sasaki, "Japan's practices and experiences of adopting UNCLOS into domestic laws", paper presented to the Conference on Practices of the UNCLOS and the Resolution of South China Sea Disputes, 3–5 September 2012, Taipei, p. 11.

[57] Michael Gagain, "Climate Change, Sea Level Rise, and Artificial Islands: Saving the Maldives' Statehood and Maritime Claims through the 'Constitution of the Oceans'", *Colorado Journal of International Environmental Law & Policy*, Vol. 23, 2012, 107.

questions that have arisen from climate change are still under discussion in academia.

It is worth mentioning that there are different terminologies which are related to the term "submerged geographic features". In addition to the term "submerged lands" as aforementioned, in the American domestic legal system, another term "tidelands" is also used, which refers to the submerged lands under the marginal sea as well as the lands covered and uncovered by the ebb and flow of the tides.[58] In that terminology, low-tide elevations are also included.

Conclusion

In conclusion, it is admitted that the claim to submerged maritime features continues to remain as a legal issue in international law and the question whether such a claim can become a clear-cut legitimate one depends upon the development of international law, in particular State practice in this respect. It is perceived that this missing link can be amended with the future development of international law supported by constant State practice.

On the one hand, we have to reconsider the principle of "land dominating the sea". It becomes a relative principle as land-locked States, no matter how big a piece of land they possess, do not enjoy any maritime zones or claim any maritime geographic features. On the other hand, when coastal States claim the continental shelf, they may not purely depend on the aforesaid principle. The legal continental shelf provided for in the LOSC deviates from the geographic definition that the continental shelf is a natural prolongation of the land mass. This relativity may have legal implications for claims to maritime geographic features in future.

---

[58] W. Page Keeton, "Federal and State Claims to Submerged Lands under Coastal Waters", *Texas Law Review*, Vol. 25, 1946–1947, 263.

Appendix 1

Regional Cooperation Agreement on Combating Piracy and Armed Robbery against Ships in Asia

The Contracting Parties to this Agreement,

Concerned about the increasing number of incidents of piracy and armed robbery against ships in Asia,

Mindful of the complex nature of the problem of piracy and armed robbery against ships,

Recognizing the importance of safety of ships, including their crew, exercising the right of navigation provided for in the United Nations Convention on the Law of the Sea of 10 December 1982, hereinafter referred to as “the UNCLOS”,

Reaffirming the duty of States to cooperate in the prevention and suppression of piracy under the UNCLOS,

Recalling “Tokyo Appeal” of March 2000, “Asia Anti-Piracy Challenges 2000” of April 2000 and “Tokyo Model Action Plan” of April 2000,

Noting the relevant resolutions adopted by the United Nations General Assembly and the relevant resolutions and recommendations adopted by the International Maritime Organization,

Conscious of the importance of international cooperation as well as the urgent need for greater regional cooperation and coordination of all States affected within Asia, to prevent and suppress piracy and armed robbery against ships effectively,

Convinced that information sharing and capacity building among the Contracting Parties will significantly contribute towards the prevention and suppression of piracy and armed robbery against ships in Asia,

Affirming that, to ensure greater effectiveness of this Agreement, it is indispensable for each Contracting Party to strengthen its measures aimed at preventing and suppressing piracy and armed robbery against ships,

Determined to promote further regional cooperation and to enhance the effectiveness of such cooperation,

Have agreed as follows:

Part I — Introduction

Article 1: Definitions

1. For the purposes of this Agreement, "piracy" means any of the following acts:
   (a) any illegal act of violence or detention, or any act of depredation, committed for private ends by the crew or the passengers of a private ship or a private aircraft, and directed:
      (i) on the high seas, against another ship, or against persons or property on board such ship;
      (ii) against a ship, persons or property in a place outside the jurisdiction of any State;
   (b) any act of voluntary participation in the operation of a ship or of an aircraft with knowledge of facts making it a pirate ship or aircraft;
   (c) any act of inciting or of intentionally facilitating an act described in subparagraph (a) or (b).
2. For the purposes of this Agreement, "armed robbery against ships" means any of the following acts:
   (a) any illegal act of violence or detention, or any act of depredation, committed for private ends and directed against a ship, or against persons or property on board such ship, in a place within a Contracting Party's jurisdiction over such offences;
   (b) any act of voluntary participation in the operation of a ship with knowledge of facts making it a ship for armed robbery against ships;
   (c) any act of inciting or of intentionally facilitating an act described in subparagraph (a) or (b).

Article 2: General Provisions

1. The Contracting Parties shall, in accordance with their respective national laws and regulations and subject to their available resources or capabilities, implement this Agreement, including preventing and suppressing piracy and armed robbery against ships, to the fullest extent possible.
2. Nothing in this Agreement shall affect the rights and obligations of any Contracting Party under the international agreements to which that Contracting Party is party, including the UNCLOS, and the relevant rules of international law.
3. Nothing in this Agreement shall affect the immunities of warships and other government ships operated for non-commercial purposes.
4. Nothing in this Agreement, nor any act or activity carried out under this Agreement shall prejudice the position of any Contracting Party with regard to any dispute concerning territorial sovereignty or any issues related to the law of the sea.
5. Nothing in this Agreement entitles a Contracting Party to undertake in the territory of another Contracting Party the exercise of jurisdiction and performance of functions which are exclusively reserved for the authorities of that other Contracting Party by its national law.
6. In applying paragraph 1 of Article 1, each Contracting Party shall give due regard to the relevant provisions of the UNCLOS without prejudice to the rights of the third Parties.

Article 3: General Obligations

1. Each Contracting Party shall, in accordance with its national laws and regulations and applicable rules of international law, make every effort to take effective measures in respect of the following:
   (a) to prevent and suppress piracy and armed robbery against ships;
   (b) to arrest pirates or persons who have committed armed robbery against ships;
   (c) to seize ships or aircraft used for committing piracy or armed robbery against ships, to seize ships taken by and under the control of pirates or persons who have committed armed robbery against ships, and to seize the property on board such ships; and
   (d) to rescue victim ships and victims of piracy or armed robbery against ships.

2. Nothing in this Article shall prevent each Contracting Party from taking additional measures in respect of subparagraphs (a) to (d) above in its land territory.

Part II — Information Sharing Center

Article 4: Composition

1. An Information Sharing Center, hereinafter referred to as "the Center", is hereby established to promote close cooperation among the Contracting Parties in preventing and suppressing piracy and armed robbery against ships.
2. The Center shall be located in Singapore.
3. The Center shall be composed of the Governing Council and the Secretariat.
4. The Governing Council shall be composed of one representative from each Contracting Party. The Governing Council shall meet at least once every year in Singapore, unless otherwise decided by the Governing Council.
5. The Governing Council shall make policies concerning all the matters of the Center and shall adopt its own rules of procedure, including the method of selecting its Chairperson.
6. The Governing Council shall take its decisions by consensus.
7. The Secretariat shall be headed by the Executive Director who shall be assisted by the staff. The Executive Director shall be chosen by the Governing Council.
8. The Executive Director shall be responsible for the administrative, operational and financial matters of the Center in accordance with the policies as determined by the Governing Council and the provisions of this Agreement, and for such other matters as determined by the Governing Council.
9. The Executive Director shall represent the Center. The Executive Director shall, with the approval of the Governing Council, make rules and regulations of the Secretariat.

Article 5: Headquarters Agreement

1. The Center, as an international organization whose members are the Contracting Parties to this Agreement, shall enjoy such legal capacity,

privileges and immunities in the Host State of the Center as are necessary for the fulfillment of its functions.

2. The Executive Director and the staff of the Secretariat shall be accorded, in the Host State, such privileges and immunities as are necessary for the fulfillment of their functions.
3. The Center shall enter into an agreement with the Host State on matters including those specified in paragraphs 1 and 2 of this Article.

Article 6: Financing

1. The expenses of the Center, as provided for in the budget decided by the Governing Council, shall be provided by the following sources:
   (a) Host State financing and support;
   (b) Voluntary contributions from the Contracting Parties;
   (c) Voluntary contributions from international organizations and other entities, in accordance with relevant criteria adopted by the Governing Council; and
   (d) Any other voluntary contributions as may be agreed upon by the Governing Council.
2. Financial matters of the Center shall be governed by a Financial Regulation to be adopted by the Governing Council.
3. There shall be an annual audit of the accounts of the Center by an independent auditor appointed by the Governing Council. The audit report shall be submitted to the Governing Council and shall be made public, in accordance with the Financial Regulation.

Article 7: Functions

The functions of the Center shall be:

(a) to manage and maintain the expeditious flow of information relating to incidents of piracy and armed robbery against ships among the Contracting Parties;

(b) to collect, collate and analyze the information transmitted by the Contracting Parties concerning piracy and armed robbery against ships, including other relevant information, if any, relating to individuals and transnational organized criminal groups committing acts of piracy and armed robbery against ships;

(c) to prepare statistics and reports on the basis of the information gathered and analyzed under subparagraph (b), and to disseminate them to the Contracting Parties;
(d) to provide an appropriate alert, whenever possible, to the Contracting Parties if there is a reasonable ground to believe that a threat of incidents of piracy or armed robbery against ships is imminent;
(e) to circulate requests referred to in Article 10 and relevant information on the measures taken referred to in Article 11 among the Contracting Parties;
(f) to prepare non-classified statistics and reports based on information gathered and analyzed under subparagraph (b) and to disseminate them to the shipping community and the International Maritime Organization; and
(g) to perform such other functions as may be agreed upon by the Governing Council with a view to preventing and suppressing piracy and armed robbery against ships.

Article 8: Operation

1. The daily operation of the Center shall be undertaken by the Secretariat.
2. In carrying out its functions, the Center shall respect the confidentiality of information provided by any Contracting Party, and shall not release or disseminate such information unless the consent of that Contracting Party is given in advance.
3. The Center shall be operated in an effective and transparent manner, in accordance with the policies made by the Governing Council, and shall avoid duplication of existing activities between the Contracting Parties.

Part III — Cooperation through the Information Sharing Center

Article 9: Information Sharing

1. Each Contracting Party shall designate a focal point responsible for its communication with the Center, and shall declare its designation of such focal point at the time of its signature or its deposit of an instrument of notification provided for in Article 18.

2. Each Contracting Party shall, upon the request of the Center, respect the confidentiality of information transmitted from the Center.
3. Each Contracting Party shall ensure the smooth and effective communication between its designated focal point, and other competent national authorities including rescue coordination centers, as well as relevant non-governmental organizations.
4. Each Contracting Party shall make every effort to require its ships, ship owners, or ship operators to promptly notify relevant national authorities including focal points, and the Center when appropriate, of incidents of piracy or armed robbery against ships.
5. Any Contracting Party which has received or obtained information about an imminent threat of, or an incident of, piracy or armed robbery against ships shall promptly notify relevant information to the Center through its designated focal point.
6. In the event that a Contracting Party receives an alert from the Center as to an imminent threat of piracy or armed robbery against ships pursuant to subparagraph (d) of Article 7, that Contracting Party shall promptly disseminate the alert to ships within the area of such an imminent threat.

Article 10: Request for Cooperation

1. A Contracting Party may request any other Contracting Party, through the Center or directly, to cooperate in detecting any of the following persons, ships, or aircraft:
   (a) pirates;
   (b) persons who have committed armed robbery against ships;
   (c) ships or aircraft used for committing piracy or armed robbery against ships, and ships taken by and under the control of pirates or persons who have committed armed robbery against ships; or
   (d) victim ships and victims of piracy or armed robbery against ships.
2. A Contracting Party may request any other Contracting Party, through the Center or directly, to take appropriate measures, including arrest or seizure, against any of the persons or ships mentioned in subparagraph (a), (b), or (c) of paragraph 1 of this Article, within the limits permitted by its national laws and regulations and applicable rules of international law.
3. A Contracting Party may also request any other Contracting Party, through the Center or directly, to take effective measures to rescue

the victim ships and the victims of piracy or armed robbery against ships.

4. The Contracting Party which has made a direct request for cooperation pursuant to paragraphs 1, 2 and 3 of this Article shall promptly notify the Center of such request.
5. Any request by a Contracting Party for cooperation involving extradition or mutual legal assistance in criminal matters shall be made directly to any other Contracting Party.

Article 11: Cooperation by the Requested Contracting Party

1. A Contracting Party, which has received a request pursuant to Article 10, shall, subject to paragraph 1 of Article 2, make every effort to take effective and practical measures for implementing such request.
2. A Contracting Party, which has received a request pursuant to Article 10, may seek additional information from the requesting Contracting Party for the implementation of such request.
3. A Contracting Party, which has taken measures referred to in paragraph 1 of this Article, shall promptly notify the Center of the relevant information on the measures taken.

Part IV — Cooperation

Article 12: Extradition

A Contracting Party shall, subject to its national laws and regulations, endeavor to extradite pirates or persons who have committed armed robbery against ships, and who are present in its territory, to the other Contracting Party which has jurisdiction over them, at the request of that Contracting Party.

Article 13: Mutual Legal Assistance

A Contracting Party shall, subject to its national laws and regulations, endeavor to render mutual legal assistance in criminal matters, including the submission of evidence related to piracy and armed robbery against ships, at the request of another Contracting Party.

Article 14: Capacity Building

1. For the purpose of enhancing the capacity of the Contracting Parties to prevent and suppress piracy and armed robbery against ships, each Contracting Party shall endeavor to cooperate to the fullest possible extent with other Contracting Parties which request cooperation or assistance.
2. The Center shall endeavor to cooperate to the fullest possible extent in providing capacity building assistance.
3. Such capacity building cooperation may include technical assistance such as educational and training programs to share experiences and best practices.

Article 15: Cooperative Arrangements

Cooperative arrangements such as joint exercises or other forms of cooperation, as appropriate, may be agreed upon among the Contracting Parties concerned.

Article 16: Protection Measures for Ships

Each Contracting Party shall encourage ships, ship owners, or ship operators, where appropriate, to take protective measures against piracy and armed robbery against ships, taking into account the relevant international standards and practices, in particular, recommendations adopted by the International Maritime Organization.

Part V — Final Provisions

Article 17: Settlement of Disputes

Disputes arising out of the interpretation or application of this Agreement, including those relating to liability for any loss or damage caused by the request made under paragraph 2 of Article 10 or any measure taken under paragraph 1 of Article 11, shall be settled amicably by the Contracting Parties concerned through negotiations in accordance with applicable rules of international law.

Article 18: Signature and Entry into Force

1. This Agreement shall be open for signature at the depository referred to in paragraph 2 below by the People's Republic of Bangladesh, Brunei Darussalam, the Kingdom of Cambodia, the People's Republic of China, the Republic of India, the Republic of Indonesia, Japan, the Republic of Korea, the Lao People's Democratic Republic, Malaysia, the Union of Myanmar, the Republic of the Philippines, the Republic of Singapore, the Democratic Socialist Republic of Sri Lanka, the Kingdom of Thailand, the Socialist Republic of Viet Nam.
2. The Government of Singapore is the depository of this Agreement.
3. This Agreement shall enter into force 90 days after the date on which the tenth instrument of notification by a State listed in paragraph 1, indicating the completion of its domestic requirements, is submitted to the depository. Subsequently it shall enter into force in respect of any other State listed in paragraph 1 above 30 days after its deposit of an instrument of notification to the depository.
4. The depository shall notify all the States listed in paragraph 1 of the entry into force of this Agreement pursuant to paragraph 3 of this Article.
5. After this Agreement has entered into force, it shall be open for accession by any State not listed in paragraph 1. Any State desiring to accede to this Agreement may so notify the depository, which shall promptly circulate the receipt of such notification to all other Contracting Parties. In the absence of a written objection by a Contracting Party within 90 days of the receipt of such notification by the depository, that State may deposit an instrument of accession with the depository, and become a party to this Agreement 60 days after such deposit of instrument of accession.

Article 19: Amendment

1. Any Contracting Party may propose an amendment to this Agreement, any time after the Agreement enters into force. Such amendment shall be adopted with the consent of all Contracting Parties.
2. Any amendment shall enter into force 90 days after the acceptance by all Contracting Parties. The instruments of acceptance shall be deposited with the depository, which shall promptly notify all other Contracting Parties of the deposit of such instruments.

Article 20: Withdrawal

1. Any Contracting Party may withdraw from this Agreement at any time after the date of its entry into force.
2. The withdrawal shall be notified by an instrument of withdrawal to the depository.
3. The withdrawal shall take effect 180 days after the receipt of the instrument of withdrawal by the depository.
4. The depository shall promptly notify all other Contracting Parties of any withdrawal.

Article 21: Authentic Text

This Agreement shall be authentic in the English language.

Article 22: Registration

This Agreement shall be registered by the depository pursuant to Article 102 of the Charter of the United Nations.

IN WITNESS WHEREOF, the undersigned, being duly authorized thereto by their respective Governments, have signed this Agreement.

Appendix 2

Agreement on the Delimitation of the Territorial Seas, Exclusive Economic Zones and Continental Shelves in the Beibu Gulf between the People's Republic of China and the Socialist Republic of Vietnam

The People's Republic of China and the Socialist Republic of Vietnam (hereinafter referred to as the Parties), in order to consolidate and develop the traditional neighboring and friendly relations between the two countries and the two peoples, to maintain and enhance the stability and development of the Beibu Gulf, based on the principles of mutual respect for each other, sovereignty and territorial integrity, non-aggression, non-interference in each other's internal affairs, equality and mutual benefit, peaceful coexistence, and in the spirit of resolving the Beibu Gulf delimitation issue through mutual compromise, friendly consultation and fairness and reasonableness, have agreed as follows:

Article 1

1. The Parties have determined the demarcation line for the territorial seas, exclusive economic zones and continental shelves of the two countries in the Beibu Gulf in accordance with the 1982 United Nations Convention on the Law of the Sea, generally accepted principles of international law and international practice, based on the full consideration of all relevant circumstances of the Beibu Gulf and on the equitable principle, and through friendly consultation.

2. The Beibu Gulf, in this Agreement, refers to the area where to the north are coasts of the land territory of China and Vietnam, to the east are coasts of China's Leizhou Peninsula and Hainan Island, to the west is the semi-enclosed bay surrounded by Vietnam's mainland coasts, and its south limit is a straight line connected by the most outer point of Yingge Zui of Hainan Island of China with the coordinates of 18°30′19″N, 108°41′17″E, through Con Co Island of Vietnam, to the seashore of Vietnam with the coordinates of 16°57′40″N, 107°08′42″E.

The Parties have decided that the above area constitutes the scope of delimitation in this Agreement.

Article 2

The demarcation line for the territorial seas, exclusive economic zones and continental shelves between the two countries in the Beibu Gulf, as agreed by the Parties, is determined by straight lines connecting the following 21 points, for which the geographic coordinates are as follows:

Point 1: 21°28′12.5″N, 108°06′04.3″E;
Point 2: 21°28′01.7″N, 108°06′01.6″E;
Point 3: 21°27′50.1″N, 108°05′57.7″E;
Point 4: 21°27′39.5″N, 108°05′51.5″E;
Point 5: 21°27′28.2″N, 108°05′39.9″E;
Point 6: 21°27′23.1″N, 108°05′38.8″E;
Point 7: 21°27′08.2″N, 108°05′43.7″E;
Point 8: 21°16′32″N, 108°08′05″E;
Point 9: 21°12′35″N, 108°12′31″E;
Point 10: 20°24′05″N, 108°22′45″E;
Point 11: 19°57′33″N, 107°55′47″E;
Point 12: 19°39′33″N, 107°31′40″E;
Point 13: 19°25′26″N, 107°21′00″E;
Point 14: 19°25′26″N, 107°12′43″E;
Point 15: 19°16′04″N, 107°11′23″E;
Point 16: 19°12′55″N, 107°09′34″E;

Point 17: 18°42′52″N, 107°09′34″E;

Point 18: 18°13′49″N, 107°34′00″E;

Point 19: 18°07′08″N, 107°37′34″E;

Point 20: 18°04′13″N, 107°39′09″E;

Point 21: 17°47′00″N, 107°58′00″E.

Article 3

1. The demarcation line from Point 1 to Point 9 as provided in Article 2 of this Agreement is the line to divide the territorial seas of the two countries in the Beibu Gulf.
2. The division of the space above and the seabed and subsoil under the territorial seas of the two countries follows the vertical direction of the demarcation line for the territorial seas of the two countries as provided in Paragraph 1 of this Article.
3. Unless otherwise agreed by the Parties, no topographical change shall change the demarcation line for the territorial seas of the two countries from Point 1 to Point 7 as provided in Paragraph 1 of this Article.

Article 4

The demarcation line from Point 9 to Point 21 as provided in Article 2 of this Agreement is the line to divide the exclusive economic zones and continental shelves of the two countries in the Beibu Gulf.

Article 5

The demarcation line to divide the territorial seas of the two countries from Point 1 to Point 7 as provided in Article 2 of this Agreement is drawn in block on the special map of the Beilun River estuary jointly surveyed and prepared by the Parties in 2000 with the scale of 1:10,000; the demarcation line for territorial seas, exclusive economic zones and continental shelves of the two countries from Point 7 to Point 21 is drawn in block on the complete map of the Beibu Gulf jointly surveyed and prepared by the Parties in 2000 with the scale of 1:500,000. The above demarcation lines are all geodesic lines.

The above special map of Beilun River estuary and the complete map of the Beibu Gulf are the maps attached to this Agreement. The ITRF-96 Coordinate System has been adopted for the above maps. Coordinates for all the demarcation points as provided for in Article 2 of this Agreement are measured on and taken from the above maps. The drawing of the demarcation line as provided for in this Agreement on the attached maps to this Agreement is only for the purpose of illustration.

Article 6

The Parties shall mutually respect their respective sovereignty over, sovereign rights to and jurisdiction in the territorial seas, exclusive economic zones and continental shelves between the two countries in the Beibu Gulf as decided by this Agreement.

Article 7

In case that any single geophysical structure of oil and gas or other mineral deposits should straddle the demarcation line as provided in Article 2 of this Agreement, the Parties shall, through friendly consultation, reach an agreement on the development of the structure or deposit in a most effective way as well as on equal sharing of the profits resulting from the development.

Article 8

The Parties have agreed to consult on matters of cooperation in respect to the rational use and sustainable development of living resources in the Beibu Gulf, and the conservation, management and use of living resources in the exclusive economic zones of the two countries in the Beibu Gulf.

Article 9

The delimitation of the territorial seas, exclusive economic zones and continental shelves of the two countries in the Beibu Gulf in accordance with this Agreement shall in no means affect or hamper the position of either Party on rules of international law in the area of the law of the sea.

Article 10

Any dispute resulting from interpretation or application of this Agreement between the Parties shall be settled through friendly consultation and negotiation.

Article 11

This Agreement shall be subject to ratification of the Parties and enter into force from the date of the exchange of ratification instruments between the Parties. The exchange of ratification instruments shall be carried out in Hanoi.

This Agreement, in duplicate and in both Chinese and Vietnamese, was signed in Beijing on 25 December 2000, and the Chinese and Vietnamese texts are equally authentic.

Representative Plenipotentiary for the People's Republic of China	Representative Plenipotentiary for the Socialist Republic of Vietnam
Tang Jiaxuan	Nguyen Dy Nien

(It is an unofficial English version translated by Zou Keyuan from the Chinese version of the Agreement.)

Bibliography

Jason Abbot and Renwick Neil, "Pirates? Maritime Piracy and Societal Security in Southeast Asia", *Pacifica Review*, Vol. 11 (1), 1999, 7–24.

W.J. Aceves, "Ambiguities in Plurilingual Treaties: A Case Study of Article 22 of the 1982 Law of the Sea Convention", *Ocean Development and International Law*, Vol. 27(3), 1996, 187–233.

Stanimir A. Alexandrov, *Self-Defense Against the Use of Force in International Law*, The Hague: Kluwer Law International, 1996.

David Anderson, "The Roles of Flag States, Port States, Coastal States and International Organizations in the Enforcement of International Rules and Standards Governing the Safety of Navigation and the Prevention of Pollution from Ships under the UN Convention on the Law of the Sea and Other International Agreements", *Singapore Journal of International & Comparative Law*, Vol. 2(2), 1998, 557–578.

David H. Anderson, "Investigation, Detention and Release of Foreign Vessels under the UN Convention on the Law of the Sea of 1982 and Other International Agreements", *International Journal of Marine and Coastal Law*, Vol. 11(2), 1996, 165–177.

Bjorn Aune, "Piracy and Its Repression under the 1982 Law of the Sea Convention", *Ocean Yearbook*, Vol. 8(1), 1989, 18–43.

Sam Bateman, "Hydrographic surveying in the EEZ: Differences and overlaps with marine scientific research", *Marine Policy*, Vol. 29 (2), 2005, 163–174.

Robert Beckman and Clive Schofield, "Moving Beyond Disputes over Island Sovereignty: ICJ Decision Sets Stage for Maritime Boundary Delimitation in the Singapore Strait", *Ocean Development and International Law*, Vol. 40(1), 2009, 1–35.

David J. Bederman, "US state boundaries — Submerged Lands Act — 1958 Geneva Convention on the Territorial Sea — straight baselines — fringing islands — ice elevations", *American Journal of International Law*, Vol. 92(1), 1998, 82–87.

Anthony Bergin, "East Asian Naval Developments-Sailing into Rough Seas", *Marine Policy*, Vol. 26(2), 2002, 121–131.

Rudolf Bernhardt, *Encyclopedia of Public International Law: Law of the Sea*, Heidelberg: Max-Planck-Institut furauslandisches offentliches Rechtund Volkerrecht,1989.

Patricia W. Birnie and Alan E. Boyle, *International Law and the Environment*, Oxford: Oxford University Press, 1994.

Patricia W. Birnie, "Piracy: Past, Present and Future", *Marine Policy*, Vol. 11 (3), 1987, 163–183.

Boleslaw Adam Boczek, "Peacetime Military Activities in the Exclusive Economic Zone of Third Countries", *Ocean Development and International Law*, Vol. 19(6), 1988, 445–468.

W.E. Butler, "Innocent Passage and the 1982 Convention: The Influence of Soviet Law and Policy", *American Journal of International Law*, Vol. 81(2), 1987, 331–347.

Michael Byers, "Policing the High Seas: The Proliferation Security Initiative", *American Journal of International Law*, Vol. 98(3), 2004, 526–545.

Lucius Caflisch (ed.), *The Peaceful Settlement of Disputes between States: Universal and European Perspectives*, The Hague: Kluwer Law International, 1998.

Antonio Cassese, "Terrorism is Also Disrupting Some Crucial Legal Categories of International Law", *European Journal of International Law*, Vol. 12 (5), 2001, 993–1001.

Jonathan I. Charney and Lewis M. Alexander (eds.), *International Maritime Boundaries*, Dordrecht: Martinus Nijhoff Publishers, 1993.

Robin Churchill, "Dispute Settlement under the UN Convention on the Law of the Sea: Survey for 2005", *International Journal of Marine and Coastal Law*, Vol. 21 (4), 2006, 377–388.

Robin Churchill, "Dispute Settlement under the UN Convention on the Law of the Sea: Survey for 2006", *International Journal of Marine and Coastal Law*, Vol. 22 (4), 2007, 463–483.

Charles Claypoole, "Access to International Justice: A Review of the Trust Funds Available for Law of the Sea-Related Disputes", *International Journal of Marine and Coastal Law*, Vol. 23(1), 2008, 77–94.

James Crawford and Donald R. Rothwell (eds.), *The Law of the Sea in the Asia-Pacific Region: Developments and Prospects*, Dordrecht: Martinus Nijhoff Publishers, 1995.

Donna Darm, "The Outward Limit of the Department of Interior's Authority over Submerged Lands — the Effect of Customary International Law on the Outer Continental Shelf Lands Act", *Washington Law Review*, Vol. 60(3), 1984–85, 673–696.

Louise de la Fayette, "ITLOS and the Saga of the Saiga: Peaceful Settlement of a Law of the Sea Dispute", *International Journal of Marine and Coastal Law*, Vol. 15 (3), 2000, 355–392.

Leticia Diaz, Barry Hart Dubner and Jason Parent, "When Is a 'Rock' an 'Island'? — Another Unilateral Declaration Defies 'Norms' of International Law", *Mich. St. Journal of International Law*, Vol. 15, 2007, 519–556.

Yoram Dinstein (ed.), *International Law at a Time of Perplexity: Essays in Honour of Shabtai Rosenne*, Dordrecht: Martinus Nijhoff Publishers, 1988.

Hasjim Djalal, Yankov Alexander and Anthony Bergin, "Draft Guidelines for Military and Intelligence Gathering Activities in the EEZ and Their Means and Manner of Implementation and Enforcement", *Marine Policy*, Vol. 29 (2), 2005, 175–183.

DUAN Jielong, "Statement on the Rule of Law at the National and International Levels", *Chinese Journal of International Law*, Vol. 6(1), 2007, 185–188.

Barry Hart Dubner, Peter Hanks and Michael Pryles, "Demarcation of Authority over Coastal Waters and Submerged Lands in the United States and Australia", *Stetson Law Review*, Vol. 8(2), 1980–81, 228–271.

John A. Duff, "A Note on the United States and the Law of the Sea: Looking Back and Moving Forward", *Ocean Development and International Law*, Vol. 35(3), 2004, 195–219.

Rene-Jean Dupuy & Daniel Vignes (eds.), *A Handbook on the New Law of the Sea*, Dordrecht: Martinus Nijhoff Publishers, 1991.

Daniel J. Dzurek, *The Spratly Islands Dispute: Who's on First?*, Durham: Department of Geography University of Durham, 1996.

Eric Ellen (ed.), *Piracy at Sea*, Paris: International Chamber of Commerce, International Maritime Bureau, 1989.

Epsey Cooke Farrell, *The Socialist Republic of Vietnam and the Law of the Sea: An Analysis of Vietnamese Behaviour within the Emerging International Oceans Regime*, The Hague: Martinus Nijhoff Publishers, 1998.

Francesco Francioni, "Peacetime Use of Force, Military Activities, and the New Law of the Sea", *Cornell International Law Journal*, Vol. 18(2), 1985, 203–226.

Erik Franckx, "American and Chinese Views on Navigational Rights of Warships", *Chinese Journal of International Law*, Vol. 10(1), 2011, 187–206.

Erik Franckx, "Innocent Passage of Warships: Recent Developments in US-Soviet Relations", *Marine Policy*, Vol. 14(6), 1990, 484–490.

Erik Franckx, "Reasonable Bond' in the Practice of the International Tribunal for the Law of the Sea", *California Western International Law Journal*, Vol. 32(2), 2001–2002, 303–342.

Michael Gagain, "Climate Change, Sea Level Rise, and Artificial Islands: Saving the Maldives Statehood and Maritime Claims Through the 'Constitution of the Oceans'", *Colorado Journal of International Environmental Law and Policy*, Vol. 23(1), 2012, 77–120.

George V. Galdorisi and Alan G. Kaufman, "Military Activities in the Exclusive Economic Zone: Preventing Uncertainty and Defusing Conflict", *California Western International Law Journal*, Vol. 32(2), 2002, 253–302.

GAO Jianjun, "Reasonableness of the Bond under Article 292 of the LOS Convention: Practice of the ITLOS", *Chinese Journal of International Law*, Vol. 7(1), 2008, 115–142.

Jack I. Garvey, "The International Institutional Imperative for Countering the Spread of Weapons of Mass Destruction: Assessing the Proliferation Security Initiative", *Journal of Conflict and Security Law*, Vol. 10(2), 2005, 125–147.

Michael Sheng-ti Gau, "Recent Decisions by the Commission on the Limits of the Continental Shelf on Japan's Submission of Outer Continental Shelf", *Chinese Journal of International Law*, Vol. 11(3), 2012, 487–504.

Tarcisio Gazzini, "The Rules on the Use of Force at the Beginning of the XXI Century", *Journal of Conflict & Security Law*, Vol. 11(3), 2006, 319–342.

Nicholas Grief, *Public International Law in the Airspace of the High Seas*, Dordrecht, Boston: Martinus Nijhoff Publishers, 1994.

Christine Gray, "A Crisis of Legitimacy for the UN Collective Security System?", *International and Comparative Law Quarterly*, Vol. 56(1), 2007, 157–170.

Douglas Guilfoyle, "Interdicting Vessels to Enforce the Common Interest: Maritime Countermeasures and the Use of Force", *International and Comparative Law Quarterly*, Vol. 56(1), 2007, 69–82.

Kay Hailbronner, "Freedom of the Air and the Convention on the Law of the Sea", *American Journal of International Law*, Vol. 77(3), 1983, 490–520.

Fox Hazel (ed.), *Joint Development of Offshore Oil and Gas: A Model Agreement for States for Joint Development with Explanatory Commentary*, London: British Institute of International and Comparative Law, 1989.

Rosalyn Higgins, "The ICJ, the ECJ, and the Integrity of International Law", *International and Comparative Law Quarterly*, Vol. 52(1), 2003, 1–20.

Seoung-Yong Hong and Jon M. Van Dyke (eds.), *Maritime Boundary Disputes, Settlement Processes, and the Law of the Sea*, Leiden: Koninklijke Brill NV, 2009.

HU Qian & CHEN Qiang, "China's Extradition Law of 2000", *Chinese Journal of International Law*, Vol. 1(2), 2002, 645–654.

Robert Jennings, "Role of the International Court of Justice", *British Yearbook of International Law*, Vol. 68(1), 1997, 1–63.

Daniel H. Joyner, "The Proliferation Security Initiative: Nonproliferation, Counterproliferation, and International Law", *Yale Journal of International Law*, Vol. 30(2), 2005, 507–548.

W. Page Keeton, "Federal and State Claims to Submerged Lands under Coastal Waters", American Bar Association, *Proceedings of the Section of Mineral and Natural Resources Law*, 29 October 1946, 56–64.

Dalchoog Kim et al (eds.), *Exploring Maritime Co-operation in Northeast Asia*, Seoul: Institute of East and West Studies, Yonsei University, 1993.

John Warren Kindt, "Dispute Settlement in International Environmental Issues: The Model Provided by the 1982 Convention on the Law of the Sea", *Vanderbilt Journal of Transnational Law*, Vol. 22(5), 1989, 1097–1118.

Natalie Klein, *Dispute Settlement in the UN Convention on the Law of the Sea*, Cambridge: Cambridge University Press, 2005.

Yue-Choong Kog, "Environmental Management and Conflict in Southeast Asia: Land Reclamation and Its Political Impact", IDSS Working Paper No. 101 (Institute of Defence and Strategic Studies, Nanyang Technological University, Singapore), January 2006.

Barbara Kwiatkowska and Alfred H. A. Soons, "Entitlement to Maritime Areas of Rocks Which Cannot Sustain Human Habitation or Economic Life of Their Own", *Netherlands Yearbook of International Law*, Vol. 31, 1990, 139–181.

Barbara Kwiatkowska, "The Australia and New Zealand v. Japan Southern Bluefin Tuna (Jurisdiction and Admissibility) Award of the First Law of the Sea Convention Annex VII Arbitral Tribunal", *International Journal of Marine and Coastal Law*, Vol. 16(2), 2001, 239–293.

Hee Kwon Park, *The Law of the Sea and Northeast Asia: A Challenge for Cooperation*, The Hague: Kluwer Law International, 2000.

Rainer Lagoni, "The Prompt Release of Vessels and Crews before the International Tribunal for the Law of the Sea: A Preparatory Report", *International Journal of Marine and Coastal Law*, Vol. 11(2), 1996, 147–164.

H. Lauterpacht, "Sovereignty over Submarine Areas", *British Yearbook of International Law*, Vol. 27, 1950, 376–433.

Eric Yong-Joong Lee, "National and International Legal Concerns regarding Recent North Korean Escapees", *International Journal of Refuge Law*, Vol. 13(1–2), 2001, 142–152.

Julia Lisztwan, "Stability of Maritime Boundary Agreements", *Yale Journal of International Law*, Vol. 37(1), 2012, 153–200.

Hugo Ignacio Llanos, "Low-Tide Elevations: Reassessing Their Impact on Maritime Delimitation", *Pace International Law Review*, Vol. 14(2), 2002, 255–272.

Michael W. Lodge, "International Seabed Authority's Regulation on Prospecting and Exploration for Polymetallic Nodules in the Area", *Journal of Energy and Natural Resources Law*, Vol. 20(3), 2002, 290–295.

Vaughan Lowe and Malgosia Fitzmaurice (eds.), *Fifty Years of the International Court of Justice: Essays in Honour of Sir Robert Jennings*, Cambridge: Cambridge University Press, 1996.

Geoffrey Marston, "Low-Tide Elevations and Straight Baselines", *British Yearbook of International Law*, Vol. 46, 1972–73, 405–424.

Thomas A. Mensah, "The Tribunal and the Prompt Release of Vessels", *International Journal of Marine and Coastal Law*, Vol. 22(3), 2007, 425–449.

Masahiro Miyoshi, "The Joint Development of Offshore Oil and Gas in Relation to Maritime Boundary Delimitation", *Maritime Briefing*, Vol. 2(5), 1999.

James B. Morell, *The Law of the Sea: The 1982 Treaty and Its Rejection by the United States*, North Carolina: McFarland & Company, 1992.

Myron H. Nordquist and John Norton Moore (eds.), *Security Flashpoints: Oil, Islands, Sea Access and Military Confrontation*, The Hague: Kluwer Law International, 1998.

Myron H. Nordquist, John Norton Moore and Kuen-chen Fu (eds.), *Recent Developments in the Law of the Sea and China*, Leiden: Koninklijke Brill NV, 2006.

Hong Thao Nguyen, "The Settlement of Disputes in Bac Bo (Tonkin) Gulf", *Vietnam Law & Legal Forum*, Vol. 7, No.77, January 2001, 15–18.

Hong Thao Nguyen, "Vietnam and Joint Development in the Gulf of Thailand", *Asian Yearbook of International Law*, Vol. 8, 2003, 137–150.

Tom Obokata, "Trafficking of Human Beings as a Crime against Humanity: Some Implications for the International Legal System", *International and Comparative Law Quarterly*, Vol. 54, 2005, 445–458.

Hal Olson, "Marine Traffic in the South China Sea", *Ocean Yearbook*, Vol. 12(1), 1996, 137–153.

Harald Haugom Olsen, "The Snake from Fujian Province to Morecambe Bay: An Analysis of the Problem of Human Trafficking in Sweated Labour", *European Journal of Crime, Criminal Law and Criminal Justice*, Vol. 16, 2008, 1–37.

Bernard H. Oxman, "Observations on Vessel Release under the United Nations Convention on the Law of the Sea", *International Journal of Marine and Coastal Law*, Vol. 10(2), 1995, 201–215.

Bernard H. Oxman, "The Third United Nations Conference on the Law of the Sea: The Ninth Session", *American Journal of International Law*, Vol. 75(2), 1981, 211–256.

Chi Young Pak, "Resettlement of the Fisheries Order in Northeast Asia resulting from the New Fisheries Agreements among Korea, Japan and China", *Korea Observer*, Vol. 30(4), 1999, 587–622.

Charles E. Pirtle, "Military Uses of Ocean Space and the Law of the Sea in the New Millennium", *Ocean Development and International Law*, Vol. 31(1–2), 2000, 7–45.

Michael Reisman and Gayl S. Westerman, *Straight Baselines in International Maritime Boundary Delimitation*, New York: St. Martin's Press, 1992.

J. Ashley Roach and Robert W. Smith, "Straight Baselines: The Need for a Universally Applied Norm", *Ocean Development and International Law*, Vol. 31(1–2), 2000, 47–80.

J. Ashley Roach and Robert W. Smith, *United States Responses to Excessive Maritime Claims*, 2nd ed, The Hague: Martinus Nijhoff publishers, 1996.

J. Ashley Roach, "Enforcement of International Rules and Standards of Navigation Safety in the Malacca and Singapore Straits", *Singapore Journal of International & Comparative Law*, Vol. 3(2), 1999, 323–336.

J. Ashley Roach, "Marine Scientific Research and the New Law of the Sea", *Ocean Development and International Law*, Vol. 27(1–2), 1996, 59–72.

J. Ashley Roach, "Today's Customary International Law of the Sea," *Ocean Development and International Law*, Vol. 45(3), 2014, 239–259.

Donald R. Rothwell, "Building on the Strengths and Addressing the Challenges: The Role of Law of the Sea institutions", *Ocean Development and International Law*, Vol. 35(2), 2004, 131–156.

Cesare P.R. Romano, *The Peaceful Settlement of International Environmental Disputes: A Pragmatic Approach*, The Hague: Kluwer Law International, 2000.

Natalino Ronzitti (ed.), *Maritime Terrorism and International Law*, Dordrecht: Martinus Nijhoff Publishers, 1990.

Natalino Ronzitti, "The Expanding Law of Self-Defence", *Journal of Conflict & Security Law*, Vol. 11 (3), 2006, 343–359.

Alfred P. Rubin, "Is Piracy Illegal?" *American Journal of International Law*, Vol. 70(1), 1976, 92–95.

Alfred Rubin, *The Law of Piracy*, Newport, RI: Naval War College Press, 1988.

Hiroko Sasaki, "Japan's Practices and Experiences of Adopting UNCLOS into Domestic Laws", paper presented to the Conference on Practices of the UNCLOS and the Resolution of South China Sea Disputes, 3–5 September 2012, Taipei.

Thomas Schindlmayr, "Sovereignty, Legal Regimes, and International Migration", *International Migration*, Vol. 41(2), 2003, 109–123.

Ivan Shearer, "Military Activities in the Exclusive Economic Zone: The Case of Aerial Surveillance", *Ocean Yearbook*, Vol. 17(1), 2003, 548–562.

Robert W. Smith, "Maritime Delimitation in the South China Sea: Potentiality and Challenges", *Ocean Development and International Law*, Vol. 41(3), 2010, 214–236.

Louis B. Sohn, "Settlement of Law of the Sea Disputes", *International Journal of Marine and Coastal Law*, Vol. 10(2), 1996, 205–217.

Yann-huei Song, "Declarations and Statements with Respect to the 1982 UNCLOS: Potential Legal Disputes between the United States and China after U.S. Accession to the Convention", *Ocean Development and International Law*, Vol. 36(3), 2005, 261–289.

Alfred H.A. Soons, *Marine Scientific Research and the Law of the Sea*, Deventer: Kluwer Law and Taxation Publishers, 1982.

Michael Stankiewicz (ed.), *Maritime Shipping in Northeast Asia: Law of the Sea, Sea Lanes, and Security*, IGCC Policy Paper No.33 (Institute on Global Conflict and Co-operation, University of California), February 1998.

Katherine R. Thomas, "The Changing Status of International Law in English Domestic Law", *Netherlands International Law Review*, Vol. 53(3), 2006, 371–398.

Gabriel V. Trinidad (ed.), *The Revised Penal Code*, 1983 Edition, Quezon City: U.P. Law Center, 1984.

Mark J. Valencia, Jenny Miller Garmendia, Jon Van Dyke, and Hasjim Djalal, *Military and Intelligence Gathering Activities in Exclusive Economic Zones: Consensus and Disagreement: A Summary of the Bali Dialogue*, Honolulu: East-West Center, 2002.

Mark J. Valencia and Jenny Miller Garmendia, *The Regime of the Exclusive Economic Zones: Issues and Responses: A Report of the Tokyo Meeting*, Honolulu: East-West Center, 2003.

Jon M. Van Dyke, Mark J. Valencia and Jenny Miller Garmendia, "The North/South Korea Boundary Dispute in the Yellow (West) Sea", *Marine Policy*, Vol. 27(2), 2003, 143–158.

Francisco Orrego Vicuña, "The International Tribunal for the Law of the Sea and Provisional Measures: Settled Issues and Pending Problems", *International Journal of Marine and Coastal Law*, Vol. 22(3), 2007, 451–462.

Paul R. William, *International Law and the Resolution of Central and East European Transboundary Environmental Disputes*, London: Macmillan Press, 2000.

Rüdiger Wolfrum (ed.), *Law of the Sea at the Crossroads: The Continuing Search for a Universally Accepted Regime*, Berlin: Duncker und Humblot, 1991.

Rüdiger Wolfrum, "Provisional Measures of the International Tribunal for the Law of the Sea", *Indian Journal of International Law*, Vol. 37(3), 1997, 420–434.

John Wong, Keyuan Zou and Huaqun Zeng (eds.), *China-ASEAN Relations: Economic and Legal Dimensions*, Singapore: World Scientific, 2006.

Michael Wood, "The International Tribunal for the Law of the Sea and General International Law", *International Journal of Marine and Coastal Law*, Vol. 22(3), 2007, 351–367.

ZHAO Lihai, *Studies on the Law of the Sea Issues*, Beijing: Peking University Press, 1996 (in Chinese).

Keyuan Zou, "Maritime Boundary Delimitation in the Gulf of Tonkin", *Ocean Development and International Law*, Vol. 30(3), 1999, 235–254.

Keyuan Zou, 'Piracy at Sea and China's Response', *Lloyd's Maritime and Commercial Law Quarterly*, Part 3, August 2000, 364–382.

Keyuan Zou "Sino-Vietnamese Fishery Agreement for the Gulf of Tonkin", *International Journal of Marine and Coastal Law*, Vol. 17(1), 2002, 127–148.

Keyuan Zou, "Sino-Japanese Joint Fishery Management in the East China Sea", *Marine Policy*, Vol. 27(2), 2003,125–142.

Keyuan Zou, "Quelling Sea Piracy in East Asia: China's Domestic Endeavour and Regional", *Journal of International Maritime Law*, Vol. 10(4), 2004, 327–342.

Keyuan Zou, *China's Marine Legal System and the Law of the Sea*, Leiden: Koninklijke Brill NV, 2005.

Keyuan Zou, "New Developments in the International Law of Piracy", *Chinese Journal of International Law*, Vol. 8(2), 2009, 323–345.

Keyuan Zou, "Navigation in the South China Sea: Why Still an Issue?" *International Journal of Marine and Coastal Law*, Vol. 32(2), 2017, 243–267.

Index

I

Lightning Source UK Ltd.
Milton Keynes UK
UKHW021825031221
394934UK00004B/298

9 789811 237058